Nuclear Magnetic Resonance Spectroscopy

Volume I

Author

Professor Pál Sohár, Ph.D., D.Sc.
Head
Spectroscopic Department
EGYT Pharmacochemical Works
Professor
Department of Organic Chemistry
Eötvös Lóránd University of Sciences
Budapest, Hungary

CRC Press, Inc.
Boca Raton, Florida

Library of Congress Cataloging in Publication Data

Sohár, Pál.
Nuclear magnetic resonance spectroscopy.

Bibliography: p.
Includes index.
1. Nuclear magnetic resonance spectroscopy.

I. Title.
QC762.S575 1983 543'.0877 82-9524
ISBN 0-8493-5632-6 (v. 1) AACR2
ISBN 0-8493-5633-4 (v. 2)
ISBN 0-8493-5634-2 (v. 3)

This book represents information obtained from authentic and highly regarded sources. Reprinted material is quoted with permission, and sources are indicated. A wide variety of references are listed. Every reasonable effort has been made to give reliable data and information, but the author and the publisher cannot assume responsibility for the validity of all materials or for the consequences of their use.

Direct all inquiries to CRC Press, Inc., 2000 Corporate Blvd., N.W., Boca Raton, Florida, 33431.

International Standard Book Number 0-8493-5632-6 (v. 1)
International Standard Book Number 0-8493-5633-4 (v. 2)
International Standard Book Number 0-8493-5634-2 (v. 3)

Library of Congress Card Number 82-9524
Printed in the United States

PREFACE

These books are an updated version of a Hungarian handbook the author published in 1976 with the same title. Since it was the first original NMR handbook in the Hungarian language, I endeavored to embrace, even if briefly, all topics of NMR spectroscopy. When the manuscript was finished, at the beginning of the 1970s, FT spectroscopy was only beginning to be studied, and thus the investigation of nuclei other than protons had much less significance than now. Therefore, many parts of the books discuss NMR phenomena primarily from the aspect of proton resonance.

With the very rapid spread of Fourier transform technique and FT instruments, the investigation of other nuclei, first of all carbon, has become feasible. In addition to chemical shifts, coupling constants and intensities are now sources of NMR information and relaxation times have become one of the starting points of structure elucidation. This progress made it necessary to deal more closely with certain theoretical and technical principlesas well as with some relationships between spectrum and chemical structure less important from the aspect of proton resonance.

Therefore, some of the problems are touched in more than one chapter. The resulting, somewhat looser structure of the books has, however, the advantage that the special principles and applications pertaining to the same nucleus may be discussed in a more concentrated and unique manner.

The encyclopedic character of the book has claimed a relatively great volume and a necessarily brief, schematic discussion of some details. But perhaps this distinguishes my book from other handbooks and justifies its publication. The books are meant primarily for chemists, undergraduates, and young spectroscopists who wish to get broad but not too detailed information on all important topics of NMR spectroscopy. I hope that scientists working in related fields, e.g., analysts, biochemists, or physicists will also find useful information in the books. At any rate, the books were written keeping in mind the demands of organic chemists who wish to use NMR spectroscopy as a tool in structure elucidation, but who also want to know, at least schematically, the theoretical foundations of the method applied.

The first chapter is a concise but comprehensive discussion, primarily from the aspect of proton resonance, of NMR theory, which assumes only minimum quantum chemical knowledge, and the mathematical apparatus is limited to the level accessible for average chemists. The main purpose of this chapter is to point out how the magnetic nuclei of molecules may be identified with the individual spin systems, and how can one determine the spectrum parameters given the key to structure elucidation, i.e., chemical shifts and coupling constants, from the measured spectrum. As the quantum theory of NMR spectroscopy is a settled, classical branch of science, and as I endeavored to review the most important principles only, the references in this chapter extend only to original publications of fundamental importance, completely neglecting more recent literature pertaining to special details.

Chapter 2 deals with the operation principles and technical problems of traditional (CW) and modern (FT) spectrometers, and also discusses special techniques of measurement. Of the latter, the measurement and use of various double resonance spectra, temperature-dependent measurements important in the investigation of dynamic phenomena, and the measurement of integrated intensities permitting quantitative analysis have the greatest weight. The theoretical foundations of Fourier transform spectroscopy, the experimental determination of relaxation times, the theory of relaxation, and relaxation mechanisms are also discussed here. In connection with relaxation times, a separate section is devoted to the cancer diagnostic application of NMR and tomography.

Chapter 3 reviews the proton resonance parameters of the most important groups of organic compounds. The huge and ever-increasing amount of data collected in this field prohibits any attempt toward completeness, and thus again I restricted myself to classical results, selecting the illustration material from the most characteristic or the first examples. Therefore, more recent literature is represented in the chapter only by some reviews or extremely interesting papers. An exception is the last section of this chapter, which deals with the medium effects, predominantly with the shift reagent technique. This part discusses the exchange phenomena and hydrogen bonds, as well as the NMR investigation of optical purity and radical reactions (CIDNP).

Chapter 4 discusses, on the basis of similar principles, the most characteristic properties of the NMR spectra of carbon nuclei, in much smaller volume those of nitrogen, and very schematically only the properties of ^{17}O, ^{19}F, and ^{31}P nuclei, with hints to the special aspects, pertaining to the given nuclei, of some theoretical problems, principles, and measurement technique.

Chapter 5 is a collection of problems. All problems are concerned with elucidation of chemical structures taken from the practice of the author. Most of the problems were real, occurring in practice. The 66 problems connected with ^{1}H NMR spectra and the 12 problems connected with carbon NMR, together with the roughly 200 spectra provide, hopefully, a good training for the chemist in his own spectrum evaluation practice.

I feel it my duty to express my gratitude to all who contributed to the publication of this book.

My first thanks should go to my students, since the subject of the book was collected first for my lectures. Inspired by their interest and questions, many ideas arose for working out certain topics and selecting the material for illustration.

To my chemist colleagues, whose names are listed in the name index, thanks are due not only for the samples synthesized partly by them and given for recording the spectra of the Problems section, but also for the preparation of model derivatives, or purification of compounds in course of structure elucidation, and also for the assistance provided by chemical information about the compounds studied.

For constructive discussions and suggestions on the completion or modification of certain parts, I am indebted to Professors G. Snatzke (Bochum), and H. Wamhoff (Bonn) and Doctors G. Jalsovszky (Budapest), I. Kövesdi (Budapest), and T. Széll (New York).

For the promotion of the English edition my thanks are due to Professor B. Csakvári and I. Kovács, as well as to Publishing House of the Hungarian Academy of Sciences; I wish to express my thanks to Dr. G. Jalsovszky for the tedious, competent, and unrewarding work of translation.

I am indebted to publishers Academic Press, Elsevier, Heyden and Son Ltd., Pergamon Press and Wiley-Interscience, personally to Professors E. D. Becker and T. C. Farrar (Washington), A. F. Casy (Alberta) and H. Wamhoff (Bonn), Doctors H. Schneiders (Bruker, Karlsruhe), D. Shaw (Oxford), C. S. Springer (New York), F. W. Wehrli (Bruker, USA), T. Wirthlin (Varian, Zug) and to firms Bruker Physik AG, JEOL, Sadtler Research Laboratories Inc., and Varian Associates for placing at my disposal illustration material or, respectively, for permitting my use of this material.

My thanks are due to the managers of my firm, EGYT Pharmacochemical Works, personally to Professor L. Pallos, Scientific Director, for ensuring the conditions for compiling this book.

It gives me great pleasure to acknowledge all my previous and present collaborators their enthusiasm, careful and competent work in all phases of the preparation of this book. I want to mention them by name: late Mrs. Dr. Zs. Méhesfalvy, Mrs. A. Csokán, Mrs. J. Csákvári, Mr. A. Fürjes, Mrs. Á. Kiss-Tamás, Mrs. M. Leszták, Mr. Gy. Mányai, Mrs. É. Mogyorósy, Mrs. Dr. K. Ósapay, Mr. I. Pelczer, Miss V. Windbrechtinger, and Mr. J. Zimonyi.

I release these books in the hope that it will benefit all those colleagues who wish to be acquainted with this singularly many faceted, still rapidly developing, in all fields of chemistry extremely efficient and again and again reviving branch of science: NMR spectroscopy.

Pál Sohár
Budapest, April 1981

THE AUTHOR

Pál Sohár, Ph.D., D.Sc. is the Head of the Spectroscopic Department of EGYT Pharmacochemical Works and Professor of the Eötvös Loránd University of Sciences, Budapest, Hungary. He graduated in 1959 at the Technical University of Budapest and obtained his Ph.D. degree in 1962 in physical chemistry with "Summa cum Laude" qualification.

Professor Sohár received the "Candidate of Sciences" degree in physical chemistry from the Hungarian Academy of Sciences in 1967 on the basis of his Thesis "Investigation of Association Structures by Infrared Spectroscopy" and the D.Sc. degree in 1973 for his research work in the application of IR and NMR spectroscopy in the structure elucidation of organic molecules. He has served as Professor at the Eötvös Loránd University since 1975. He was the Head of the Spectroscopy Department of the Institute of Drug Research. He began his career in this Institute in 1959 and received promotions to scientific assistant in 1962, to senior assistant in 1969, and to scientific counselor from 1974 to 1980. He assumed his present position in 1980.

Professor Sohár is a member of the Committees of Physical and Inorganic Chemistry, Spectroscopy and Theoretical Organic Chemistry of the Hungarian Academy of Sciences and the member of ISMAR (International Society of Magnetic Resonance, Chicago, Illinois). He is the Secretary of the Committee on Molecular and Material Structure of the Hungarian Academy of Sciences.

Professor Sohár is the author of more than 180 scientific papers and has been the author or coauthor of six books, among them the first monographs in Hungarian on infrared spectroscopy and nuclear magnetic resonance. He is the coeditor of the series *Absorption Spectra in the Infrared Region* published by Akadémiai Kiadó, Budapest (Publishing House of the Hungarian Academy of Sciences). He has given more than 150 scientific presentations or invited lectures, and was several times invited lecturer of postgraduate courses at the Technical Universtiy of Budapest.

His current major research interests include structure elucidation of organic compounds by nuclear magnetic resonance and infrared spectroscopy.

TABLE OF CONTENTS

Volume I

Chapter 1
Theory of Nuclear Magnetic Resonance Spectroscopy

Chapter 2

NMR Spectrometers, Recording Techniques, Measuring Methods

Volume II

Chapter 3
Proton Resonance Spectroscopy

Chapter 4
The Resonance Spectra of Nuclei Other Than Hydrogen

TABLE OF CONTENTS

Volume III

Chapter 1

THEORY OF NUCLEAR MAGNETIC RESONANCE SPECTROSCOPY

INTRODUCTION

Spectroscopy deals with interactions between electromagnetic radiation and matter. The different branches of spectroscopy study phenomena emerging in the molecule absorbing the different frequency components of the spectrum — which is the frequency-resolved representation of the electromagnetic radiation. Depending on the frequency of electromagnetic radiation absorbed, the UV and IR (vibration) spectra appear at higher frequencies, whereas far IR (rotational) and nuclear magnetic resonance (NMR) spectra fall into the range of microwaves and radiowaves (see Figure 1).

The appearance of NMR spectra is therefore connected with the absorption of radiowaves. The absorption of radiation induces transitions among discrete energy levels — just like in other branches of spectroscopy. The energy levels are determined by the magnetic properties of the nuclei in the sample. During energy transitions, the orientation of the magnetic moment of atomic nuclei relative to the external magnetic field is changed.

The absorption of the radiowaves occurs in a resonance fashion, at discrete frequency values, and the frequency of resonance is specific to the individual nuclei, provided the external magnetic field is the same. The frequency, ν, is therefore of fundamental significance in NMR spectroscopy. Its unit is the hertz (Hz), and its dimension is $[s^{-1}]$.

It has to be noted that in IR spectroscopy, ν denotes wavenumbers $[cm^{-1}]$. As

$$\nu = c/\lambda \tag{1}$$

where λ [m] is the wavelength and c is the velocity of light ($\sim 3.10^8$ m/s), a wavelength λ of 1 m corresponds to 3.10^8 Hz = 3.10^2 MHz frequency. The interval of radiofrequencies associated with nuclear magnetic resonance is 10^3 — 10^{-2} MHz.

NMR is a relatively young science. Studying the hyperfine structure in high-resolution electronic spectra of atoms, Pauli concluded in 1924[1098] that certain atomic nuclei possess angular momentum and therefore magnetic moments as well; in other words they behave as spinning charged particles in classical mechanics.

Rabi and co-workers (1939) were the first to prove[1157] experimentally the correctness of Pauli's assumption in their studies on atomic beams. Later they also determined the magnitude of the nuclear angular and magnetic momenta of a few nuclei. Using the hypothesis of Goudsmit and Uhlenbeck,[1435] Stern and Gerlach (1921) were able to show via classical experiments that the intrinsic angular momentum (spin) of the electron is quantized.[551] Following this argument, Dennison in 1927 assumed that the nuclear angular momentum is also quantized and the different quantum states can be characterized by the spin quantum number I.[351] Substantiation of Dennison's assumption required resonance experiments in which the resonance-like absorption of the radiation associated with the transition of nuclei between different spin quantum states is induced by irradiating the nuclei with a radiation of appropriate energy.

Following unsuccessful experiments[572,573] of Gorter in 1936 and 1942, the first resonance signals were observed in 1945 independently by Bloch et al.[150] and Purcell et al.[1153] Bloch studied the water molecule, while Purcell investigated paraffin.

It was Knight in 1949,[774] then Proctor and Yu,[1145] Dickinson,[364] Lindström,[868] and Thomas[1401] in 1950, who established that for metals and for ^{14}N, ^{19}F, and ^{1}H isotopes of inorganic and organic compounds, the magnetic moment of the nuclei depends on their chemical envi-

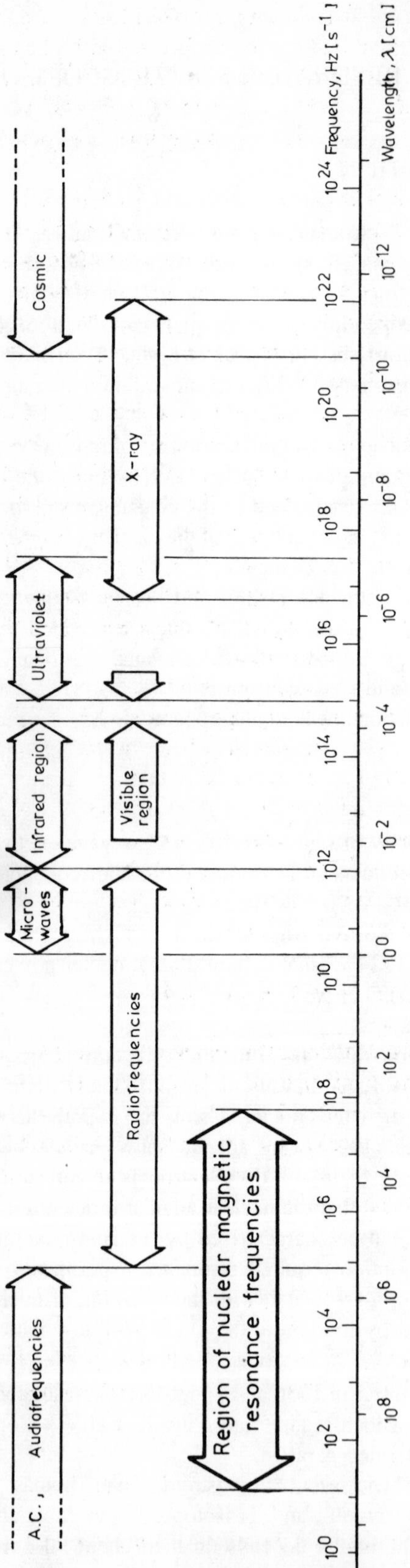

FIGURE 1. The electromagnetic spectrum with the range of NMR frequencies.

ronment, in other words, on molecular structure. Resonance absorption of identical nuclei in different molecules appears therefore at different frequencies or fields. Similarly the identical nuclei of the same molecule in different structural environments may have different resonance frequencies. This was first demonstrated in 1951 by Arnold et al.,[58] who were able to record that the ^{1}H NMR signals of the methyl, methylene, and hydroxyl protons of alcohol appeared separately.

Consequently, the position of the resonance signal can be correlated with molecular structure. The extremely fast development of the NMR method to become one of the powerful methods of structural research can be basically ascribed to this discovery. NMR spectra have proved to be one of the richest sources of information about molecular structure, and they have made possible the solution of many problems that were either impossible or very difficult to solve earlier. In general, the structure of NMR spectra is much simpler than that, for example, of IR spectra, thus their interpretation is easier.

IR and NMR spectroscopy are rather fortunately complementary to one another. IR bands reflect the characteristics of polar atomic groupings and bonds (the intensity of an IR band increases proportionally to the bond or group polarity), whereas ^{1}H NMR spectra provide information mainly about nonpolar C–H bonds — which possess in the IR spectrum rather uncharacteristic absorptions.

NMR is an efficient tool in almost every area of organic chemistry; its use extends from identification, which can be made when the spectrum of the unknown compound is compared to spectra of known ones, through studies of hydrogen bonds and substituent effects, structural elucidations, and stereochemical, conformational, or kinetic studies.

The fast expansion of the method has been greatly aided by modern electronics and transistor techniques, allowing the commercial production of high-performance NMR spectrometers. These enable, in spite of the very exacting technical requirements of the method, the fast, operationally simple, and dependable recording of resonance spectra.

The use of superconducting magnets gave a new impulse to the development of NMR spectroscopy, which made possible one-order increase of previously accessible magnetic fields and the application of Fourier transform measuring technique, which has resulted more orders of increase in sensitivity of NMR spectrometers, and it made possible routine investigation of insensitive atomic nuclei, primarily the ^{13}C isotope.

1.1 THE MAGNETIC RESONANCE SPECTRUM

1.1.1. Magnetic Properties of Atoms and Molecules; Magnetic Moment and Susceptibility

Under the effect of magnetic field, a magnetic moment opposite to the field (diamagnetic) is induced in atoms and molecules. The magnetic moment **M** produced in unit volume of matter, i.e., magnetization, is a sum of individual magnetic moment vectors $\boldsymbol{\mu}_i$, induced in the atoms or molecules. If there are *N* magnetic species in unit volume,

$$\mathbf{M} = \sum_{i=1}^{N} \mu_i \tag{2}$$

When the magnetic field is created by electric current, it is useful to define field **H**, which is the total conduction current, I_v, encompassed by Curve l (conductor or line of force of length l):

$$\oint \mathbf{H}\, dl = I_v \tag{3}$$

It is a good approximation for most substances that magnetic moment **M** is proportional to the field inducing it:

$$\mathbf{M} = \chi\mathbf{H} \tag{4}$$

and by means of Equation 3, **H** is easy to determine from the readily measurable quantity I_v. Its SI unit is $m^{-1}A$, the CGS unit is oersted $[cm^{-1/2}g^{1/2}s^{-1}]$. Proportionality factor χ is magnetic susceptibility, a dimensionless number, which gives the magnitude of magnetic moment induced in 1 mol of substance by unit field **H**. In diamagnetic materials, χ is a small negative number in the order of 10^{-5}. In classical physics and up to the last years in NMR literature, too, vector **H** was termed as "magnetic field strength". This name is, however, misleading, since they are identical only in vacuum, in the case of magnetic field induced by a free magnetic pole or in the theory treating magnetism in the magnetic pole approximation. Therefore, this terminology should be avoided.[1279]

The magnetic field which has changed in the atomic or molecular medium and which is of fundamental importance in the theory of electromagnetic field is characterized by vector **B** (magnetic flux density, magnetic induction), defined by

$$\mathbf{B}dl = \mu_0(I_v + I_m) \tag{5}$$

which takes into account both types of current, conduction current, I_v, and current, I_m, due to magnetization **M** of the medium. In Equation 5, μ_0 is the permeability of vacuum, and its magnitude is, by definition, $\mu_0 = 4\pi \cdot 10^{-7}$. Its SI unit is $m^{-1}sA^{-1}V = mkgs^{-2}A^{-2}$.

Magnetic field **B** and **H** are connected by the following relationship of general validity:

$$\mathbf{B} = \mu_0(\mathbf{H} + \mathbf{M}) \tag{6}$$

or, with regard to Equation 4,

$$\mathbf{B} = \mu_0(1 + \chi)\mathbf{H} = \mu\mathbf{H} \tag{7}$$

Equation 7, similar to Equation 4, is an approximation, since χ and μ are not constant, being functions of **B**. Since, however, in most cases — disregarding of paramagnetic materials — χ is very small, the use of Equations 4 and 7 instead of Equation 6 of general validity causes negligible error.

Paramagnetic atoms or molecules have permanent magnetic moments independently of the magnetic field. This permanent moment is generally several orders higher than the moment induced by the field. The thermal motion of paramagnetic atoms or molecules (e.g., iron, ferromagnetic materials) and the simultaneous presence of magnetic field produce a macroscopic magnetic moment of the same direction as the field (paramagnetic). For paramagnetic materials $\chi > 0$, whereas for diamagnetic ones, $\chi < 0$. Equation 7 shows that in vacuum (of susceptibility $\chi = 0$) $\mu = \mu_0$, and thus the substitution of μ for μ_0 in the above equations means that the medium is taken into account.

The SI unit of magnetic field **B** is $[kgs^{-2}A^{-1} = m^{-2}sV]$, the tesla (T). Its use is, however, inconvenient, since the unit is too big in comparison with magnetic fields occurring in practice. Its CGS analogue is, therefore, still used: 1 G $[cm^{-1/2}g^{1/2}s^{-1}]$ is equivalent to 10^{-4} T (but not identical, since the dimensions of **B** are not the same in the SI and CGS systems!). In CGS the magnetic permeability, μ of Equation 7 is dimensionless, and its magnitude in the "nonrationalized" form is unity ($\mu' = 1$). In the SI system, rationalized permeability, μ, is used: $\mu = 4\pi\mu'$. This is the reason why both the units and the numerical values of **B** and **H** $[cm^{-1/2}g^{1/2}s^{-1}]$ are identical in the CGS system. In turn, it follows from Equation 6 that the units of **M** and **H** are identical in both the SI and the CGS systems.

Starting from the fact that **M** = 0 in vacuum (no magnetization is possible) and thus $B = \mu_0 H$, it is easy to obtain the conversion factors between the SI and CGS units of **B** and **H**. Taking into account Equation 3, a straight conductor of infinite length in which a current of 1 A flows induces a magnetic field of $B_{SI} = \mu_0 I/2\pi r = 4\pi \cdot 10^{-7}/2\pi = 2 \cdot 10^{-7}$ T at a distance of 1 m. Using CGS units, $B_{CGS} = 4\pi\mu' I/2\pi r = 2I/r = 2 \cdot 10^{-1}/10^2$, since I_{CGS}: [biot], r_{CGS}: [cm] and 1 A = 10^{-1} biot and 1 m = 10^2 cm, hence $B_{CGS} = 2 \cdot 10^{-3}$ G. Therefore, 1 T $\triangleq$ 10^4 G. From this, the conversion factor of H_{SI} and H_{CGS} may also be determined. 1 H_{SI} [Am^{-1}] corresponds to μ_0 T (B_{SI}):$B_{SI} = 4\pi \cdot 10^{-7}$ T $\simeq 4\pi \cdot 10^{-3}$ G. $H_{CGS} = B_{CGS} = 4\pi \cdot 10^{-3} = 0.01256$ G, oersted, and 1 Oe $\triangleq$ 1/0.01256 $m^{-1}A$ = 79.6 $m^{-1}A$.

From atomic physics it is well known that the magnetic moment of the electrons of an atom or of a molecule is given by:

$$\mu_e = g_e \beta\sqrt{J(J + 1)} \tag{8}$$

where g_e is the nuclear g, or Lande factor, and J is the quantum number representing angular momentum, while β is the Bohr magneton:

$$\beta = eh/4\pi m_e c = 9.2732 \cdot 10^{-24} \ [JT^{-1}] \tag{9}$$

where e is the electronic charge, and h is the Planck constant, while m_e is the electronic mass.

Nuclear magnetic moment is given by a similar expression:

$$\mu_N = g_N (m_H/m_N)\, \beta_N \sqrt{I(I + 1)} \tag{10}$$

where β_N is the nuclear magneton, expressed in proton mass units ($\mu_H = 1.41049 \cdot 10^{-26}$ JT^{-1}), while I is the so-called *spin quantum number* associated with the total angular momentum, **P**, of the nucleus.

If paramagnetic nuclei of moment μ_N are immersed into an external magnetic field, these become oriented relative to the direction of the field and the magnitude of the resultant susceptibility is described by the Langevin-Brillouin relationship known from the classical theory of magnetism:

$$\chi_N = N\mu_N^2/3kT \tag{11}$$

where N is number of magnetic nuclei in unit volume, k is the Boltzmann factor, and T is the absolute temperature. Equation 11 is not valid for low temperatures and high fields since it is merely the first term of an expansion!

It follows that both magnetic moment and susceptibility of atomic nuclei are much smaller than those of electrons. Since g_e and g_N have the same magnitude and $\beta_N/\beta = 1/1836$, the paramagnetic susceptibility of atoms is smaller by a factor of about 1836^2 than diamagnetic susceptibility resulting from electrons. This is the reason why the static measurement of paramagnetism is very difficult, even at temperatures in the neighborhood of absolute zero. This also explains why electron spin resonance (ESR) spectroscopy developed faster, since to induce and detect this kind of resonance is experimentally incomparably easier.

1.1.2. The Quantized Nature of the Angular and Magnetic Moments of Atomic Nuclei

As mentioned in the introduction, Pauli's assumption about the existence of *angular momentum* of certain atomic nuclei was verified by many experiments. The total angular momentum, **P**, results from the *spin* and *orbital angular momenta* of the nucleons. For an even number of nucleons, the maximum projection value of **P** in a given direction is always an integral multiple of $h/2\pi = \hbar$, and when in addition also the number of protons and neutrons is even, this projection vanishes, whereas for odd numbers of nucleons, it is a half-integral multiple of $\hbar$.

These rules have important practical consequences. Those nuclear species, namely, whose spin quantum number is zero, have according to Equations 10 and 12 neither nuclear nor magnetic angular moments. In other words, they are inactive in NMR spectroscopy. Such are the nuclei of ^{12}C, ^{16}O, and ^{32}S. This circumstance simplifies the ^{1}H NMR spectra of organic compounds, since no magnetic interaction (the so-called spin-spin interaction) is expected between protons and the nuclei ^{12}C, ^{16}O, and ^{32}S. The resonance absorption of O, C, and S is observed only for isotopic species ^{17}O, ^{13}C, and ^{33}S (''magnetic active'' or simply ''magnetic'' nuclei) of low abundance.

The absolute value of the angular momentum vector is

$$\mathbf{P} = \hbar\sqrt{I(I+1)} \tag{12}$$

The largest component taken in arbitrary direction x is then:

$$(P_x)_{max} = \hbar I \tag{13}$$

In addition P_x can also assume — for independent particles with equal probability — the values $\hbar(I - 1)$, $\hbar(I - 2)$, . . . $\hbar(-I + 1)$, $\hbar(-I)$, altogether $(2I + 1)$ different values. A free nucleus therefore can be found in $(2I + 1)$ different magnetic states. These states are degenerate, i.e., (in the absence of external magnetic field) their energy is identical and they differ only in the relative position of the angular momentum vector and the direction of measurement. The experimentally determinable magnetic moment $\boldsymbol{\mu}$ is always parallel or antiparallel to the angular momentum vector **P**, or zero if the latter vanishes. Their ratio, γ, is the *gyromagnetic ratio* and is characteristic of the individual nuclei:

$$\mu/\mathbf{P} = \gamma \; [s^{-1}T^{-1}] \tag{14}$$

For the hydrogen is $\gamma_H = 2.67519 \cdot 10^8 \; s^{-1}T^{-1}$.

Similarly to **P** the magnetic moment $\boldsymbol{\mu}$ is also quantized:

$$\mu = \gamma\hbar\sqrt{I(I+1)} \tag{15}$$

The quantized nature means that the components of $\boldsymbol{\mu}$ measurable in an arbitrary direction — being for example, when the nucleus is studied in the presence of a magnetic field the direction of the external field, $\mathbf{B}_0$ — can only assume distinct values:

$$\mu_{\mathbf{B}_0} = m\mu/I = \gamma\hbar m \tag{16}$$

where $\mu = (\mu_{B_0})_{max} = \gamma\hbar I$ is the value of the maximum component along the direction of the field $\mathbf{B}_0$ and m can again assume the values $I, I - 1, \ldots -I + 1, -I$. The μ is therefore the maximum component of the vector $\boldsymbol{\mu}$.

According to the theory of Dirac, the magnitude of the magnetic moment of the proton is $\beta N\sqrt{I(I + 1)} = (eh/4\pi m_H c)\sqrt{I(I + 1)}$. The experimentally determined value is, however, about 2.79 times of the theoretical one; this difference is taken into account by the Lande factor, g. In fact the magnetic moment $\boldsymbol{\mu}$ is not collinear with the angular moment **P**, but — using the classical picture for describing the motion — rotates about **P**, tracing out a conical surface, and experimentally one measures the time average in the direction of **P**. Furthermore, the true magnitude of the magnetic moment vector $\boldsymbol{\mu}$ defined by Equations 10 and 15 cannot even in principle be equal to the maximal component along a given spatial direction, as demanded by the Heisenberg uncertainty principle.

Thus, the number m defining the angular and magnetic moments in a certain direction is

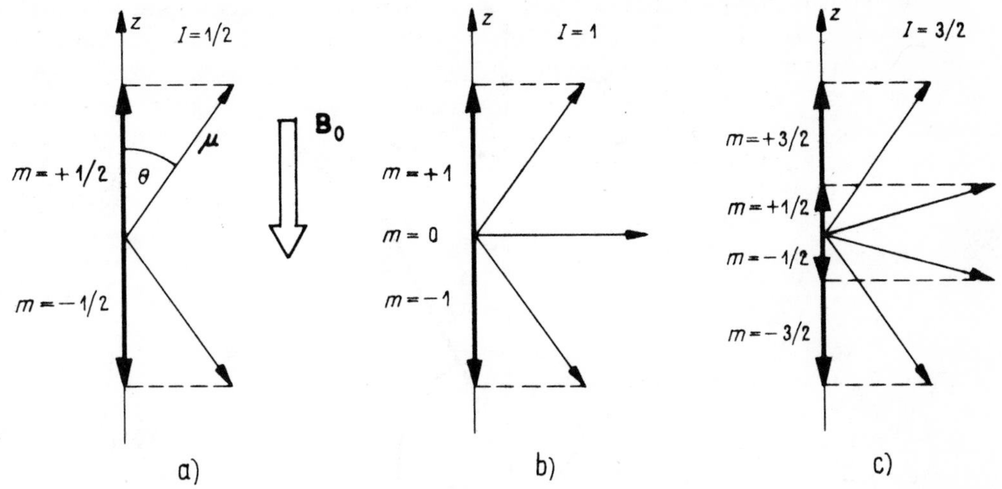

FIGURE 2. Possible orientations of the magnetic moment relative to the external magnetic field $\mathbf{B}_0$ for nuclei of (a) $I = 1/2$, (b) $I = 1$, and (c) $I = 3/2$.

$\pm$ 1/2 for the proton, $\pm$ 1 and 0 for ^{14}N, $\pm$ 3/2 and $\pm$ 1/2 for ^{7}Li, since the spin quantum numbers for these nuclei are 1/2, 1, and 3/2, respectively.

Therefore, when the nucleus is placed into an external magnetic field, the magnetic moment vector can be aligned in certain directions only. The value of the angle θ between the external field $\mathbf{B}_0$ and $\boldsymbol{\mu}$ can only assume such values that $\mu \cos \theta$ (the projections of $\boldsymbol{\mu}$ along the direction of the field $\mathbf{B}_0$) be integral and half-integral multiples of $\gamma\hbar = \mu/I$, i.e., $\cos \theta = m/\sqrt{I(I + 1)} \approx m/I$ (see Figure 2).

1.1.3. The Motion of Nuclei in an External Magnetic Field; The Larmor Precession

Nuclei have angular momenta arising from the motion of the nucleons and therefore also possess magnetic momenta, in other words they can be regarded as spinning charged particles. Any rotating electrically charged particle generates a magnetic field of definite magnitude and orientation. Moving nuclei can thus be treated as small electromagnets associated with a magnetic moment vector $\boldsymbol{\mu}$.

In order to see what happens when the moving nuclei having a magnetic moment $\boldsymbol{\mu}$ are immersed into a magnetic field $\mathbf{B}_0$, we have to examine the motion of the nuclei in the magnetic field. A rotating particle retains its angular velocity and axis of rotation until an external force acts upon it. When any rotating particle is put into an external field, its angular velocity remains the same, but there is a change in the orientation of its rotational axis, and this axis will itself rotate about the direction of the field. For the nuclei in the field $\mathbf{B}_0$, this means that vectors $\mathbf{P}$ and $\boldsymbol{\mu}$ are rotating about the direction of the external field $\mathbf{B}_0$. This complex motion — including the rotation of the nucleus about $\mathbf{P}$ and $\boldsymbol{\mu}$ — is called Larmor precession (see Figure 3).

During precession, a torque $\mathbf{L}$ is exerted upon the particle of magnetic moment $\boldsymbol{\mu}$, which — according to the Newton's laws of motion — is equal to the change of the angular momentum during the precessional motion:

$$\mathbf{L} = d\mathbf{P}/dt = \dot{\mathbf{P}} \tag{17}$$

It is known from the elementary theory of magnetism that

$$\mathbf{L} = \mu \times \mathbf{B}_0 \tag{18}$$

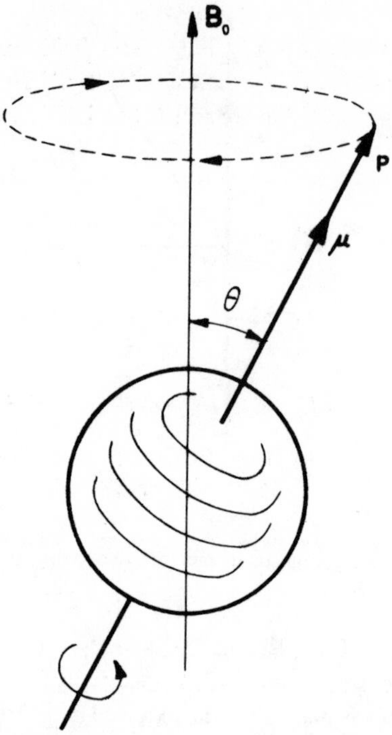

FIGURE 3. The precession of a nucleus having magnetic **μ** and angular momentum **P** in an external magnetic field $\mathbf{B}_0$.

Comparing the above formula with Equations 14 and 17, we obtain:

$$\dot{\mathbf{P}} = \gamma \mathbf{P} \times \mathbf{B}_0 \tag{19}$$

This equation of motion describes the precession of **P** about the magnetic field $\mathbf{B}_0$. The angular velocity of this motion is given by

$$\omega_0 = \dot{\mathbf{P}}/\mathbf{P} = \gamma \mathbf{B}_0 \tag{20}$$

This is the Larmor equation which gives the frequency of the precession by

$$\nu_0 = \omega_0/2\pi = \gamma \mathbf{B}_0/2\pi = \gamma\!\!\!/ \mathbf{B}_0 \tag{21}$$

where $\gamma\!\!\!/$ is the value of gyromagnetic ratio as divided by 2π.

It follows from Equation 21 that the precessional frequency ν_0 is independent of the magnitude of the θ angle spanned by the magnetic moment **μ** and the direction of the external magnetic field $\mathbf{B}_0$ and that its value (which depends on the gyromagnetic ratio) is characteristic for individual nuclei. The value of ν_0 is given in Table 1 [434,1379] for the nuclei most important in organic chemistry.

1.1.4. Magnetic Levels and the Energy of Transitions Between Them

In the absence of an external magnetic field, the orientation of the magnetic moments of nuclei is arbitrary. In the presence of some external field $\mathbf{B}_0$, however, there can be only $(2I + 1)$ definite orientations relative to the field.

Different orientations of the magnetic moment are associated with different energy states:

Table 1
SPIN QUANTUM NUMBER, NATURAL ABUNDANCE, RELATIVE NMR SENSITIVITY, AND RESONANCE FREQUENCY IN POLARIZING FIELDS OF 1.00, 1.41, 2.34, 5.87, AND 11.74 T, RESPECTIVELY, FOR SOME NUCLEI[434,1379]

Isotope	Spin quantum number, I_z	Natural abundance (%)	Relative NMR sensitivity	Resonance frequency (MHz) 1.00 T	1.41 T	2.34 T	5.87[b] T	11.74[b] T
^{1}H	1/2	99.985	1.00	42.577	60.000	100.000	250.000	500.000
^{2}H	1	$1.5 \cdot 10^{-2}$	$9.65 \cdot 10^{-3}$	6.536	9.2104	15.351	38.376	76.753
^{3}H	1/2	0.00[a]	1.21	45.414	63.997	106.664	266.658	533.317
^{6}Li	1	7.43	$8.51 \cdot 10^{-3}$	6.265	8.8286	14.775	36.789	73.578
^{7}Li	3/2	92.58	0.293	16.547	23.317	38.862	97.158	194.317
^{10}B	3	19.58	$1.99 \cdot 10^{-2}$	4.575	6.4479	10.746	26.866	53.732
^{11}B	3/2	80.42	0.165	13.660	19.250	32.084	80.209	160.419
^{12}C	0	98.892	0.00	—	—	—	—	—
^{13}C	1/2	1.108	$1.59 \cdot 10^{-2}$	10.705	15.087	25.144	62.860	125.721
^{14}N	1	99.63	$1.01 \cdot 10^{-3}$	3.077	4.3343	7.2238	18.059	36.118
^{15}N	− 1/2	0.37	$1.04 \cdot 10^{-3}$	4.315	6.0798	10.133	25.332	50.664
^{16}O	0	99.96	0.00	—	—	—	—	—
^{17}O	− 5/2	$3.7 \cdot 10^{-2}$	$2.91 \cdot 10^{-2}$	5.772	8.134	13.56	33.892	67.784
^{19}F	1/2	100.0	0.83	40.055	56.446	94.077	235.192	470.385
^{29}Si	− 1/2	4.70	$7.84 \cdot 10^{-2}$	8.460	11.919	19.865	49.662	99.325
^{31}P	1/2	100.0	$6.63 \cdot 10^{-2}$	17.235	24.288	40.481	101.202	202.404

[a] Radioactive isotope.
[b] Data from "Almanac 1982" from Bruker AG, Bruker Analytic GMBH, Karlsruhe, Rheinstetten, West Germany.

$$\mathbf{E} = -\mu \cdot \mathbf{B}_0 = \mu B_0 \cos\theta = \mu B_0 m/I \qquad (22)$$

Energy levels are numbered following decreasing magnitude: $E_1 > E_2 > \ldots > E_n > E_{n+1}, \ldots$, etc. At the lowest energy level, the magnetic moment is oriented parallel to the field $\mathbf{B}_0$; its projection is therefore $-\mu$. (Provided that the direction of $\mathbf{B}_0$ is taken to coincide, as customary with the negative z axis.) Level E_1 there corresponds to an orientation of the magnetic moment antiparallel to the field $\mathbf{B}_0$ and to the spin quantum number $m = + I$, and so to a $+ \mu$ projection component.

In the case of $I = 1/2$ therefore, $E_1 = \mu B_0$ (if $m = + 1/2$) and $E_2 = - \mu B_0$ (if $m = - I/2$). For $I = 1$, $E_1 = \mu B_0$, $E_2 = 0$ and $E_3 = - \mu B_0$ (provided m has the values $+1$, 0, and -1, respectively). Finally for the case $I = 3/2$, $E_1 = \mu B_0$, $E_2 = \mu B_0/3$, $E_3 = - \mu B_0/3$, and $E_4 = -\mu B_0$ (for m values $+ 3/2$, $+ 1/2$, $- 1/2$, and $- 3/2$).

Considering Equations 13 and 17, the energy can be given as:

$$E = \gamma \hbar B_0 m \qquad (23)$$

where $m = I, I - 1, \ldots, - I + 1, - I$.

Any two neighboring energy levels are different from one another by $\gamma\hbar B_0$. The higher B_0, the greater the separation of the energy levels.

Transitions are possible between certain levels accompanied by uptake or release of energy. Later it will be shown by quantum-mechanical arguments that only transitions between neighboring energy levels are possible or "allowed", i.e., in reality only such transitions occur. Naturally this problem is immaterial with a single proton since there are only two

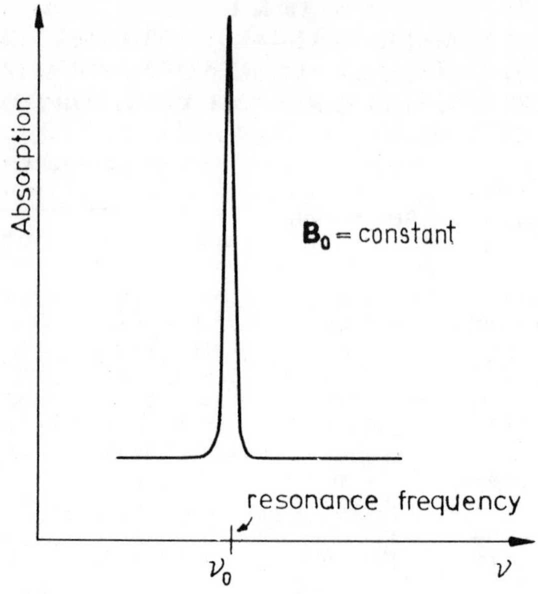

FIGURE 4. Absorption as a function of frequency in constant external magnetic field.

levels (according to the values $m = -1/2$ and $m = +1/2$). Transition energy for two vicinal levels is

$$\Delta E = \hbar\nu = \gamma h B_0 \tag{24}$$

From which

$$\nu = \gamma B_0 \tag{25}$$

In other words, transitions between adjacent energy levels are possible only when the nucleus in field $\mathbf{B}_0$ absorbs electromagnetic radiation of the frequency as in Equation 25 or when it emits radiation of the same frequency. With magnetic fields in the order of 1 T, the transitions between magnetic levels are brought about by radiation in the RF range.

As it can be seen, the Larmor frequency, ν_0, is equal to the frequency, ν, of transition between two adjacent energy levels. If radiation of such $\nu = \nu_0$ frequency interacts with the nucleus, a change of state is effected while the precession angle θ (the angle between $\mathbf{B}_0$ and $\boldsymbol{\mu}$) is altered. When $\mathbf{B}_0$ is constant and the frequency is changed continuously, a maximum of absorption or emission will be observed at the value corresponding to the Larmor frequency, i.e., magnetically active nuclei, when placed in a magnetic field, absorb or emit the electromagnetic radiation in a resonance fashion (see Figure 4).

1.1.5. Interpretation of Transitions Between Magnetic Energy Levels

It was mentioned that when the nucleus is irradiated with a frequency satisfying the resonance condition given by the Larmor equation, absorption takes place encountered and the nucleus is excited to a magnetic level of higher energy. Simultaneously, the orientation of the nuclear magnetic moment in the $\mathbf{B}_0$ field is altered. Let us now consider, using concepts of classical mechanics, the change of the magnetic quantum state (energy) of nuclei upon the absorption of radiowaves.

It was shown that in an external magnetic field $\mathbf{B}_0$, the nucleus undergoes precession about the field axis (see Figure 3), while $\mathbf{B}_0$ exerts a torque upon it, trying to force the axis of rotation towards the direction of the field. In order to produce a transition between the magnetic levels, a $\mathbf{B}_1$ magnetic field — perpendicular to the $\mathbf{B}_0$ and rotating about its direction in phase with the nucleus — should be superimposed upon $\mathbf{B}_0$. Field $\mathbf{B}_1$ exerts a torque upon the nucleus, resulting in a precession and thereby in a transition between the two magnetic levels corresponding to the two different orientations of the moment $\boldsymbol{\mu}$. This effect can be explained as follows.

Let the direction of field $\mathbf{B}_1$ be perpendicular to the direction of field $\mathbf{B}_0$ that lie in a plane defined by $\boldsymbol{\mu}$ and $\mathbf{B}_0$ (see Figure 5a). For the moment let us disregard the effect of $\mathbf{B}_1$. Then, according to the previous discussion, the nucleus rotates about $\boldsymbol{\mu}$ and simultaneously $\boldsymbol{\mu}$ also rotates about the $\mathbf{B}_0$. If now $\mathbf{B}_1$ rotates together with moment $\boldsymbol{\mu}$, then the precession about $\mathbf{B}_0$ can be left out of consideration from the point of view of the relative position of $\mathbf{B}_1$ and $\boldsymbol{\mu}$ (see Figure 5b) and thus the precession is reduced to rotation about $\boldsymbol{\mu}$. Let us now consider field $\mathbf{B}_1$ as well, which induces precession about its own directions (see Figure 5c). Evidently during such a motion the direction of $\boldsymbol{\mu}$ is changed so that its positions in the two half-periods correspond precisely to orientations associated with the two possible energy levels. The rotating field $\mathbf{B}_1$ superimposed upon $\mathbf{B}_0$ then induces indeed transition between the energy levels.

$\mathbf{B}_0$ and $\mathbf{B}_1$ may be denoted as *polarizing* and *excitation* fields, respectively, because the former effects about the ordering of the magnetic moments, whereas the latter triggers spin flipping from an orientation of lower energy into one of higher energy.

1.1.6. Population of the Magnetic Energy Levels

Population of energy levels — as known from thermodynamics — decreases exponentially with increasing energy. It is expected therefore that in a magnetic field, the magnetic moment of the majority of nuclei is oriented in the direction of the field.

For protons there are two energy levels, according to the values + 1/2 and − 1/2 of m, and the partition of the nuclei among these states is defined by the Boltzmann distribution law. Considering Equation 22 as well, we obtain:

$$N''/N' = e^{\Delta E/kT} = e^{\gamma \hbar B_0/kT} = e^{2\mu B_0/kT} \approx 1 + 2\mu B_0/kT \tag{26}$$

where N'' and N' denote the number of nuclei at the lower and higher levels, ΔE denotes the energy difference of the two levels, k denotes the Boltzmann constant, and T denotes the absolute temperature.

The ratio N''/N' for protons at room temperature and a magnetic field of 1 T is very close to unity: 1.0000066. The reason for this is the small energy difference (ΔE) between the two levels the energy of thermal motion being sufficient to raise nuclei to the higher level.

In contrast to the energy changes in usual chemical reactions whose magnitude is $\sim 5 \cdot 10^4$ J/mol, the radiation energy sufficient to raise 1 mol of protons to the higher magnetic niveau is only about $2 \cdot 10^{-2}$ J. The energy of thermal motion is of magnitude $5 \cdot 10^2$ J/mol at room temperature.

The above relationship explains also why it is necessary to apply very high polarizing fields in magnetic resonance experiments. The population difference of the higher and lower levels is, namely, proportional to the energy difference ΔE which increases proportional to the magnitude of B_0. The condition for resonance absorption is a population difference of the levels concerned. Unless other phenomena would not reestablish the original distribution continuously, absorption of energy would be observable only for a short time, until the number of emitting atoms becomes equal to the number of absorbing ones.

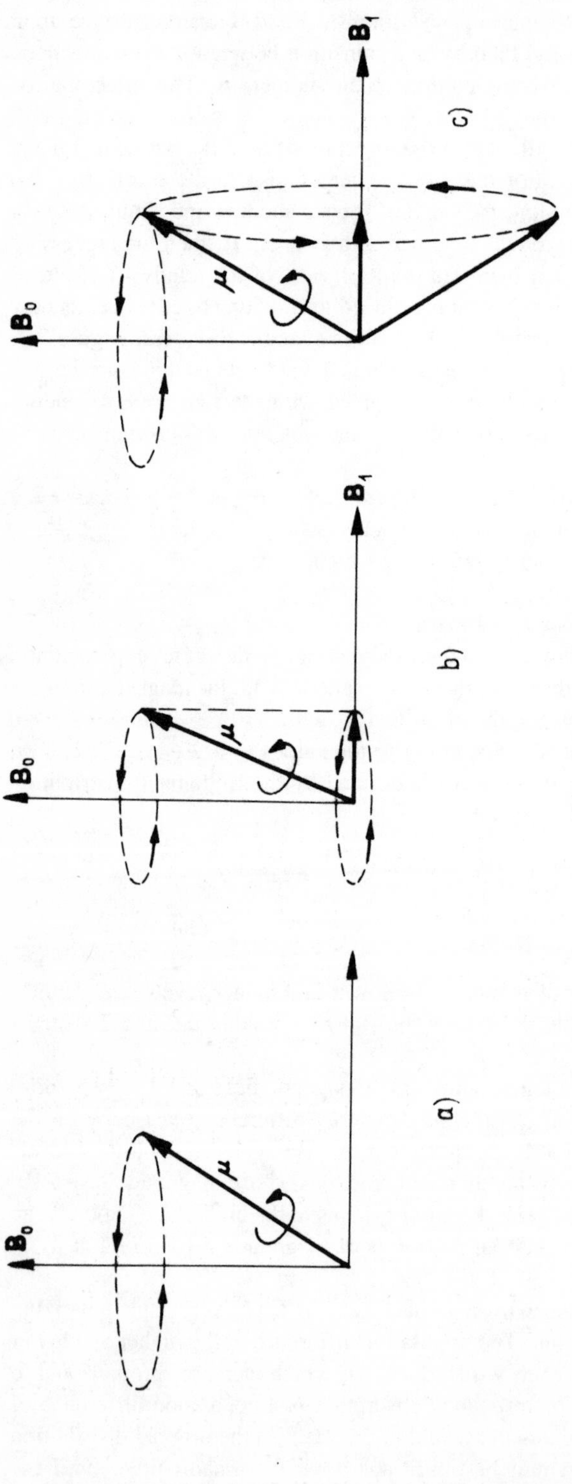

FIGURE 5. Precession of the magnetic moment in the rotating excitation field, $\mathbf{B}_1$ perpendicular to polarizing field $\mathbf{B}_0$. (a) Precession around $\mathbf{B}_0$ in the absence of $\mathbf{B}_1$, (b) rotation of $\mathbf{B}_1$ around $\mathbf{B}_0$ with frequency of precession of the magnetic moment, and (c) precession of the magnetic moment around $\mathbf{B}_1$.

1.1.7. Macroscopic Magnetization of Spin Systems and the Spin Temperature[434]

The macroscopic magnetization of an ensemble of magnetic nuclei induced by a polarizing field depends on the distribution of the nuclei among the magnetic levels. Using the above considerations concerning the population of the individual levels, the macroscopic magnetization induced by the external field $\mathbf{B}_0$ in a system of nuclei of spin quantum number I can be calculated. On applying the polarizing field, the magnetic moments of the spin system examined become ordered relative to the direction of $\mathbf{B}_0$. Consequently, the nonzero macroscopic magnetization $\mathbf{M}$ appears in the direction $\mathbf{B}_0$; its magnitude being proportional to B_0 (compare Equation 6).

The appearance of $\mathbf{M}$ is a consequence of the fact that the population of different spin states is not uniform; therefore in an external magnetic field, the nuclei possess a resultant paramagnetic susceptibility χ_0. Its approximate magnitude is given by Equation 4.

Macroscopic magnetization can thus be described by the following expression:

$$\mathbf{M} = \sum_{m=-I}^{+I} \frac{N_m m\mu}{I} \tag{27}$$

where N_m is the population of the level characterized by quantum number m:

$$N_m = \frac{N}{2I+1}\left(1 - \frac{m\mu B_0}{IkT}\right) \tag{28}$$

Comparing the two expressions above, one obtains

$$M = N\mu^2(I+1)B_0/3kTI \tag{29}$$

For protons and other nuclei of spin quantum number 1/2, $\chi_0 = N\mu^2/kT$. For a nonquantized (classical) system, the classical Langevin-Brillouin expression (see Equation 11) can be recovered by leaving out from Equation 29 the factor $(I + 1)/I$, containing spin quantum number I.

Macroscopic magnetization is a paramagnetic phenomenon originating from atomic nuclei and depending on the temperature. The value of χ_0 is $3 \cdot 10^{-10}$ for 1 mol of water protons. Since diamagnetic susceptibilities are in the order of 10^{-6}, no static measurement is possible for nuclear paramagnetism at room temperature. Even at 2 K the contribution of paramagnetism to the total susceptibility of hydrogen is only 20%.

In a field $\mathbf{B}_0$, the ratio of processes requiring energy, i.e., those which excite nuclei from a lower level to a higher one, to those which result in emission, in an unperturbed nuclear system, depends on the term populations. In the absence of $\mathbf{B}_0$ and in equilibrium at temperature T, every magnetic level is populated equally, as in the absence of a polarizing field, the energy of the levels and, therefore, Boltzmann factor determining their population is identical. If the nuclear system is transferred into a magnetic field $\mathbf{B}_0$, the originally degenerate levels are split to $2I + 1$ levels of different energy. This is Zeeman splitting. The number of nuclei on the m-th energy level is given by Equation 28.

In these expressions, the value of T is not necessarily the actual ambient temperature, only when the system has already reached thermal equilibrium after the application of $\mathbf{B}_0$. The establishment of thermal equilibrium has, however, a finite rate (see later); thus in the moment of applying the polarizing field and for a short time thereafter, the distribution of the nuclei among the levels does not correspond to the equilibrium values. Such distributions can be characterized by a fictitious value of T, the *spin temperature*. Thus, situations when there is an excess of nuclei at the higher level are characterized by negative spin temperatures; such spin systems can be experimentally realized (e.g., in masers).

1.1.8. Spin-Lattice Relaxation

As we have already seen, the original uniformity of the population of the Zeeman levels is upset when a polarizing field ($\mathbf{B}_0$) appears. Until, the establishment of the equilibrium distribution given by Equation 28, there is a constant change in the populations. When the spin of the nuclei $I = 1/2$ (for example for protons), the population of the two levels given by

$$N'' = \frac{N}{2}\left(1 + \frac{\mu B_0}{kT}\right) \text{ and } N' = \frac{N}{2}\left(1 - \frac{\mu B_0}{kT}\right) \tag{30}$$

Let the probability of transition from the lower level to the upper one be W'' (absorption) and let W' be the probability of the reverse process (emission). In equilibrium the number of transitions in opposite directions are identical: $W''N'' = W'N'$, from which:

$$W'/W'' = N''/N' \approx 1 + 2\mu B_0/kT \tag{31}$$

This expression can also be regarded as the definition of spin temperature. Since the difference of N'' and N' changes always by two units, one obtains

$$\dot{n} = 2(N'W' - N''W'') \tag{32}$$

where $n = N'' - N'$. The population difference in equilibrium is

$$n_e = (N'' + N')\,\mu B_0/kT = N\mu B_0/kT \tag{33}$$

Expressing the value of W'' and W' transition probabilities by their average W, we obtain $W'' = W - W\mu B_0/kT$ and $W' = W + W\mu B_0/kT$. Upon substituting these expressions into Equation 32 and using the notations n and n_e,

$$\dot{n} = 2W(n_e - n) \tag{34}$$

Accordingly, it follows that the farther the system is from equilibrium, the faster is the change in the term populations. Upon integration of Equation 34 gives:

$$n_e - n = (n_e - n)_0\, e^{-2Wt} = (n_e - n_0)\, e^{-t/T_1} \tag{35}$$

where the index zero refers to initial values of n_e and n, and

$$T_1 = (2W)^{-1} \tag{36}$$

is the *spin-lattice relaxation time*. The T_1 is defined as the time in which the difference $(n_e - n)$ drops to its *e*-th portion.

The spin-lattice relaxation time T_1 (which is inversely proportional to the transition probability!) can also be regarded as the average lifetime of a spin state. Since it represents the component of the resultant vector of magnetic moment in the direction of $\mathbf{B}_0$ and describes its change in time, it is called "longitudinal relaxation". The ratio of establishing thermal equilibrium is characterized therefore by T_1. The adjective "spin-lattice" of T_1 refers to the fact that it gives the rate of reaching thermal equilibrium between, on the one hand, the nuclear or spin system and, on the other, the medium (crystal lattice in the case of solids). For this reason, the magnitude of T_1 depends much on the medium. In the liquid phase, its value is between 10^{-2} and 10^2 s; for solids longer relaxation times can be encountered, extending up to a few days. T_1 values for various nuclear species are rather different. With

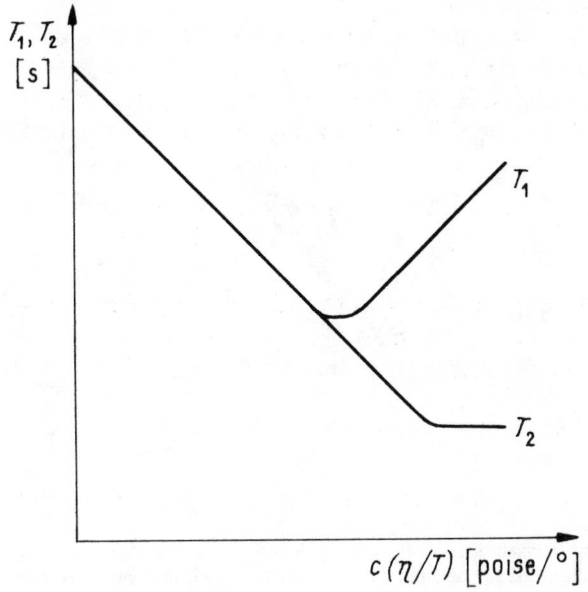

FIGURE 6. The spin-lattice (T_1) and spin-spin relaxation (T_2) times as a function of the ratio of viscosity and absolute temperature.[154]

$I \geqslant 1$, relaxation times are much shorter. Such nuclei possess electric quadrupole moments, which promote further energy exchange with the lattice. The presence of paramagnetic ions in the lattice has similar effects and in such cases T_1 may drop to a level of about 10^{-4} s.

Corresponding to Equation 28 or 33, only at infinitely high temperatures would populations of magnetic levels have the same probabilities. For statistical reasons, however, before the application of the polarizing field, all possible orientations of nuclear moments are equally probable. Consequently, at the moment of switching on the polarizing field, spin temperature can be considered as infinitely high; the system "cools off" until reaching thermal equilibrium. This happens via energy exchange between nuclei of the same molecule or of different ones. This energy exchange cannot occur by collision since the nuclei concerned are bound in a molecule (isolated adiabatically). Therefore, different mechanisms of energy exchange should be considered. The most important relaxation mechanism is the dipole-dipole interaction of the spin system and the lattice (quantitative treatment of this problem and of other mechanisms will be given in Section 2.2.6).

In the vicinity of the nucleus under study, the neighboring molecules are in a state of rapid thermal motion and their dipoles establish quickly changing local magnetic fields. This tumbling may contain components which are perpendicular to $\mathbf{B}_0$ and rotate with the Larmor frequency. These can lead to spin transitions, just like the excitation field $\mathbf{B}_1$. Evidence for this is the viscosity dependence of the relaxation time T_1. Minimum T_1 values are observed at medium viscosity. At low viscosities because of averaging of local fields due to fast tumbling, T_1 increases (see Figure 6). Conversely for high viscosities it is the smaller number of possible local fields (among them those leading to relaxation) which results in the increase of T_1. For similar reasons, T_1 is large for solids. If T_1 is great enough (> 10 s), relaxation processes can be observed easily and the intensity of the resonance signal increases continuously from the moment of the onset of fields $\mathbf{B}_0$ and $\mathbf{B}_1$ until thermal equilibrium is reached (see below).

1.1.9. Spin-Spin Relaxation

In addition to energy exchange with the lattice, nuclei may exchange energy with neigh-

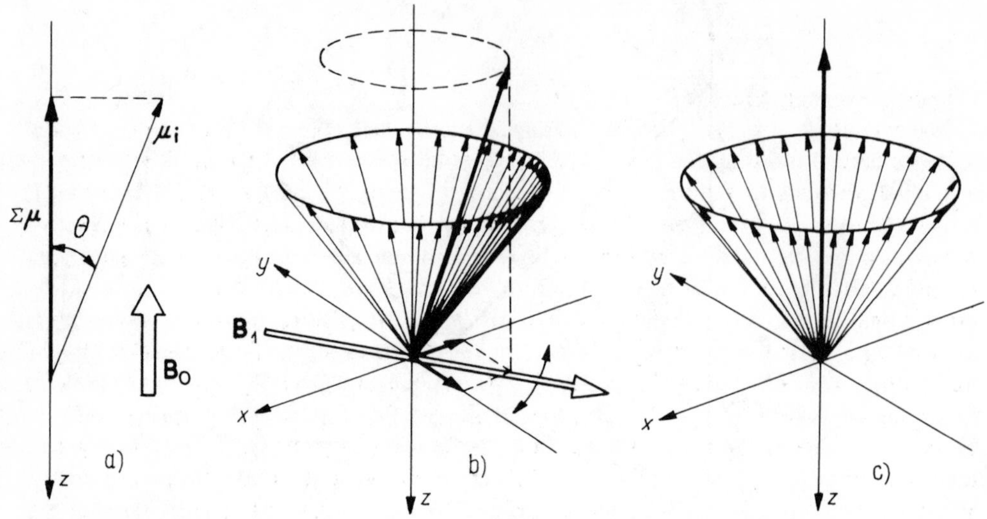

FIGURE 7. Spin-spin relaxation (T_2), the phase memory. (a) The macroscopic magnetization (the resultant magnetic moment) in the polarizing field $\mathbf{B}_0$, (b) the resultant magnetic moment and its components in the excitation field $\mathbf{B}_1$, and (c) reestablishment of the resultant moment to the direction of $\mathbf{B}_0$ (statistical distribution of the components) as a result of the spin-spin relaxation.[1379]

boring nuclei of the same kind. Precession of identical, and only of identical nuclei, can lead to an oscillating field at each other's location with the Larmor frequency, and this may effect a mutual flipping of the spins of the interacting nuclei, while their total energy remains constant. As a result of this process called *spin-spin* or *transversal relaxation*, the M_x and M_y components of the macroscopic magnetization **M** perpendicular to the field $\mathbf{B}_0$ are reduced. The time-constant of the process T_2 is the spin-spin relaxation time.

This phenomenon can be explained as follows. When the excitation field $\mathbf{B}_1$ is switched on, the nuclear moments become aligned in the direction of $\mathbf{B}_1$. This gives rise to such a macroscopic magnetization in the spin system, which has components perpendicular to the direction $\mathbf{B}_0$ as well. Let us, for the sake of simplicity, assume that all magnetic moment vectors originate from the same point. In the presence of $\mathbf{B}_0$, all moments form an angle θ with the direction of $\mathbf{B}_0$ and precess with a random, but steady phase around it. The magnetic moments are statistically uniformly distributed over the precession cone, and the resultant magnetic moment vector **M** is collinear with $\mathbf{B}_0$ (see Figure 7a). The effect of the presence of the rotating excitation field $\mathbf{B}_1$ is to tip the **M** vector towards the direction of $\mathbf{B}_1$ in the plane defined by $\mathbf{B}_0$ and $\mathbf{B}_1$. At the same time, the distribution of the moments $\boldsymbol{\mu}$ over the precession cone is no longer uniform, but they become concentrated around the instantaneous direction of $\mathbf{B}_1$. Precession about $\mathbf{B}_1$ can be neglected here, since it is much slower (because the angular velocity is proportional to the magnitude of the field; compare Equation 20). The **M** vector which is precessing just like the $\boldsymbol{\mu}$ moments of the individual nuclei about the direction of $\mathbf{B}_0$ (the *z*-axis), but it acquires nonzero M_x and M_y components (see Figure 7b). If now for some reason the individual moments do not preserve their phase coherence, the moments become after a certain time again statistically distributed over the precession cone. **M** then returns to the direction of $\mathbf{B}_0$, and its *x* and *y* components disappear (see Figure 7c). The time over which the magnetization in the direction of $\mathbf{B}_1$ is reduced by a factor of *e* is called *transversal relaxation time* or *phase memory*, and its symbol is T_2. "Transversal" refers to the fact that T_2 describes the time dependence of M_x and M_y which are perpendicular to $\mathbf{B}_0$. The term "phase memory" on the other hand points out that the magnitude of T_2 determines the average time over which the phase of nuclear precession is preserved. Thus, the change of the components M_x and M_y of the macroscopic magnetization within a unit time is determined by the magnitude of T_2:

$$\dot{M}_{x,y} = -M_{x,y}/T_2 \qquad (37)$$

(The negative sign refers to the decrease of $M_{x',y'}$.)

The magnitude of T_2 is dependent on several factors. One of them is the already-mentioned pairwise mutual transition of spins to the opposite level, actuated by the local field built up at mutual positions as a result of their precession. During this process, the M_z component remains constant, whereas the phase of precession is changed. Thus, spin-spin relaxation reduces T_2. Hence the term: spin-spin relaxation time for T_2. Since this type of interaction requires that the nuclei be in one another's vicinity for a longer time, T_2 is inversely proportional to viscosity (see Figure 6). At low viscosities, the probability of relaxation is reduced as a consequence of rapid thermal motions, and therefore T_2 increases. It follows that for liquids of low viscosity, $T_1 \approx T_2$, while for viscous liquids $T_2 \ll T_1$ and T_2 lowers to a minimal values of $\sim 10^{-6}$ s. T_2 depends on spin-lattice relaxation processes, i.e., on T_1 as well, since the phase of precession of the nucleus may change in each spin-lattice relaxation transition. T_2 is greatly reduced by the inhomogeneity of the magnetic field $\mathbf{B}_0$. When $\mathbf{B}_0$ shows inhomogeneities across the total volume of the spin-system (sample), the precession frequencies are different as a result of the different fields. This means that even if at a given instant all nuclei are in the same phase, the differences in rotational frequencies lead in a short time to a dephasing; hence T_2 is reduced to minimal values. Solid particles in the solution may also cause magnetic inhomogeneities, thus dephasing. Slowly changing local magnetic fields generated by tumbling of dipole molecules in solids and viscous liquids exert the same effect and also explain line broadening in polar solvents, e.g., dimethyl sulfoxide.

1.1.10. Saturation

According to the foregoing analysis, there are two competing processes in NMR experiments. Upon absorption of radiofrequency (RF) radiation, the number of excited nuclei exceeds those returning via emission into the ground state. Consequently the population difference between the two levels becomes smaller. This change is opposed by nonradiative relaxation processes which take the nuclei from the excited state back to the ground state. The two processes must be in equilibrium. When more nuclei are excited than relaxed in unit time, the population difference between the levels quickly vanishes, leading to the disappearance of the resonance signal. This phenomenon is termed *saturation*, and it can be eliminated by lowering the power of the radiofrequency.

The quantitative picture of saturation can be outlined as follows. Before the application of RF field $\mathbf{B}_1$ — considering in the first approximation only T_1 relaxation — the change of the number of ground-state nuclei n in unit volume is given (compare Equations 34 and 36) by

$$\dot{n} = (n_e - n)T_1^{-1} \qquad (38)$$

where n_e is the value of n for thermal equilibrium.

Since the saturation occurs when the population between the ground- and excited-state magnetic levels is equalized, the rate of T_1 relaxation has foremost significance in this respect. T_1 determines the rate of establishing or reestablishing the population difference of the levels, the change in time of M_z component of the macroscopic magnetization. Saturation is, however, also effected, though to a smaller extent, by T_2. As we shall see, the line width of the resonance signal is determined by T_2, when the signal is broad, saturation (disappearance of the signal) sets in earlier.

Following the application of the RF field $\mathbf{B}_1$, resonance-absorption brings about a decrease of $2nW_a$ in the value of n in unit time, provided the transition probability in unit time is given by W_a. Considering this we obtain

$$\dot{n} = (n_e - n)/T_1 - 2nW_a \qquad (39)$$

When the two opposing processes get into equilibrium, the spin-system assumes a stationary state, i.e., $n = 0$, therefore,

$$\frac{n_{st}}{n_e} = \frac{1}{1 + 2W_aT_1} = Z \qquad (40)$$

where Z is the so-called saturation coefficient, expressing the deviation of n from the thermal equilibrium value (the index of n_{st} refers to the stationary state).

The value of transition probability W_a for $I = 1/2$ and a RF field $\mathbf{B}_1$ is

$$W_a = \frac{1}{4}\gamma^2 B_1^2 g(\nu) \qquad (41)$$

Here $g(\nu)$ is a function defining the shape (natural width) of the resonance line, which is related to the relaxation time T_2 via:

$$g(\nu)_{max} = 2T_2 \qquad (42)$$

The spin-system saturates most quickly for a maximum value of $g(\nu)$. Substituting $g(\nu)$ by $2T_2$ gives:

$$\frac{n_{st}}{n_e} = \frac{1}{1 + \gamma^2 B_1^2 T_1 T_2} = Z_0 \qquad (43)$$

where Z_0 is the saturation factor referring to the maximum value of $g(\nu)$. The value of Z_0 is 1 at thermal equilibrium and 0 for saturation. It then follows:

$$Z_0 = T/T_s \qquad (44)$$

where T_s is the spin-temperature (in thermal equilibrium $T_s = T$; for saturation the spin-temperature is infinite). Evidently the larger are T_1 and T_2, the shorter is the time in which the system reaches the saturation. The magnitude of the saturation factor depends also on B_1. When B_1 is too low, the resonance signal disappears, whereas if it is too high, saturation sets in and signal intensity again drops to zero.*

1.1.11. Natural Line Width of Nuclear Magnetic Resonance Lines

The NMR absorption signal always possesses a certain line width. Line broadening is dependent on several factors. The natural line width of any spectral line is determined by the lifetime of the excited state. This is the outcome of the Heisenberg uncertainty principle, which, as is well known, states that two independent canonically conjugate parameters of a corpuscular system cannot be simultaneously determined with arbitrary accuracy. Energy and lifetime are constituting such a pair and for them it applies that:

$$\Delta t \cdot \Delta E = \Delta t h \Delta\nu \gtrsim h/2\pi \qquad (45)$$

thus,

$$\Delta\nu \gtrsim (2\pi\Delta t)^{-1} \qquad (46)$$

Therefore, the shorter is the average lifetime of the excited state, the greater is the

* Compare p. 22 and Equation 60.

broadening of the resonance signal. Compared with electronic and IR spectroscopy, lifetimes in NMR are rather long. From Equation 46, it follows that for an average lifetime of 10^2 s, the line width is smaller than 10^{-2} Hz. In practice, however, line widths are always greater. Under experimental conditions, there is always some magnetic inhomogeneity in the external field and inside the sample. Not even the best spectrometer can provide a perfectly homogeneous magnetic field, and due to inhomogeneities, lines sharper than 10^{-1} Hz cannot even in principle be expected. In inhomogeneous magnetic fields, the relaxation times and, therefore, the average lifetimes are sharply reduced, sometimes even down to about 10^{-4} s, resulting in very broad lines (with line width of 10^4 Hz), which cannot experimentally be observed any more. This is the reason why the NMR spectra of viscous liquids, solids, and compounds possessing quadrupole moments, as well as samples containing paramagnetic ions, cannot be at all recorded, or can be only using very sophisticated methods. Since for hydrogens $T_1 > 10^{-1}$ s, the line width of in ^{1}H NMR spectra depends on T_2 only.

1.1.12. The Bloch Equations[434]

The first quantitative description for the behavior of magnetic nuclear moments in varying magnetic fields was provided by Bloch in his famous equations describing and defining the time dependence of the total magnetic moment and its components in a unit volume. The solution of these equations originally proposed as hypothetical ones proved to provide an unambiguous and satisfactory explanation for a diversity of phenomena encountered in magnetic resonance, especially for transient and time-dependent effects.

Let us consider first the problem of a system of nuclei having a spin of 1/2 in a constant polarizing field $\mathbf{B}_0$. Let z be the orientation of the field. Then the time dependence of the resultant macroscopic magnetic moment is given by

$$\dot{\mathbf{M}} = \gamma(\mathbf{M} \times \mathbf{B}) \tag{47}$$

which is obtained from the comparison of Equations 2, 14, and 19. The components of $\dot{\mathbf{M}}$ are obtained from the following determinant:

$$\begin{vmatrix} \mathbf{i} & \mathbf{j} & \mathbf{k} \\ M_x & M_y & M_z \\ B_x & B_y & B_z \end{vmatrix} \tag{48}$$

where **i, j,** and **k** are unit vectors and M_x, M_y, and M_z are components of magnetization, while B_x, B_y, and B_z are field components. The motion of the nuclear system in a polarizing field $B_z = B_0$ is then:

$$\dot{M}_x = \gamma(M_y B_z - M_z B_y) = \gamma B_0 M_y \tag{49a}$$

$$\dot{M}_y = -\gamma(M_x B_z - M_z B_x) = -\gamma B_0 M_x \tag{49b}$$

$$\dot{M}_z = \gamma(M_x B_y - M_y B_x) = 0 \tag{49c}$$

since $B_x = B_y = 0$.

The solution of these equations leads to the result that **M** is precessing about direction z with the Larmor frequency. Let us now consider the time-dependent change of **M** due to relaxation processes.

The change of the component M_z in time is determined by relaxation time T_1:

$$M_z = (N'' - N') = \mu n \tag{50}$$

and from Equations 34 and 36,

$$\dot{n}\mu = \dot{M}_z = 2W\mu(n_e - n) = 2W(M_e - M_z) = (M_e - M_z)/T_1 \tag{51}$$

where M_e is the macroscopic magnetization belonging to thermal equilibrium. The other two components have already been defined in the discussion of spin-spin relaxation (see Equation 37).

The Bloch equations are derived by taking into account the components γ ($\mathbf{M} \times \mathbf{B}_1$) originating from the presence of excitation field $\mathbf{B}_1$ perpendicular to $\mathbf{B}_0$ and rotating with an angular velocity, ω:

$$(B_1)_x = B_1 \cos \omega t \tag{52a}$$

$$(B_1)_y = -B_1 \sin \omega t \tag{52b}$$

$$(B_1)_z = 0 \tag{52c}$$

Upon substituting the field components (see Equation 52) into the determinant (see Equation 48) and considering the relaxation processes as well, the Bloch equations are obtained:

$$\dot{M}_x = \gamma(M_y B_0 + M_z B_1 \sin \omega t) - M_x/T_2 \tag{53a}$$

$$\dot{M}_y = \gamma(M_z B_1 \cos \omega t - M_x B_0) - M_y/T_2 \tag{53b}$$

$$\dot{M}_z = \gamma(M_x B_1 \sin \omega t + M_y B_1 \cos \omega t) - (M_z - M_e)/T_1 \tag{53c}$$

The solution for the Bloch equations in the case of a constant or slowly changing RF field, provided the absorption of RF energy keeps always exactly in balance with the energy released from the nuclei to the lattice (by the spin-lattice relaxation), i.e., if $\dot{M}_z = 0$, is the following:

$$M_x = \frac{1}{2} \chi_0 \omega_0 T_2 \frac{2B_1 \cos \omega t \cdot \Delta\omega T_2 + 2B_1 \sin \omega t}{1 + (\Delta\omega)^2 T_2^2 + \gamma^2 B_1^2 T_1 T_2} \tag{54a}$$

$$M_y = \frac{1}{2} \chi_0 \omega_0 T_2 \frac{2B_1 \cos \omega t - 2B_1 \sin \omega t \cdot \Delta\omega T_2}{1 + (\Delta\omega)^2 T_2^2 + \gamma^2 B_1^2 T_1 T_2} \tag{54b}$$

$$M_z = \chi_0 B_0 \frac{1 + (\Delta\omega)^2 T_2^2}{1 + (\Delta\omega)^2 T_2^2 + \gamma^2 B_1^2 T_1 T_2} \tag{54c}$$

where $\Delta\omega = \omega_0 - \omega$.

As it can be seen from the above expressions, the macroscopic magnetization has a component also in the direction of the RF field $\mathbf{B}_1$ ($M_{x,y} \neq 0$), and it is this fact that enables the absorption of the RF energy, the appearance of nuclear resonance. The term $\gamma^2 B_1^2 T_1 T_2$, occurring in the saturation factor, too (see Equation 43) is found in the expression of all three components. NMR experiments can therefore be performed with success only if the experimental conditions are chosen so that the saturation factor is kept within the range required for convenient magnitudes of the components M_x, M_y, and M_z.

It is advantageous to introduce complex quantities in the examination of the Bloch equations. Magnetic susceptibility as written up as a complex quantity is

$$\chi = \chi' - i\,\chi'' \tag{55}$$

For a complex oscillating field of magnitude $2B_1e^{i\omega t}$, the complex magnetization is given by

$$M = (\chi' - i\,\chi'')\,2B_1e^{i\omega t} = 2B_1(\chi'\cos\omega t + \chi''\sin\omega t) \tag{56}$$

$$- 2iB_1(\chi''\cos\omega t - \chi'\sin\omega t)$$

If now the RF field applied in the study of nuclear resonance is defined as the real part of the complex field $2B_1e^{i\omega t}$, the real part of the magnetization M_x is given by:

$$M_x = 2B_1(\chi'\cos\omega t + \chi''\sin\omega t) \tag{57}$$

The experimentally observable interaction between the susceptibility of the nuclear system and the RF field is obviously only related to the real component of M_x. The first term of Equation 57 is the magnetization component in phase with the RF field, whereas the second term represents the out-of-phase component (shifted by $\pi/2$).

A comparison of Equations 57 and 54a gives for the Bloch susceptibilities:

$$\chi' = \frac{1}{2}\chi_0\omega_0T_2\frac{T_2\Delta\omega}{1 + (\Delta\omega)^2T_2^2 + \gamma^2B_1^2T_1T_2} \tag{58a}$$

and

$$\chi'' = \frac{1}{2}\chi_0\omega_0T_2\frac{1}{1 + (\Delta\omega)^2T_2^2 + \gamma^2B_1^2T_1T_2} \tag{58b}$$

As can be seen, both susceptibility components are frequency dependent and contain the dimensionless term $\gamma^2B_1^2T_1T_2$.

When $\mathbf{B}_1$ is small enough so that saturation can be neglected, i.e., $\gamma^2B_1^2T_1T_2 \ll 1$, Equations 58a and 58b are simplified:

$$\chi' = \frac{1}{2}\chi_0\omega_0T_2\frac{T_2\Delta\omega}{1 + (\Delta\omega)^2T_2^2} \tag{59a}$$

and

$$\chi'' = \frac{1}{2}\chi_0\omega_0T_2\frac{1}{1 + (\Delta\omega)^2T_2^2} \tag{59b}$$

Plotting the values of χ' and χ'' as a function of $T_2\Delta\omega$ (see Figure 8), curve χ'' shows the resonance character of the absorption, while the curve χ' belongs to the dispersion corresponding to the absorption.

The electronic system of NMR spectrometers allows the recording of both absorption and dispersion curves, simply by means of shifting the phase of the excitation field B_1 relative to that induced the detector coil. Usually it is the absorption curve that is recorded, but sometimes* recording of the dispersion curve is more practical.

The examination of Equations 54a and 58b shows the influence of T_1, T_2, and B_1 on the magnetization component M_x and the susceptibility χ'' and thereby upon the intensity of the absorption maximum. It is obvious and convenient to study this effect when $\Delta\omega$ =

* Compare to Volume II, p. 146 and p.259.

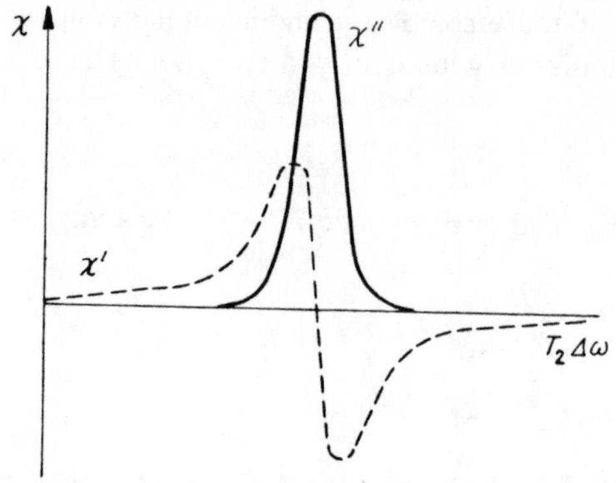

FIGURE 8. Frequency dependence of the Bloch susceptibilities. Absorption (χ'') and dispersion (χ') curve.[434]

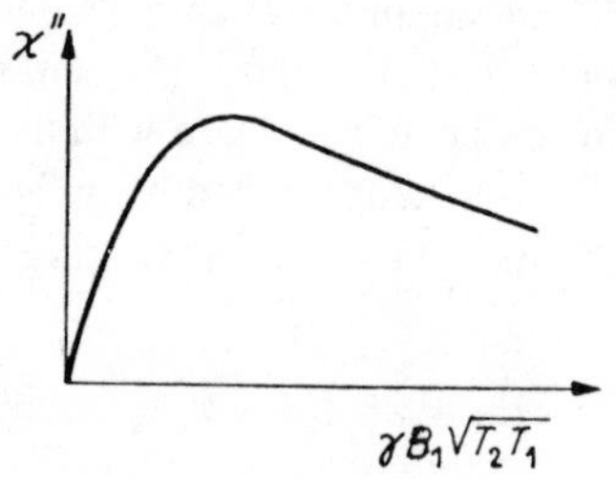

FIGURE 9. The susceptibility χ'' (signal intensity) as a function of B_1 for a given spin system ($\gamma\sqrt{T_1T_2}$ = constant).[1375]

0 ($\omega_0 = \omega$), i.e., at resonance. It follows from Equation 58b that when T_1 is very large, the χ'' susceptibility component disappears because there is no resonance signal due to saturation. Similarly no absorption can be observed when the T_2 is very small because in this case phase coherence cannot be maintained, the *x*, and *y* components of magnetization are vanishing quickly.

From the practical point of view, it is especially important to ascertain the way M_x and thus signal intensity depends on the excitation field B_1, since the latter is the only factor that can easily be changed experimentally. When B_1 is small the denominator of Equation 54a is approximately 1, thus M_x is proportional to B_1. If B_1 is sufficiently large, the unity in the denominator can be neglected and an inverse proportionality is found to exist between M_x and B_1. The RF field B_1, therefore, goes through an optimum from the point of view of signal intensity (see Figure 9), which can be obtained by determining the extremum $(dM_x/dB_1)_{\omega_0=\omega}$. Taking then $\gamma^2 B_1^2 T_1 T_2 = 1$, we find that

$$(B_1)_{\text{opt}} = (\gamma\sqrt{T_1 T_2})^{-1} \tag{60}$$

Considering that for protons in the liquid phase T_1 and T_2 is in the range of 10^{-1} to 10^{1} s, the optimal value for B_1 lies in the range 10^{-7} to 10^{-9} T. A value of B_1 giving highest signal intensity is associated, however, with significant signal broadening. The width of the signal can be characterized by $\Delta\omega_{1/2}$ i.e., the width belongs to the half of the maximum of the function χ'' defined by Equation 58b, $(\chi'')_{max}/2$ and is given by

$$\Delta\omega_{1/2} = 2(\omega_0 - \omega) = 2\Delta\omega \tag{61}$$

As the maximum of χ'' is obtained at resonance ($\Delta\omega = 0$), we have (see Figure 8)

$$(\chi'')_{max} = \frac{1}{2}\,\chi_0\,\omega_0\,T_2\,\frac{1}{1 + \gamma^2 B_1^2 T_1 T_2} \tag{62}$$

Upon comparing Equations 62 and 58b, it can be seen that for the values $(\chi'')_{max}/2$ the following relation holds:

$$T_2^2 \Delta\omega^2 = 1 + \gamma^2 B_1^2 T_1 T_2 \tag{63}$$

and from this:

$$\Delta\omega_{1/2} = 2\Delta\omega = \frac{2}{T_2}\sqrt{1 + \gamma^2 B_1^2 T_1 T_2} \tag{64}$$

Provided the RF field B_1 has a small value, the signal width is given by $\Delta\omega_{1/2} \approx 2/T_2$.

Considering the average magnitude of T_2, the signal width is of the order of 10 to 10^{-1} Hz, if the frequency corresponding to the polarizing field B_0, which lies between 1 and 100 MHz. This difference of five to nine orders of magnitude is a proof for the resonance character of the absorption. This follows also from Equations 54a and 58b, giving the magnitude of M_x and χ'', since M_x and χ'' are large only for small values of $\Delta\omega$, the quantities M_x and χ'' being inversely proportional to the square of the frequency difference $\Delta\omega$. In other words, the absorption signal has finite intensity only in the immediate vicinity of the resonance frequency ω_0; the resonance signal is sharp.

Substitution of the optimum of B_1 obtained from Equation 60 into Equation 64 gives $\Delta\omega_{1/2} = 2\sqrt{2}/T_2$, i.e., in this case lines are approximately one and a half times broader. Higher resolution (sharper signals) is obtained only when the value of B_1 is taken smaller than the optimum with respect to signal intensity.

In the case of thermal equilibrium:

$$\chi'' = \omega_0\,\chi_0\,g(\nu)/4 \tag{65}$$

and from this

$$g(\nu) = \frac{2T_2}{1 + (\Delta\omega)^2 T_2^2} \tag{66}$$

As can be established from Equation 66, the maximum of $g(\nu)$ is $2T_2$, in accordance with Equation 42.

The intensity of the resonance signal (the area under the peak) depends besides χ'' and B_1 also on the value of ω_0. The relationship is linear, and the intensity of the line is given by:

$$A \sim 4\,\omega_0\,\chi'' B_1^2 \tag{67}$$

From the above expression, the important conclusion can be drawn that the line intensity increases with the second power of the measuring frequency, because in Equation 67 the square of frequency, ν_0, is found ($\omega_0 = 2\pi\nu_0$ and ω_0 figure also in the susceptibility component χ'', compare Equation 58b). For example, at 100 MHz, the intensity of the lines is about 2.16 times greater than at 60 MHz. This value corresponds only to $(100/60)^{3/2}$ because in reality there is linear increase of noise with frequency as well, and the true intensity of the signals is determined by the signal-to-noise ratio, ζ.* Actually ζ increases proportionally to power 3/2 of frequency.

1.2. CHEMICAL SHIFT

1.2.1. The Phenomenon of Chemical Shift

Proctor and Yu observed[1145] in 1950 that the resonance signal of nitrogen in ammonium nitrate, NH_4NO_3, is a doublet, and model experiments showed that this splitting of about 10^{-4} T was not due to the isotope ^{15}N. It was then concluded that the frequency of the NMR signals depends not only on the external field, but is influenced also by the magnetic field in the close surrounding of the nucleus. The electrons moving around the nucleus induce a weak magnetic field which — according to the Lenz law — is opposite to the external magnetic field $\mathbf{B}_0$ and therefore $\mathbf{B}_l$, the resultant local field around the nucleus, is somewhat lower than $\mathbf{B}_0$. Since in the ammonium ion the distribution of the bonding electrons surrounding the nitrogen is different from that in the nitrate ion, the local magnetic field $\mathbf{B}_l$ is of different magnitude, thus the resonance frequencies of the two ionic groups are different. This phenomenon — the dependence of the position of the signal of the same isotope on the nature of the bonds connecting the nucleus to its neighbors and on the structure of the adjacent atomic groupings, in short, on the chemical environment — is called *chemical shift*.

From this interdependence of chemical shift and chemical environment follows that *it is possible to draw conclusions concerning the chemical environment of a nucleus based on the value of chemical shift*. This recognition is the basic of applicability of NMR in chemical structural investigations.

The dependence of the NMR spectrum on chemical structure is a consequence of local magnetic fields induced by electrons orbiting about the nucleus as well as by those around the adjacent nuclei. The resonance lines of nuclei in a molecular context are shifted as compared with that of "ideal" nuclei, isolated from the molecular surroundings. Nuclei having different chemical environments give rise to different resonance signals. Consequently there may be as many resonance lines in the NMR spectra of molecules of different structure as there are nuclei structurally different from one another. Thus, e.g., ethanol, *p*-xylene, and acetone should give rise to three, two, and one proton-resonance signal, according to the presence of methyl-, methylene-, and hydroxy-protons in ethanol, methyl-, and aromatic protons in *para*-xylene and six methyl-protons in acetone; e.g., in the ^{1}H NMR spectrum of 2,6-di-*t*-butyl-4-methyl-phenol (**1**) there are indeed four lines, (see Figure 10), according to the presence of *t*-butyl-, methyl-, hydroxyl-, and aromatic ring protons at δ = 1.24, 2.25, 4.90, and 6.36 ppm.** The 18 *t*-butyl, the 3 methyl-, and the 2 aromatic hydrogens are equivalent within each group; therefore, their shifts are identical. The assignment of the signals to the different kinds of protons is done on the basis of their relative intensity, as, in the above order, they correspond to 18, 3, 1, and 2 hydrogens, and the signal intensity is proportional to the number of corresponding nuclei.*** For the signal height, this statement is only approximately true, as the width of the individual lines is different.† The accurate intensity ratios are obtained by comparing the areas under the signal curves. NMR spectrometers have built-in integrating facilities, which record the *integrated spectrum*, on which steps of different height indicate the integral values at positions corresponding to the signals. The integral value, i.e., *the step height, is proportional to the number of protons* (compare Figure 10). The first organic compound in the ^{1}H NMR spectrum of which structurally different hydrogen nuclei could be observed as separate signals was ethanol.[58] The signals

* Compare p. 140 and Equation 304.

** As to the definition and units of chemical shifts see p. 41-44 and Equation 74.

***Compare p. 142.

† Compare p. 142.

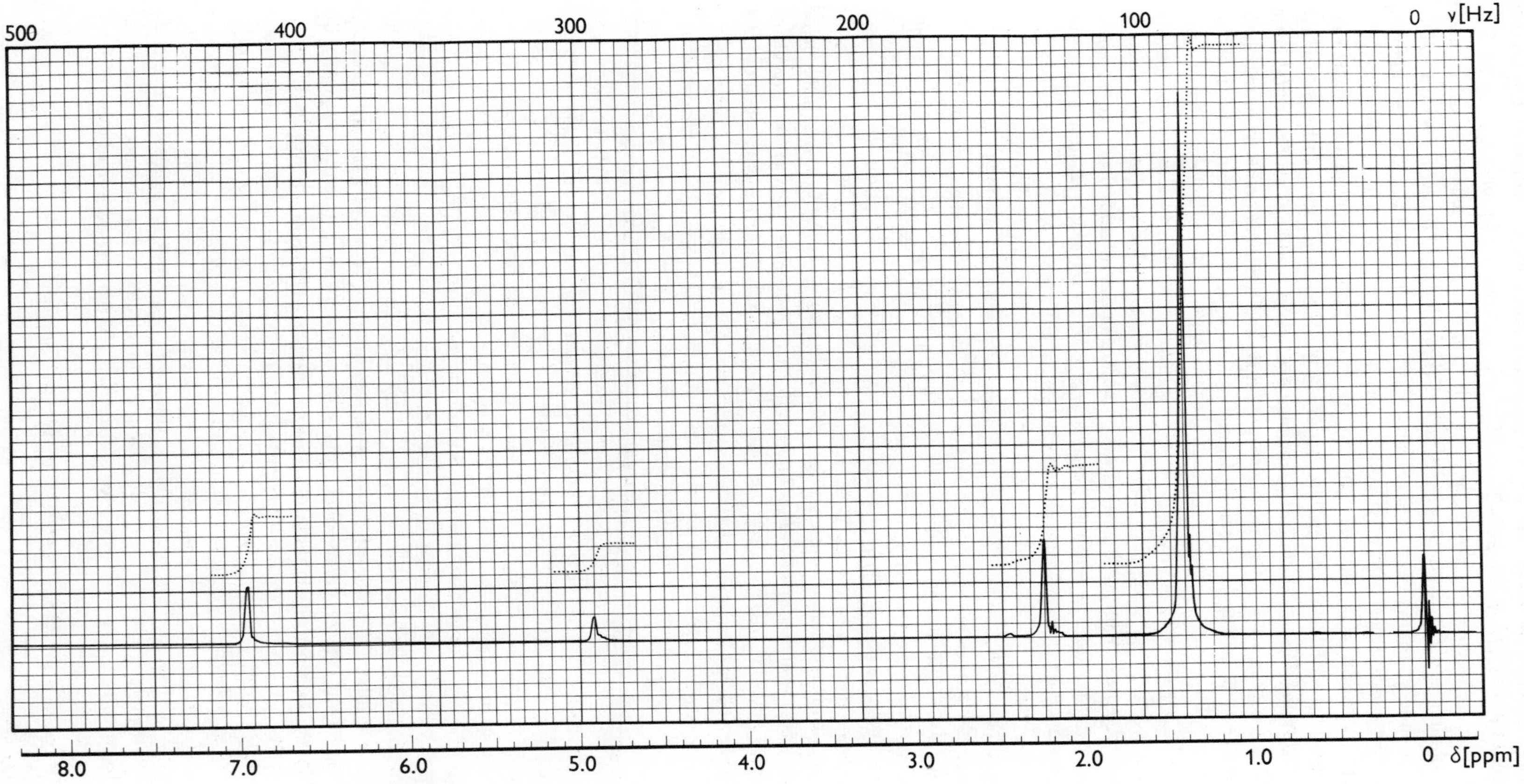

FIGURE 10. The ^{1}H NMR spectrum of 2,6-*t*-butyl-4-methylphenol (**1**) with the integrals. (If not otherwise indicated, the ^{1}H NMR spectra recorded in $CDCl_3$ solution at 60 MHz will be shown.)

could be easily assigned to the hydroxyl-, methylene-, and methyl-protons on the ground of the intensity ratio 1:2:3. A magnetic field of 0.7 T was applied in this experiment. The distance between the signals was about 3 μT; therefore in order to be able to detect separate signals, a resolution of at least $1{:}10^6$ was necessary. The resolution of present-day instruments is at least three orders of magnitude better, about $1{:}10^9$, which means that the measurement of chemical shift can be done with high accuracy as high as 10^{-7} T $\triangleq$ 0.1 ppm. Chemical shifts fall into a rather narrow interval, 10^{-3} to 10^{-4} times the external magnetic field $\mathbf{B}_0$, and their magnitude depends on the nucleus under study. Thus, e.g., for protons, this range is of order of 20 ppm, while for fluorine resonance spectra, it can be as much as 600 ppm.

The difference between the external magnetic field $\mathbf{B}_0$ and the local one $\mathbf{B}_l$ is proportional to the magnitude of $\mathbf{B}_0$ acting upon the molecule:

$$\Delta B = B_0 - B_l = \sigma B_0 \quad (68)$$

where σ is the screening factor, a dimensionless number. For a constant field $\mathbf{B}_0$ the resonance frequency ν_ℓ corresponding to the altered local field $\mathbf{B}_l$ is obviously different from the resonance frequency of an isolated nucleus:

$$\nu_\ell = \gamma B_l = \gamma B_0 (1 - \sigma) \quad (69)$$

Knowing σ, the chemical shift can be calculated:

$$\Delta\nu = \nu_0 - \nu_\ell = \sigma\gamma B_0 = \sigma\nu_0 \quad (70)$$

In other words, the magnitude of the chemical shift is proportional to the external magnetic field $\mathbf{B}_0$, or to the resonance frequency ν_0 (compare Figure 11 showing the schematic spectrum of toluene at 60 and 100 MHz, where the intensity enhancement of a factor of about 2.1 can also be seen clearly).*

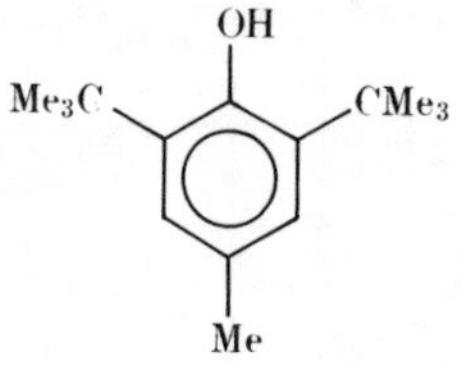

1

1.2.2. Diamagnetic Shift

In order to determine the value of the chemical shift, it is necessary to know the value of the shielding factor σ. For a given spherically symmetrical (neutral) nucleus, about which the distribution of the electrons is naturally also spherically symmetric, it holds:

$$\sigma = \frac{4\pi e^2}{3m_e c^2} \int_0^\infty r\rho(r)\mathrm{d}r \quad (71)$$

This is the Lamb formula[807] in which e and m_e stand for the charge and mass of electron, respectively, while ρ(r) is the electronic density per one electron at a distance r from the nucleus. For hydrogen e.g., σ is 18 ppm, for carbon it is 260 ppm, and for lead it is 10^{-2}.[365]

* Compare p. 24.

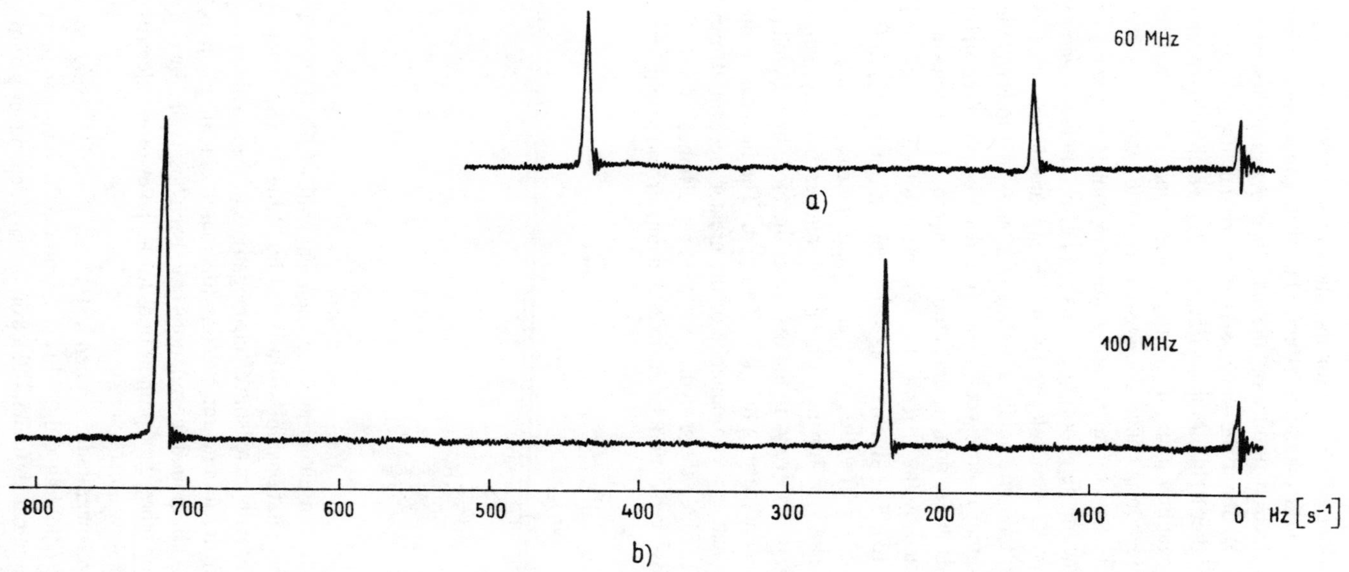

FIGURE 11. The schematic spectrum of toluene (a) at 60 and (b) at 100 MHz.

The more electrons orbit about the nucleus, i.e., the higher the electron density, the greater is the shielding effect and the larger is σ.

In real molecules, however, the situation is more complex.[753] There is no spherical symmetry and electron density about the nuclei is also a function of the nature of the chemical bonds. For this reason, σ can only be calculated for a very few simple cases.[1163]

Numerous methods have been developed[554,1132,1231] for the qualitative assessment of the relative magnitude of σ, by factorizing of the shielding effect. One of its components is the diamagnetic shielding provided by the electron cloud about the nucleus in question, described by the Lamb formula, reduces $\mathbf{B}_0$ to $\mathbf{B}_l$ about the nucleus. For a constant frequency ν_0, therefore, resonance appears only at higher field (compare Equation 21) which compensates for the diamagnetic shielding. This entails a shift of the resonance signal in the upfield direction (see Figure 12). This phenomenon is called *diamagnetic (upfield) shift*.

Diamagnetic shielding effect is influenced by neighboring atoms or groups by means of altering electron density. Electron-attracting neighboring groups ($-I$ effect) reduce electron density and thereby the shielding as well, while electron-withdrawing substituents lead to an opposite shift. It follows, e.g., that the more acidic a hydrogen atom is, the lower is the field at which the proton resonance signal is observed. This can indeed be observed upon comparing the spectra of acids and alcohols, amides and amines, or those of amines and their salts. The latter case is exemplified by the spectra of 2-amino-thiazole (**2**) and its HCl salt in DMSO-d_6 solution (see Figure 13). The signal of the salt is shifted downfield, to the left (see Figure 13a); hence the chemical shift value, δ, of all signals increases in the case of the salt.* The H-4 signal is at 7.03 and 7.35 for the base and the salt, respectively, while the H-5 signal is at 6.62 and 7.04 ppm, respectively. In the first case, therefore, the chemical shift difference is 0.30 ppm; for the second case in turn it is 0.42 ppm. The increase of the chemical shifts in the spectrum of the salt is a consequence of the greater electron affinity ($-I$ effect) of the substituent at C-2. Since these hydrogens are at a distance of several bonds from the substituent, the shift observed cannot be attributed to anisotropic neighboring group effect (compare next section).

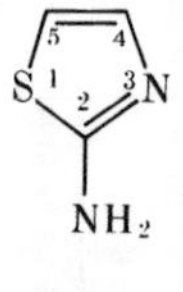

2

The electron density around the hydrogen nuclei and the chemical shift of the ^{1}H NMR signal are therefore closely related; this relationship is illustrated by Table 2, which demonstrates that the chemical shift increases with increasing electronegativity of the substituents, i.e., with the decrease of electron density around the protons (decreased shielding). It is understandable that for a great variety of compounds correlations have been observed between chemical shifts and electron density, or other data proportional to the latter (e.g., electronegativity, the Hammet and Taft constants, etc.).

For the determination of the relative magnitude of chemical shifts, it is, however, not sufficient to consider the diamagnetic shielding only. This is borne out by contradictions often encountered when correlating chemical shifts with electron density-dependent parameters (compare Section 3 also), e.g., chemical shifts of acidic and aldehyde protons are nearly identical, although the former is much more acidic.

* Compare with Figure 21.

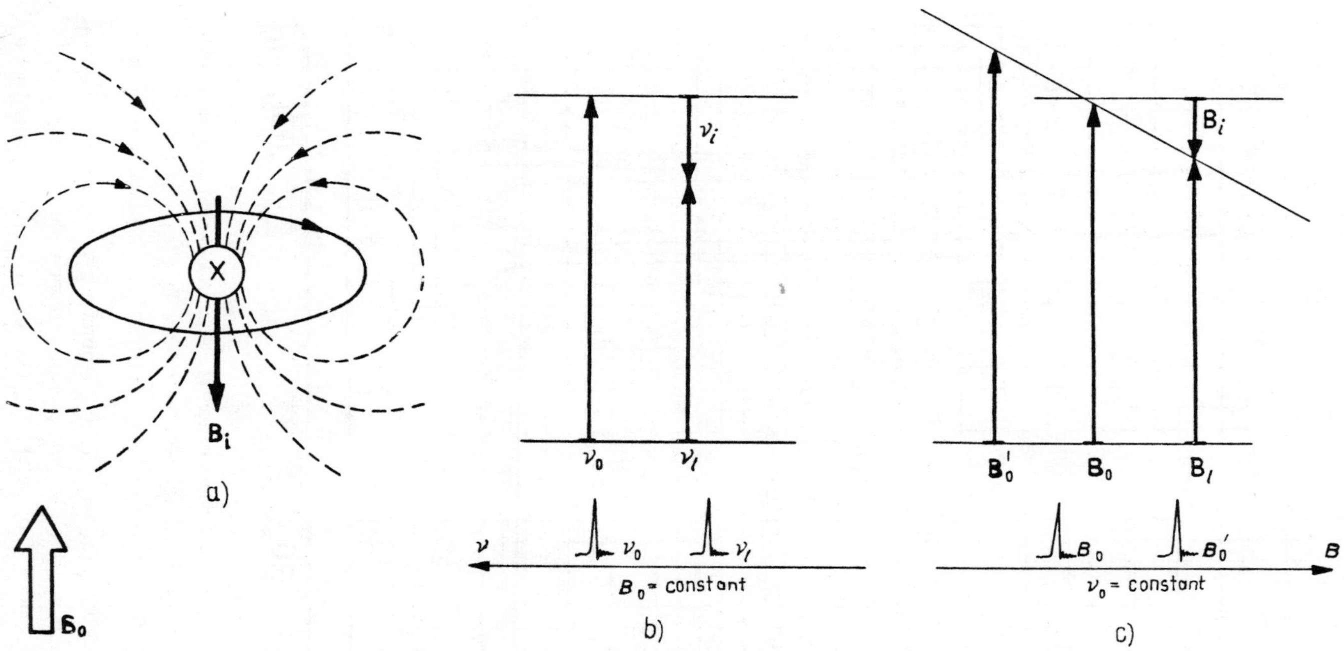

FIGURE 12. Diamagnetic shielding. (a) The induced magnetic field, $\mathbf{B}_i$, (b) the shift of the spectral line at constant polarizing field $\mathbf{B}_0$, and (c) at constant radiofrequency, ν_0. $\mathbf{B}_l$ is the local field around the shielded nucleus, $\mathbf{B}_0$ is the resonance field of the shielded nucleus, and ν_0 and ν_l are the resonance frequencies of the nonshielded and the shielded nucleus, respectively, at constant polarizing field $\mathbf{B}_0$, and ν_i is the decrease of resonance frequency caused by shielding.

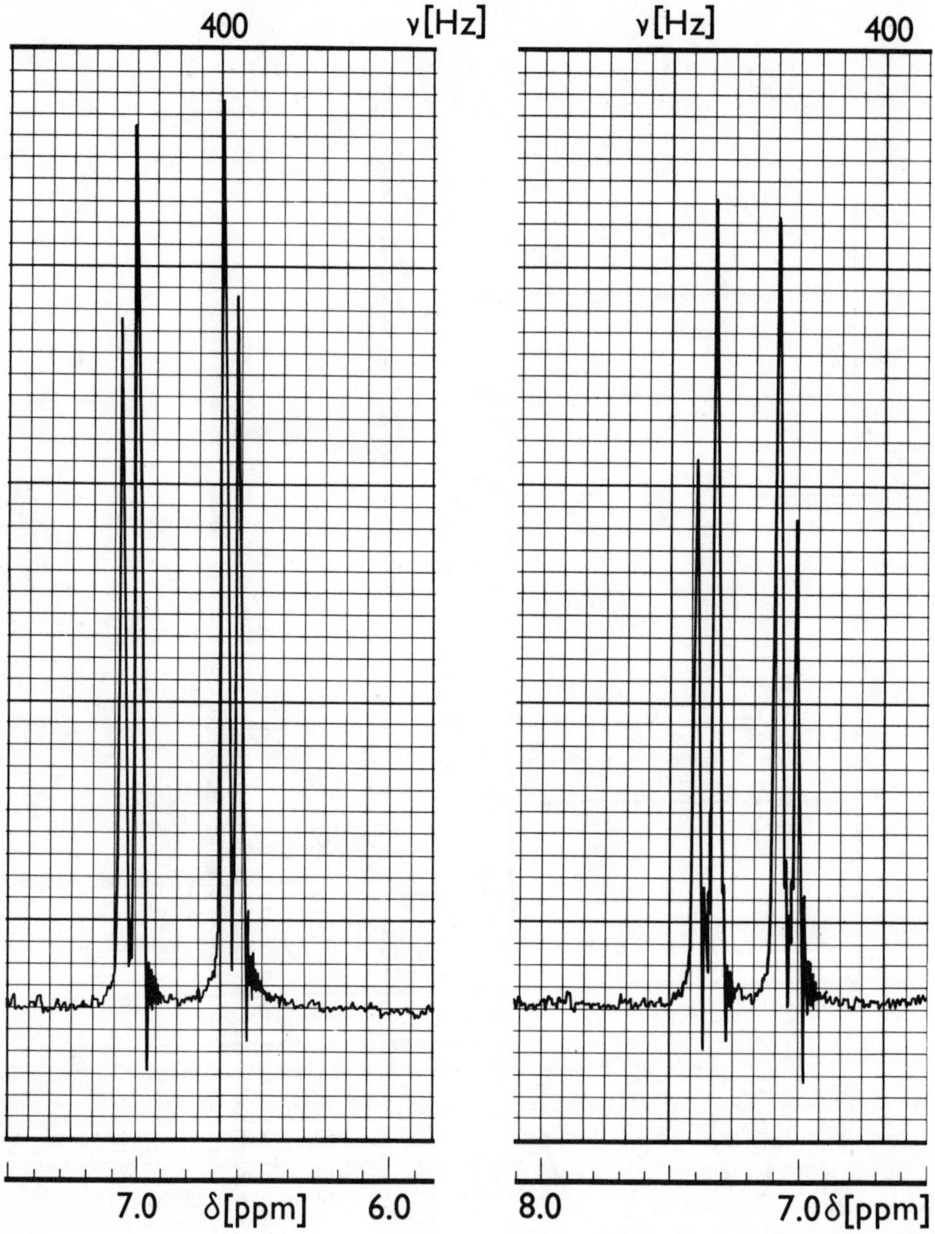

(a) (b)

FIGURE 13. The signals of ring protons of (a) 2-aminothiazole (**2**) and (b) its HCl salt in aqueous DMSO-d_6 solution.

1.2.3. Paramagnetic Shift

Another factor contributing to the resultant shielding around a given nucleus is the temperature-independent paramagnetism which originates from the nonspherical electron distribution.* This results in a dependence of shielding upon the mutual orientation of the molecule and the polarizing magnetic field. When a linear molecule is oriented parallel to the field direction, there is no additional effect. When, on the other hand, the molecular axis is perpendicular to $\mathbf{B}_0$ — or, generally, $\mathbf{B}_0$ is asymmetric in relation to molecular

* This is not to be mistaken for the temperature-dependent paramagnetism found in the case of molecules with odd electron spin.

Table 2
CORRELATION BETWEEN THE ELECTRONIC DENSITY AND THE ^{1}H NMR CHEMICAL SHIFT (PPM) IN A FEW SIMPLE COMPOUNDS[31,434,649]

Compound	ppm	Compound	ppm	Compound	ppm
CH_4(gas)	0.13	CH_4	0.23	$[C_5H_5]^-$	5.64
$C(CH_3)_4$	0.94	CH_3I	2.16	C_6H_6	7.37
CH_3SCH_3	2.08	CH_3Br	2.68	$[C_7H_7]^+$	9.24
$N(CH_3)_3$	2.13	CH_3Cl	3.05		
$N^+(CH_3)_4Cl^-$	3.20	CH_3F	4.26		
CH_3OCH_3	3.24				

symmetry — certain electron transitions, which are otherwise forbidden become allowed for the paramagnetic excited electronic states, and this is manifested in deshielding of the involved nucleus — in *paramagnetic (downfield) shift* of its signal. The excited electrons induce a local field parallel to the perturbing field $\mathbf{B}_0$, resulting in a deshielding effect. This effect is the more important, the closer the levels of excited and ground states (the smaller the perturbation energy), and it could only be observed in a pure form by means of a model capable of existing both as a covalent molecule and a purely ionic pair (XY and X^+Y^-, respectively). In the ionic pair there is no possibility for the paramagnetic effect since there are no easily excitable bonding electrons. Since the hydrogen atom does not possess excited electronic states close to the ground state, paramagnetic effects can be neglected in ^{1}H NMR spectroscopy, but for other nuclei, as ^{13}C, ^{19}F, etc., it can be decisive.* For the hydrogen molecule, theoretical calculations show that the paramagnetic effect reduces the diamagnetic shielding by a factor of 5/6 only.

1.2.4. The Magnetic Anisotropy Effect of Neighboring Groups

The third factor contributing to the total shift comes from the electron currents of the neighboring atoms. In the case of a given bond X–Y, when the bond is parallel to the field $\mathbf{B}_0$, the induced field due to the electrons of the adjacent atom X reduces the field about the atom Y, while the opposite is true for a perpendicular orientation (see Figure 14). Experimentally one observes, of course, an average value, and it depends on the magnetic susceptibility, χ, of the atom X, whether the opposite effects cancel each other or there is a net change of the field at atom Y. If χ is isotropic (spherically symmetric), there is no change of field at the site of nucleus Y. If in turn it is anisotropic and the field is greater along the bond X–Y than perpendicularly to it (a case frequently found with linear molecules), the shielding increases, and the signal of Y appears at higher field as it would be expected considering the distribution of electrons alone. Unlike the diamagnetic shielding effect, which is transferred through bonds, the s.c. *neighboring group anisotropic effect* is acting *through space*.

The directional dependence of the shielding factor, σ, is given by the McConnell equation[958] or by similar expressions:

$$\Delta\sigma \sim \frac{1 - 3\cos^2\theta}{r^3}\,\Delta\chi \qquad (72)$$

where r and θ define the mutual positions of the nucleus of interest and the neighboring groups or bonds of axial symmetry (regarding the latter in the point-mass approximation), while $\Delta\chi = \chi_\parallel - \chi_\perp$ is the anisotropy of the magnetic susceptibility (see Figure 15).

It can be seen from this expression that the change in shielding ($\Delta\sigma$) caused by the

* Compare Volume II, p. 150, 239, 261, and 262.

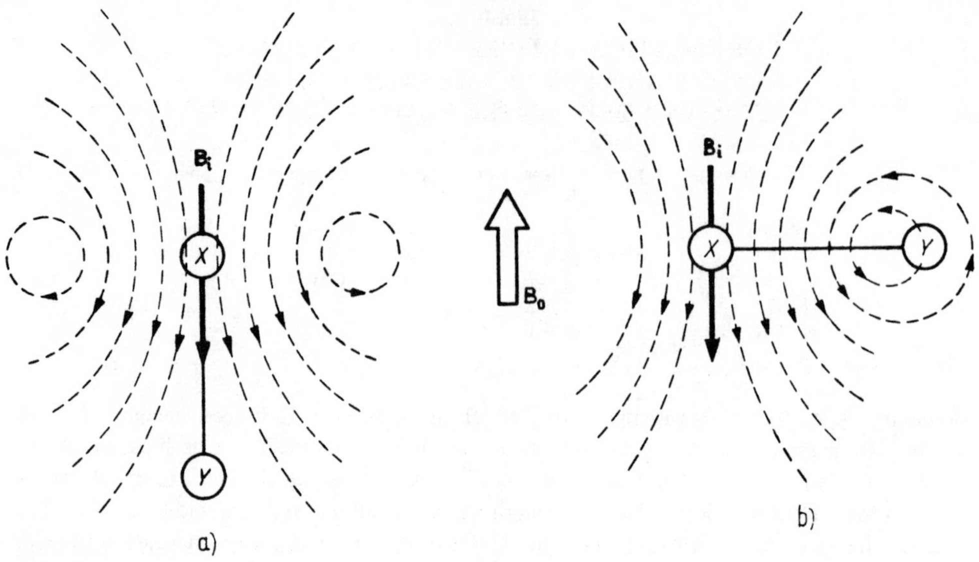

FIGURE 14. The induced magnetic field for (a) parallel and (b) perpendicular orientations of a linear molecule XY relative to $\mathbf{B}_0$.

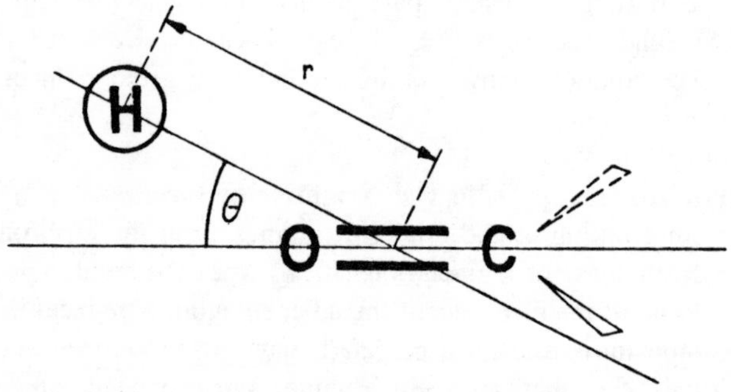

FIGURE 15. The geometry-dependence of anisotropic shielding effect of the carbonyl group.

anisotropy of susceptibility is zero for a given value of the angle θ ($\cos^2 \theta = 1/3$), where it changes sign. This means that the neighboring group effect is a function of geometry, both with regard to sign and magnitude. Regarding the effect over the entire space reveals that the certain (mostly chromophoric) groups possess a *shielding cone* (see Figure 16). When nucleus comes within the shielding cone a diamagnetic shift of its signal is experienced, while for a position outside the shielding cone, a shift of opposite sense appears. $\Delta\chi$ and thereby $\Delta\sigma$ can be calculated theoretically; these calculations are, however, inaccurate. One reason for this is that the groups are treated in the point-mass approximation; the other is that this effect cannot be separated from other factors influencing the shielding. The shielding cones of the various functional groups have, therefore, been determined empirically by comparison of model compounds. The values thus obtained are very useful in structure determination.

From a study of the series of aldehydes of the general formula R–CH_2–CHO, where R = Me, Et, Pr, *i*Pr, *n*Bu, and *t*Bu, was deduced the actual form of the shielding cone of a carbonyl group (see Figure 16). Among the rotamers **3a** and **3b** of compounds RCH_2CHO, the stability of the latter is higher with bulkier R groups. From a simultaneous increase of

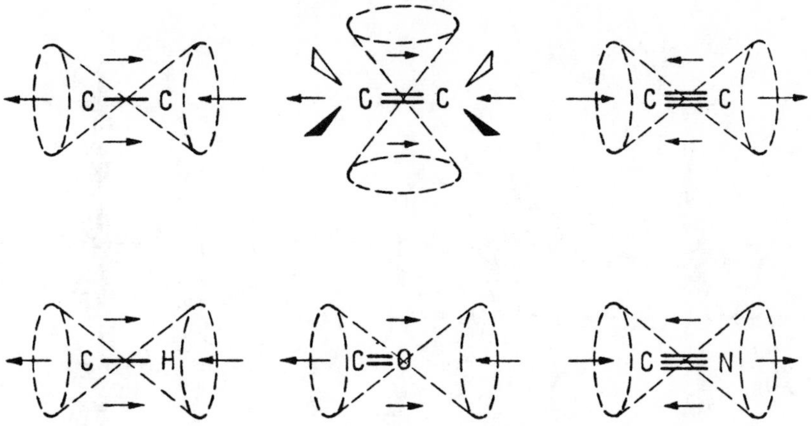

FIGURE 16. The "shielding cone" of a few types of bonds.

the diamagnetic shift of the H_α signal with the size of R, it follows that the cone of anisotropy has the form as given in Figure 16.

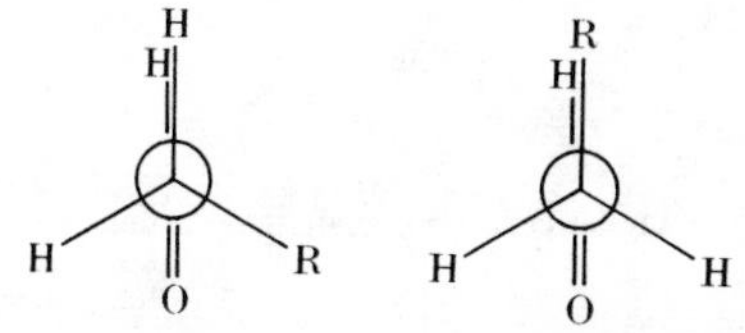

3a **3b**

A good example for neighboring group anisotropy is acetylene.[1125] In molecules perpendicular to the field $\mathbf{B}_0$ (see Figure 17a) due to perturbation of certain σ-electronic states, a paramagnetic dipole is created at the carbon atoms which reduces the field around the protons. In parallel oriented molecules, no dipole is induced by the σ-electrons; π-electrons provide diamagnetic shielding on both carbon and hydrogen (see Figure 17b), which is possible only in this latter orientation. The protons are then more shielded in both cases, and therefore an upfield shift is expected for the ^{1}H NMR signal. Indeed a diamagnetic shift of 3.83 ppm relative to ethylene was found for the proton signal in acetylene.[1128]

Anisotropy effect explains the similar chemical shifts of aldehyde and carboxylic protons despite the evidently lower electron density at the acidic protons. The aldehyde hydrogen, however, is subject to the anisotropy effect of the carbonyl group which causes a downfield shift because the hydrogen lies outside the shielding cone.

Other functional groups may also show significant anisotropy effects. That of the hydroxyl group, for example, is borne out by the shifts of the methyl and 4-methyne signals of compounds **4a** to **4c**.[336,692]

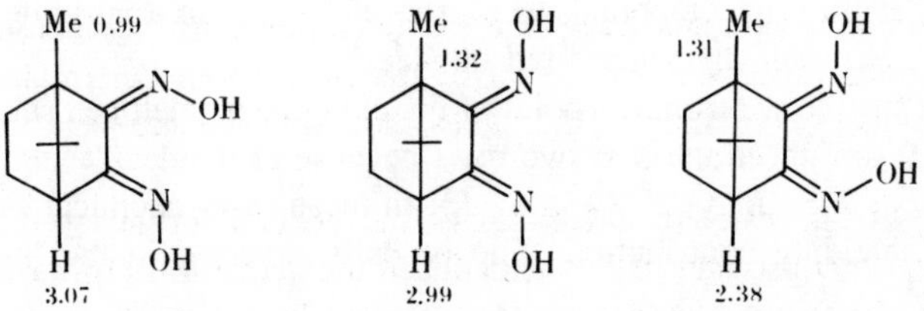

4a **4b** **4c**

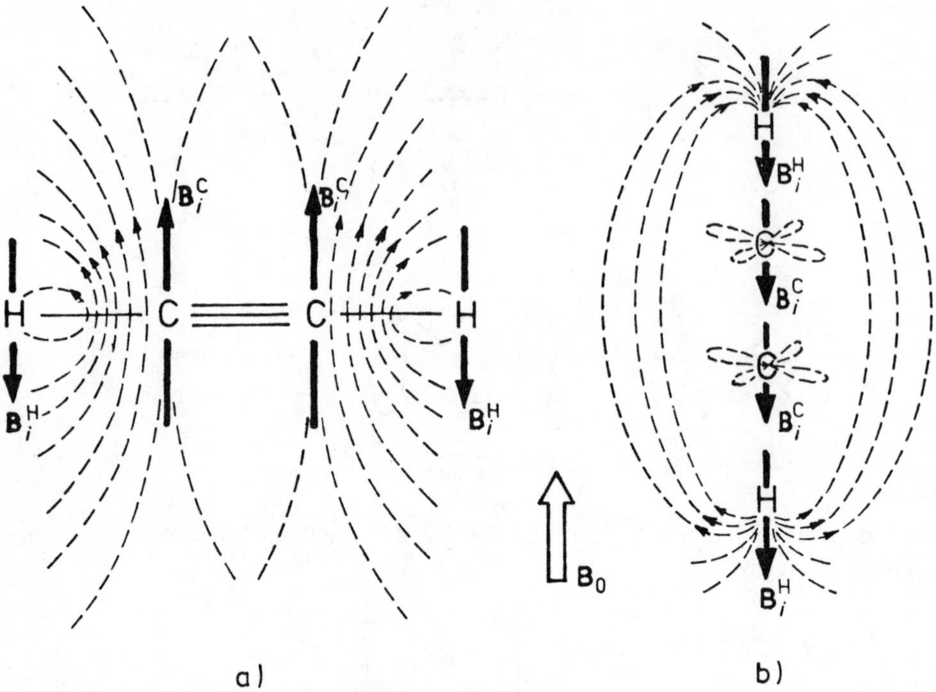

FIGURE 17. The induced (a) paramagnetic and (b) diamagnetic dipole in acetylene for perpendicular and parallel orientations relative to $\mathbf{B}_0$.

The sulfoxyde group gives also rise to rather significant anisotropy shifts of signals of a nearby hydrogen. This enables the simple determination of the configuration of the S → O bond in rigid systems, as in **5a** to **5c**,[1137] where above all H–3,3′,6,7 signals are shifted markedly (see Formulas).

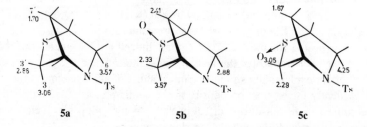

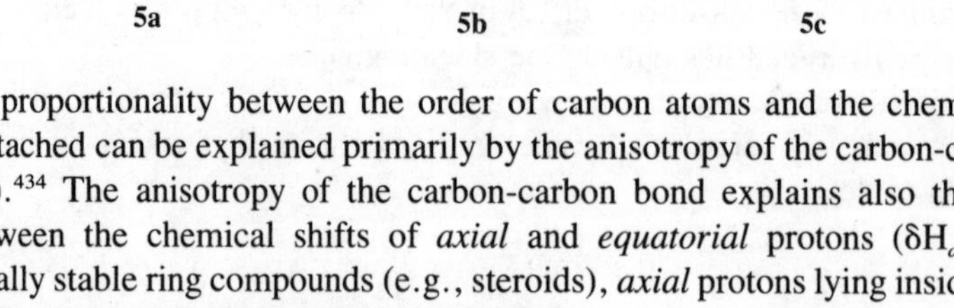

5a **5b** **5c**

The direct proportionality between the order of carbon atoms and the chemical shift of the protons attached can be explained primarily by the anisotropy of the carbon-carbon bonds (see Table 3).[434] The anisotropy of the carbon-carbon bond explains also the difference observed between the chemical shifts of *axial* and *equatorial* protons ($\delta H_a < \delta H_e$) in conformationally stable ring compounds (e.g., steroids), *axial* protons lying inside the shielding cone, while *equatorial* hydrogens fall outside.*

Table 3 is also instructive from the point of the effect of the halogen substituent. δ-values decrease from left to right in the first two rows because of the decreasing $-I$ effect of the substituents ($-I_{OH} > -I_{Cl} > -I_{Br} > -I_I$). In these cases chemical shifts are governed by diamagnetic shielding contribution. The order is reversed in the last row (*iso*propyl

* Compare Volume II, p. 27.

Table 3
CHEMICAL SHIFTS OF ANALOGOUS HYDROGENS ATTACHED TO PRIMARY, SECONDARY, AND TERTIARY CARBON ATOMS[434]

X	OH	Cl	Br	I
CH_3X	3.38	3.05	2.68	2.16
CH_3CH_2X	3.59	3.57	3.36	3.15
$(CH_3)_2CHX$	3.85	4.13	4.20	4.25

derivatives). This can be only explained assuming that the halogens influence shifts not only through reducing the diamagnetic shielding by their $-I$ effect, but also by the anisotropic neighboring group effect. This latter increases in the order Cl < Br < I because due to the increasing atomic radius, the halogen atom approaches the adjacent hydrogens and gives rise to a paramagnetic shift. The net shift is the vectorial sum of the two effects. For methyl- and ethyl-halogenides, diamagnetic shielding predominates, while for the *iso*propyl series, the anisotropy effect takes over and reverses the order of shifts.

1.2.5. Effect of Intramolecular Ring Currents

Components of the resultant shielding discussed in the foregoing section originate from electron currents localized at a single atom or bond. With ring systems having continuous conjugation, first of all with aromatics, a special case for the anisotropic neighboring group effect comes into being which is a consequence of electron delocalization in the ring, called *intramolecular ring currents*. For aromatic compounds, ring currents play a decisive role among factors influencing resultant shielding of protons.

In benzene oriented perpendicularly to the polarizing field $\mathbf{B}_0$, the movement of the electrons parallel to the plane of the ring and around the main symmetry axis induces a magnetic field that reduces $\mathbf{B}_0$ along the molecular axis, while increasing it about the protons (see Figure 18).

The aromatic proton signals experience — relative to, for example, the olefinic protons of cyclohexadiene — a downfield shift of about 1.5 ppm. According to theoretical calculations,[1483] the shielding effect vanishes with increasing distance from the ring along the molecular axis. In the plane of the ring, the shielding effect first drops upon moving away from the center of the ring, reaches zero at a distance of about 2 Å (at about 1.3 times ring radius), and then it is replaced by deshielding zone. This deshielding effect has a maximum at about 2.25 Å and falls off quickly at larger distances (see Figure 19a).

Diamagnetic shielding in the direction of the molecular axis is responsible for the specific effect of aromatic solvents causing upfield shift of solutes.* If points in the plane perpendicular to the benzene ring experiencing the same shielding are connected, the picture given in Figure 19b is obtained. The solid line represents the planar intersection of the shielding cone points, i.e., the points of zero shielding contributions.

The effect of the aromatic ring currents is well illustrated in 1,4-polymethylene-benzenes (**6**). The protons attached to methylene groups around the middle of the chain suffer an upfield shift of about 1 ppm (~0.7 ppm) because they are situated in the shielding cone of the benzene ring, compared to the rest of methylene protons (~1.4 ppm). The latter have shifts characteristic of cycloalkanes, except for the two methylene groups attached directly to the aromatic ring which have δ-values (~2.5 ppm) similar to that (~2.7 ppm) of ethylbenzene.[1483]

* See p. 40.

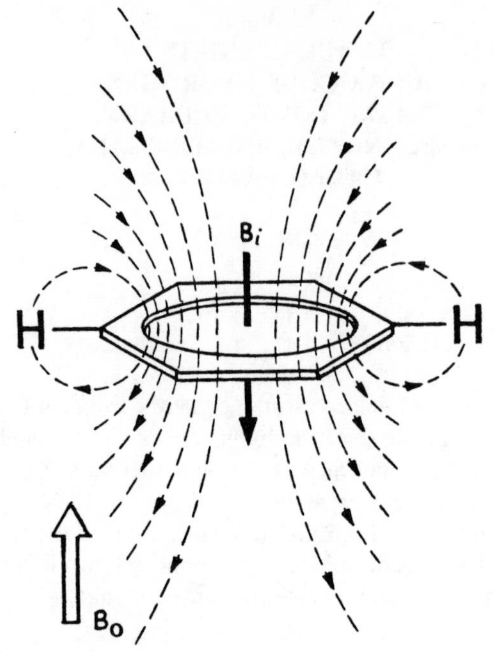

FIGURE 18. Magnetic field induced by ring currents in the benzene rings.

The methyl signal of compound **7** is at 1.77 ppm, whereas the methyl groups of 9,10-dimethyl-anthracene (**8**) give rise to a signal at 3.08 ppm. The diamagnetic shift of 1.31 ppm can be ascribed to the anisotropic effect of the aromatic rings.[363]

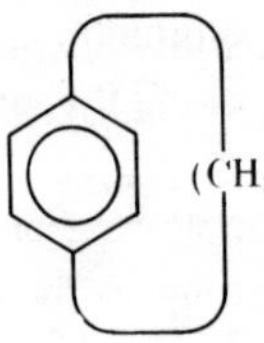

6

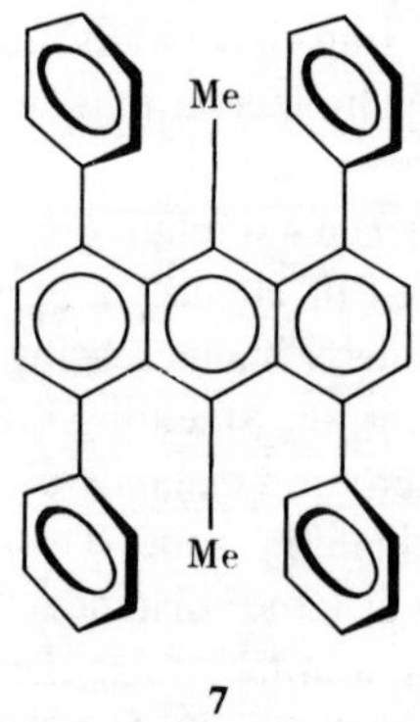

7

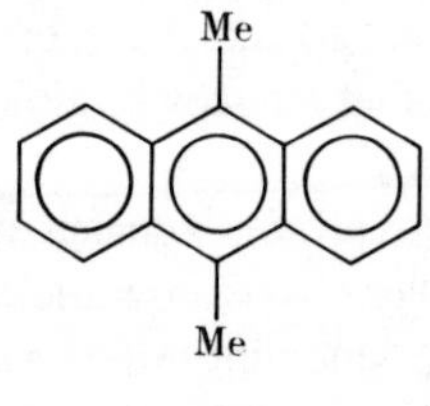

8

In the spectrum of *syn*-paracyclonaphthane **9a**, the H_A and H_B signals are shifted upfield (7.15 ppm) as compared to the corresponding resonance signal (7.50 ppm) of the *anti* isomer (**9b**) or that (7.55 ppm) of 1,4-dimethylnaphthaline (**10**). The δH_C signal shows an upfield shift in **9b** (δH_C 6.72, and 7.00 ppm in the case of **9b, 9a**, and **10**, respectively[1475]). The negative chemical shifts (δ = − 1.8 ppm, δ = − 3.9 ppm) of "internal" (CH and NH) protons of aromatic annulenes,[706] e.g., **11**,* and porphyrins, e.g., **12****[5,101,1556] may also be explained by the diamagnetic effect of ring currents. The signals of "external" protons appear at higher chemical shifts (8.9 and 9.96 ppm) similar to the signals of aromatic protons.

* Compare Volume II, p. 60 and 61.
** Compare Volume II, p. 91.

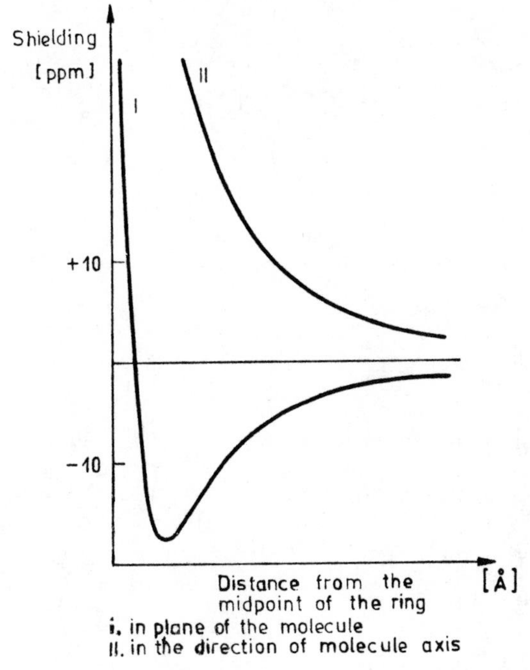

(a)

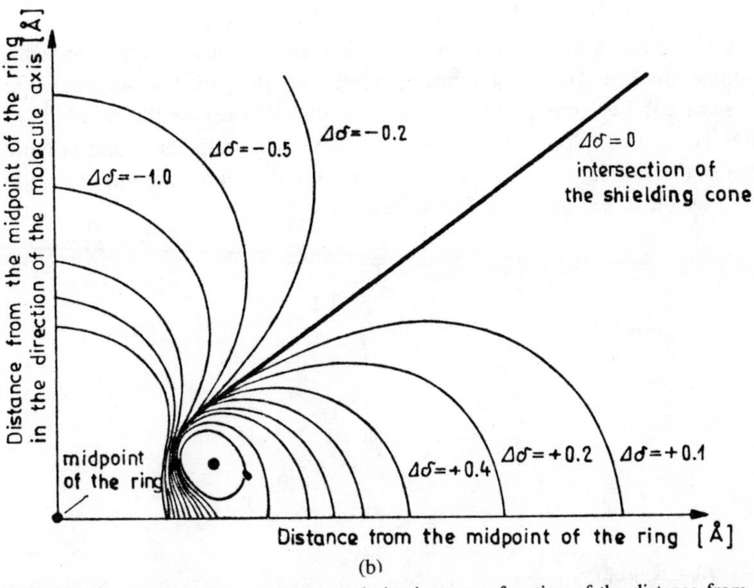

(b)

FIGURE 19. (a) Shielding in benzene derivatives as a function of the distance from the centrum of the ring, in the molecule plane (Curve 1), and perpendicular to it along the molecular axis (Curve II) and (b) isochron curves connecting identically shielded points in a plane perpendicular to the ring currents. Intersection line of the shielding cone (thick line).[724]

The methyl signal of the annulene derivative **13**, which originates from groups situated inside of the ring system causing diamagnetic shift, also appears at negative δ-value (−4.25 ppm). In the spectrum of the anionic form $[\mathbf{13}]^{2-}$ of this compound, the same signal shifts downfield by about 25 ppm: δ = 21.24 ppm.[165,981] This is the largest chemical shift measured so far in ^{1}H NMR spectroscopy.

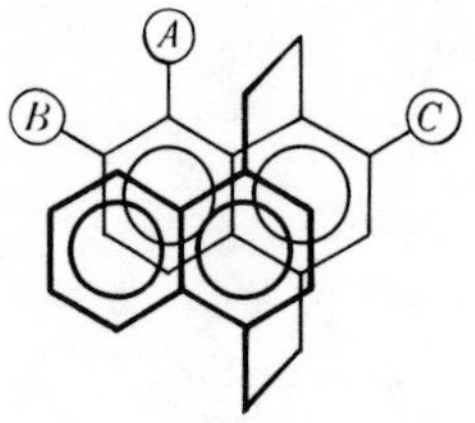

9a

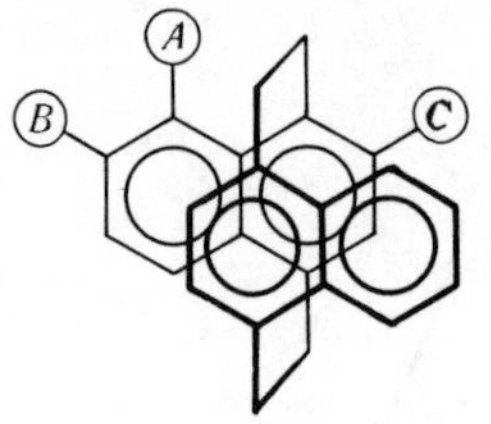

9b

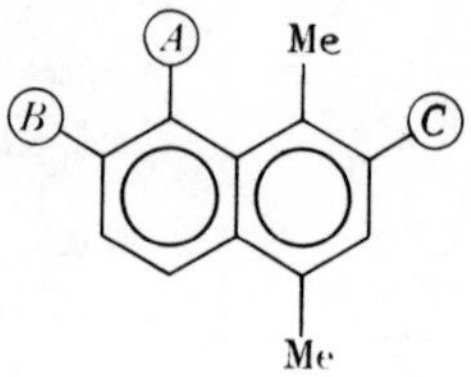

10

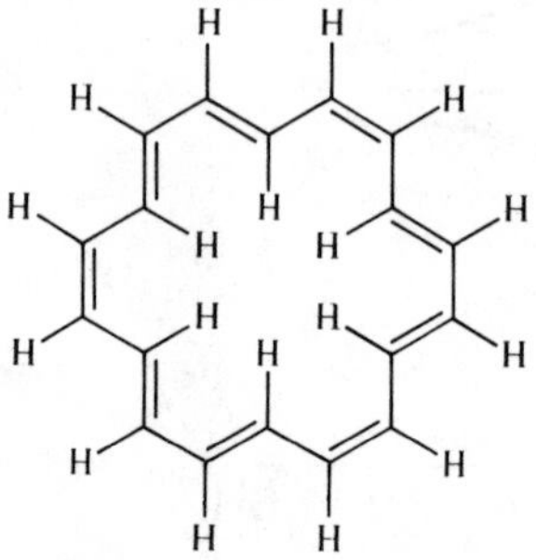

11

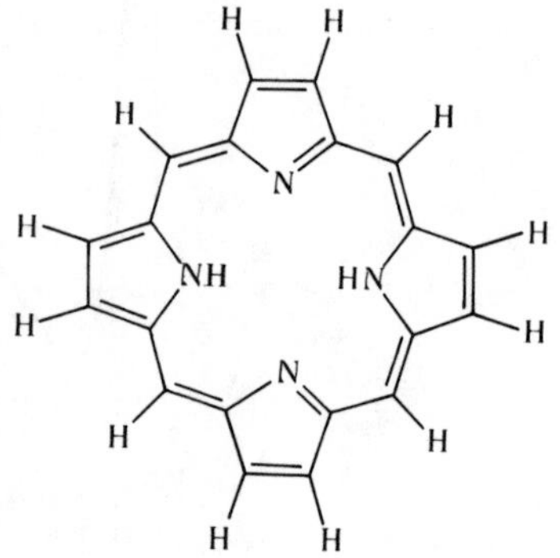

It can be proved quantum chemically[1133] that in continuously conjugated (plane polygon) ring systems, the free electronic motion along the ring means negative contribution to magnetic susceptibility (induction of a diamagnetic dipole), if the number of delocalized π-electrons is $4n + 2$. The contribution is positive (the induced dipole is paramagnetic) if the number of the π-electrons is $4n$. This explains the very large paramagnetic shift* between **13** ($4n + 2$ system) and its anion ($4n$ system).

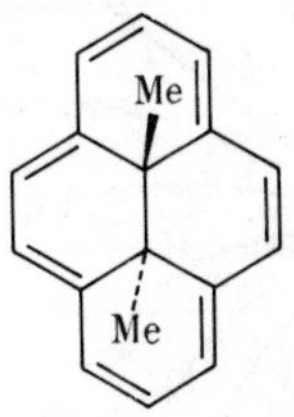

13

1.2.6. Solvent Effects[672]

Chemical shift may be strongly influenced by magnetically anisotropic solvents. Solvation leads to definite mutual orientation between solvent and solute molecules (s.c. collision complexes are formed). Thus, no averaging takes place among shielding differences coming from anisotropy. For this reason the aromatic (disk-like) solvents having diamagnetic shielding in the direction of their axes cause diamagnetic shifts relative to values obtained in isotropic solvents. This is illustrated in Figure 20 which depicts possible mutual arrangements of the benzene and solute A. In arrangement (a) one would expect increased shielding, while for (b) deshielding is likely. The first case is, however, more probable in accordance with

* Compare Volume II, p. 60.

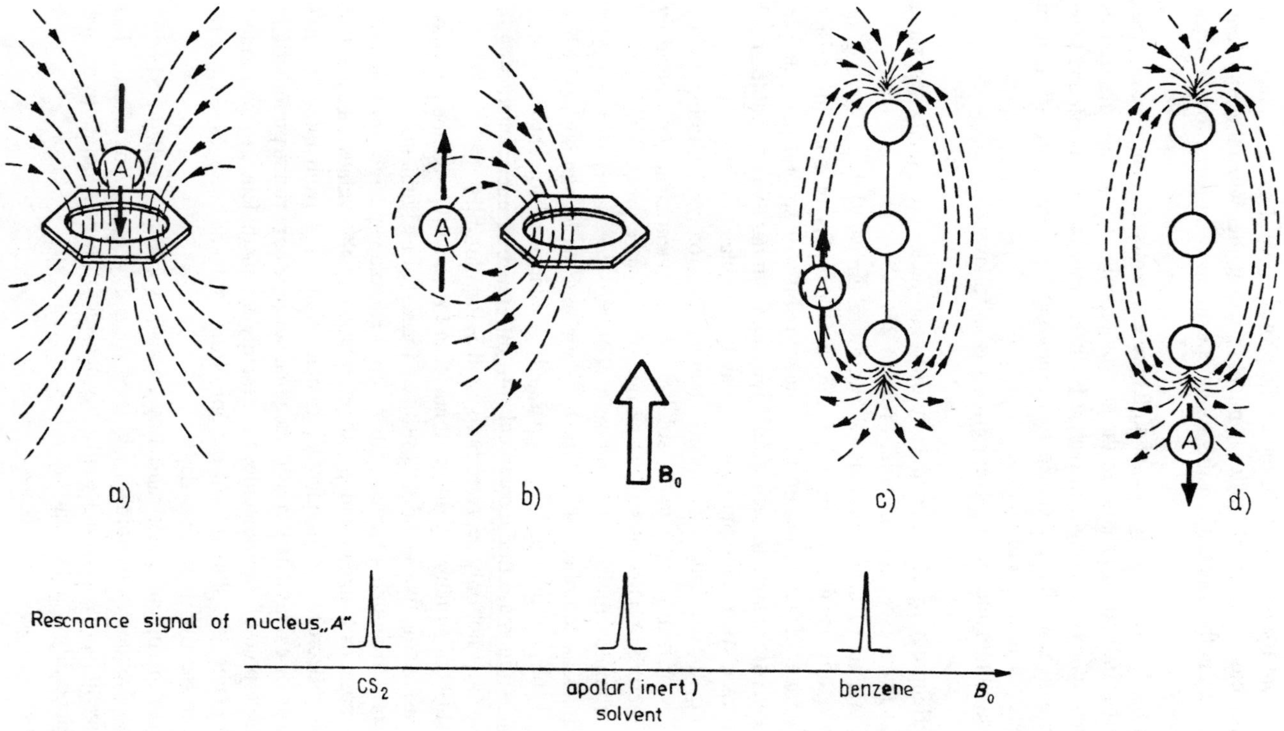

FIGURE 20. Change of the chemical shift in solution for different orientation of the solute and plate- or rod-shaped solvent molecules.

experimental results, showing a diamagnetic shift of about 1.3 ppm.* For linear solvent molecules, where diamagnetic shielding is expected in the direction of the molecular axis, arrangement (c) is favored. Therefore downfield shifts are predicted and found (~ 0.5 ppm).[226]

Solvent effects can be used extensively in structure elucidation work. With unknown compounds it is advisable to take the spectra in at least two solvents of different character.

In apolar solvents, except for the signals of the acidic protons (OH, NH, SH, etc.), there is usually no significant change in the ^{1}H NMR chemical shifts, neither when these solvents are compared among one another, nor when the solute spectra are compared to gas-phase or pure liquid state spectra.

On going over to aromatic solvents significant shifts are encountered, and these changes can be rather different for individual protons in the same molecule. These shift differences are usually called *solvent effects*, although, in fact, it is differences of solvent effects. (The true solvation effects could have been obtained by comparing chemical shifts in solution with those the gas-phase or pure liquid phase.)

The main information obtainable from solvent effects is as follows:

1. Accidentally overlapping signals of chemically different nuclei are usually separable using a different solvent (see, for example, Problem **25**).
2. Identification of signals of acidic protons. While the chemical shifts of hydrogens attached to carbon are usually different by only a few tenths of ppm in various solvents,** the signals of acidic proton may undergo shifts of several ppm, and in addition, a characteristic change in their shape, ranging from hardly observable broad signals to sharp peaks. Also spin-spin splitting may be eliminated or brought, respectively, by changing the solvent, varying pH, temperature, or concentration, or by adding a few drops of heavy water to the solution. These phenomena are caused by changes in the structure of associates or in the rate of exchange processes.
3. Interpretation of complex multiplets. Spin-spin coupling (see later) may cause to appear certain signals as complicated multiplets. Their analysis is only possible when the components of the multiplet are well separated. A change of solvent may cause such a separation. An increase in the chemical shift difference of nuclei interacting in spin-spin coupling may not only separate but simplify the multiplets.***
4. Specific solvent shifts of structurally different nuclei of the same type can be predicted theoretically at least qualitatively and thus utilized for signal assignment.

   The α-methyl proton signals of cyclic ethers, for example, undergo upfield shifts by 0.05 to 0.38 ppm in benzene and pyridine, relative to the values measured in CCl_4.[1521] This shift decreases with increasing size of the ring from three- to six-membered. This can be explained by considering that benzene or pyridine preferentially solvates the positive pole of the molecule, i.e., mainly the methylene groups farthest from the oxygen. Thus, the shielding of the anisotropic solvent becomes smaller on the H_α atoms when the ring is increased.[1523]

   This special case of solvent effect caused by aromatic solvents is called ASIS.[815]† Mostly C_6D_6 (less frequently pyridine-d_5) is used for studying solvent effects. The corresponding solvent pair is CCl_4, or less often (when solubility considerations require it) it is $CDCl_3$. Generally it holds that:

* Compare p. 41.

** The only exception is the acetylenic proton of monosubstituted acetylenes, which shows significant solvent shifts (Compare Volume II, p. 47 and Problem **34**).

***Compare e.g., Figure 44.

† Aromatic solvent induced shift.

a. In benzene solutions, shifts are diamagnetic (sometimes as much as 1.5 ppm) relative to values in CCl_4 or $CDCl_3$.
b. The largest shifts are observed for signals of nuclei closest to the positive pole of the dissolved dipolar molecule and decreases gradually towards the negative pole.
c. The magnitude of the shifts increases proportionally to the dipole moment of the molecule dissolved.

Many empirical correlations have been elaborated on the above ground between chemical structure and ASIS. Among these the Bhacca-Williams rule[1521] states that if a plane is set perpendicularly to the C=O double bond at the carbon atom, the ASIS (the shift $\Delta\delta_{CDCl_3 - C_6D_6} = \delta_{CDCl_3} - \delta_{C_6D_6}$) is negative for hydrogens on the side towards the oxygen atom, while positive for signals of hydrogens on the other side. In steroids, for example, the configuration of the methyl groups can be established using this rule. In C_6D_6 the signals of *axial* methyl protons vicinal to the carbonyl group show a diamagnetic shift [$\Delta\delta Me(a) = 0.2 - 0.3$ ppm], whereas the *equatorial* ones undergo a smaller paramagnetic shift [$\Delta\delta Me(e) = -0.05 - -0.15$ ppm].[131,183] For example, the C-4 methyl groups of Structure **14** give in $CDCl_3$ at 1.06 ppm overlapping signals, which are split in C_6D_6 ($\delta Me(a)$: 0.83 ppm, $\delta Me(e)$: 1.16 ppm).[131] The same is found for the 2-methylene-protons, too.[131] This empirical rule helps often the determination of relative configuration in groups rigid ring systems.

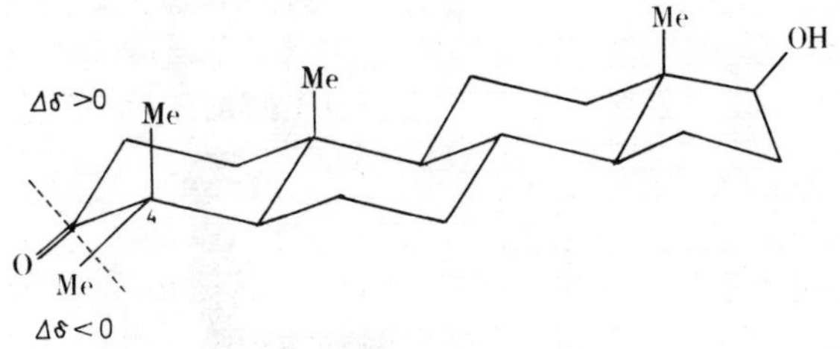

14

5. Corresponding signals of enantiomeric solutes are coincident in achiral solvents, but may split in a chiral solvent.* From the ratio of their intensity, optical purity of optically active compounds can be determined. Moreover, NMR spectra taken in optically active solvents may allow conclusions about absolute configurations.

In addition to those discussed in the previous sections (e.g., $-I$ effect, ring current, etc.), sometimes further structural factors influencing chemical shifts should be taken into consideration. These effects can, however, be separated only with difficulty both from one another and from the ones discussed before. Although their influence in ^1H NMR is usually small, their existence is one of the reasons why it is so difficult to calculate theoretically, even approximately, chemical shifts. These factors are connected to inter- or intramolecular interactions (e.g., hydrogen bonding, solvent-solute interactions, etc.) or to the physical state of the sample (phase, temperature, concentration).[1108] Electric fields created by electric di- or quadrupoles[935] or van der Waals' forces[687,936] may influence the nuclear shielding, too.

1.2.7. Additivity and Determination of Chemical Shifts; Correlation Tables; Reference Substances

Comparison of shifts in different compounds provides possibility for deducing shielding factors for different substituents, structural units, groups, and atoms. These substituent effects

* Compare Volume II, p. 114-115.

Table 4
CHARACTERISTIC PROTON CHEMICAL SHIFT RANGES FOR VARIOUS FUNCTIONAL GROUPS

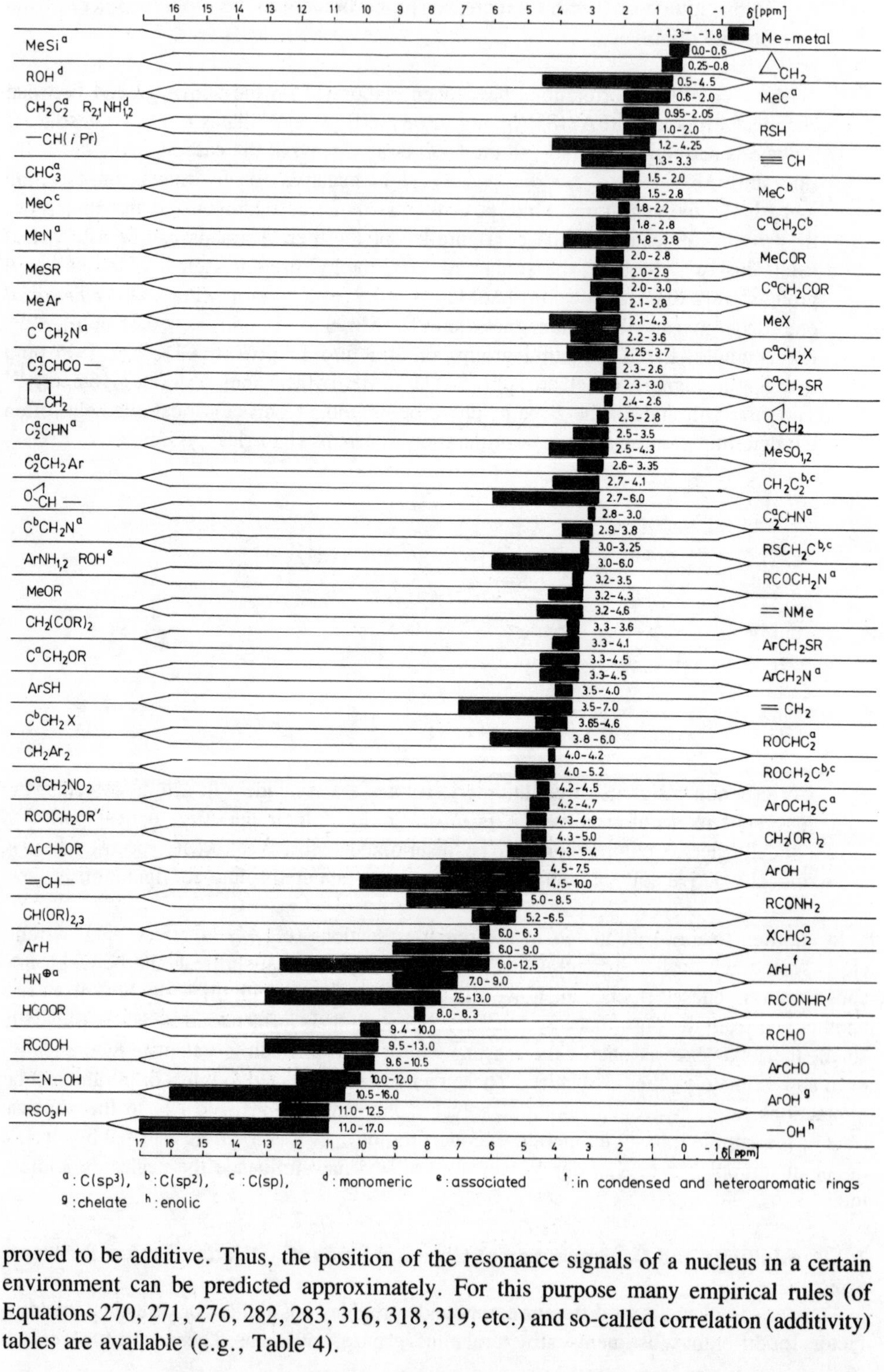

[a]: C(sp³), [b]: C(sp²), [c]: C(sp), [d]: monomeric [e]: associated [f]: in condensed and heteroaromatic rings
[g]: chelate [h]: enolic

proved to be additive. Thus, the position of the resonance signals of a nucleus in a certain environment can be predicted approximately. For this purpose many empirical rules (of Equations 270, 271, 276, 282, 283, 316, 318, 319, etc.) and so-called correlation (additivity) tables are available (e.g., Table 4).

The chemical shifts are relative values. In order to determine the absolute chemical shifts, we have to know the accurate values of B_0 and ν_0. A sufficiently exact determination of the latter (10^{-8} to 10^{-9}), however, would be rather difficult. It is much simpler to refer the chemical shifts to some standard, instead of the resonance of an idealized, nonshielded nucleus. The most generally used *reference* substance in ^{1}H and ^{13}C NMR spectroscopy — for nonaqueous solutions — is *tetramethylsilane,* $SiMe_4$ (TMS), the H and C atoms of which, apart from a few exceptions, are the most shielded sorts of these nuclei. Therefore on the scale chosen arbitrarily to have a zero chemical shift for TMS, the shifts of most proton and carbon types have identical signs (positive).

Since TMS is practically insoluble in water, in aqueous solutions the sodium salt of 3,3-dimethyl-2-silapentane-5-sulfonic acid, $Me_3Si(CH_2)_3SO_3Na$ is used (DSS). Chemical shifts referred to DSS or TMS agree within about 0.02 ppm. The sharp singlet signal of the methyl-groups of DSS is not influenced by pH changes — the multiplet of methylene groups are insignificant* for DSS concentrations $< 1\%$.

The TMS resonance signal is very sharp, and — according to the 12 equivalent hydrogens — also very intense on a molar basis; thus the determination of its exact position is easy, even at low concentrations. It is soluble in most NMR solvents. A further advantage of this standard is that it is nonpolar, quite inert, and does not interact with either solvents or the solute. Therefore it can be applied as an internal standard,** since when added to solution of the sample, it does not influence the chemical shifts of the latter. Because of its volatility (bp 27°C), it can easily be removed from the sample following the measurements.

The definition of chemical shift as referred to TMS is

$$\delta = (\nu_S - \nu_{TMS})\, 10^6/\nu_0 \text{ (ppm)} \tag{73}$$

and from the above, since by definition $\delta_{TMS} = 0$,

$$\delta = (\nu_S/\nu_0)\, 10^6 \text{ (ppm)} \tag{74}$$

where ν_s and ν_{TMS} are the resonance frequencies of the sample and the standard, respectively, in Hz, and ν_0 is the measuring frequency in Hz. The factor 10^6 is used only for convenience; thus usually δH values between 1 and 15 are obtained. Recording δ-values offer also the advantage that this quantity is independent of the measuring frequency, thus of the instrument used. The chemical shift values obtained according to Equation 70 depend, however, on the magnitude of ν_0. For protons, whose shielding coefficient is larger than that of TMS, δ-values are negative. Although δ-values are more convenient, the chemical shifts are sometimes given in τ units: $\tau = 10 - \delta$.[1406] While ^{1}H NMR shifts greater than $\delta = 10$ ppm (and $\tau < 0$) are quite frequent for organic compounds (e.g., enols, acids, aldehydes, amides, etc.), negative δ-values are extremely rare (aromatic annulenes, porphyrines, certain metallo-organic compounds). In Figure 21 the different shift scales are given, indicating also the position of the resonance signal of a few simple compounds.

* Compare Figure 85.

** When the standard for some reason cannot be mixed with the sample, it is placed into the probe either in a sealed capillary tube or by applying a double-walled probe, with the standard in the mantle. In such cases the local magnetic field is different in the sample and in the standard, and this may lead to significant differences compared to chemical shifts measured in the presence of internal standards. The change of the local field is due to the different volume susceptibility of the solution. The error from this source can reach 1 ppm, thus values referred to external standards should be corrected. The corresponding formula and the correction factors therein (volume susceptibilities) are available in the literature for the usual solvents.[1379]

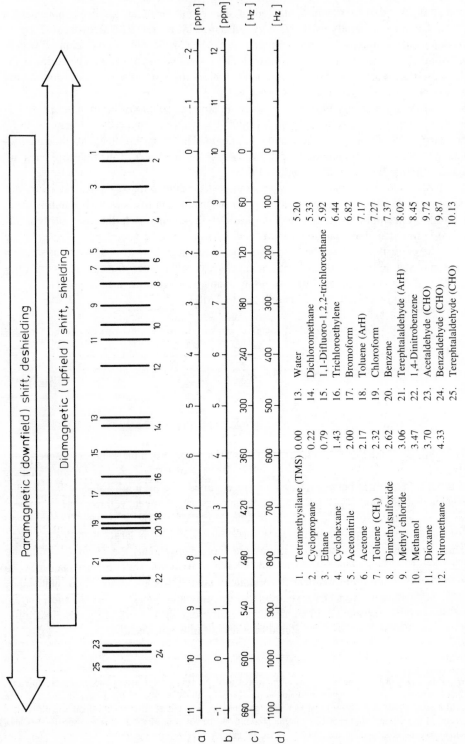

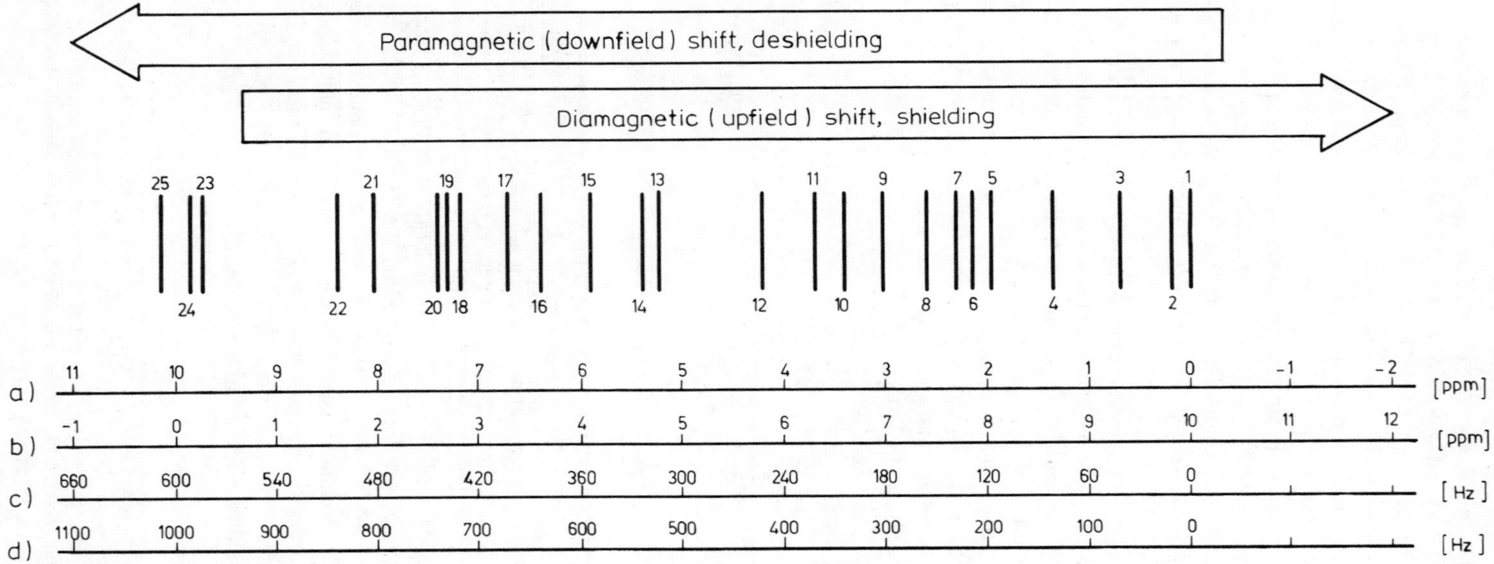

1.	Tetramethysilane (TMS)	0.00	13.	Water	5.20
2.	Cyclopropane	0.22	14.	Dichloromethane	5.33
3.	Ethane	0.79	15.	1,1-Difluoro-1,2,2-trichloroethane	5.92
4.	Cyclohexane	1.43	16.	Trichloroethylene	6.44
5.	Acetonitrile	2.00	17.	Bromoform	6.82
6.	Acetone	2.17	18.	Toluene (ArH)	7.17
7.	Toluene (CH_3)	2.32	19.	Chloroform	7.27
8.	Dimethylsulfoxide	2.62	20.	Benzene	7.37
9.	Methyl chloride	3.06	21.	Terephtalaldehyde (ArH)	8.02
10.	Methanol	3.47	22.	1,4-Dinitrobenzene	8.45
11.	Dioxane	3.70	23.	Acetaldehyde (CHO)	9.72
12.	Nitromethane	4.33	24.	Benzaldehyde (CHO)	9.87
			25.	Terephtalaldehyde (CHO)	10.13

FIGURE 21. The chemical shifts of NMR solvents referred to TMS on the (a) δ- and (b) τ- scale, as well as the frequency scale at (c) 60 and (d) 100 MHz.

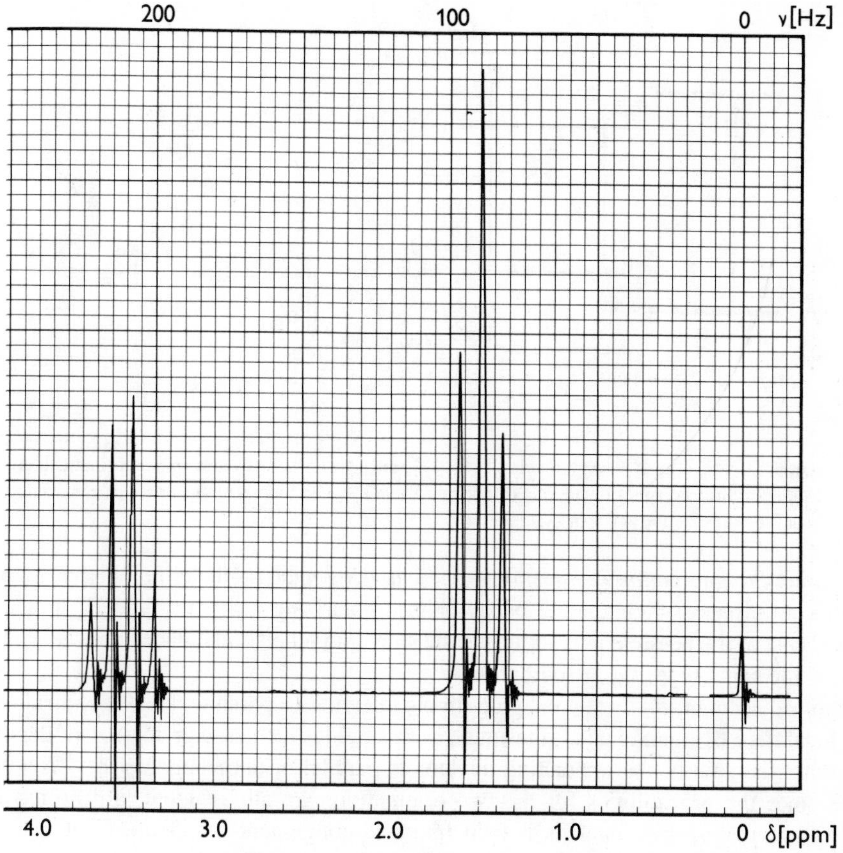

FIGURE 22. The spectrum of ethyl chloride.

1.3. SPIN-SPIN COUPLING: QUALITATIVE INTERPRETATION

1.3.1. The Phenomenon of Spin-Spin Coupling

In Section 1.2 we have shown that *the number of signals is given by the number of structurally different nuclei in the molecule*. In accord with this statement, the ^{1}H NMR spectrum of, e.g., 2,6-di-*t*-butyl-4-methylphenol (**1**), is composed of four sharp maxima, corresponding — in the order of increasing chemical shifts — to hydrogens of the *t*-butyl-, methyl-, hydroxyl-, and aromatic groups. However, many of the simplest compounds give much more complicated spectra, which cannot be interpreted without further considerations.

The methyl signal of ethyl chloride (CH_3–CH_2Cl) is composed of three closely spaced lines of intensity ratio 1:2:1, while the methylene signal is split into four components (see Figure 22), with an intensity ratio of 1:3:3:1. The ratio of the total intensities of the line groups (3:2) shows that the triplet belongs to the methyl group, while the quartet belongs to the methylene hydrogens. Similar spectra are obtained for many monosubstituted ethane derivatives, like ethyl bromide and iodide, nitroethane, propionitrile, propionic acid, propionyl chloride, etc. Similarly, most of the simple compounds have ^{1}H NMR spectra *composed of more lines than expected on the basis of the chemical shift theory*. It is, however, true *that the lines can be grouped into as many sets, as there are structurally different protons in the molecule*. For the complete interpretation of such spectra, in addition to the chemical shift, one needs another parameter, the *coupling constant*.

1.3.2. The Multiplicity of Spin-Spin Splitting and the Relative Intensity of Lines

The splitting of the signals into multiplets described above can be interpreted by taking

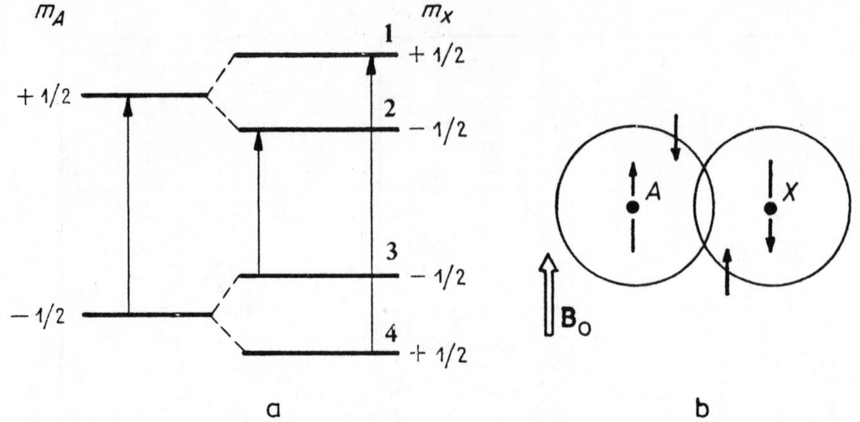

FIGURE 23. (a) The *A* energy levels of the *AX* spin system split by the spin-spin interaction (according to the two spin orientations of nucleus *X*) and (b) the mutual arrangement of electronic and nuclear spins for the *AX* spin system.

into account *the interaction of the spins of neighboring nuclei*. Nuclei in the vicinity of one another can influence their mutual behavior in a polarizing magnetic field $\mathbf{B}_0$ in a variety of ways. One of the possible interactions is the spin-spin relaxation induced between adjacent nuclei of the same type, discussed earlier.

Among adjacent nuclei there are dipole-dipole interactions possible, leading to an alteration of the local field $\mathbf{B}_l$ around nuclei, inasmuch as this field is increased or decreased, depending on whether the spin of the interacting nucleus is parallel or antiparallel to $\mathbf{B}_0$. As a consequence since the probabilities for the two orientations are almost identical, the resonance signal of the neighboring nucleus is split into two components of identical intensity. The magnitude of splitting depends on the distance and mutual orientation of the nuclei. Since these parameters undergo constant change due to the fast tumbling and the local fields are averaged out, the dipole-dipole interaction or coupling plays a role only for solids and viscous liquids.

There is, however, a third mode for coupling between nuclei, transferred by the bonding electrons. The simplest possible case is when the coupling of two nonequivalent magnetic nuclei of 1/2 spin (*A* and *X*) is examined, and it is supposed that the difference of their chemical shifts $\Delta\delta AX$ is much larger than the coupling constant characterizing the interaction between them (see below). The orientation of spin *X* parallel to the external field $\mathbf{B}_0$ reduces the local field about nucleus *A*. (Similar to Figure 14b, where for *A* and *X* stand *X* and *Y*, respectively.) Since the separation between the energy levels of atom *A* is proportional to the magnitude of the local field $\mathbf{B}_l$ (compare Equation 24), the distance of the *A* levels is then reduced. For an antiparallel orientation of the spin *X*, on the other hand, $\mathbf{B}_l$ about nucleus *A* and the separation of the *A* levels is also increased (see Figure 23a). These hold as well for exchanging the role of nuclei *A* and *X*; therefore, corresponding to the parallel or antiparallel orientation of the nucleus *A*, the energy levels of nucleus *X* are split also.[1166] This can be interpreted as follows: there is a magnetic interaction between nucleus *X* and its bonding electrons, and due to this coupling, the electron spin efforts to take antiparallel orientation to the nuclear spin. The spins of the bonding electrons must, however, be oriented also antiparallel, according to the Pauli principle. Consequently, the electron spin of atom *A* has the same orientation as that of nuclear spin *X*. The former then has such as influence upon the nuclear spin *A* that it tends to be oppositely oriented, this latter orientation being also opposite to that of the nuclear spin of *X* (see Figure 23b).

Consequently, the antiparallel orientation of the two nuclear spins is more favorable

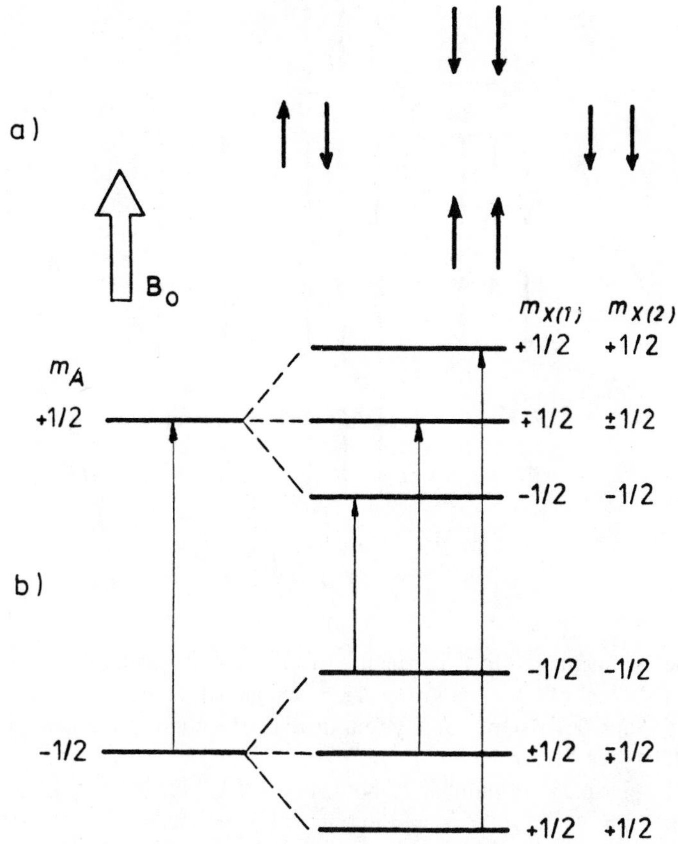

FIGURE 24. (a) The possible orientations for the spins of two equivalent nuclei in the polarizing magnetic field and (b) the *A* energy levels of the AX_2 spin system split by the spin-spin interaction according to the two spin orientations of the nuclei *X*.

energetically and manifests itself in a lowering of the corresponding levels **2** and **4**. The levels (**1** and **3**) attached to parallel orientation of spins *A* and *X* on the other hand are higher (see Figure 23a).

The two possible mutual orientation of spins *A* and *X* have almost identical probabilities, therefore, the signal is split into two components of the same intensity, situated at equal distance from the original line position.

When proton *A* has two equivalent neighbors of spin 1/2 (AX_2 spin system, e.g., protons in 1,1,2-trichloroethane), the spins of nuclei *X* may form three different combinations. Both protons can have either parallel or antiparallel orientations relative to $\mathbf{B}_0$, and also may have mutually opposite orientations. The latter combination can be realized in two ways (see Figure 24a). In the first case, the magnetic field is about proton *A*, consequently the distance of the corresponding energy levels is decreased; in the second it is increased, while in the third case it remains unchanged (see Figure 24b). This leads to the appearance of three transitions of different energy; one of these has the frequency of the coupling-free case. Three equidistant spectral lines are observed at these three absorption frequencies with the intensity ratio of 1:2:1, corresponding to the statistical weight of the transitions. Naturally the orientation of proton *A* can also influence the energy levels of protons *X*, thus the signal of the latter is also split, to 1:1 doublet (compare Figure 28). The magnitude of splitting is the same, since the coupling is transferred through the same bonding electrons.

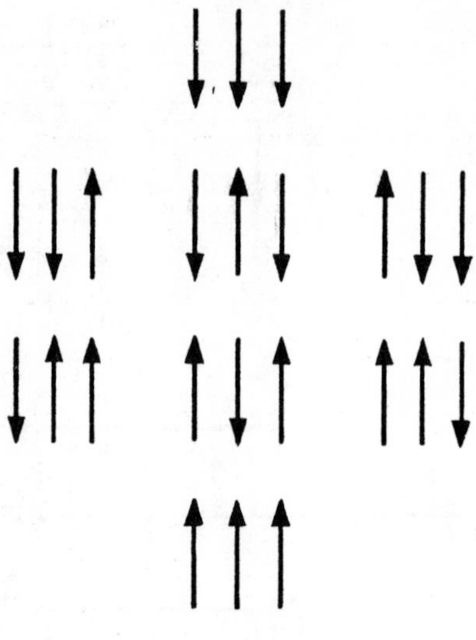

FIGURE 25. Possible arrangement of three equivalent spins in the polarizing magnetic field.

By the same arguments the triplet structure of the methyl signal and the quartet splitting of the methylene signal of the ethyl group can be explained (compare Figure 22). The spins of the methyl protons can assume namely four distinct orientations; the ratio of the statistical weights is 1:3:3:1 (see Figure 25).

Provided the spin quantum number of both A and X is 1/2, the resonance signal of the nucleus A in the spin system A_mX_n is split into $n + 1$ components. The situation is the same when the n number of atoms X do not belong to the same group, as for example the six methyl hydrogens of *iso*propyl-derivatives (see Figure 26), but are structurally identical. Generally, therefore, *the splitting of the lines is* $(n + 1)$*-fold*, if there are n equivalent protons or other nuclei of spin 1/2 in the neighborhood of the group giving rise to the multiplet. This rule applies also for nuclei of spin $I \neq 1/2$; in such cases the multiplet consists of $2nI + 1$ equidistant lines.

When there are i groups of different chemical shifts interacting with the nucleus under study, composed of $n_1 \ldots n_i$ equivalent nuclei of 1/2 spin, the splitting is given by:

$$(n_1 + 1)(n_2 + 1) \ \ldots \ (n_i + 1) \tag{75}$$

or, alternatively, for nuclei having $I_1 \ldots I_i \neq 1/2$:

$$(2n_1I_1 + 1)(2n_2I_2 + 1) \ \ldots \ (2n_iI_i + 1) \tag{76}$$

The relative intensity of the lines (A_k) for nuclei of spin 1/2 is

$$A_k = \binom{n}{k} = \frac{n!}{(n-k)!k!} \tag{77}$$

where $0! = 1$. The above form of spin-spin coupling is called *first-order coupling* (see below).

Spin-spin coupling was first observed by Proctor and Yu[1146] in the study of ^{121}Sb resonance of sodium hexafluoro antimonate ($NaSbF_6$). The resonance signal shows — in accord with the $(n + 1)$ rule — seven equidistantly spaced lines with the intensity ratio of 1:6:15:20:15:6:1 (compare the similar symmetrical septet of *iso*propyl iodide in Figure 26).

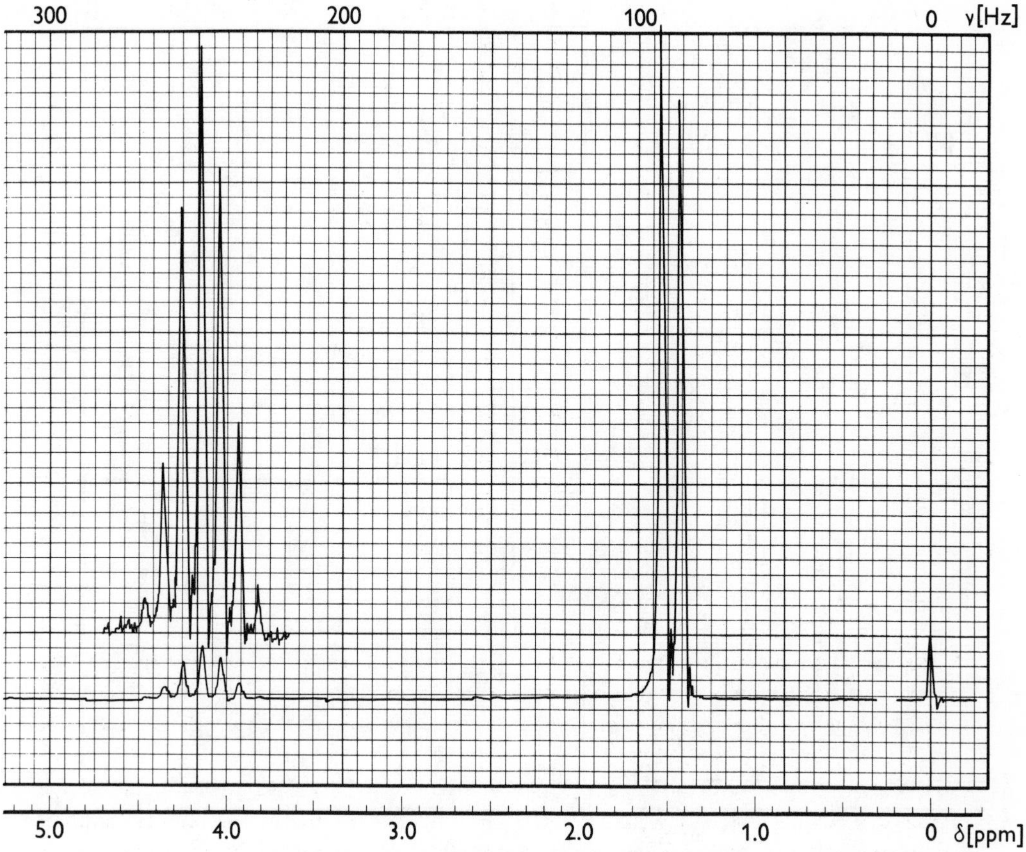

FIGURE 26. The spectrum of 2-iodopropane (Me_2CHI).

1.3.3. First-Order and Higher-Order Spin-Spin Coupling

The physical basis of spin-spin coupling is as shown the interaction of nuclear spins. This can be easily demonstrated by the spectra of such ethane derivatives in which the protons are successively replaced by noncoupling substituents, e.g., by chlorine or bromine. Except for fluorine the halogens behave as magnetic inactive nuclei* because though, possessing magnetic moments ($I_{Cl} = 3/2$, $I_{Br} = 3/2$, $I_I = 5/2$), their quadrupole moment highly accelerates the relaxation processes, resulting in the averaging of the spin states, i.e., the average lifetime of individual spin states becomes too short for the observation of the corresponding coupling in the spectra.** These haloethanes have the general formula A_nX_m ($n = 1, 2$; $m = 1, 2, 3$), where nuclei A and X may become equivalent upon symmetrical 1,2-disubstitution ($A_nX_m \rightarrow A_{n+m}$).

When one or two methylene hydrogens of the ethyl chloride molecule are replaced by chlorine, the methyl triplet (see Figure 22) is simplified to an 1:1 doublet or collapses to a singlet. In the former case, the methylene signal is still an 1:3:3:1 quartet, but its total intensity drops to one third of the intensity of the methyl doublet (see Figure 27). Upon substituting one or all three hydrogens of the methyl group by chlorine, the spectrum becomes of a single singlet: the four or two hydrogens, respectively, being equivalent. When two hydrogens are chlorine-substituted, i.e., when 1,1,2-trichloroethane is considered, the spectrum is an 1:1 doublet and an 1:2:1 triplet (see Figure 28). These correspond to the methylene and methyne protons, respectively; their intensity ratio is 2:1. The spectrum of 1,1,2,2-

* Compare p. 73.

** Compare Volume II, p. 97, 238, and 251.

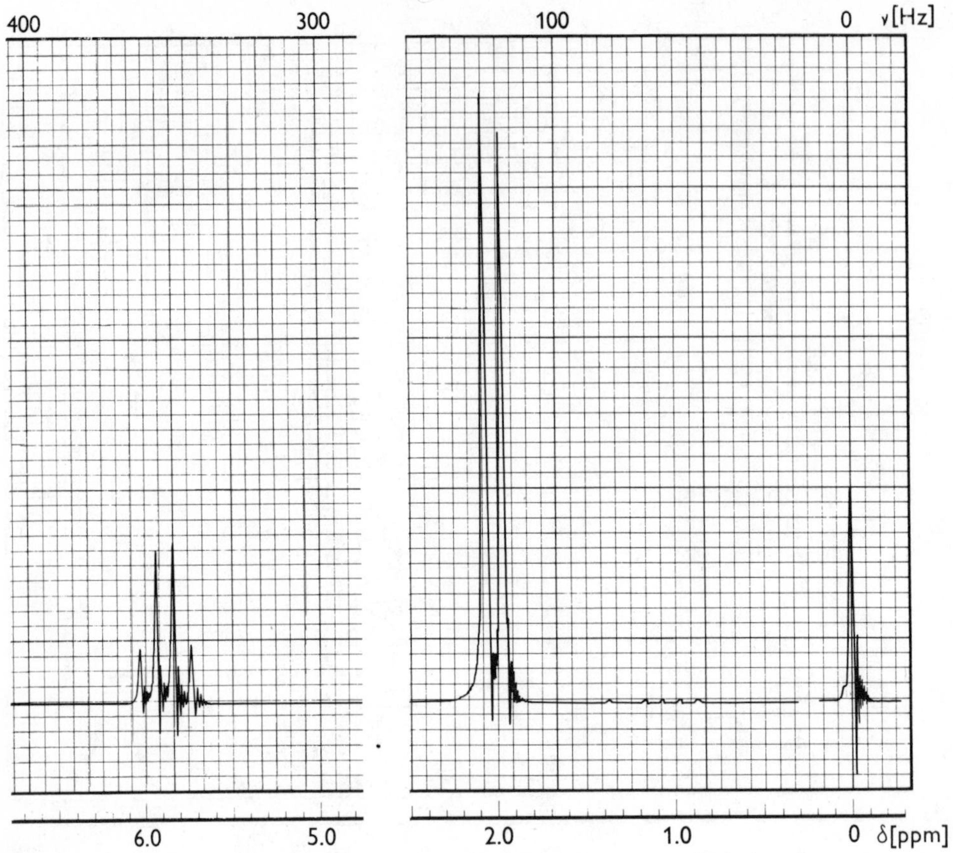

FIGURE 27. The spectrum of 1,1-dichloroethane (Me–$CHCl_2$).

tetrachloro- and pentachloroethane is a singlet, due, in the first case, to the molecular symmetry. In the case of asymmetrical 1,2- and 1,1,2,2-substitution, there are two-two triplet and doublet in the spectrum. An example for the first case is Figure 53a and for the latter an example is Figure 40.

The above examples follow the $(n + 1)$ rule and also obey Equation 77 in respect of the relative intensity of the lines. These rules apply for more complex cases, too.

For example, in the spectrum of crotonic aldehyde (**15**), the aldehyde proton gives rise to a doublet, while eight lines represent the β-hydrogen due to interaction with the methyl group and the α-proton, respectively. The signal of the α-hydrogen consists of 16 lines, as not only the *vicinal* (β-) olefinic and aldehyde protons, but also the methyl hydrogens participate in splitting: $2 \cdot 2 \cdot 4 = 16$ (see Figure 29). The coupling with the methyl group is the smallest; that caused by the β-proton is in turn the largest. Accordingly the methyl hydrogens coupled both to the α- and β-protons give rise to four lines. Thus, the example of crotonic aldehyde also exemplifies that spin-spin coupling can occur not only between *vicinal* nuclei, but also between those farther removed from each other (see later).

The methyne signal of RCH_2–$CH(CH_3)$–CH_2R' type compounds consists theoretically of $3 \cdot 3 \cdot 4 = 36$ lines, showing that if a nucleus is adjacent to many different kinds of atoms, its signal can be rather complex. In such cases lines are usually not separated, and — depending on the resolution — a more or less structured broad multiplet is observed.

As opposed to chemical shift and signal intensity, the *spin-spin coupling* (fine structure) is *field and frequency invariant*.[607] This is clear from Figure 30a and b, showing the ethyl multiplets of ethylbenzene at 60 and 100 MHz. While the triplet and the quartet are much

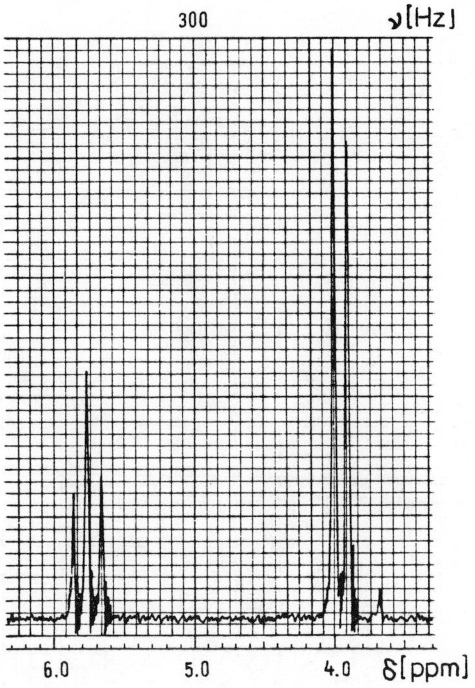

FIGURE 28. The spectrum of 1,1,2-trichloroethane ($CH_2Cl–CHCl_2$).[1044]

more separated (on the hertz scale!), at 100 MHz the distance of the lines of the multiplets remains unaltered.

The triplets and quartets of some ethyl derivatives show further splitting at 60 MHz (see Figure 30c), which is absent at 100 MHz (see Figure 30d). This fine structure is therefore frequency dependent.

There are molecules in the spectra of which the multiplets of chemically different protons are not separated. For example, in the spectrum (see Figure 31c) of acrylonitrile (**16**), line-components cannot be assigned to the protons *A*, *B*, and *C*. The multiplet structure is frequency dependent and becomes simpler at higher measuring frequencies. Already at 100 MHz (see Figure 31b), the lines of one of the protons are clearly distinguished; the other two give still overlapping multiplets. At 220 MHz (see Figure 31c), line groups for all protons are well separated. It should also be noted that there is no apparent regularity in the relative intensities of the lines.

The simple splitting, which is frequency invariant and follows the $(n + 1)$ rule, is called *first-order splitting*, while the frequency-dependent splittings, leading to the fine structure of the first-order multiplets and to more complicated spectra, as for example of acrylonitrile, are called *higher-order spin-spin splitting*. The order of splitting depends (see later) on the ratio $J/\Delta\nu$, where $\Delta\nu$ is the chemical shift difference of the interacting nuclei (in hertz) and J is the coupling constant.

1.3.4. The Coupling Constant

The parameter characterizing quantitatively spin-spin couplings, i.e., the interaction among the spin, is the coupling constant J, usually given in hertz units, because J is frequency invariant. J is, in addition to chemical shift δ, the other fundamentally important information which can be obtained from the NMR spectrum and used in structure elucidation.

Couplings are always mutual; two atomic groups, A_n and X_m, coupled to one another show

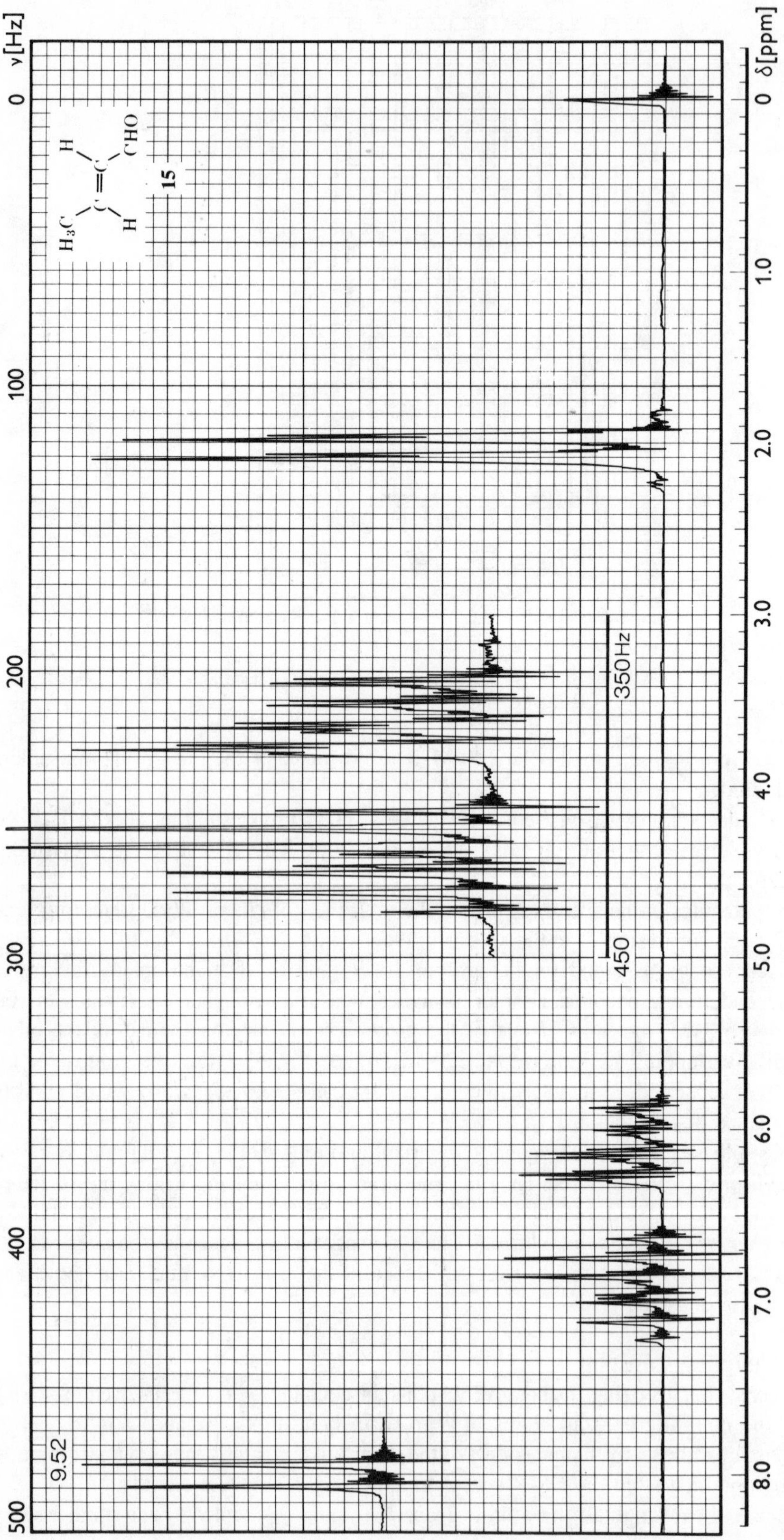

FIGURE 29. The spectrum of crotonaldehyde (**15**).

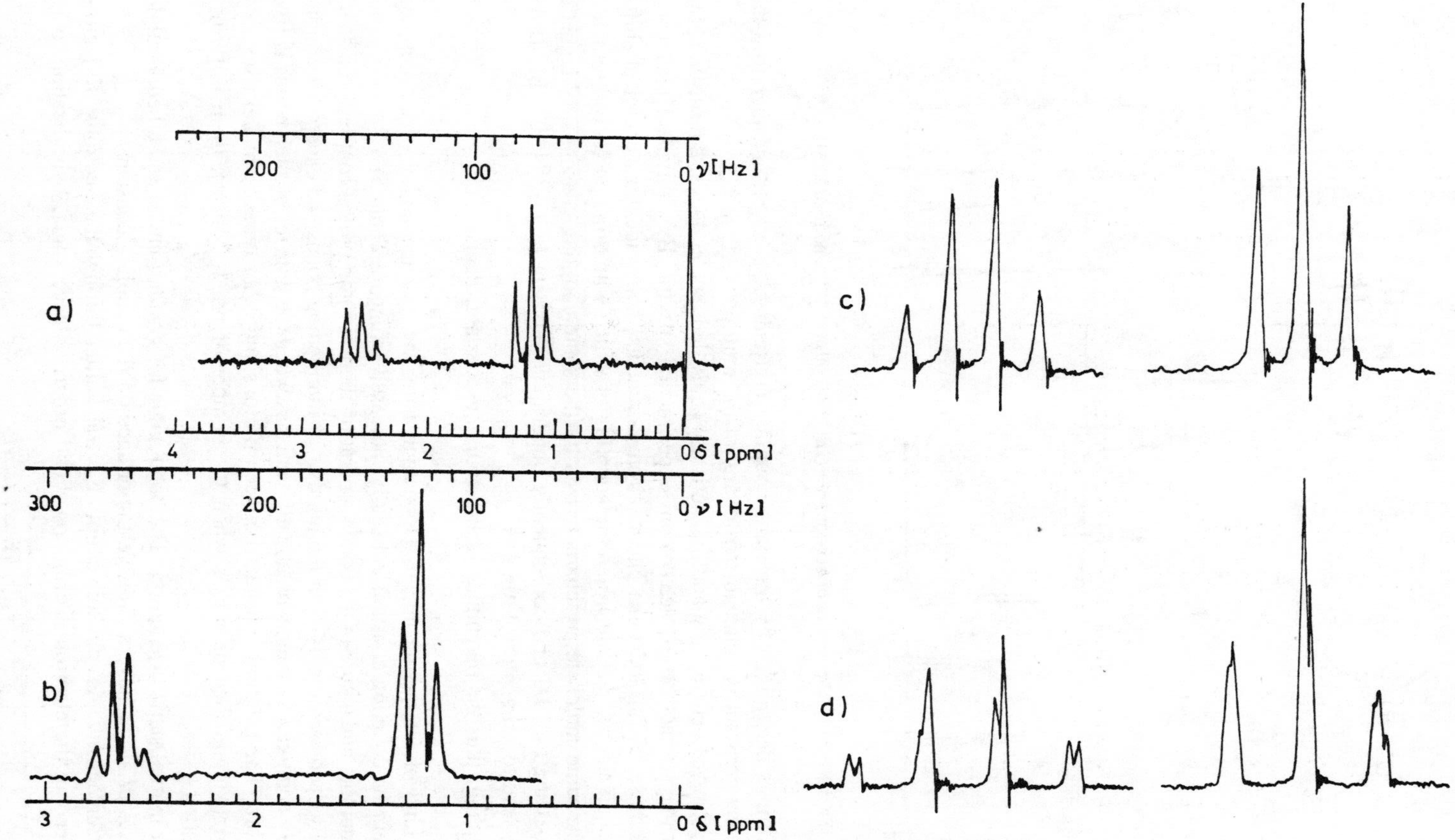

FIGURE 30. The ethyl signals of ethyl benzene and ethyl thiocyanate (a) and (b) at 60 MHz as well as (c) and (d) at 100 MHz.[720]

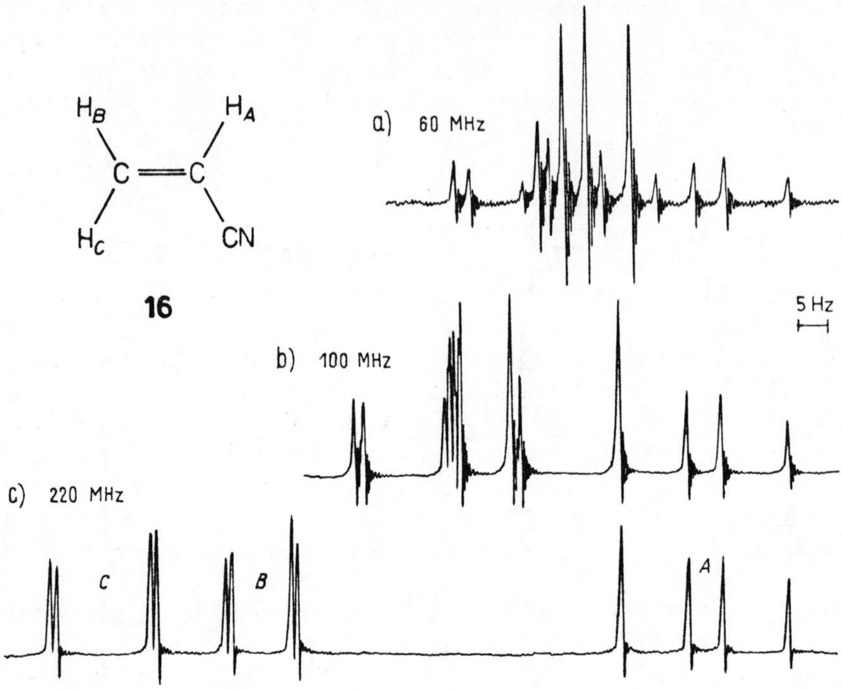

FIGURE 31. The spectra of acrylonitrile (**16**) at (a) 60, (b) 100, and (c) 220 MHz.[1443]

always splittings determined by the same coupling constant ($J_{AX} \equiv J_{XA}$), and therefore the same coupling constant is obtainable from any of the multiplets, A_n or X_m.

Coupling between chemically and magnetically equivalent nuclei does not result in splitting, thus, they cannot be extracted from experimental spectra. By equivalent nuclei we understand those having the same chemical environments, hence identical chemical shifts (chemical equivalence, see later), and the coupling of which with other nuclei or groups of nuclei in the same molecule are characterized by the same coupling constants.* Magnetic nuclei possessing $I > 1/2$ and consequently quadrupole moment (e.g., ^{14}N, Cl, Br, I) do not usually lead to spin-spin splitting.**

The coupling, thus the magnitude of the coupling constant, depends on:

1. The kind of interacting nuclei. Upon increasing the atomic number of the coupled nuclei, the coupling constant is usually increased (compare Table 5).
2. The number and nature of the bonds separating them. Proton-proton coupling constants usually decrease quickly with the number of intervening bonds and except for continuously conjugated systems practically vanishes when the number of such bonds is five or more. The coupling constant between protons and other types of nuclei rises first to a maximum then decreases when the number of bonds between them is further increased.
3. Their mutual steric orientation. The bond angle for *geminal* nuclei and dihedral angle for *vicinal* atoms have a decisive importance on the coupling constant.
4. Electron density. As spin-spin coupling is transmitted through the electrons, all factors influencing the electron distribution (bond order, valence angle, hybridization, con-

* Magnetic equivalence; compare p. 70.

** Compare p. 49 and Volume II, p. 97.

jugation, inductive and mesomeric effects, etc.) have an effect upon the magnitude of J. The coupling constant is independent of temperature, concentration, and physical state of the sample. It does change as a function of these parameters is indicative of conformational equilibria or other alterations influencing molecular structure.

Solvents usually by H-bonding have various degrees of effect up the coupling constants. They usually decrease when the electronic density around the coupled nuclei is decreasing as a consequence of solvation or association.

For protons coupling constants are between 0 and 25 Hz, whereas for other nuclei they may exceed 1000 Hz. Some characteristic intervals of J, values between different nuclei are given in Table 5.

Coupling constants can be negative. Their sign does not influence spectra of spin systems characterized by a single constant only or first-order spectra of spin-systems composed of sets magnetically equivalent nuclei only. Thus, from such spectra only their absolute value can be obtained. For more complicated spin systems, spectra do depend on the sign of the coupling constants,* but even in such cases the analysis of the spectra gives only their relative signs.

If the number of σ-bonds between the interacting protons in saturated open chain hydrocarbons is an odd number (the coupled spins are antiparallel), the sign of the coupling constant is positive by definition.[753] For parallel spin-spin alignment (even bonds between the protons) $J < 0$. The coupling constant of *geminal* hydrogens [$J^{gem} \equiv {}^2J(\mathrm{H,H})$] is accordingly negative and that of *vicinal* ones [$J^{vic} \equiv {}^3J\ (\mathrm{H,H})$] is positive. Since spin-spin coupling is transmitted not only through the σ-electrons, it is not always possible to determine the sign of J simply by counting the bonds. Various structural factors determining the magnitude of couplings will be discussed in connection with proton-proton interactions, but the coupling constants of other nuclei are influenced similarly by these factors.

1.3.5. Geminal Couplings

Geminal (^2J) coupling constants are generally negative in saturated open chain hydrocarbons. The *geminal* coupling of methyl protons can not be observed since the signal of the equivalent methyl protons is a singlet.** It can be, however, determined indirectly, e.g., when one of the hydrogens is substituted by deuterium. (Another possibility will be mentioned in Section 4.1.3.2.) In the ^{1}H NMR spectrum in the absence of other couplings, the group CDH_2 gives rise to three lines of equal intensity as the X_2 part of the AX_2 system. The multiplicity is $2nI_D + 1$ (compare Formula 76), where $n = 1$, $I_D = 1$. The A portion of the AX_2 spectrum could naturally be observed only in the deuterium resonance spectrum, which is a triplet of 1:2:1 intensity. From the coupling constant $^2J(\mathrm{D,H})$, the $^2J(\mathrm{H,H})$ value of the methyl protons can be calculated*** by:[125]

$$J(\mathrm{H,H}) = 6.55\, J(\mathrm{D,H}) \qquad (78)$$

The above empirical relationship for converting coupling constants applies for any type of group.

The signals of isolated methylene groups in open chain compounds generally do not split due to *geminal* coupling, too, apart from special cases to be discussed later, since as a result of free rotation the two hydrogens become equivalent. In ring compounds, however, *geminal* coupling can be frequently observed.

* Compare p. 124, compare also Sections 2.1.6.2 (p. 149) and 4.1.3.2 (Volume II, p 209-210).

** Compare p. 51 and 91-92.

***Compare Equation 324.

Table 5
CHARACTERISTIC COUPLING CONSTANTS FOR DIFFERENT PAIRS OF NUCLEI IN VARIOUS TYPES OF MOLECULES

Coupling	Group	J /Hz/	Note
^{1}J(H, H)	H – H	280[a]	
^{2}J(H, H)	CH_4	12.5[a]	
	H – C – H	10 – 18[b]	
		3 – 9	X : CH_2
		4 – 7	X : O
		1 – 2	X : NR
		< 1	X : S
	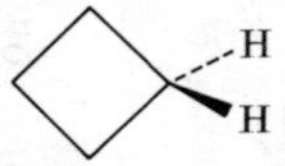	16 – 20	$n \geq 1$
	X = C(H)(H)	0 – 4[b]	X : C
		7 – 17	X : N
		42[a]	X : O
^{3}J(H, H)	$CH_3 - CH_3$	8[a]	
	>CH – CH<	2 – 9[c]	hindered rotation
		6 – 7[c]	free rotation
	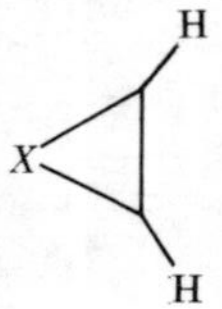	7 – 13	c, X : CH_2
		4 – 10	t, X : CH_2
		2 – 5	c, X : O
		1 – 3	t, X : O
		~6	c, X : NR
		~3	t, X : NR
		6 – 7	c, X : S
		5 – 6	t, X : S
^{3}J(H, H)	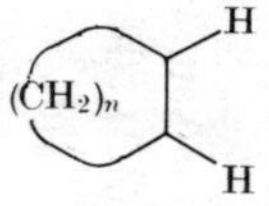	~10	c (n = 2)
		~5	c (n = 3)
			t (n = 2, 3)
		8 – 12	a, a (n = 4)
		3 – 5	a, e (n = 4)
		2 – 3	e, e (n = 4)
	>CH – CH = X	1 – 3	X : O
		4 – 10	X : =CH–
	– CH = CH –	5 – 14[b]	Z
		12 – 19[b]	E
	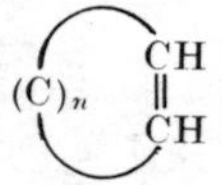	0 – 2	n = 1
		2 – 4	n = 2
		5 – 7	n = 3
		8 – 11	n = 4
		10 – 13	n = 5
	= CH – CH =	9 – 13	
	= CH – CHO	5 – 8	
	>CH – OH	~5	split only in dry DMSO-d_6
	CH ≡ CH	9.8[a]	
^{4}J(H, H)	>CH – C ≡ CH	2 – 3	
	– CH = C = CH –	5 – 6	
	>CH – C = CH –	0 – 2	$\lvert J^{cd} \rvert > \lvert J^{td} \rvert$

$^{n}\underline{J}$(H, H)	benzene ring (H, H)	7 – 10	$\underline{o}$ ($^{3}\underline{J}$)
		1 – 3	$\underline{m}$ ($^{4}\underline{J}$)
		0 – 1	$\underline{p}$ ($^{5}\underline{J}$)
	five-membered ring (X)	2 – 5	$^{3}\underline{J}$
		1 – 3	$^{4}\underline{J}$
$^{1}\underline{J}$(X, H)	D – H	1 – 2	
	B – H	30 – 150	
	C – H	125	sp^3
	C = H	170	sp^2
	C ≡ H	250	sp
	^{15}N – H	50–100	
	O – H	70	
	P – H	200–500	
	OP – H	500–700	
$^{2}\underline{J}$(X, H)	>N – C – H	0 – 3	^{14}N (sp^3) C (sp^3)
		1 – 10	^{15}N (sp^3)
	>N = CH –	10 – 20	
	= N – CH<	10 – 20	
	^{15}N – CHO	20	
	= CH – N<	1 – 6	N (sp^3)
		40 – 80	C (sp^3)
	F—C – H	70 – 90	C (sp^2)
	P C – H	10 – 20	C (sp^3)

$^{5}\underline{J}$(H, H)	>CH – C = C – CH<	1 – 2	$\underline{J}^{cd}$
		~0	$\underline{J}^{td}$
	>CH – C – C = CH –	~1	
	>CH – C ≡ C – CH<	0 – 2	
$^{1}\underline{J}$(X, Y)	B – F	10 – 80	
	B – P	~50	
	C – C	30 – 60	sp^3
	C = C	~ 70	sp^2
	C ≡ C	~170	sp
	C – ^{15}N	0 – 20	sp^3, sp^2, sp
	C – F	220–350	
	C – Si	~50	
	C – P	10 – 60	
	^{15}N = ^{15}N	~15	
	^{14}N – F	100 – 250	
	Si – F	350 – 500	
	P – F	700–1500	
	P – P	~100	
	OP – PO	~500	
$^{2}\underline{J}$(X, Y)	⩾C – C ≡ ^{15}N	0 – 5	$^{2}\underline{J}$(N, C)
	⩾C – C – F	30 – 100	$^{2}\underline{J}$(F, C)
		150–230	C (sp^3)
	F – C – F	290–305	cyclohexanes
		6 – 90	C (sp^2)

Table 5 (continued)
CHARACTERISTIC COUPLING CONSTANTS FOR DIFFERENT PAIRS OF NUCLEI IN VARIOUS TYPES OF MOLECULES

^3J(X, H)	>N – C = CH –	1 – 6	
	>CH – C ≡ ^{15}N	1 – 2	
	>CH – CF<	5 – 20	
	– CH = CF –	1 – 20	c
		10 – 40	t
^4J(F, H)	>CH – C – CF<	0 – 5	
^nJ(F, H)	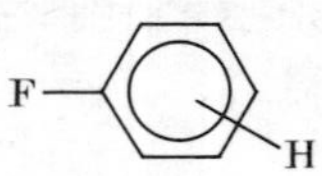	6 – 10	o, ^3J
		5 – 8	m, ^4J
		2 – 3	p, ^5J

^3J(F, F)	>CF – CF<	0–20	
		~ 27	a, a
		~ 9	e, e
		~ 1	a, e
	– CF = CF –	10– 60	c
		110–130	t
^4J(F, F)	>CF – C – CF<	7– 10	C (sp^3)
^nJ(F, F)	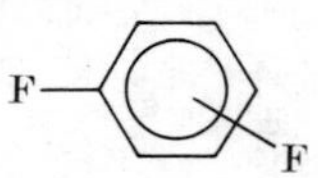	17– 21	o, ^3J
		2 – 4	m, ^4J
		15 –15	p, ^5J

[a] Determined from the J(H, D) coupling constant of the partially deuterated molecule
[b] The coupling constant depends on the electronegativity of the neighbouring substituents
[c] The coupling constant depends on the dihedral angle
[d] The coupling constant depends on the rotation about the C – C bond

c : cis, t : trans, cd : cisoid, td : transoid, o : ortho, m : meta, p : para

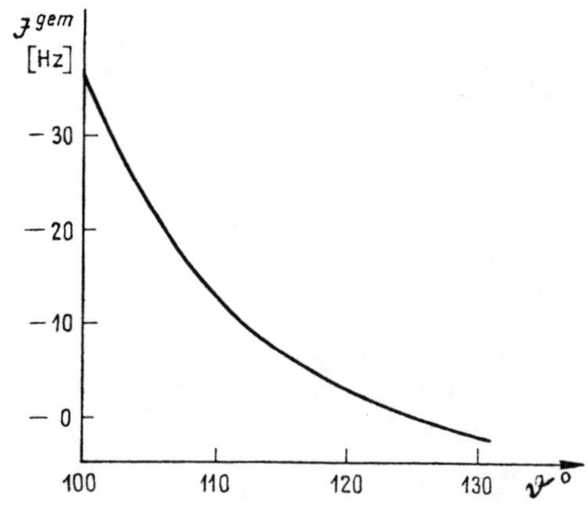

FIGURE 32. Dependence of J^{gem} coupling constant of *geminal* hydrogens upon the dihedral angle.[605]

The magnitude of ^2J(H,H) coupling constants is a function of molecular structure. It has been theoretically shown that J^{gem} increases with increasing ϑ_{HCH} bond angle; since generally $J^{gem} < 0$, its absolute value drops (see Figure 32). Thus, with cyclohexane derivatives ($\vartheta = 109°$), J^{gem} falling into the range of 12 to 18 Hz, whereas in cyclopropanes ($\vartheta = 114°$), their value is found between 4 to 9 Hz. For the terminal ethylene groups ($\vartheta = 120°$), the splittings of 0 to 3 Hz are found and J^{gem} is frequently positive. The agreement between the experimental and theoretical data is generally rather poor in consequence, that valance angle is only one of the factors determining the coupling constants.

Electron-withdrawing substituents on a carbon atom carrying the hydrogens increase (with regard to the absolute value decrease) the value of *geminal* coupling constants.[1129] J^{gem} (CH_4) = − 12.4, J^{gem} (CH_3F) = − 9.6 Hz. A significant change is found in the case of the cyclopropane derivatives [J^{gem} = (− 3) − (− 9) Hz], as compared to the oxiranes [J^{gem} = (+ 4) − (+ 7) Hz]. The − *I* effect of a *vicinal* substituent leads to a change of opposite sense, similarly as in the case of *vicinal* couplings. Compounds **17a** and **17b**, for example, give J^{gem} values of − 6.8 and − 9.7 Hz, respectively.[1526]

A linear correlation was found between the number of *vicinal* π-bonds and J^{gem}. The values of J^{gem} for methane (number of π-bonds: 0), toluene (one *vicinal* π-bond), acetonitrile (two π-bonds), and malonodinitrile (four π-bonds) are 12.4, 14.5, 16.2, and 20.3 Hz, respectively.[85] The dependence in ring compounds of J^{gem} on the dihedral angle, φ, between the π-bond and the CH bonds is shown on Figure 33.[85]

In accordance with this, cyclohexanones (φ ≈ 75°) and cyclopentanones (φ ≈ 20°, 40°) can be characterized by J^{gem} of 12 to 13 and 16 to 17 Hz, respectively. The corresponding J^{gem} value of cyclopentenedione (**18**) is 21.8 Hz, as a result of dihedral angles ~20 and 40° and the cumulated effects of two π-bonds. This, s.c. π-contribution can be useful in structural elucidation. J^{gem} is very sensitive not only to structural changes, but also to the physical state of the sample. J^{gem} values of the methylene hydrogens neighboring the oxygen in a CH_2O group of $MeCH(OCH_2R)_2$-type acetals show solvent and concentration dependence: they decrease upon increasing dielectric constant of the solvent.[1172]

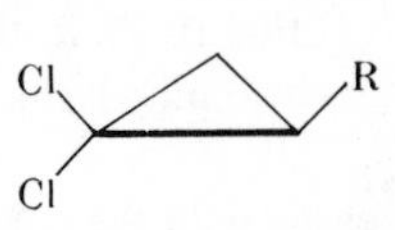

17a R=COOH
17b R=OAc

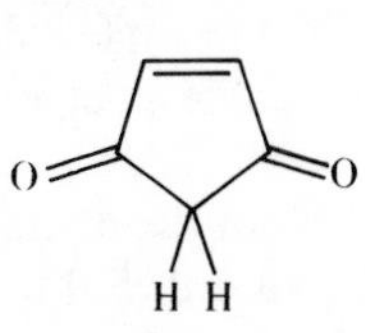

18

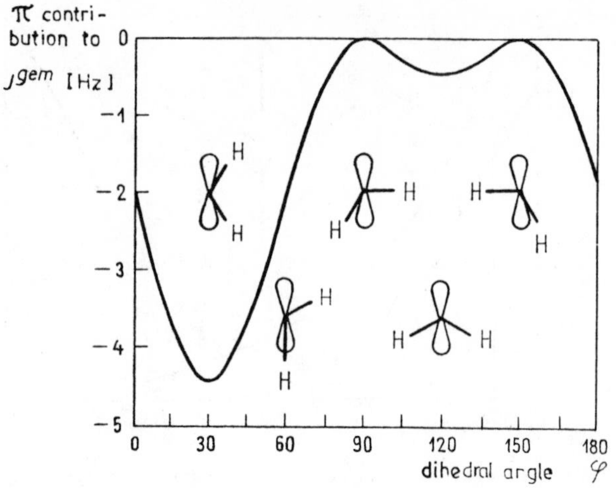

FIGURE 33. The dependence of J^{gem} of *geminal* methylene hydrogens *vicinal* to an unsaturated carbon atom, upon the dihedral angle, φ, between the C–H and π-bonds.[85]

1.3.6. Vicinal Couplings

The most important factor (after bond order) influencing *vicinal* coupling constants ($J^{vic} \equiv {}^3J$) of protons is the dihedral angle of the *vicinal* CH bonds and is described by the theoretically derived Karplus equation:[751]

$$J^{vic} = a \cos^2\varphi - b \tag{79}$$

where the value of the constants *a* and *b* is different for the ranges $0° < \varphi < 90°$ and $90° < \varphi < 180°$. In order to avoid this inconvenience, instead of Equation 79, for molecules such as ethane and its $CHR_2–CHR'_2$ type derivatives, Equations 80a or 80b have been proposed,[588] with different constants to be deduced empirically.

$$J^{vic} = A + B\cos\varphi + C\cos^2\varphi \tag{80a}$$

$$J^{vic} = A' + B'\cos\varphi + C'\cos 2\varphi \tag{80b}$$

Equations 80a and 80b are not independent, they are interconvertible, substituting the following relationships for the constants: $A = A' + C'/2$, $B = B'$, and $C = C'/2$. According to Equations 79, 80a, or 80b, J^{vic} depends on φ as shown in Figure 34a. The experimental data are in qualitative agreement with the calculated results (see Figure 34b).

Although Equation 79 is only of qualitative nature, and therefore the measured value of the coupling constants cannot be used for the calculalion of the dihedral angles, it is of much use for distinguishing *axial* or *equatorial* and *cis* or *trans* orientation of *vicinal* protons in saturated six-membered rings and in olefins, respectively. It follows from Equations 79, 79a, and 79b, namely that the coupling constant of *diaxial* protons ($\varphi = 180°$) is substantially larger than that of *axial-equatorial* or *diequatorial* ones ($\varphi = 60°$). Similarly, J^{vic} of *trans* olefin protons ($\varphi = 180°$) is expected to be larger than that of their *cis* counterparts ($\varphi = 0°$). Besides the dihedral angle φ, other structural factors also play a role in the establishment of coupling.

In cycloolefins ^3J(H,H) of the olefinic protons increases with the ring size. Its expectable value is 0.5 to 2.0, 2.5 to 4.0, 5.0 to 7.0, and 8.5 to 10.5 (larger values may occur, too) Hz for three-, four-, five-, and six-membered rings. Since the dihedral angle is constant throughout this series ($\varphi = 0°$), evidently *J* depends not only on φ, but also on the angle

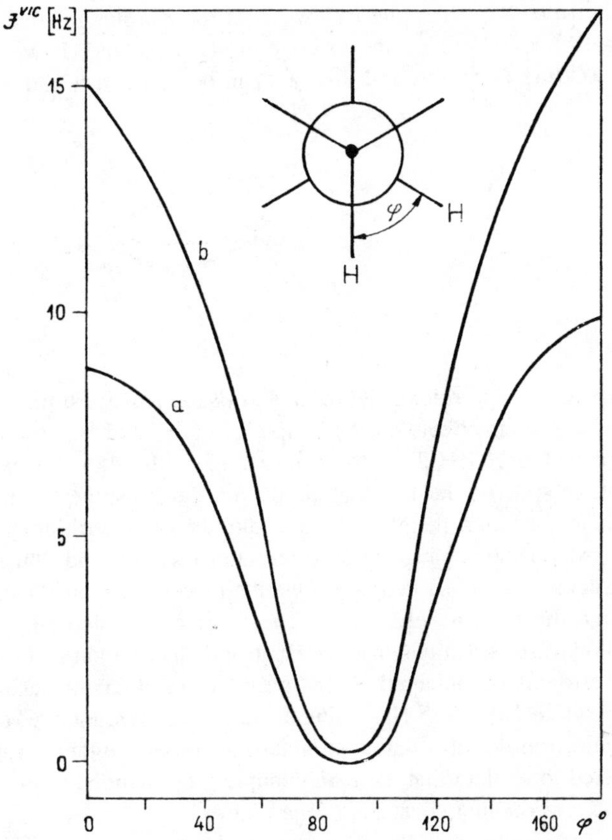

FIGURE 34. The (a) calculated and (b) measured dependence of coupling constants of *vicinal* protons upon the dihedral angle.[751]

δ defined by the HCC bonds (*vicinal* couplings can be characterized by two such bond angles, which are identical in cycloolefins).

In accordance with Equation 79, for cyclopropanes J^{cis}_{vic} ($\varphi = 0°$) $> J^{trans}_{vic}$ ($\varphi \approx 144°$) always holds ($^3J^{cis} \approx 7$ to 13 Hz, $^3J^{trans} \approx 4$ to 10 Hz), and the same is true for oxiranes ($J^{cis} \approx 2$ to 5 Hz, $J^{trans} \approx 1$ to 2.5 Hz). Thus, a significant difference exists for both types of compounds between the magnitude of both $^3J^{cis}$ and $^3J^{trans}$, although neither φ nor ϑ are significantly different.

In oxiranes the heteroatom reduces both *vicinal* couplings, evidently through its $-I$ effect. By oversimplification we may say that the heteroatom reduces couplings by withdrawing the electrons transmitting the interaction. Electronegative substituents generally lower the coupling constants in a linear manner:[8,306,834,894,1526]

$$J_0 - J = \Delta J = K\Delta E = K(E_0 - E_H) \qquad (81)$$

where ΔJ is the difference of the coupling constant relative to J_0 of an unsubstituted compound, while ΔE is a quantity proportional to the difference in electronegativity[691] between hydrogen (E_H) and the substituent (E_0). The *vicinal* coupling constants of ethane, ethyl chloride, and 1,1,2-trichloroethane are, for example, about 8,[894] 7,[8] and 6 Hz, respectively.

Due to this phenomenon, bridgehead hydrogens in condensed oxiranes do not couple their *vicinal* partners.[1522] In 3,4-epoxy-tetrahydrofurane (**19**), e.g., the H-3 signal is a singlet, i.e., $J^{cis}_{2,3} \approx J^{trans}_{2,3} \approx 0$,* and therefore the H-2,2′ protons give rise to a s.c. *AB* spectrum.

* Compare Problem **63** and Volume III, p. 232.

The absence of the *trans* ($J'_{2,3}$) coupling follows from the Karplus relationship ($\varphi = 90°$), but due to the small dihedral angle, the *cis* coupling ($J_{2,3}$) should, however, be nonzero. Nevertheless this splitting is absent, and this fact can be ascribed to the $-I$ effect of the heteroatom.[1522]

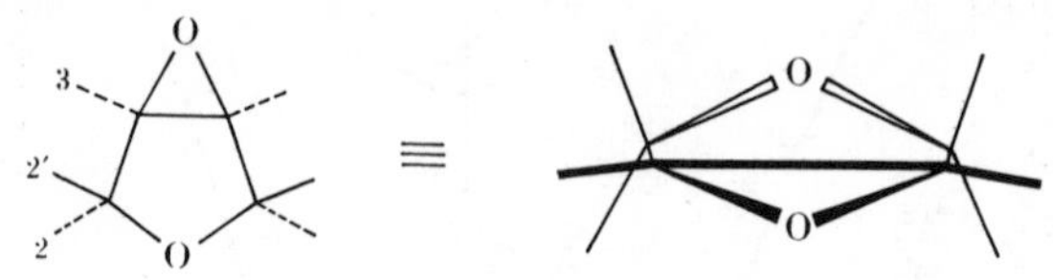

19

In rigid systems (e.g., in steroids), the *axial* and *equatorial* substituents can be distinguished on the basis of their effects on the couplings of H_α and H_β atoms. The coupling constants ^3J(H,OH) and ^3J(H-1,H-2) of the isomeric pair **20a, 20b**, for example, are significantly different, in spite of the fact that φ, ϑ, and the substituent are identical in the isomers.[1160] This difference must therefore be related to the steric position of the substituent. One of the effects which may change in such structures as **20a** and **20b** is the "through space" spin-spin interaction among hydrogens via their close electron clouds, in contrast to transmission through the bonding electrons.* The other effect, depending also upon the mutual steric position of the substituents is the electronic distribution and the change of the bondlengths. This effect is encountered in its "pure" form in naphthalene derivatives,[1522] where $J_{1,2} \approx 8$ Hz, while $J_{2,3} \approx 6.5$ Hz. Using again a simplified picture, the shorter bond involved a higher electron density, thus a stronger interaction (higher J value). The same argument can be used to explain that the *ortho* coupling constants of aromatic compounds are smaller than the *cis* coupling constants of the olefins.

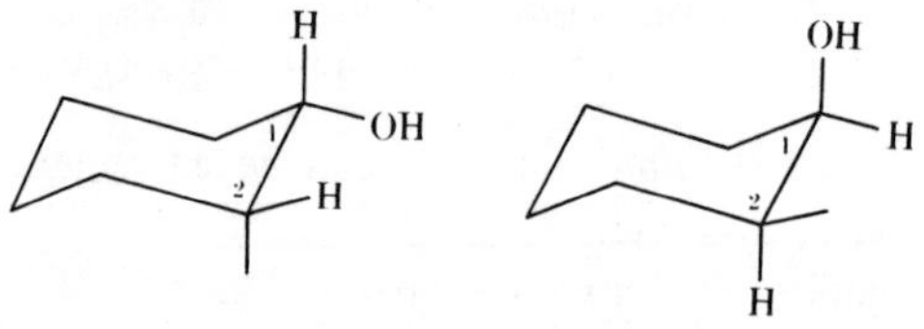

20a **20b**

1.3.7. Long-Range Couplings

In saturated compounds significant coupling can only be observed between hydrogens separated by only two or three bonds. With high-resolution measurements it is also possible to observe the s.c. "long-range" couplings between nuclei separated by more than three bonds ($^nJ = 0.1$ to 3.0 Hz, $n \geq 4$). In unsaturated compounds — due to the presence of a delocalized π-electron system — protons separated by five, six, or even more bonds can be coupled with each other.

1.3.7.1. Long-Range Couplings in Unsaturated Systems: Allylic and Homoallylic Coupling

The frequent long-range coupling between hydrogens of RR′ CH–C=CH– group is called the allylic coupling: $^4J = 0$ to 3 Hz. *Cisoid* hydrogens (**21a**) possess[1368] usually a larger coupling constant than their *transoid* isomers (**21b**). In cycloolefins ^4J lowers with increasing ring size.[1312] Substituents can have a significant influence and may cause the insignificance or even a reversal of sign of the difference between the *cisoid* and *transoid* couplings. ^4J

* Compare p. 68.

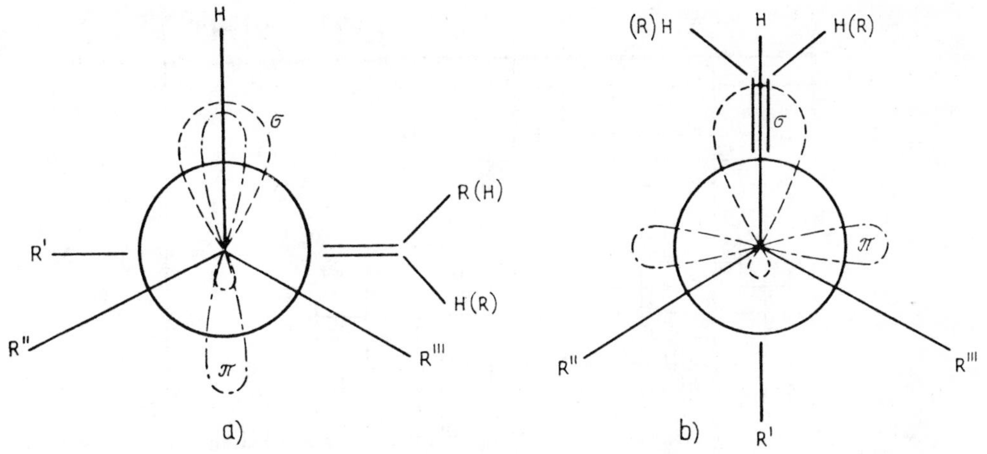

FIGURE 35. The overlap of σ- and π-orbitals transmitting long-range spin-spin interactions in compounds of type RHC=CR'–CHR''R'''.

may be altered not only by the electronegativity of the substituents, but also by its bulky, too. The larger ^4J is, the greater the overlap between the σ and π electrons transmitting the spin-spin interaction. As a consequence, ^4J and the conformation are interrelated. Thus, for example, between the two rotamers shown in Figures 35a and 35b, the former should exhibit a larger long-range coupling. ^4J may be solvent dependent, since in different solvent the ratio of the rotamers may be different.

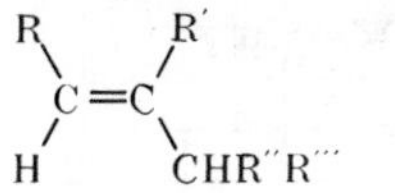

21a

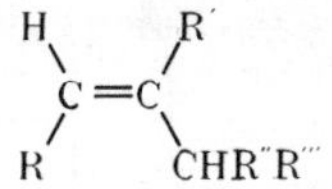

21b

Dependence of allylic coupling on geometry can be studied in rigid compounds. Thus, for example, in the steroid epimers **22a, 22b** (X = OAc, Me, etc.), the H-4 signal is a doublet for **a** (^4J = 1.5 to 2 Hz) and a singlet for **b** because of the large overlap between the σ-orbitals of the C_4-H bond and π-orbitals of the C-4,5 double bond in **a** epimers (as in Figure 35a), while in configuration **b** the arrangement shown in Figure 35b is realized.[1552]

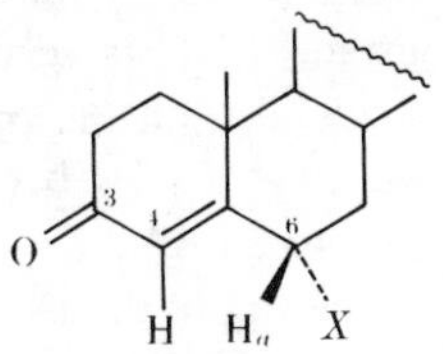

22a

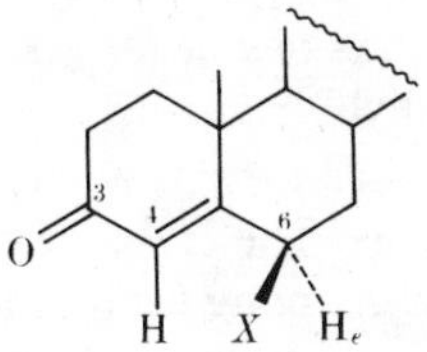

22b

In the spectrum of the analogue unsubstituted at C-6 (**22**, X = H), the H-4 signal is a broad line of about 3.5 Hz width. The reason for this is that the *axial* methylene proton H-6*a* coupled to H-4 interacts also both with its *equatorial* pair H-6*e* and with the H-7 atoms. Therefore there are several different energy states of H-6*a*, each of which interacts with those of H-4, causing the further splitting of its signal. This phenomenon is called *virtual coupling*[1015] and may lead to a splitting of the signal of H-4 even when the H-4,H-6*e* coupling

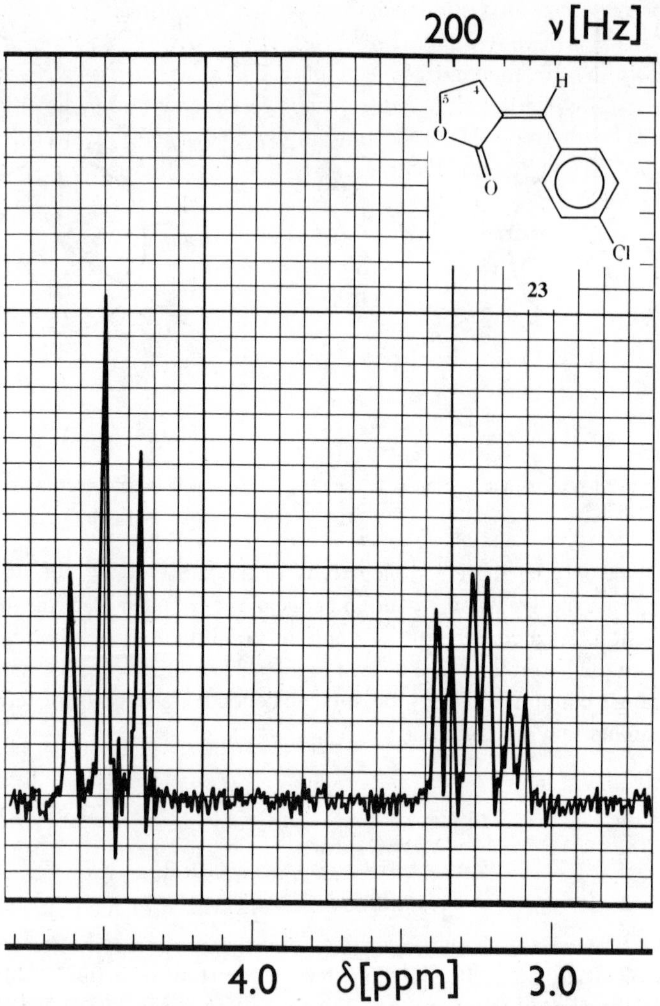

FIGURE 36. The multiplets of methylene protons of Compound **23**. (Courtesy of Professor H. Wamhoff, Friedrich Wilhelm University, Institute of Organic and Biochemistry, Bonn.)

is practically zero. In fact, signal H-4 is the *A* part of an *ABCDX* multiplet, which consists of a number of lines when one of the coupling constants (J_{AX}) is very small.*

The magnitude of allylic coupling constant depends also on the presence of further π-electrons. In Compound **23** there is a long-range coupling of 3 Hz between the olefinic and C-4 methylene hydrogens. The H-4,5 atoms give rise to an *AA'XX'* spectrum of two 1:2:1 triplets, and the ^4J-type coupling leads to a further doubling of the upfield (H-4) triplet (see Figure 36). The relative positions of the σ-orbitals of the *cisoid* H-4 atoms are identical in respect of the π-orbitals. This permits significant overlap and explains the large and identical splitting for both hydrogens.

Allylic type ^4J coupling can be frequently observed in cyclic olefins, as well as in aromatic and heteroaromatic compounds between olefinic or aromatic hydrogen and the protons of a *vicinal* methyl group (H–C=C–CH_3). In such cases interaction usually results only in a broadening not in splitting of the methyl signals, manifested in a reduction of the height of the maximum. This is especially evident when there is another methyl group in the molecule,

* Compare Section 1.5.5.2.

which does not participate in a similar interaction and gives therefore a much sharper and higher singlet, making thereby direct comparison possible (compare Problems **25, 46,** and **60**). On the spectrum of Compound **24** (see Figure 37) the acetyl-methyl signal at 2.95 ppm is sharp, while the C-4-methyl signal at 2.42 ppm is broader and lower, due to the small allylic splitting. The quartet of H-3 (the *A* part of an AX_3 system) appears as a broad signal at 6.18 ppm. The *AX* doublets of H-5,6 come at 6.90 and 7.68 ppm, respectively (J = 9 Hz); the signal of the chelated hydroxyl is found at 13.5 ppm.

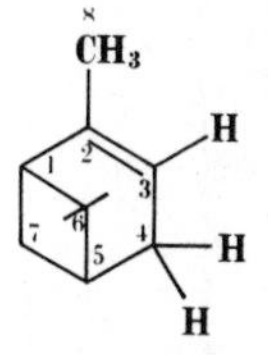

25

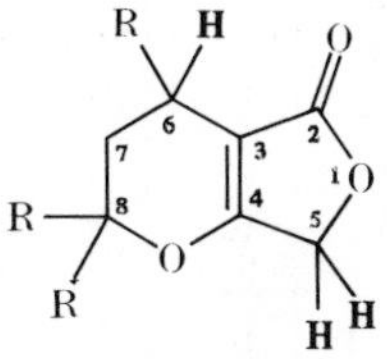

26 R=H
27 R=Me

The long-range coupling ^5J between hydrogens of >CH–C=C–CH< groups is called *homoallylic coupling*. ^5J = 0 to 3 Hz and are dependent on valence angles.[1114] Arrangements allowing better overlap between the σ and π electrons correspond to larger values of ^5J. It has been shown that ^5J has a maximum when the angle between the CH bond and the planes of the C=C bond is 90° (maximum overlap between the σ and π orbitals), while around 0° it has a minimum.[752]

In contrast to allylic couplings $^5J^{transoid} > {}^5J^{cisoid}$.[239] Thus, for example, the methyl group on the unsaturated C-2 in α-pinene (**25**) has an 1:3:3:1 quartet because the constants corresponding to allylic and homoallylic coupling with H-1, H-3, and one of the two β-protons, H-4, respectively, are practically the same ($^4J \approx {}^5J \approx 2$ Hz).

Homoallylic coupling is operative in Compound **26**; the C-5 methylene signal is a triplet, as a result of coupling ^5J with H-6 atoms. Accordingly, in the analogue **27**, the 5-methylene protons give a doublet. In both cases ^5J = 1.5 Hz.

Interactions between hydrogens in the *meta*- and *para*-positions of benzene derivatives as well as the non*vicinal* hydrogens of five- and six-membered heteroaromatics can consider as long-range couplings (^4J and ^5J, respectively). In benzene derivatives, the $J^o > J^m > J^p$ relation always holds. The *para*-coupling is < 1 Hz and cannot usually be observed, while J^m = 1 to 3 Hz. In the case of heteroaromatic compounds, however, exceptions are rather frequent. Thus, in pyrazine (**28a**) $J^o \approx J^p \approx 2$ Hz and $J^m \approx 0$ Hz.[307] Generally, π-electrons favor the emergence of significant long-range couplings. For example, in **28b**, the ^7J coupling between the protons of the two methyl groups is 0.7 Hz.[307]

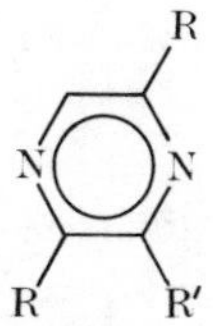

28a R, R=H
28b R=Me, R′=Cl

Interaction between the hydrogen on the sp carbon atom of the monosubstituted acetylenes (CH≡CR) and the α-protons is characterized by ^4J = 2.5 to 3 Hz. In conjugated systems, hydrogens separated by several bonds may interact. As an example the spectrum of compound

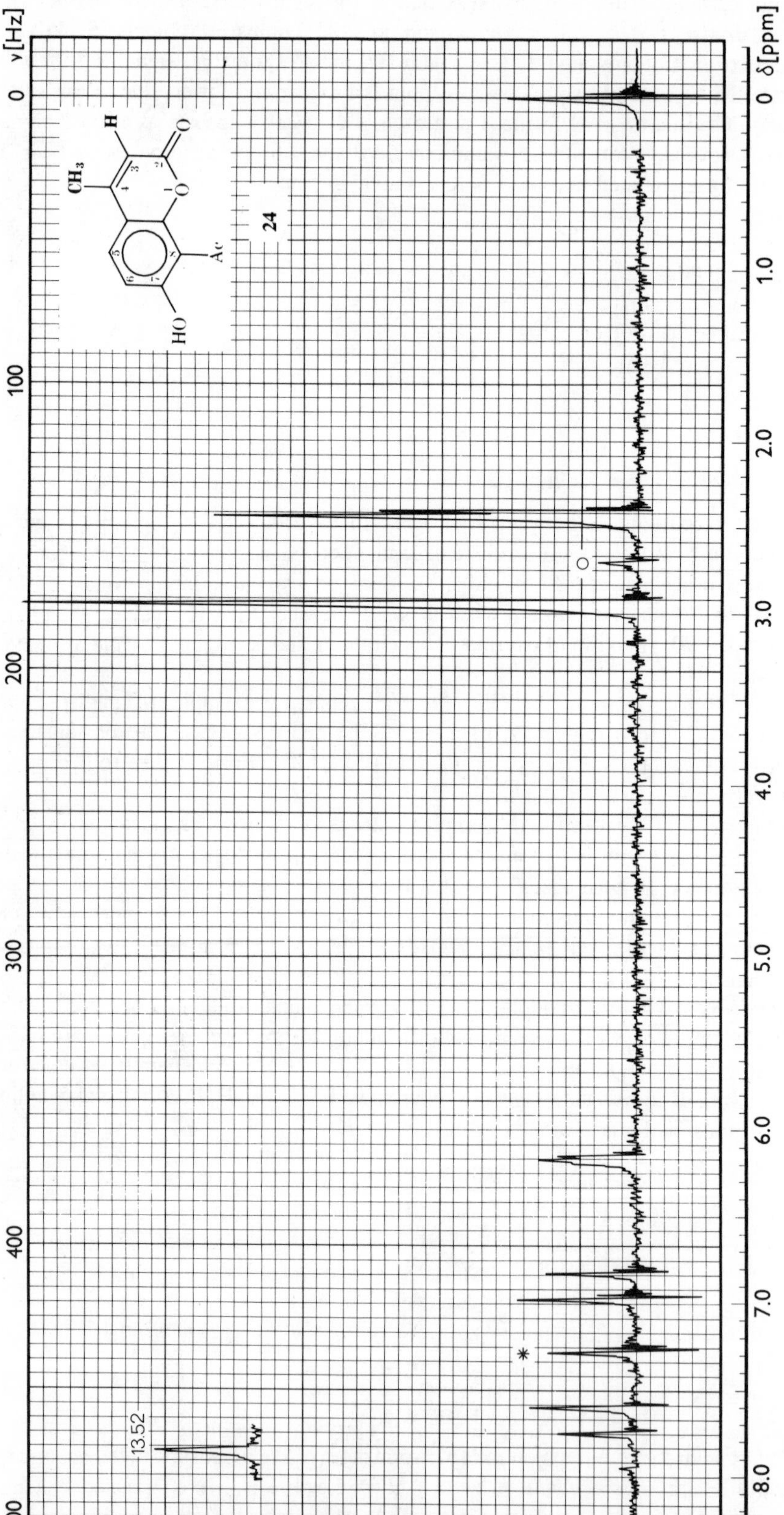

FIGURE 37. The spectrum of 4-methyl-7-hydroxy-8-acetylchromone-2 (**24**). (Signals caused by solvents are marked with asterisks and those due to impurities by dots, compare Volume III, p. 1, last paragraph.)

$H_3C\text{–}(C{\equiv}C)_3\text{–}CH_2OH$ is quoted showing ^9J = 0.4 Hz long-range coupling between the methyl and methylene protons.[1316]

The ^4J couplings of type CH=C=CH of the allenes are about the same in their absolute value as the ^3J *vicinal* couplings of type >C=C=CH–CH< (^3J, ^4J: 6 to 7 Hz). Similarly to olefins, $J^{gem} \leqslant 2$ Hz and the ^5J long-range couplings of type >CH–C=C=CH– are of approximately same magnitude.

1.3.7.2. Long-Range Couplings in Saturated and Fluorine-Containing Compounds

All long-range couplings discussed so-far were transmitted by π-electrons. Exceptionally, if certain structural conditions are met, long-range coupling can also be observed in saturated compounds, though the splitting is, usually, very small.

For the dependence of long-range couplings of protons separated by four bonds, the following relationship is valid:[236]

$$^4J = K\cos^2\varphi^{1,3}\cdot\cos^2\varphi^{2,4} - C \qquad (82)$$

where K and C are constants, while $\varphi^{1,3}$ and $\varphi^{2,4}$ are the dihedral angles between bonds 1 and 3 and 2 and 4, respectively.

According to this relationship, it is understandable that the value of ^4J in the zig-zag (''W''-arrangement; S-*trans* conformation) is greater than in the S-*cis* conformation, in which coplanarity of the four bonds is less likely due to the steric repulsion of the substituents and for which therefore the value of $\cos^2\varphi$ is smaller. This is the s.c. ''W-mechanism'' (''W-pattern'') of long-range couplings[974] (compare Problems **63** and **66**). Thus, for example, the coupling constants ^4J for the *cis* (**29a**) and *trans* (**29b**) anellated decalines and *N*-methylquinolisidinium derivatives are different; in fact, only hydrogens in ''W''-arrangement coupled the methyl protons. Thus, in **29a** the methyl protons are coupled to only one hydrogen, whereas in **29b**, where there are three such protons, the methyl protons are coupled to all three.[1294,1525] Accordingly, the value of ^nJ ($n > 3$) is influenced mainly by configuration and conformation.

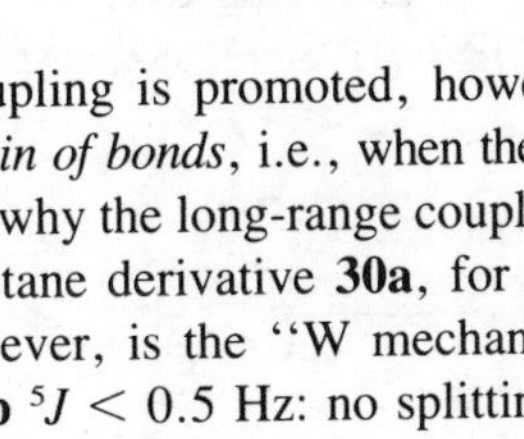

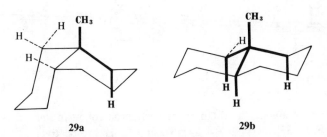

29a

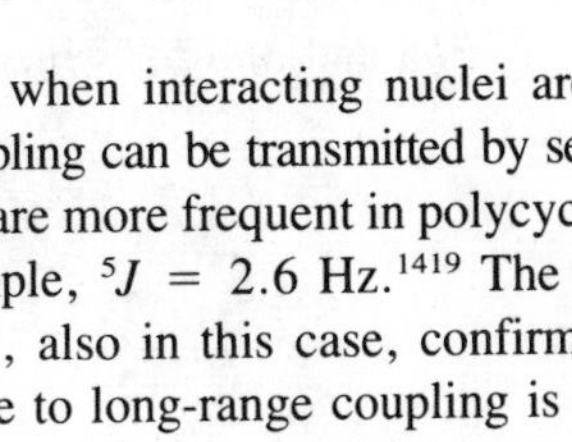

29b

Long-range coupling is promoted, however, when interacting nuclei are connected by *more than one chain of bonds*, i.e., when the coupling can be transmitted by several routes.[1389] This is the reason why the long-range couplings are more frequent in polycyclic compounds. For the tricyclooctane derivative **30a**, for example, ^5J = 2.6 Hz.[1419] The most important contribution, however, is the ''W mechanism'', also in this case, confirmed by the fact, that in isomer **30b** ^5J < 0.5 Hz: no splitting due to long-range coupling is observable.

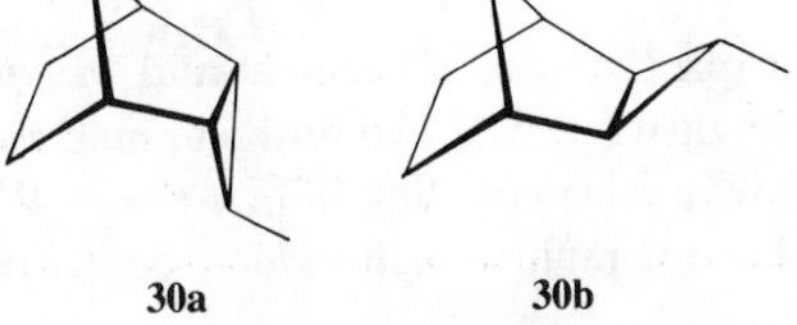

30a **30b**

The incorporation of heteroatoms is also favorable for long-range couplings (compare Problem **63**). In Compounds **31** the ^5J long-range coupling between H-1 and H-4 is 1.6 Hz.[1555]

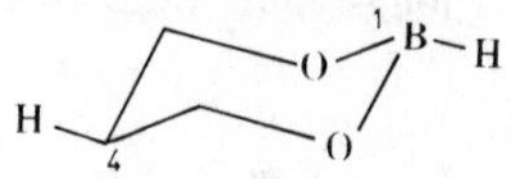

31

Instructive examples are provided by some steroids, aromatic in ring A (**32**): coupling $^5J_{6,9}$ is significant, whereas $^4J_{6,8}$ transmitted through a single chain of bonds, containing no double bonds, cannot be observed.[726]

"*Through space*" interactions may occur between nuclei in steric vicinity and may be much stronger than expected considering only the number of separating bonds. Extremely large ^4J long-range coupling constants can be measured, e.g., in the spectra of fused ring systems, which has been attributed to the overlap of the C–H bonding orbitals. The coupling between the anellated hydrogens of Compound **33** is ^4J = 18 Hz,[1078] due to the through-space interaction.

Significant through-space coupling is rare in proton resonance. The atomic radius of fluorine being larger than that of hydrogen, and also the J(F,F) couplings are greater by one or two orders of magnitude, this phenomenon can be observed frequently in fluorine compounds. Through-space interaction explains, e.g., in compounds of type **34**, that the coupling constant of F-8 and the protons in R is gradually decreasing in the series R = XMe_n (X = O, S, N, and C, and n = 1, 1, 2, and 3), R=Me and R=H, i.e., remarkably, $^6J > {}^5J > {}^4J$.[199] The plausible reason is that the fluorine is the closest to the methyl hydrogens of R=XMe_n groups, while the $\overline{\text{HF}}$ distance is the largest when R=H.

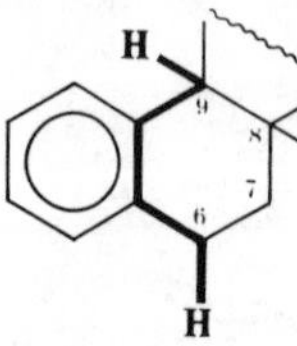

32

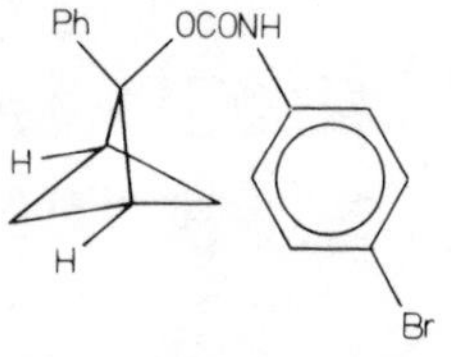

33

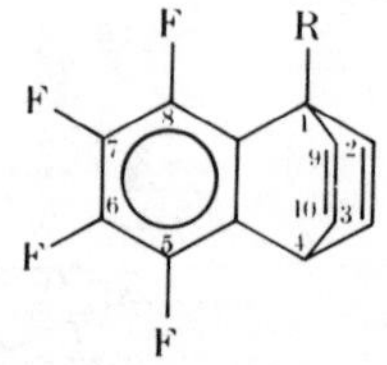

34

In the same way among the rotamers of perfluoroacrolein in the S-*trans* and S-*cis* conformation (**35a** and **35b**) $^4J(F_a,F_b)$ = 84.5 Hz and <2 Hz, respectively.[200] A further example is the *cis-trans* pair **36a** and **36b** for which ^5J(F,F) is 12 and 0 Hz, respectively.[1038]

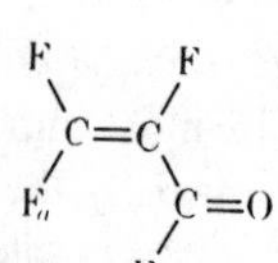

35a

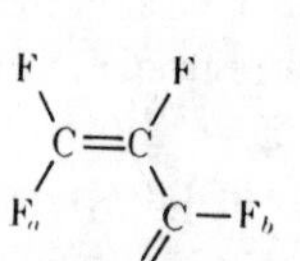

35b

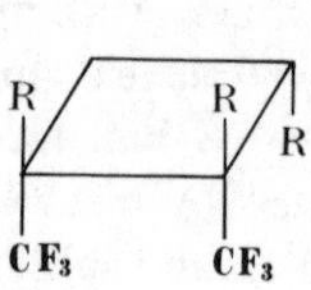

36a

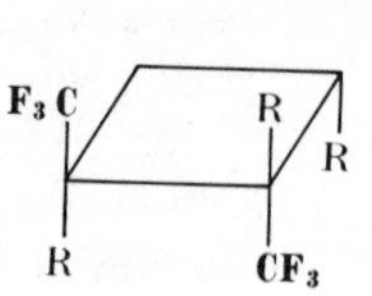

36b

1.3.8. Classification of Spin Systems. Chemical and Magnetic Equivalence[570,979]

When the chemical shift difference of two different magnetic nuclei is much greater than the constant of their coupling, this coupling is first order. When chemical shift difference and coupling constant are comparable, a high-order coupling results. The former spin system is denoted by the symbol *AX*, while the latter is denoted by *AB*, thereby using the adjacent or remote letters of the alphabet in order to indicate whether the chemical shifts of the two interacting nuclei are similar or very different. Coupled spin systems consisting of more than two nuclei can be characterized similarly. Spin systems of the same type result in analogous spectra, therefore this kind of systematization has great practical significance. When a coupled spin system is recognized to belong to a certain type, theoretically derived relationships for general cases can be applied for determination of chemical shifts and coupling constants from the observed spectrum. For the classification of spin systems, it is necessary to introduce the concepts of *chemical and magnetic equivalence*.

Two nuclei of the same isotope are *chemically equivalent* (and *homotopic*) in a molecule possessing an n-fold symmetry axis C_n if upon transformation by C_n (rotation by $2\pi/n$ about the symmetry axis) the two nuclei are interchanged and the resultant configuration is indistinguishable from the original one. Therefore, protons in the following molecules are chemically equivalent: CH_4, CH_3Cl, CH_2Cl_2, benzene, *p*-dichlorobenzene, 1,3,5-trichlorobenzene.

Hydrogens in molecules of type CH_2RR' are chemically nonequivalent. Although the two hydrogens can be transformed into one another by reflecting them through the CRR′ plane, these molecules lack an axis of symmetry. This type of *chemical non-equivalence* called also *enantiotopy* can, however, be observed in NMR only in the presence of a chiral agent (e.g., an optically active solvent or shift reagent). Atoms or groups like the two hydrogens in the present case are called *enantiotopic*, since they have the same mutual relationship as that of the enantiomers.

Thus, the two *vicinal* hydrogens of *meso*-1,2-dichloro-1,2-dibromoethane (**37**) are enantiotopic because in the molecule there is no symmetry axis, only a center of symmetry in the given conformation.

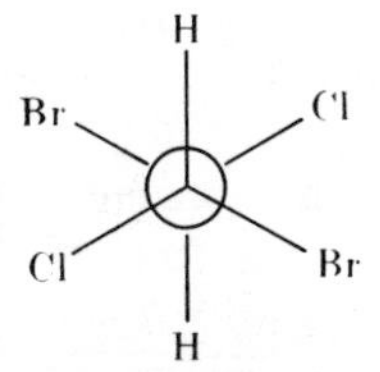

37

The chemical shift of chemically equivalent nuclei is always identical, i.e., they are *isochronous*. Isochrony of two nuclei may be accidental and is therefore a necessary, but not a sufficient condition for chemical equivalence. For example, the methylene hydrogens of CH_2RR'-type molecules are isochronic in achiral solvents, but may show anisochrony in chiral ones. In achiral solvents all such atomic sets of any molecules are isochronous which can be interchanged by any symmetry operation.

Those nuclear pairs that cannot be interchanged by any symmetry operation are called *diastereotopic* pairs. Diastereotopic pairs of nuclei are always chemically nonequivalent and are potentially anisochronic. Accordingly, homotopic, enantiotopic and diastereotopic nuclei constitute a series of decreasing symmetry. Any of two hydrogens of chloroethylene, $H_2C{=}CHCl$, forms a pair of diastereotopic atoms. Although the molecule possesses a symmetry element, the molecular plane, this cannot serve to interchange any two protons. Any nuclear pair of the same isotope in asymmetric molecules constitutes a diastereotopic pair, although this cannot always be observed due to accidental isochrony or to averaging caused by molecular motions.

In the ^1H NMR spectrum of methylacetylene ($HC{\equiv}CCH_3$) taken in deuterochloroform at 60 MHz, the signals of the acetylenic and methyl protons coincide at 1.80 ppm[1044] due to *accidental anisochrony*. In the same way the signals for the two methylene groups of β-cyano methylpropionate (NC–CH_2–CH_2–COOMe) is a singlet at 2.68 ppm.[1044] Toluene gives only two singlets, corresponding to the methyl and to the accidentally isochronous aromatic hydrogens. Accidental isochrony can, however, be overcome in most cases by changing solvent or measuring frequency (see Problem **25**).

Magnetically equivalent (isogam) nuclei are sets of chemically equivalent nuclei when their spin-spin interactions with each of the other chemically equivalent groups of nuclei in the molecule can be described by coupling constants of the same sign and magnitude. Not all chemical equivalent nuclei are, however, necessarily magnetically equivalent. Although atoms H and H*, and similarly F and F* in 1,1-difluoroethylene (**38**) are chemically equivalent, the heteronuclear coupling between nuclei *trans* to each other is evidently different from those between nuclei in a *cis* arrangement accordingly, $J(\text{H,F}) \equiv J(\text{H*, F*}) \neq J(\text{H,F*}) \equiv J(\text{H*,F})$.

In asymmetrically *para-* (**39**) and symmetrically *ortho*-disubstituted benzenes (**40**), H_A and H_B atoms are, of course, chemically equivalent, but, for example the coupling $J(H_A,H_B) \equiv J^o$ is evidently different from the interaction $J(H_A,H_B{*})$, because the former is established through three bonds, while the latter is transmitted through five or four bonds, respectively: J^p (in **39**), J^m (in **40**).

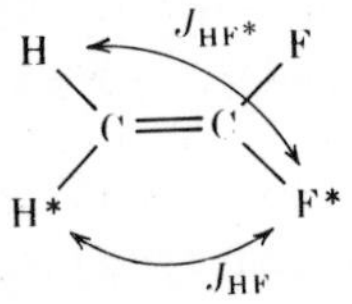

38

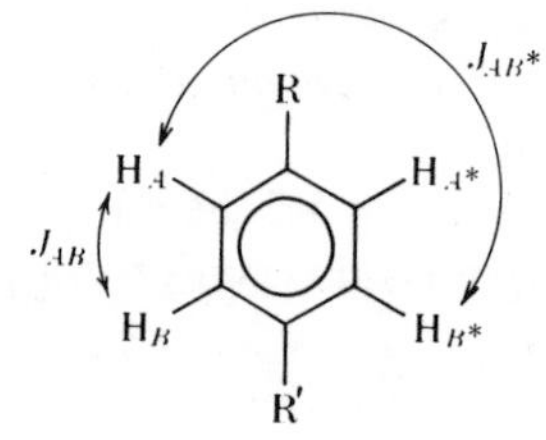

39

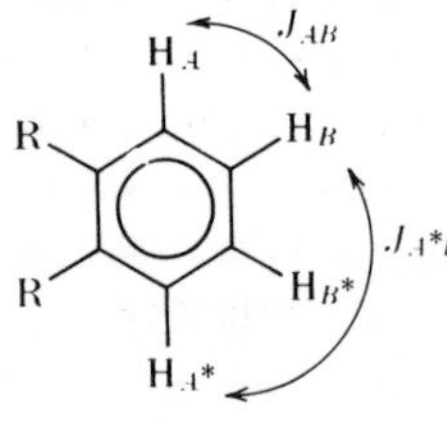

40

Experimental spectra show often apparent magnetic equivalence for nonequivalent nuclei. Thus, for propane derivatives of type R–CH_2^A–CH_2^B–CH_2^C–R′, the signal of the central (*B*) methylene group is, in many cases, a regular 1:4:6:4:1 quintet (e.g., 1-chloro-3-bromopropane).[1044] It follows that in these cases, despite that R ≠ R′, $J_{AB} = J_{BC}$ accidentally.

This phenomenon is therefore termed *accidental magnetic equivalence*. Increasing the measuring frequency or the resolving power or the application of more indifferent solvents causing lesser line width may eliminate accidental magnetic equivalence.

According to conventions proposed by Bernstein, Pople, and Schneider,[123] spin systems are distinguished from one another in the following way: different sets of chemically nonequivalent nuclei, the *basis-groups*, are denoted by different capitals of the alphabet. When the chemical shift difference between the basis groups is small, the first letters of the alphabet are taken in succession, A, B, and C. When there is a large chemical shift difference, then letters from the beginning and the end of the alphabet are chosen, e.g., *A* and *X*. Thus, the symbol *ABXY* represents a four-spin system of chemically nonequivalent nuclei, in which chemical shifts of the pairs *A* and *B* as well as *X* and *Y* similar between each other, but very different between the pairs. *AMX* stands for three nonequivalent nuclei, where all interaction between the possible pairs are of first order. Chemically equivalent nuclei are denoted by the same capital letter. The number of them is given in the lower index, e.g., A_2, A_2B_3, or AB_2X_2. Chemically equivalent, but magnetically nonequivalent nuclei are distinguished by priming, e.g., the spin systems representing the H and F atoms of Compound **38** is labeled as *AA′XX′* (ΔδHF is large) and *AA′BB′* stands for hydrogens in **39** and **40,** where the value of $\Delta\delta H_AH_B$ is usually small.

Temperature-independent anisochrony — Similar to diastereomers, the scalar physical and chemical properties of *diastereotopic* pairs of nuclei are different, and they are in principle, always anisochronous. This type of inherent anisochrony is independent of the temperature and cannot be equalized by any type of molecular motion. The *cis* and *trans* *vicinal* and *geminal* hydrogen pairs of $H_2C{=}CHCl$ molecule are, for example, always chemically nonequivalent and potentially anisochronous.

Temperature-dependent nonequivalence — In the course of certain intra- and intermolecular motions, a mutual exchange of chemical environments can occur for a given pair of nuclei. When this exchange process is fast enough as compared to the average lifetime of spin states, it leads to the isochrony of nuclei, which would by the molecular symmetry otherwise be anisochronous. In other words the spectrometer ''perceives'' an average of the originally different magnetic shielding of the nuclei, and thereby the original nonequivalence is eliminated. The two nuclei, therefore, give rise to a common signal with an averaged chemical shift. The rate of the exchange processes is temperature dependent and for this reason anisochrony observed at lower temperatures is replaced by isochrony at higher ones.

In the spectra of *N,N*-dimethyl-amides (**41**) appear, for example, two methyl signals at room temperature (compare Problems **51**) due to the *cis* and *trans* positions of the methyl groups relative to the carbonyl oxygen. As a consequence of the mesomerism (**a** ↔ **b** ↔ **c**) in amides, the CH bond has a double bond character, therefore the rotamers **b** and **c** cannot freely interconvert (*hindered rotation*). At higher temperatures, the rotation about the CN bond becomes very fast, and the splitting of the methyl signal disappears.*

Besides internal rotation,** another kind of intramolecular motion which can eliminate anisochrony is *inversion of pyramidal atoms* as N, P, etc.,*** *ring inversion*, † *tautomerism*,‡ valence isomerization,‡‡ etc. The equivalence of the 12 protons in cyclohexane is caused, for example, by a fast ring inversion.+

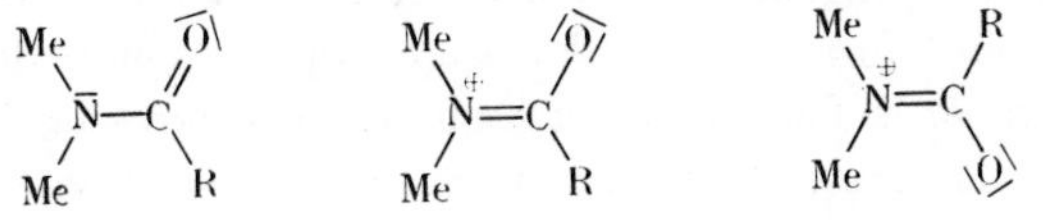

41a **41b** **41c**

Acidic protons of the solute may participate in fast inter- and intramolecular exchange involving the solute itself, the acidic protons, or the water content of the solvent and give rise to a common signal, at an average shift value.++ The rate of exchange depends on the temperature, the solvent, the solute concentration, and on the pH value as well.

Anisochrony in ethane derivatives — In the ethane derivatives of type RCH_2–CR′R″R‴, (where R′ may be hydrogen or a substituent equal to R), the anisochrony of methylene hydrogens has both temperature-dependent and temperature-independent components.

At lower temperatures, the statistical weight in the conformation equilibrium of the three stable rotamers (**42a** to **42c**) is not identical due to the different energy contents of the rotamers. It can easily be seen that the chemical environments of the two methylene protons H_A and H_B in the rotamers **42a** to **42c** are different, thus their magnetic shielding values are

* Compare Volume II, p. 92.

** See also Volume II, p. 66, 90-91, 107-108, etc., also Figure 148, p. 93.

*** Compare Volume II, p. 16-17, 43-46, and Volume I, p. 72.

† Compare Volume II, p. 19, 20, 24, 25, 43-46, 56, 91, etc.

‡ Compare Volume II, p. 88 and Problem **56**.

‡‡ Volume II, p. 57-59, 60-61, 75 and Problem **50**.

\+ Compare Volume II, p. 25, 100, 106, etc.

++ Compare Volume II, p. 105.

also different. This holds also for all intermediate conformations, including the eclipsed ones as well. Upon increasing the temperature, the anisochrony ($\Delta\delta AB$) of the protons H_A and H_B is gradually reduced and approaches asymptotically a constant value. This happens because, at higher temperatures, fast free rotation occurs and the relative probabilities of the conformers become identical. The anisochrony originating from unequally rotameric population becomes smaller. This constitutes the temperature-dependent part of the anisochrony.

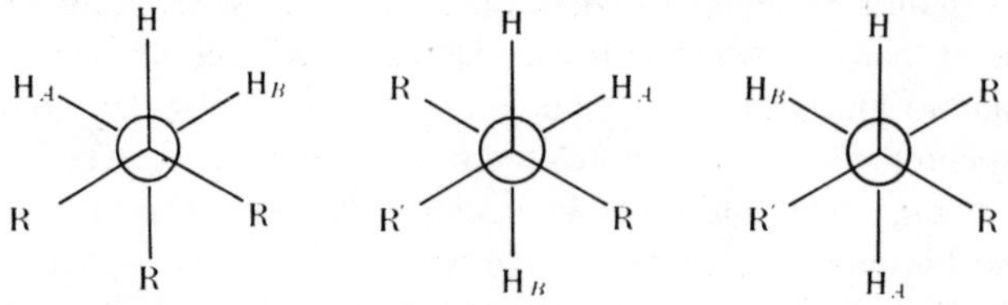

42a **42b** **42c**

Averaging, however, is incomplete. It can be seen, namely, that H_B can never be brought into exactly the same environment as that of H_A in any of the rotamers. Thus, for example, H_A in rotamer **a** has the same position as H_B in rotamer **c**: substituent R″ is in *trans* position and the immediate neighbors (in a clockwise direction) are H and R′ on the projection. The total environment for H_A in **a** is still different from that of H_B in **c** because the second neighbors are different, i.e., those of H_A are H and *R* in **a**, while they are *R* and H for H_B in **c**. In other words, the shielding effect of the group –CHR′R″ is anisotropic in respect of the methylene hydrogens, which thus are inherently nonequivalent and anisochronic in molecules of type $CH_2R–CR'R''R'''$.

This kind of symmetry persists for free and fast internal rotation also, and the anisochrony due to it is therefore independent of the temperature. An example is the molecule CF_2Br CHBrCl, whose *geminal* fluorine atoms are diastereotopic and anisochronous.[601,602] For this reason, an *ABX* multiplet is found even at higher temperatures in the fluorine resonance spectrum. At low temperatures, the conformational equilibrium are frozen in,[1035,1036] giving three superimposed *ABX* multiplets with intensity corresponding to the ratio of the three different staggered forms.

Similarly, the methylene protons of acetaldehyde diethyl acetal [MeCH(OEt)$_2$] and of *N*-benzyl-*N*-methyl-1-phenyl-ethylamine [PhCH$_2$–NMe–CH(Me)Ph], as well as the olefinic protons of Compound **43**,[1280,1299] are anisochronous. In compound PhCH$_2$–NMe–OMe, a condition for diastereotopy is the slow inversion. It has indeed been found that methylene protons anisochronous at low temperature becomes isochronous upon increasing of the temperature and thereby the rate of inversion. This is therefore an example for temperature-dependent diastereotopy.[586]

Anisochrony does not necessarily involve slow internal rotation, but may be associated with a nonuniform population of rotameric states. This means that in spite of free internal rotation monosubstituted ethanes may give rise to *AA′BB′C* multiplet (**44a**) at lower temperatures, but A_2B_3-type spectra (**44b**) will be obtained at higher temperatures.

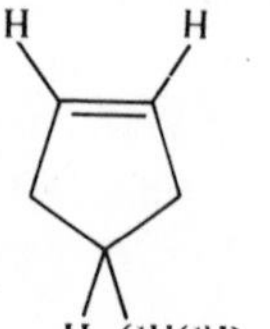

43

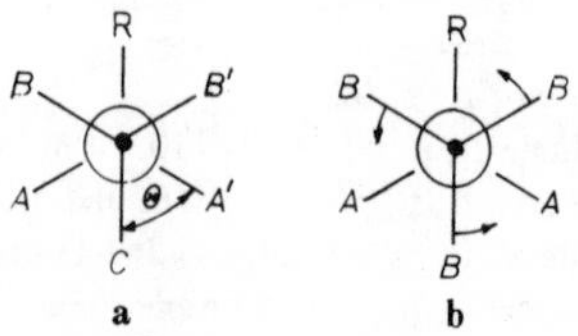

44

Examples for various types of spin systems are found in Table 6. It has to be noted that the ratio $J/\Delta\nu$ cannot be predicted easily, thus classification of the spin system should always be based on the spectrum itself. Thus, for example, the monosubstituted oxiranes may belong to systems *AMX*, *ABX*, or *ABC*, ethyl-derivatives may belong to systems A_2X_3 or to A_2B_3, while five-membered heterocyclic systems as pyrrol or thiophene can be assigned to types A_2X_2, A_2B_2, $AA'XX'$, or $AA'BB'$. Heteronuclear systems, however, are usually of first order because $J/\Delta u$ for such cases is small (chemical shifts for different kind of atoms are very different). In the assignment of the spin systems, one can disregard nonmagnetic nuclei of spin $I = 0$, e.g., ^{12}C, ^{16}O, etc. (furthermore , atoms Cl, Br, and I).*

1.4 THE QUANTUM-MECHANICAL TREATMENT OF MAGNETIC RESONANCE SPECTRA[300,434,1211]

The theoretical interpretation of first-order spectra is simple. Higher-order NMR spectra can, however, be interpreted only invoking quantum mechanics. Therefore it is necessary to review the most important features of the quantum-mechanical interpretation of the NMR phenomena.

We have seen that in the presence of an external (polarizing) field $\mathbf{B}_0$ constant in time a system of magnetic nuclei can be characterized by discrete energy levels. The distribution of nuclei among these levels follows Boltzmann's law. The only change a rotating excitation field $\mathbf{B}_1$ brings about are transitions among these levels, i.e., some nuclei are excited from ground levels to higher ones. At frequencies, ν, corresponding to the excitation energies $\Delta E = E' - E'' = h\nu$ (E'' and E' stand for the energies of ground state and excited state), resonance absorption of the RF radiation inducing field $\mathbf{B}_1$ takes place in the sample. An NMR spectrum is the assembly of such absorption maxima.

The theoretical interpretation of spectra means the solution of the following problems:

1. Calculation of the energy levels, E_i, split by the field $\mathbf{B}_o$ for a system of any type of magnetic nuclei.
2. Selection of levels between which transitions induced by the excitation field $\mathbf{B}_1$ are allowed.
3. Determination of the energy, ΔE_{ij}, of the allowed transitions and of their relative probability.
4. Defining magnetic interactions of nuclei and the extent to which energy levels and transitions altered by these spin-spin interactions.

The result of these calculations is a set chemical shifts, ν_i, and coupling constants, J_{ij}. These parameters determine the resonance spectrum of a given spin system.

Having these data one can obtain the number of spectral lines of the spin system under study, their distance and intensity ratios, and hence, the theoretical spectrum. Theoretical spectra give the key to calculate coupling constants and chemical shifts from experimental spectra.

The application of quantum mechanics enabled to achieve of all these goals. It has become possible to reproduce the experimental spectra to their finest details and provide a quantitative treatment of all phenomena related to nuclear magnetic resonance.

1.4.1. The Schrödinger Equation and Hamilton Operator of Spin Systems

The classical equations of motion given by Newton are not valid for atomic systems. Instead, the motion of a particle or of a system of particles is described by the Schrödinger

* See footnote on p. 49.

Table 6
CLASSIFICATION OF SOME MOLECULES AND GROUPS OF COMPOUNDS TO VARIOUS SPIN SYSTEMS

Spin system	Compound
AX	$CHCl_2F$; $R_2CH-CHO$; $Cl_2CH-CFCl_2$; 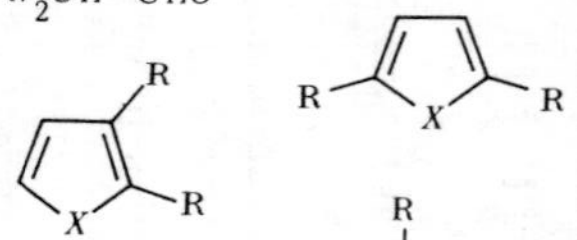
AB	$ClCH=CHBr$; $H_2C=CClBr$; 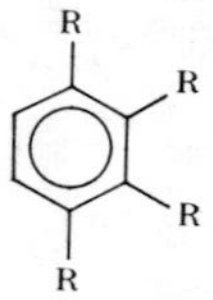
A_2	H_2; CH_2Cl_2; $Cl_2CH-CHCl_2$; 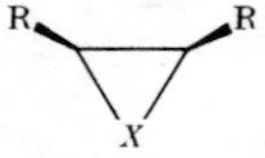
AX_2	CH_2FCl; $RCH_2-CHR'R''$; $^{13}CH_2Cl_2$; 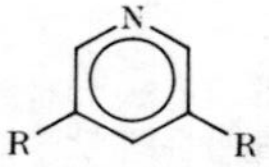; 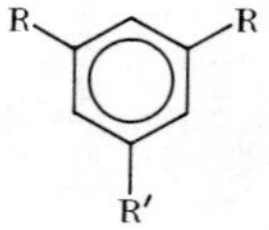
AB_2	$ClCH=C=CH_2$; $CHCl_2-CH_2Cl$
A_3	NH_3; CH_3R; ; 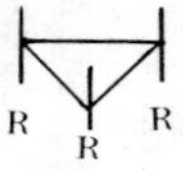

Spin system	Compound
A_2X_2	CH_2F_2; $N\equiv{}^{13}C-CH_2-{}^{13}C\equiv N$; $R-CH_2-CH_2-R'$
A_2B_2	SF_4Cl_2; $ClCH_2-CH_2OH$
AA'XX'	HOOC(H)C=C(H)–C(H)=C(H)COOH; ; 
AA'BB'	HOOC(H)C=C(H)–C(H)=C(H)COOH; 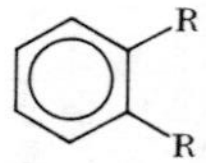; 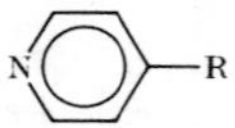
A_4	CH_4; $R-CH_2-CH_2-R$; $H_2C=CH_2$; $H_2C=C=CH_2$; ; 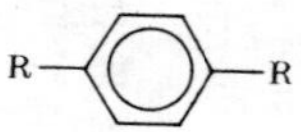
AX_3	CH_3F; CHF_3; $^{13}CH_3Cl$; CH_3CHRR'
AB_3	CH_3SH; CH_3CHO; $HC\equiv C-CH_3$; $CH_3-CR=CHR'$
A_2X_3	$CH_3-CH_2NO_2$; CH_3-CH_2R
A_2B_3	$(CH_3-CH_2)_2Hg$
A_6	CH_3-CH_3; ; 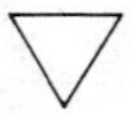

AMX	$H_2C=CH{-}OCOR$ $F_2C=CFCl$ $^{13}CHFCl_2$ $HC{-}C{-}CH=CHOMe$	$H_2C=CHR$
ABX	$H_2C=CFCl$ $H_2C=CHOR$ 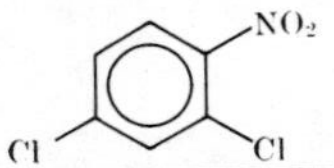	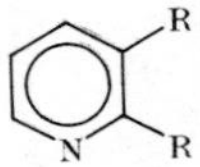
ABC	$H_2C=CHCN$ 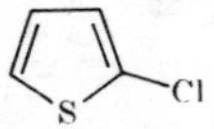	
AMPX	$^{13}CH{\equiv}C{-}CHFCl$	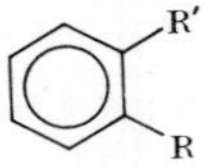
ABMX	$^{13}CH_2=CFCl$	
ABCX	$H_2C=CFH$	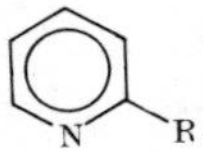 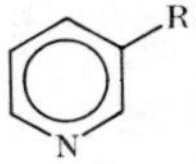
ABCD	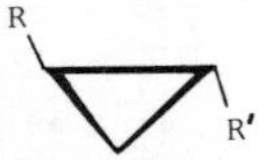	
AMX_2	$H_2C=CBr{-}CH_2Br$	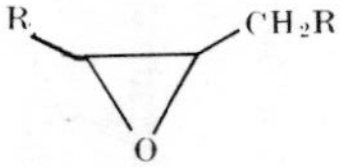
ABX_2	$H_2C=CCl{-}CH_2Cl$	
AB_2X		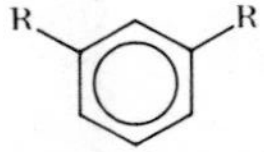
ABC_2	$H_2C=CR-CH_2OR$	

AB_4	SF_5Cl
AM_2X_2	$^{13}CH_2F_2$
A_6B_2	$CH_3{-}CH_2{-}CH_3$
AM_2X_3	$CF_3{-}CH_2{-}OH$
A_9	$BrC(CH_3)_3$

A_{12} 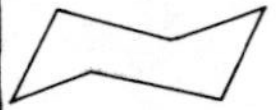 $C(CH_3)_4$ 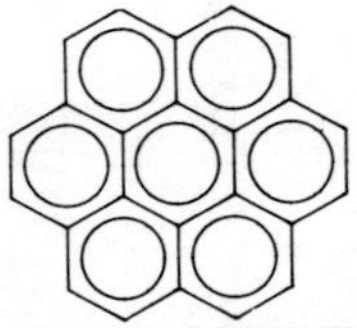

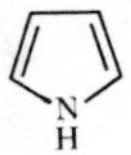 $AA'MM'X$

 $AA'BB'X$

 $AA'BXX'$

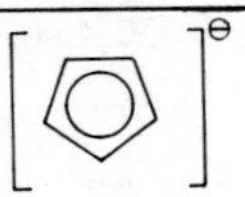 A_5

 $ABCC'DD'KK'MM'X$

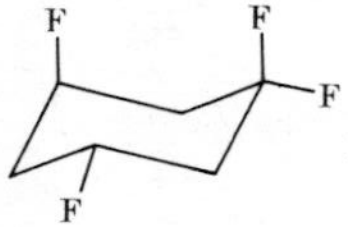 $AA'BB'CC'MM'XX'YY'$

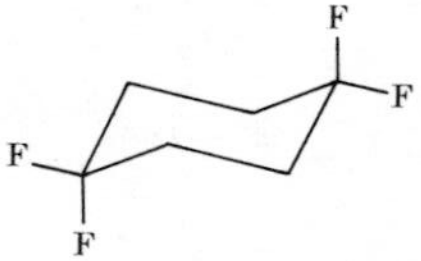 A_8X_4

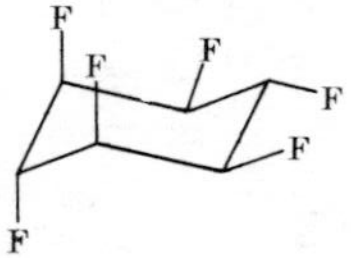 A_6X_6

equation. In the case of problems of height complexity, it is advisable to set up the Schrödinger equation — utilizing the close analogies among the quantities of classical mechanics and quantum mechanics. In classical mechanics, the sum, H, of the kinetic energy, E_k, and the potential energy, E_p, for motions of a system in a conservative field of force is equal to a constant, E:

$$H = E_k + E_p = E \tag{83}$$

Here H is the s.c. Hamilton-function which can generally be expressed by means of canonically conjugate pairs of quantities (e.g., momenta and coordinates or energies and times). If Equation 83 is ‘‘multiplied’’ by the stationary wavefunction, ψ of the corpuscule or a corpuscular system, the time-independent Schrödinger equation* of the corpuscule or corpuscular system is obtained:

$$\mathcal{H}\psi = E\psi \tag{84}$$

where $\mathcal{H}$ is the so-called *Hamilton operator*.

The possible values of a given physical quantity are obtained in quantum theory as the solutions of wavefunctions of the following form:

$$\mathcal{O}\psi = \lambda\psi \tag{85}$$

which are obtained as the eigenvalues, λ_i, of the wavefunction, ψ. Here $\mathcal{O}$ is the operator representing the physical quantity in question. The solution of the above equation furnish eigenfunctions, ψ_i, as well.

By the solution of the Schrödinger Equation 84, one obtains the eigenfunctions, ψ_i, and the discrete energy levels, the eigenvalues E_i, belonging to the possible state of the corpuscular system. In order to solve the wave equation one has to know the form of the Hamiltonian, $\mathcal{H}$.

The Hamiltonian of the magnetic moment $\boldsymbol{\mu}$, belonging to an isolated magnetic nucleus, i.e., for a single spin, is obtainable from the expression of the classical energy, in case of a polarizing field $\mathbf{B}_0$ (compare Equations 16 and 22):

$$\mathcal{H} = -\mu B_0 = -\gamma\hbar I_z B_0 \tag{86}$$

where I_z can have the values $m = I, I - 1, \ldots -I$; therefore the allowed energies of the system (compare Equation 23) are

$$E_i = m\gamma\hbar B_0 = mh\nu_0 \tag{87}$$

In the case of more than one nuclei, the Hamiltonian can be given by:

$$\mathcal{H} = \sum_i \gamma_i B_i \mathcal{I}_{zi} + \sum_{i} \sum_{< j} J_{ij}\, \mathcal{I}_i\, \mathcal{I}_j \tag{88}$$

where γ_i is the gyromagnetic factor of the i-th nucleus, $B_i = B_0(1 - \sigma_i)$ is the resonance field of the i-th nucleus (the direction of the field, B_i, is the negative z axis), $\mathcal{I}_i$ and $\mathcal{I}_j$ are the spin angular momentum operators for the i-th and j-th nucleus, and $\mathcal{I}_{zi}$ is the z component of $\mathcal{I}_i$, while J_{ij} is the spin-spin coupling constant between the i-th and j-th nucleus. From

* Since under the conditions mentioned in the introduction nuclear systems are time invariant, the use of the much more complicated time-dependent Schrödinger equation is not necessary for our problem.

Equation 88, the eigenvalues of $\mathcal{H}$ are obtained in frequency units (hertz), i.e., $\mathcal{H}$ here plays the role of a *frequency operator*.

The use of the Hamilton operator in these two senses (as energy or frequency operator) will be found throughout in the discussions; it only involves, however, a multiplication or division by h. (Accordingly eigenvalues are obtained either in energy or in frequency units.) In the following $\mathcal{H}$ will mean either the energy or the frequency operator, depending on the logic of the context.

Upon substituting the resonance frequency using Equation 21 into Equation 88:

$$\mathcal{H} = \sum_i \nu_i \mathcal{I}_{zi} + \sum_{i<j}\sum J_{ij} \mathcal{I}_i \mathcal{I}_j \tag{89}$$

The above equation is usually abbreviated as:

$$\mathcal{H} = \mathcal{H}^0 + \mathcal{H}' \tag{90}$$

where the first term ($\mathcal{H}^o$) is representing the interaction between the external magnetic field and the spins, while $\mathcal{H}'$ accounts for internal spin-spin interactions.

1.4.2. Spin Angular Momentum Operators and Their Properties

For the construction of the Hamilton operator, it is necessary to know the properties of the spin angular momentum operators. The *spin angular momentum* operator $\underline{\mathcal{I}}$, representing the angular momentum of a given magnetic nucleus is a vector operator, i.e., it is defined by three linearly independent operator components:

$$\underline{\mathcal{I}} = \mathcal{I}_x, \mathcal{I}_y, \mathcal{I}_z \tag{91}$$

The operator components unlike the total angular momentum operator $\underline{\mathcal{I}}$, are measurable quantities. In keeping with the Heisenberg uncertainty relationship, however, only one component can be determined to an arbitrary accuracy. In other words, the magnitude and direction of the angular momentum of corpuscular systems cannot be determined simultaneously.* When therefore, as usual, the magnitude of the discrete component, $\hbar I_z$, in the direction of the polarizing field, $\mathbf{B}_0$ is given exactly, automatically the components I_x and I_y become indeterminates. This is the reason that the maximum value of $\hbar I_z$ is smaller than the angular momentum vector, $\hbar\sqrt{I(I+1)}$, or that its orientation in the z direction is forbidden. This means, namely, that we would be able to determine simultaneously the exact values of both the magnitude and orientation of the three vector components (this would be equivalent to $I_x = I_y = 0$).

By the help of the operator components, a fourth operator $\mathcal{I}^2$, called *the square of the spin vector*, can also be defined and the allowed values of which add up to the square of the spin angular momentum are

$$\mathcal{I}^2 = \underline{\mathcal{I}} \cdot \underline{\mathcal{I}} = \mathcal{I}_x^2 + \mathcal{I}_x^2 + \mathcal{I}_y^2 \tag{92}$$

To the nucleus of spin I therefore belongs a series of the spin angular momentum operator $\mathcal{I}^2$, $\underline{\mathcal{I}}$, $\mathcal{I}_z$, $\mathcal{I}_x$, $\mathcal{I}_y$. Let us take $\psi_{I,m}$ as the eigenfunction of a spin state of a given magnetic nucleus; then having performed upon it the operations defined by the operators, the eigenvalues of the operator are obtained:

$$\mathcal{I}^2 \psi_{I,m} = I(I+1)\,\psi_{I,m} \tag{93}$$

$$\mathcal{I}_z \psi_{I,m} = m\psi_{I,m} \tag{94}$$

* Compare p. 6.

Here I is the spin quantum number, while m can assume the usual values, $I, I - 1, ..., -I$.

The operators $\mathcal{I}^2$ and $\mathcal{I}_z$ therefore do not alter the eigenfunctions, ψ, i.e., the function ψ is a solution for both equations, $\mathcal{I}^2\psi = \lambda\psi$ and $\mathcal{I}_z\psi = \lambda_z\psi$, where λ_i and λ_{zi} are the eigenvalues of operators $\mathcal{I}^2$ and $\mathcal{I}_z$. On the other hand, the operator $\mathcal{I}_x$ and $\mathcal{I}_y$ do not leave the functions ψ unchanged. They are, therefore, not solutions of the equations $\mathcal{I}_x\psi = \lambda_x\psi$ and $\mathcal{I}_y\psi = \lambda_y\psi$.

The Hamilton operator, $\mathcal{H}$, and the z component of the angular momentum operator possess simultaneous eigenfunctions and eigenvalues. This becomes obvious from Equation 86 divided by h (writing $\mathcal{H}$ as frequency operator) and substituting the frequency ν_0 in place of γB_o according to Equation 21:

$$\mathcal{H} = \nu_0 \mathcal{I}_z \tag{95}$$

Expression 95 is the simplest form of Equation 89 as applied to a single magnetic nucleus ($J = 0, i = 1$).

The physical meaning of the above considerations, also derived from the classical picture, is that different orientations of the spin angular momentum with respect to $\mathbf{B}_0$ — corresponding to the eigenvalues of the angular momentum operator — are associated with different energies (frequencies), i.e., different eigenvalues of the Hamiltonian.

Operators $\mathcal{I}_z$ and $\mathcal{H}$ are then commutative (interchangeable), i.e.,

$$\mathcal{H}(\mathcal{I}_z\psi) = \mathcal{I}_z(\mathcal{H}\psi) \tag{96}$$

The total spinvector operator, $\underset{\sim}{\mathcal{I}}$, does not commute either with $\mathcal{H}$ or $\mathcal{I}_z$, while the square operator, $\mathcal{I}^2$, possess simultaneous eigenfunctions and eigenvalues:

$$\mathcal{H}(\mathcal{I}^2\psi) = \mathcal{I}^2(\mathcal{H}\psi) \tag{97}$$

$$\mathcal{I}_z(\mathcal{I}^2\psi) = \mathcal{I}^2(\mathcal{I}_z\psi) \tag{98}$$

It is the above relationship which justifies the introduction of the operator, $\mathcal{I}^2$, whose eigenvalues are determinable ($\underset{\sim}{\mathcal{I}} \cdot \underset{\sim}{\mathcal{I}} = \mathcal{I}^2$ yields a scalar quantity, just like $\mathcal{I}_z$ as opposed to the case of the operators $\underset{\sim}{\mathcal{I}}$, $\mathcal{I}_x$, and $\mathcal{I}_y$.

The operations symbolized by operators $\mathcal{I}_x$ and $\mathcal{I}_y$ acting upon the eigenfunctions $\psi_{I,m}$ can be carried out by the application of the operators $\mathcal{I}^+$ and $\mathcal{I}^-$ defined by

$$\mathcal{I}^{\pm} = \mathcal{I}_x \pm i\, \mathcal{I}_y \tag{99}$$

therefore:

$$\mathcal{I}_x = \frac{1}{2}(\mathcal{I}^+ + \mathcal{I}^-) \tag{100}$$

and

$$\mathcal{I}_y = -\frac{\mathcal{I}^+ - \mathcal{I}^-}{2i} = \frac{i}{2}(\mathcal{I}^- - \mathcal{I}^+) \tag{101}$$

and the eigenvalues are

$$\mathcal{I}^{\pm}\psi_{I,m} = [\sqrt{(I \mp m)(I \pm m + 1)}]\psi_{I,m\pm 1} \tag{102}$$

The s.c. raising operator, $\mathcal{I}^+$ (see also Equations 102 and 113), transforms the ground state wavefunction, ψ_m, into excited state wavefunction, ψ_{m+1} therefore also called *excitation* or *absorption operator*. In the same way (see Equations 102 and 114), the lowering operator, $\mathcal{I}^-$, can be termed *emission operator*.

1.4.3. Eigenfunctions of Spin Systems

To spin states of a given spin system, there belongs a set of eigenfunctions. These eigenfunctions, just like other ones belonging to other quantized motions of molecules, are *orthogonal* and *normalized*. This means that

$$\int_{-\infty}^{+\infty} \psi_m \psi_n d\tau = \int_{-\infty}^{+\infty} \psi_n \psi_m d\tau = 0 \quad (103)$$

provided $n \neq m$ and

$$\int_{-\infty}^{+\infty} \psi_m \psi_m d\tau = \int_{-\infty}^{+\infty} \psi_n \psi_n d\tau = 1 \quad (104)$$

The functions obeying conditions 103 and 104 are called orthogonal and normalized, respectively.

For nuclei with $I = 1/2$, m can only assume the values $\pm 1/2$; there are then only two eigenfunctions. In case of a single nucleus, these so-called basis functions of the "system" are $\psi_{1/2,1/2} \equiv \alpha$ and $\psi_{1/2,\,-1/2} \equiv \beta$. These wavefunctions are naturally also orthonormal, therefore

$$\int_0^{2\pi} \alpha\beta d\varphi = \int_0^{2\pi} \beta\alpha d\varphi = 0 \quad (105)$$

$$\int_0^{2\pi} \alpha\alpha d\varphi = \int_0^{2\pi} \beta\beta d\varphi = 1 \quad (106)$$

Here the integral over $d\varphi$ involves the consideration of all values of the wavefunction product belonging to all possible spin orientations. Such integrals as well as those of the Equations 103 and 104, and similar ones occur quite often in quantum mechanics. For the sake of simplicity, we use the following symbolism introduced by Dirac:

$$\langle \psi_m | \psi_n \rangle = \langle \psi_n | \psi_m \rangle = 0 \quad (107)$$

$$\langle \psi_m | \psi_m \rangle = \langle \psi_n | \psi_n \rangle = 1 \quad (108)$$

and similarly:

$$\langle \alpha | \beta \rangle = \langle \beta | \alpha \rangle = 0 \quad (109)$$

$$\langle \alpha | \alpha \rangle = \langle \beta | \beta \rangle = 1 \quad (110)$$

In these expressions, the integrals $\int_{-\infty}^{+\infty} \ldots d\tau$ and $\int_0^{2\pi} \ldots d\varphi$ are replaced by the brackets (< and >), and multiplication is designated by a vertical line.

It follows from Equations 93, 94, and 102 that upon performing the operations symbolized by the spin angular momentum operators on the eigenfunctions α and β, the following results are obtained:

$$\mathcal{I}^2 \alpha = \frac{3}{4} \alpha \qquad \mathcal{I}^2 \beta = \frac{3}{4} \beta \quad (111)$$

$$\mathcal{I}_z \alpha = \frac{1}{2} \alpha \qquad \mathcal{I}_z \beta = -\frac{1}{2} \beta \quad (112)$$

$$\mathcal{I}^{+}\alpha = 0 \qquad \mathcal{I}^{+}\beta = \alpha \tag{113}$$

$$\mathcal{I}^{-}\alpha = \beta \qquad \mathcal{I}^{-}\beta = 0 \tag{114}$$

$$\mathcal{I}_{x}\alpha = \frac{1}{2}\beta \qquad \mathcal{I}_{x}\beta = \frac{1}{2}\alpha \tag{115}$$

$$\mathcal{I}_{y}\alpha = \frac{i}{2}\beta \qquad \mathcal{I}_{y}\beta = -\frac{i}{2}\alpha \tag{116}$$

1.4.4. Multispin Systems

The spin angular momentum operators of systems of N nuclei can be given by:

$$\underset{\sim}{\mathcal{I}}_{\mathrm{T}} = \sum_{j=1}^{N} \underset{\sim}{\mathcal{I}}_{j} \tag{117}$$

$$\mathcal{I}_{w\mathrm{T}} = \sum_{j=1}^{N} \mathcal{I}_{wj} \tag{118}$$

where $w = x, y,$ or z:

$$\mathcal{I}_{\mathrm{T}}^{\pm} = \sum_{j=1}^{N} \mathcal{I}_{j}^{\pm} = \sum_{j=1}^{N} \mathcal{I}_{xj} \pm i \sum_{j=1}^{N} \mathcal{I}_{yj} \tag{119}$$

$$\mathcal{I}_{\mathrm{T}}^{2} = \sum_{j=1}^{N} \mathcal{I}_{j}^{2} + 2\sum_{i<j}\sum \underset{\sim}{\mathcal{I}}_{i}\underset{\sim}{\mathcal{I}}_{j} \tag{120}$$

The basis functions of a spin system of N nuclei, provided the eigenfunctions of the individual nuclei are mutually independent, i.e., there is no spin-spin coupling, can be written up in the following form:

$$\psi_{k} = \psi_{k}^{1}\psi_{k}^{2} \cdots \psi_{k}^{N} = \prod_{j=1}^{N} \psi_{k}^{j} \tag{121}$$

where ψ_k^j denotes the k-th individual eigenfunction of the j-th nucleus. When the spin quantum number of all atoms is 1/2, then there are 2^N, whereas if the spin quantum number is I, there are $(2I + 1)^N$ product eigenfunctions, ψ_k, (basis functions) of the spin system.

The m_{T} eigenvalues of the resultant operator $\mathcal{I}_{z\mathrm{T}}$, of spin systems of N nuclei are $(m_{\mathrm{T}})_{\max}$ $(m_{\mathrm{T}})_{\max} - 1, \ldots - (m_{\mathrm{T}})_{\max} + 1, - (m_{\mathrm{T}})_{\max}$, where

$$I_{z\mathrm{T}} = (m_{\mathrm{T}})_{\max} = \sum_{j=1}^{N} I_{zj} = \sum_{j=1}^{N} (m_{j})_{\max} \tag{122}$$

For nuclei of spin 1/2, $(m_{\mathrm{T}})_{\max} = N/2$.

There are therefore altogether $2(m_{\mathrm{T}})_{\max} + 1$ eigenvalues. The number of basis functions pertinent to identical values of m_{T} is

$$\frac{N!}{\left(\frac{N}{2} - m_{\mathrm{T}}\right)!\left(\frac{N}{2} + m_{\mathrm{T}}\right)!} \tag{123}$$

Accordingly, when $I = 1/2$ and $N = 5$, the number of basis functions is $2^5 = 32$, and to each of the eigenvalues $m_{\mathrm{T}} = \pm 5/2, \pm 3/2,$ and $\pm 1/2$, there belong one, five, and ten of product functions, respectively. Eigenvalues $m_T = \pm 5/2$ correspond to the basis functions ααααα and βββββ respectively, the eigenvalue + 1/2 to functions αααββ, ααβαβ, αβααβ, βαααβ, ααββα, αβαβα, βααβα, αββαα, βαβαα, and ββααα. Here and in the following discussions in the indexes of the previous product functions, the serial numbers of the nuclei have been omitted, e.g., instead $\alpha_1\alpha_2\alpha_3\alpha_4\alpha_5$, we wrote ααααα etc. Thus, for example, the product $(\alpha_1\beta_2 + \beta_1\alpha_2)(\alpha_3\beta_4 + \beta_3\alpha_4)$ can simply be written as αβαβ + βααβ + αββα + βαβα. It has to be borne in mind that index numbers are listed from left to right, starting with 1. Furthermore, indexes always start anew from 1 following the product sign (perpendicular line) or, in the case of new terms, following the + and − signs. Thus: $\alpha|\alpha \equiv \alpha_1\alpha_1$ and $\alpha\alpha|\alpha\alpha \equiv (\alpha_1\alpha_2)(\alpha_1\alpha_2)$, etc.

The eigenvalues of the operator $\mathcal{I}_T^2$ are $I_T(I_T + 1)$, where, I_T, the resultant spin quantum number, can assume the same values as $|I_{zT}|$, i.e., for N nuclei of spin 1/2 its magnitude is $N/2$, $N/2 - 1$, . . . 1/2 or 0, depending on whether N is odd or even. The number of the basis functions for a given eigenvalue I_T is

$$\frac{N!\,(2I_T + 1)}{\left(\frac{N}{2} + I_T + 1\right)!\left(\frac{N}{2} - I_T\right)!} \tag{124}$$

It is noted at this point that upon performing operations on product functions of multispin systems, one follows the same route as in the differentiation of product functions. Each transformed product function is composed of as many terms as the number of factors in the original function, and in each term, the operation under consideration is performed upon just one factor leaving the rest of them unchanged. Thus,

$$\mathcal{I}^2|\alpha\alpha\rangle = |\frac{3}{4}\alpha\alpha\rangle + |\frac{3}{4}\alpha\alpha\rangle = |\frac{3}{2}\alpha\alpha\rangle$$

$$\mathcal{I}^2|\beta\alpha\rangle = |\frac{3}{4}\beta\alpha\rangle + |\frac{3}{4}\beta\alpha\rangle = |\frac{3}{2}\beta\alpha\rangle$$

$$\mathcal{I}_z|\alpha\alpha\rangle = |\frac{1}{2}\alpha\alpha\rangle + |\frac{1}{2}\alpha\alpha\rangle = |\alpha\alpha\rangle$$

$$\mathcal{I}_z|\alpha\beta\rangle = |\frac{1}{2}\alpha\beta\rangle - |\frac{1}{2}\alpha\beta\rangle = 0$$

$$\mathcal{I}_z|\beta\beta\rangle = -|\frac{1}{2}\beta\beta\rangle - |\frac{1}{2}\beta\beta\rangle = -|\beta\beta\rangle$$

$$\mathcal{I}^+|\alpha\alpha\rangle = |0\alpha\rangle + |\alpha 0\rangle = 0 = \mathcal{I}^-|\beta\beta\rangle$$

$$\mathcal{I}^+|\alpha\beta\rangle = |0\beta\rangle + |\alpha\alpha\rangle = |\alpha\alpha\rangle = \mathcal{I}^+|\beta\alpha\rangle$$

$$\mathcal{I}^+|\beta\beta\rangle = |\alpha\beta\rangle + |\beta\alpha\rangle = \mathcal{I}^-|\alpha\alpha\rangle$$

$$\mathcal{I}^-|\alpha\beta\rangle = |\beta\beta\rangle + |\alpha 0\rangle = |\beta\beta\rangle = \mathcal{I}^-|\beta\alpha\rangle$$

$$\mathcal{I}_x|\beta\alpha\rangle = |\frac{1}{2}\alpha\alpha\rangle + |\frac{1}{2}\beta\beta\rangle$$

$$\mathcal{I}_y|\beta\beta\rangle = -|\frac{i}{2}\alpha\beta\rangle + |\frac{i}{2}\beta\alpha\rangle$$

and similarly for three factors:

$$\mathcal{I}^2 |\alpha\beta\alpha\rangle = |\frac{3}{4}\alpha\beta\alpha\rangle + |\frac{3}{4}\alpha\beta\alpha\rangle + |\frac{3}{4}\alpha\beta\alpha\rangle = |\frac{9}{4}\alpha\beta\alpha\rangle$$

$$\mathcal{I}^+ |\beta\beta\alpha\rangle = |\alpha\beta\alpha\rangle + |\beta\alpha\alpha\rangle + |\beta\beta 0\rangle = |\alpha\beta\alpha\rangle + |\beta\alpha\alpha\rangle$$

The above expressions mean therefore the sum

$$\sum_{i=1}^{N} \psi^1 \psi^2 \ldots \mathcal{I}_i \psi^i \ldots \psi^N$$

for the system of N magnetic nuclei, i.e., for the product function composed of N factors.

When an "operator product" is found in the expression (the operator "factors" refer only to the identically indexed components of the product wavefunction), upon performing the operating symbolized by the operator indexed correspondingly one by one on each factor in the product function, the other factors remain unaltered, and the product of the expressions obtained this way give the final result. For example,

$$\mathcal{I}_{x1} \mathcal{I}_{x2} |\alpha_1 \alpha_2 \alpha_3\rangle = |\frac{1}{4}\beta\beta\alpha\rangle$$

$$\mathcal{I}_1^+ \mathcal{I}_2^+ |\beta_1 \alpha_2 \beta_3\rangle = |\alpha 0 \beta\rangle = 0$$

When the eigenfunctions of the individual nuclei of the spin system composed of N nuclei are not mutually independent — when there exists spin-spin coupling — then instead of $K = 2^N$ product functions of form given in Equation 121 for nuclei with spin 1/2, only the basis functions ψ_K obtained by the linear combination of the former:

$$\psi_K = \sum_{j=1}^{K} C_j \varphi_j = C_1\varphi_1 + C_2\varphi_2 + \cdots + C_K\varphi_K \qquad (125)$$

satisfy the Schrödinger equation of the system. Here the normalized nature of the basis functions means

$$\sum_{j=1}^{K} |C_j|^2 = 1 \qquad (126)$$

The product functions φ_j, are also orthonormal: $\langle\varphi_i|\varphi_j^*\rangle = \delta_{ij}$, where φ_j^* is the conjugate complex* of the function φ_j,δ_{ij} the Krönecker delta is always zero, except for $i = j$ when $\delta_{ij} = 1$.

The Schrödinger equation of the system (compare Equation 84) is

$$\mathcal{H} \sum_{j=1}^{K} C_j\varphi_j = E \sum_{j=1}^{K} C_j\varphi_j \qquad (127)$$

Upon multiplying by φ_i^* both sides of Equation 127 and integrating over the whole space — for all possible spin orientations — we obtain:

$$\left\langle \varphi_i^* \mathcal{H} \sum_{j=1}^{K} C_j\varphi_j \right\rangle = E \left\langle \varphi_i^* \sum_{j=1}^{K} C_j\varphi_j \right\rangle \qquad (128)$$

* The aim of using conjugate complexes is to get real quantities as the product of the eigenfunctions.

Denoting the integrals on the left-hand side $<\varphi_i^*|\mathcal{H}|\varphi_j>$ — the matrix elements of the Hamilton operator — by symbol $\mathbf{H}_{ij}$ after rearrangement:

$$\sum_{j=1}^{K} C_j(\mathbf{H}_{ij} - E\delta_{ij}) = 0 \tag{129}$$

The above expression leads to a system of homogeneous linear equations in C_j whose nontrivial solutions (when $C_j \neq 0$) are obtained when the condition $|\mathbf{H}_{ij} = \delta_{ij}\mathrm{E}| = 0$ is satisfied. This expression is the determinant of a K-th order secular equation system which is completely analogous to, e.g., the determinant characteristic to the vibrational motion of polyatomic molecules, playing an important part in the theory of IR spectroscopy.[671] Particularized the determinant can be written as:

$$\begin{vmatrix} \mathbf{H}_{11} - E & \mathbf{H}_{12} & \mathbf{H}_{13} & \cdots & \mathbf{H}_{1K} \\ \mathbf{H}_{21} & \mathbf{H}_{22} - E & \mathbf{H}_{23} & \cdots & \mathbf{H}_{2K} \\ \cdot & \cdot & & & \cdot \\ \cdot & \cdot & & & \cdot \\ \cdot & \cdot & & & \cdot \\ \cdot & \cdot & & & 1 \\ \mathbf{H}_{K1} & \mathbf{H}_{K2} & \mathbf{H}_{K3} & \cdots & \mathbf{H}_{KK} - E \end{vmatrix} = 0 \tag{130}$$

where the terms $\mathbf{H}_{ij}$ are the matrix elements of Equation 129. The solution of the determinant (Equation 130) yields a series of E values, satisfying the equations system. Upon substituting them back to Equation 129, the coefficients C_j are also obtained.

The K-th order determinant has K solutions. This means that the spin system of N nuclei possessing $K = 2^N$ basis functions can have K energy levels, provided all nuclei have spin $I = 1/2$. Some of the levels can be degenerate.

For solution of the given problem — for the construction of the theoretical nuclear magnetic resonance spectrum — first of all the matrix elements $\mathbf{H}_{ij}$ should be given, expressed by the frequency values, ν, and coupling constants, J. For this purpose Equation 89 suffices. Considering the properties of the operators, $\mathcal{H}$, and eigenfunctions, φ, many terms in the determinant vanish and the determinant can be factorised into subdeterminants. Because of commutativity of operators, $\mathcal{I}_{zT}$ and $\mathcal{H}$, and orthogonality of the functions, φ_i, e.g., all terms in the determinant becomes zero for which $j \neq i$ in $\mathbf{H}_{ij}$ further φ_i and φ_j are product functions not belonging to the same eigenvalue I_{zT}.

It can also be shown that in the off-diagonal elements the term $\mathbf{H}°$ always vanishes, as the product functions φ_i and φ_j are always eigenfunctions of $\mathcal{H}°$. Therefore the operation $\mathcal{H}°$ does not influence φ_j, in other words, there is an integral factor in the term $\mathbf{H}°$ of $<\varphi_i|\varphi_j>$ which is zero due to the orthogonality. Therefore:

$$\mathbf{H}^{0}_{ij} \equiv <\varphi_i|\mathcal{H}^{0}_{ij}|\varphi_j> = 0 \tag{131a}$$

and

$$\mathbf{H}^{0}_{ii} \equiv <\varphi_i|\mathcal{H}^{0}_{ii}|\varphi_i> = \sum_{k=1}^{N} \nu_k m_k \tag{131b}$$

where the coefficients m_k, for the nuclei are the eigenvalues of operators $\mathcal{I}_z$ (compare Equations 89 and 122). Furthermore, for nuclei of spin 1/2:

$$\mathbf{H}'_{ij} \equiv \langle\varphi_i|\mathcal{H}'_{ij}|\varphi_j\rangle = \frac{1}{2}\sum_{k} \sum_{<l} J_{kl}U_{kl} \tag{131c}$$

where $U_{kl} = +1$ when the spin of the k-th and l-th nuclei is different in both basis functions φ_i and φ_j. For all other cases $U_{kl} = 0$; furthermore,

$$\mathbf{H}'_{ii} \equiv \langle\varphi_i|\mathcal{H}'_{ii}|\varphi_i\rangle = \frac{1}{4}\sum_{k} \sum_{<l} J_{kl}V_{kl} \tag{131d}$$

where $V_{kl} = \pm 1$, depending on the mutual orientation of the k-th and l-th nuclei; whether it is parallel or antiparallel. Equations 131c and 131d can be condensed to a more general common expression, which is more amenable. The second term in Hamiltonian

$$\sum_{i} \sum_{<j} J_{ij}\, \mathcal{I}_i\, \mathcal{I}_j = \sum_{i} \sum_{<j} J_{ij}(\mathcal{I}_{xi}\mathcal{I}_{xj} + \mathcal{I}_{yi}\mathcal{I}_{yj} + \mathcal{I}_{zi}\mathcal{I}_{zj}) \tag{132}$$

Using Equations 112, 115, and 116, it can be shown that if we let the operation in Equation 132 to be carried out on any possible pair of product functions, $\varphi_i\varphi_j$, the result is the following:

$$\frac{J_{ij}}{4}(2\varphi_i^{\mathcal{P}}\varphi_j^{\mathcal{P}} - \varphi_i\varphi_j) \tag{133}$$

where $\varphi_i^{\mathcal{P}}\ \varphi_j^{\mathcal{P}}$ is the so-called permutated product function. This means that Equation 132 can be given in the following simpler form:

$$\frac{1}{4}\sum_{i} \sum_{<j} J_{ij}(2\mathcal{P}_{ij} - 1) \tag{134}$$

Here $\mathcal{P}$ is the so-called *permutation operator* which leads to the permutation of the components when applied to a product wavefunction:

$$\mathcal{P}_{kl}|\cdots\varphi_j\varphi_k\varphi_l\varphi_m\cdots\rangle = |\cdots\varphi_j\varphi_l\varphi_k\varphi_m\cdots\rangle \tag{135}$$

Thus, for example,

$$\mathcal{P}_{1,2}|\alpha\beta\rangle = \mathcal{P}_{1,2}|\alpha_1\beta_2\rangle = |\alpha_2\beta_1 = \beta_1\alpha_2\rangle = |\beta\alpha\rangle$$

$$\mathcal{P}_{1,2}|\alpha\beta\alpha\rangle = |\beta\alpha\alpha\rangle\,;\ \mathcal{P}_{2,3}|\alpha\beta\alpha\rangle = |\alpha\alpha\beta\rangle,\ \text{etc.}$$

The above discussion will be now illustrated by the determinant written for the simplest spin system A_2 built from two nuclei of spin 1/2.

The 4 × 4 determinant is factorized into two 1 × 1 and one 2 × 2 subdeterminant, and in this case the solution is very simple. The combined eigenfunctions are

$$\psi_{I_T,m} = C_1\varphi_1 + C_2\varphi_2 + C_3\varphi_3 + C_4\varphi_4 \tag{136}$$

where φ_1, φ_2, φ_3, and φ_4 are the product functions of form $\alpha_1\alpha_2$, $\alpha_1\beta_2$, $\beta_1\alpha_2$, and $\beta_1\beta_2$. The indexes can be neglected, thus they will be $\alpha\alpha$, $\alpha\beta$, $\beta\alpha$, and $\beta\beta$. The determinant is

$$\begin{vmatrix} \mathbf{H}_{11} - E & \mathbf{H}_{12} & \mathbf{H}_{13} & \mathbf{H}_{14} \\ \mathbf{H}_{21} & \mathbf{H}_{22} - E & \mathbf{H}_{23} & \mathbf{H}_{24} \\ \mathbf{H}_{31} & \mathbf{H}_{32} & \mathbf{H}_{33} - E & \mathbf{H}_{34} \\ \mathbf{H}_{41} & \mathbf{H}_{42} & \mathbf{H}_{43} & \mathbf{H}_{44} - E \end{vmatrix} = 0 \qquad (137)$$

Upon expanding the determinant, the energy values, E_i, are obtained and resubstituting them yields the coefficients, C_i. As the operators $\mathcal{H}$ and $\mathcal{I}^2$ possess simultaneous eigenvalues, a completely analogous secular equation can be set up for the calculation of the eigenvalues $I_T(I_T + 1)$. The only change to be introduced is that in place of the energy eigenvalues E_i the eigenvalues λ_i of the operator $\mathcal{I}^2$ are introduced. Having calculated these eigenvalues, coefficients C_i are obtained in the same way as in the case of the energy equation. The calculation of the energy values will be mentioned later.* In the present case, it is more practical to calculate the eigenvalues of the operator $\mathcal{I}^2$, since these are the same for any system of two nuclei of spin 1/2, whereas the energies depend on the type of nuclei and their chemical environment.

The determinant written for $\mathcal{I}^2$ is the following:

$$\sum C_j | \mathcal{I}^2_{ij} - \lambda_i \delta_{ij} | = 0 \qquad (138)$$

where $\mathbf{I}^2_{ij} = <\varphi^*_i|\mathcal{I}^2|\varphi_j>$ means the matrix elements containing the $\mathcal{I}^2$ operator. Considering Equations 131a to 131d, there are several zero elements as shown below:

$$\begin{vmatrix} \mathbf{I}^2_{11} - \lambda & 0 & 0 & 0 \\ 0 & \mathbf{I}^2_{22} - \lambda & \mathbf{I}^2_{23} & 0 \\ 0 & \mathbf{I}^2_{32} & \mathbf{I}^2_{33} - \lambda & 0 \\ 0 & 0 & 0 & \mathbf{I}^2_{44} - \lambda \end{vmatrix} = 0 \qquad (139)$$

This fourth-order determinant can be decomposed into two first-order and one second-order subdeterminant. Therefore φ_1 and φ_4 are eigenfunctions of the operator $\mathcal{I}^2$, while the eigenvalues are I^2_{11} and I^2_{44}. According to Equation 120:

$$\mathcal{I}^2 |\varphi_1> = (\mathcal{I}^2_1 + 2\,\mathcal{I}_1\,\mathcal{I}_2 + \mathcal{I}^2_2)|\alpha\alpha> = [\,\mathcal{I}^2_1 + 2(\mathcal{I}_{x1}\mathcal{I}_{x2} + \mathcal{I}_{y1}\mathcal{I}_{y2} + \mathcal{I}_{z1}\mathcal{I}_{z2}) + \mathcal{I}^2_2\,]|\alpha\alpha> \qquad (140)$$

because

$$2\,\mathcal{I}_1\,\mathcal{I}_2|\alpha\alpha> = 2(\mathcal{I}_{x1}\mathcal{I}_{x2} + \mathcal{I}_{y1}\mathcal{I}_{y2} + \mathcal{I}_{z1}\mathcal{I}_{z2})|\alpha\alpha> \qquad (141)$$

From the determinant (Equation 139), it follows that $\mathbf{I}^2_{11} = <\alpha\alpha|\mathcal{I}^2|\alpha\alpha> = \lambda$. Performing the operations defined by Equations 111, 112, 115, and 116 from Equation 140, we obtain

* Compare p. 88-92, Section 1.5.1.

that λ, i.e., the eigenvalue of the operator $\mathcal{I}^2$, equals $I_{11}^2 = I_T(I_T + 1) = 2$, therefore $I_T = 1$. (The solution $I_T = -2$ is physically meaningless, as the square operator of the spin vector cannot possess negative eigenvalues.* In a similar fashion $I_{44}^2 = 2$, and also $I_T = 1$.

In the same way for all matrix elements of the second-order subdeterminant, we get the same eigenfunctions: $\varphi_2 + \varphi_3 = \alpha\beta + \beta\alpha$. On dividing by this quantity, the second-order subdeterminant assumes the form:

$$\begin{vmatrix} (1-\lambda) & 1 \\ 1 & (1-\lambda) \end{vmatrix} = 0 \tag{142}$$

from which $\lambda^2 - 2\lambda = 0$, therefore $\lambda = 0$ or 1 thus $I_T = 0$ or 1. When the value of λ is already known, the coefficients C_2 and C_3 in Equation 136 can be calculated according to Equations 126 and 138:

$$C_2(1-\lambda) + C_3 = 0 \tag{143}$$

$$C_2 + C_3(1-\lambda) = 0 \tag{144}$$

and

$$C_2^2 + C_3^2 = 1 \tag{145}$$

Substituting the values of λ, we obtain that $C_2 = \pm\, C_3 = 1/\sqrt{2}$, and from the above equations:

$$\psi_{0,0} = \frac{1}{\sqrt{2}}(\alpha\beta - \beta\alpha) \tag{146}$$

$$\psi_{1,0} = \frac{1}{\sqrt{2}}(\alpha\beta + \beta\alpha) \tag{147}$$

1.4.5. Transition Probabilities and the Relative Intensities of Spectral Lines; Selection Rules

In NMR spectroscopy there is no need for measurement of absolute intensities which are very difficult to carry out, since for the analysis of the spectra the relative intensities of the spectral lines suffice, provided that these are measured under strictly identical experimental conditions.

The relative intensity of the spectral lines is proportional to *transition probability*. The determination of the latter, therefore, enables the calculation of the relative intensity of the spectral lines theoretically.

The probability of a given transition — similar to the case of, for example, IR spectroscopy[671] — is proportional to the square of the matrix elements of the s.c. *transition moment*. Components of the matrix of the transition moment for a single magnetic nucleus are

$$\mathbf{R}_w = \langle \psi_{m'}^* | \mathcal{I}_w | \psi_{m''} \rangle \tag{148}$$

and for a spin system of N nuclei

$$\mathbf{R}_w = \langle \psi_{m'}^* | \sum_{i=1}^{N} \mathcal{I}_{wi} | \psi_{m''} \rangle \tag{149}$$

* Compare p. 81.

where $\psi_{m''}$ is the wavefunction of th initial state, $\psi^*_{m'}$ is the conjugate complex of the final state, and $w = x, y$, or z. When the values of the above expression for $\mathbf{R}_w$ are calculated utilizing Equations 112, 115, and 116, we always find that $\mathbf{R}_z = 0$. The operator, $\mathcal{I}_z$, leaves namely the wavefunctions, ψ, unchanged; therefore having performed the operation symbolized by $\mathcal{I}_z$, it still is true that $\psi^*_{m'} \neq \psi_{m''}$ thus the integral in Equation 149 is zero, due to the orthogonality of the basis functions. This is in accord with the fact that the polarizing field $\mathbf{B}_0$ of the direction z cannot induce transitions among the magnetic levels. Since the x and y components of $\mathbf{B}_1$ and so of the transition moment are equivalent, it is enough to examine only one of the matrix elements, e.g., $\mathbf{R}_x$ for the calculation of relative intensities.

Transition probabilities, W thus relative intensities are proportional to the square of the components of the transition moment. The use of the operator $\mathcal{I}^+$ instead of $\mathcal{I}_x$ in the expression of $\mathbf{R}^2_x$ is much more illuminating (compare Equation 113), since the latter alters the spin wavefunctions, indeed according to the transitions $m'' \rightarrow m'$ ($\beta \rightarrow \alpha$). The commutativity of the operators follows from Equations 113 and 115. The relative intensity is, therefore, proportional to the expression:

$$W \sim [\langle\psi'|\ \mathcal{I}^+|\psi''\rangle]^2 \qquad (150)$$

where ψ' and ψ'' are the wavefunctions of the excited and the ground states, respectively. Upon performing the operation $\mathcal{I}^+$ on the eigenfunction ψ'', the value of m'' changes by $+1$, therefore, when m' (in case of multinuclear systems the eigenvalue I_{zT}) is greater by one unit than m'' we obtain a nonzero value, in spite of the orthogonality, for W. Therefore only transitions obeying the condition

$$\Delta I_{zT} = \pm 1 \qquad (151)$$

are *allowed*, because only these have nonzero probabilities. All other combinations are *forbidden*, in other words, the probability of the corresponding transition is zero. Equation 151 is a s.c. *selection rule* also occurring in optical spectroscopy[671] and is strictly valid for any spin system.

The selection rule (Equation 151) for protons and for other single-nuclear "systems" of spin 1/2 follows by itself, since there are only two levels. As for molecular vibrations, molecular symmetry plays a decisive role in the study of the spin systems.

The eigenfunction ψ possesses symmetry properties invariant to the operations $\mathcal{I}_x$ and $\mathcal{I}^+$ or $\mathcal{I}_y$ and $\mathcal{I}^-$. Expressions 148 to 150 are therefore nonzero, only when the symmetry species of the functions ψ' and ψ'' are identical. This is therefore another selection rule, according to which *only eigenfunctions of the same symmetry species can be combined.* The experimental finding to be illustrated later is that all transitions among the symmetric and antisymmetric levels ($s \leftarrow\!\!|\!\!\rightarrow a$) are forbidden.*

When the spin system under study has symmetry elements, the theoretical treatment becomes simpler, since certain elements of the secular determinant (see Equation 130) vanish or become identical. By denoting the total spin angular momentum of the groups of chemically equivalent nuclei by I_G, it can be shown that

1. Only those product functions can combine with belong to identical eigenvalues I^2_G.
2. Only such transitions are allowed for which the selection rule $\Delta I^2_G = 0$ is valid.

Condition 1 is equivalent, for example, to the statement that the two third-order subdeterminants of the ABC system in the case of an AB_2 spin-system (where due to the molecular

* Compare e.g., p. 91.

symmetry the two nuclei, B, are chemically equivalent) decompose into a first-order and second-order determinant each. Because of the selection rule under Point 2, certain transitions, allowed for an ABC system, become forbidden; the structure of the spectrum becomes simpler. In the subsequent discussion more examples will be given to illustrate these statements.

1.5. THE QUANTITATIVE STUDY OF SPIN SYSTEM

The quantum-mechanical interpretation of NMR phenomena provides all the theoretical tools necessary for the calculation of even the most complicated spectra. In the simpler (first-order) cases, of course, it is not necessary to invoke quantum mechanical considerations, and they can be tackled satisfactorily without calculations merely by simple logics. In the determinant (see Equation 130) of a spin system involving only first-order coupling, only the diagonal elements have nonzero values; the K-th order determinant can be decomposed into K first-order subdeterminants. This is where the term ''first-order coupling'' comes from. The product functions of Equation 121 do not combine with each other in this case.

When the condition for first-order coupling is not satisfied, i.e., when chemical shift differences (in Hz) are not greater by about one order of magnitude than the corresponding coupling constants, however, the study of spin systems becomes more complex. The study of even the simplest case — that of the A_2 spin system (e.g., the H_2 molecule) — can only be accomplished using quantum mechanics.

1.5.1. The A_2 Spin System

Let us take first the simplest possible case, that of a single nucleus of spin $I = 1/2$, with $m = +1/2$, i.e., having basis function α. There is, of course, no possibility for interaction. The quantum-mechanical treatment yields the energy of the system.

The time-independent Schrödinger equation of an isolated nucleus A is (compare Equation 84):

$$\mathcal{H}\alpha = E\alpha \tag{152}$$

where α is the stationary spin wavefunction, and E is the energy of the nucleus in a static polarizing field $\mathbf{B}_0$. In order to determine the possible energy, function should be extended to include all possible values of the wavefunction α, depending on the coordinate φ (for all possible spin orientations):

$$\int_{\varphi=0}^{360} \alpha|\mathcal{H}|\alpha d\varphi = E\int_{\varphi=0}^{360} \alpha^2 d\varphi \tag{153}$$

From this equation it follows (by applying previously introduced abbreviations):

$$E = \frac{\langle\alpha|\mathcal{H}|\alpha\rangle}{\langle\alpha|\alpha\rangle} = \langle\alpha|\mathcal{H}|\alpha\rangle \tag{154}$$

as the wavefunction is normalized ($\langle\alpha|\alpha\rangle = 1$).

To obtain the value of E, the Hamilton operator (see Equation 88) should be multiplied by h and then substituted into Equation 154. Thus, the result is obtained in energy units. The second term in Equation 88 is zero — since there is no interaction. Therefore,

$$E = h\gamma B_l(\langle\alpha_1|\mathcal{I}_{z1}|\alpha_1\rangle) \tag{155}$$

The product $h\gamma B_l$ can be taken out since it is not influenced by the operation $\mathcal{I}_z$. Here

B_l is the local field around the nucleus under study, which depends on B_0 as described by Equation 68. According to Equation 112, $\mathcal{I}_z|\alpha\rangle = \alpha/2$; thus:

$$E = \frac{1}{2} h\gamma B_l(\langle\alpha|\alpha\rangle) = \frac{1}{2} h\gamma B_l = \frac{1}{2} h\gamma B_0(1-\sigma) \qquad (156)$$

obeying the normalization condition. This expression is, in fact, the energy defined by Equation 23 for the case $m = 1/2$, also taking nuclear shielding into account. The result (see Equation 156) can also be obtained by multiplying Equation 95 by h and substituting the eigenvalue of $\mathcal{I}_z$ $(+1/2)$.

One nucleus naturally possesses two energy levels (2^N, N = 1), according to the eigenvalues $m = +1/2$ and $m = -1/2$ for I_z. When $m = -1/2$, the basis function β should be substituted into Equations 152 to 156, and in this case one obtains energy of the same absolute value, but of opposite signs: $E(\alpha) = -E(\beta)$. Two-spin systems possess $2^2 = 4$ energy levels, which are obtained from the possible combinations of the individual eigenvalues I_z of the nuclei ($m_1 = m_2 = +1/2$, $m_1 = +1/2$, and $m_2 = -1/2$; $m_1 = -1/2$ and $m_2 = +1/2$; finally $m_1 = m_2 = -1/2$.

Let us first examine the total energy of two, chemically equivalent nuclei of spin $+1/2$ (the resultant spin quantum number is $I_{zT} = 1$), assuming, for the time being, that there is no interaction between them. The common wavefunction of the two nuclei is the product of the individual wavefunctions: $\psi = \alpha_1\alpha_2 = \alpha\alpha$. This way,

$$E = h\gamma B_l(\langle\alpha_1\alpha_2|\mathcal{I}_{z1} + \mathcal{I}_{z2}|\alpha_1\alpha_2\rangle) =$$

$$h\gamma B_l\,[(\langle\alpha_1|\mathcal{I}_{z1}|\alpha_1\rangle) + (\langle\alpha_2|\mathcal{I}_{z2}|\alpha_2\rangle)] \qquad (157)$$

since the operator $\mathcal{I}_{z1}$ and $\mathcal{I}_{z2}$ does not influence the wavefunction α_2 and α_1, respectively, B_l can be taken out of the expression (the nuclei are chemically equivalent). As there is no interaction, the energy is therefore the sum of the individual energy values of the two nuclei: $E_{1\,2} = E_1 + E_2 = h\gamma B_0(1 - \sigma)$.

When $m = -1/2$ for the two nuclei, the value of $E_{1,2}(\beta\beta)$ is obtained completely analogously by using the basis function $\psi = \beta\beta$, for which it holds $E_{1,2}(\alpha\alpha) = -E_{1,2}(\beta\beta)$.

In the case of states in which $I_{zT} = 0$ (for one nucleus $I_z = +1/2$, and for the other, $-1/2$), the product functions αβ and βα, respectively, as were shown* cannot be used anymore. These contain namely the impossible statement that the spins I_z of the two equivalent nuclei are not identical (i.e., $+1/2$ and $-1/2$). The wavefunctions of the state are obtained through the linear combination of the former. The wavefunctions so obtained should satisfy the following conditions: they should be symmetric (invariant) or antisymmetric (different only in sign) in respect to the interchange of the nuclei; furthermore, they should be orthonormal. These conditions are satisfied — as it has already been shown (compare Equations 146 and 147) — by the wavefunctions, $\psi = 1/\sqrt{2}\,(\alpha\beta \pm \beta\alpha)$.

One of the two functions is invariant towards the exchange of the nuclei, while the other changes sign. The former is therefore the symmetric wavefunction, while the other is the antisymmetric one. Their orthogonality can be seen from the following:

$$\frac{1}{2}\,[\langle(\alpha\beta + \beta\alpha)|(\alpha\beta - \beta\alpha)\rangle] = \frac{1}{2}\,[(\langle\alpha\beta|\alpha\beta\rangle) + (\langle\beta\alpha|\alpha\beta\rangle) - (\langle\alpha\beta|\beta\alpha\rangle) - (\langle\beta\alpha|\beta\alpha\rangle)] =$$

$$\frac{1}{2}\,[(\langle\alpha|\alpha\rangle)(\langle\beta|\beta\rangle) + (\langle\beta|\alpha\rangle)(\langle\alpha|\beta\rangle) - (\langle\alpha|\beta\rangle)(\langle\beta|\alpha\rangle) - (\langle\beta|\beta\rangle)(\langle\alpha|\alpha\rangle)] = 0$$

Due to the normalization condition of α and β, the integrals in square brackets are one in the first and last terms, while the two middle terms are zero, corresponding to their or-

* Compare p. 86.

Table 7
BASIS FUNCTIONS, RESULTANT SPIN QUANTUM NUMBERS, SYMMETRY, SYMBOLS, AND ENERGIES OF THE A_2 SPIN SYSTEM WITHOUT (J = 0) AND WITH SPIN-SPIN INTERACTION ($J \neq 0$)

Basis function				Quantum state		
					Energy	
Symbol	**Form**	I_{zT}	**Symmetry**	**Symbol**	$J = 0$	$J \neq 0$
ψ_1	$\alpha\alpha$	+1	s	s_1	$\gamma hB_0(1-\sigma)$	$\gamma hB_0(1-\sigma) + hJ/4$
ψ_2	$(\alpha\beta + \beta\alpha)/\sqrt{2}$	0	s	s_0	0	$+hJ/4$
ψ_3	$(\alpha\beta - \beta\alpha)/\sqrt{2}$	0	a	a_0	0	$-3hJ/4$
ψ_4	$\beta\beta$	−1	s	s_{-1}	$-\gamma hB_0(1-\sigma)$	$-\gamma hB_0(1-\sigma) + hJ/4$

thogonality. Hence the whole expression is zero. The combined basis functions are normalized, since $<\psi|\psi> = 1$:

$$\frac{1}{2}[(<\alpha\beta|\alpha\beta>) \pm 2(<\alpha\beta|\beta\alpha>) + (<\beta\alpha|\beta\alpha>)] =$$

$$\frac{1}{2}[(<\alpha|\alpha>)(<\beta|\beta>) \pm 2(<\alpha|\beta>)(<\beta|\alpha>) + (<\beta|\beta>)(<\alpha|\alpha>)] = 1$$

The integrals in the first and last terms in square bracket are again one, while the term in the middle are zero due to the normalization and orthogonality of α and β, respectively.

In the absence of spin-spin interaction between the spins of the two nuclei, the energy levels $E(\alpha\beta)$ and $E(\beta\alpha)$ are both equal to zero:

$$E(\alpha\beta) = E(\beta\alpha) = E(\alpha) + E(\beta) = 0 \tag{158}$$

The above are summarized in Table 7, where the notation of the levels refers to the symmetric (s) or antisymmetric (a) nature of the energy levels, while the lower index refers to the values of I_{zT}. It can easily be seen that, according to the selection rule $\Delta I_{zT} = \pm 1$ (see Equation 151), the energy of any transition of the A_2 spin system is the same: $|\Delta E| = \gamma hB_0(1 - \sigma)$; therefore, the spectrum is a single line. The interaction energy can be very simply calculated using Equation 134 for any level. The contribution of the interactions calculated by Equation 134 is expressed with the coupling constants J and therefore is obtained in frequency units. Therefore, the energy is obtained from these upon multiplication by h. For example, for the level corresponding to the eigenfunction, $(1/\sqrt{2})(\alpha\beta + \beta\alpha)$:

$$<(1/\sqrt{2})(\alpha\beta + \beta\alpha)|\frac{J_{1,2}}{4}(2P - 1)|(1/\sqrt{2})(\alpha\beta + \beta\alpha)> =$$

$$\frac{J_{1,2}}{8}[<(\alpha\beta + \beta\alpha)|(2\beta\alpha + 2\alpha\beta - \alpha\beta - \beta\alpha)>] =$$

$$\frac{J_{1,2}}{8}[<(\alpha\beta + \beta\alpha)|(\alpha\beta + \beta\alpha)>] = \frac{J_{1,2}}{8}[1 + 0 + 0 + 1] = J_{1,2}/4$$

The change due to spin-spin coupling in the other three energy levels can be calculated in a similar fashion. The energy level diagram of the A_2 spin system — compared to the interaction-free case — is found in Figure 38.

The change of the energies (see Table 7) as a consequence of spin-spin coupling means that all s states are destabilized in the same degree, whereas the state a is stabilized to a much greater extent.

Table 7 shows that energy of transitions $s_{-1} \to s_0$ and $s_0 \to s_1$ both for the interacting system and for the interaction-free case are equivalent to each other. Therefore, when only these transitions are allowed, the spectrum of the A_2 spin system is composed of only one line, even in case of spin-spin coupling. When the transitions $s_{-1} \to a_0$ and $a_0 \to s_1$ become also allowed, one can expect the splitting of the lines since the transition energy differs by $\pm J$ from previous values (compare Table 7).

Let us then examine the relative probability of transitions. There are formally six transitions possible: $s_{-1} \to a_0$, $s_{-1} \to s_0$, $s_{-1} \to s_1$, $a_0 \to s_0$, $a_0 \to s_1$, $s_0 \to s_1$.

Using the basis functions given in Table 7, the relative probability of the transitions of the A_2 spin system can be deduced according to Equation 150:

$s_{-1} \to a_0$:

$$W \sim \frac{1}{2}\,[<(\alpha\beta - \beta\alpha)|\,\mathcal{I}^+|\beta\beta>]^2 = \frac{1}{2}\,[<(\alpha\beta - \beta\alpha)|(\alpha\beta + \beta\alpha)>]^2 =$$

$$\frac{1}{2}\,[<\alpha\beta|\alpha\beta> - <\beta\alpha|\alpha\beta> + <\alpha\beta|\beta\alpha> - <\beta\alpha|\beta\alpha>]^2 = \frac{1}{2}\,[1 - 0 + 0 - 1]^2 = 0$$

$s_{-1} \to s_0$:

$$W \sim \frac{1}{2}\,[<(\alpha\beta + \beta\alpha)|\,\mathcal{I}^+|\beta\beta>]^2 = \frac{1}{2}\,[<(\alpha\beta + \beta\alpha)|(\alpha\beta + \beta\alpha)>]^2 = \frac{1}{2}\,[1 + 0 + 0 + 1]^2 = 2$$

$s_{-1} \to s_1$:

$$W \sim \frac{1}{2}\,[<\alpha\alpha|\,\mathcal{I}^+|\beta\beta>]^2 = [<\alpha\alpha|(\alpha\beta + \beta\alpha)>]^2 = [0 + 0]^2 = 0$$

$a_0 \to s_0$:

$$W \sim \frac{1}{4}[<(\alpha\beta + \beta\alpha)|\,\mathcal{I}^+|(\alpha\beta - \beta\alpha)>]^2 = \frac{1}{4}\,[<(\alpha\beta + \beta\alpha)|\,(\alpha\alpha - \alpha\alpha)>]^2 = 0$$

$a_0 \to s_1$:

$$W \sim \frac{1}{2}\,[<\alpha\alpha|\,\mathcal{I}^+|(\alpha\beta - \beta\alpha)>]^2 = \frac{1}{2}\,[<\alpha\alpha|(\alpha\alpha - \alpha\alpha)>]^2 = 0$$

$s_0 \to s_1$:

$$W \sim \frac{1}{2}\,[<\alpha\alpha|\,\mathcal{I}^+|(\alpha\beta + \beta\alpha)>]^2 = \frac{1}{2}\,[<\alpha\alpha|(\alpha\alpha + \alpha\alpha)>]^2 = \frac{1}{2}\,[1 + 1]^2 = 2$$

The transition probability of the transitions: $s_{-1} \to a_0$, $a_0 \to s_0$, $a_0 \to s_1$ are zero according to the forbidden nature of the symmetric ↔ antisymmetric transitions,* while that of the transition $s_{-1} \to s_1$ is zero according to the selection rule given by Equation 151. For the two allowed transitions ($s_{-1} \to s_0$ and $s_0 \to s_1$), Equation 151 holds, and the relative probabilities are the same in both cases, i.e., 2.

In fact, the observed resonance spectrum of any A_2 spin systems is a single line. In a similar way, the A_n-type spin system composed of n equivalent nuclei (where n can be any small integral number) was shown by experiment to consist — in accord with theory — a single line. Singlet spectra of A_n systems are called *zero-order* spectra. (In practice it is often found that chemically nonequivalent nuclei give rise to a singlet as a result of accidental

* Compare p. 87.

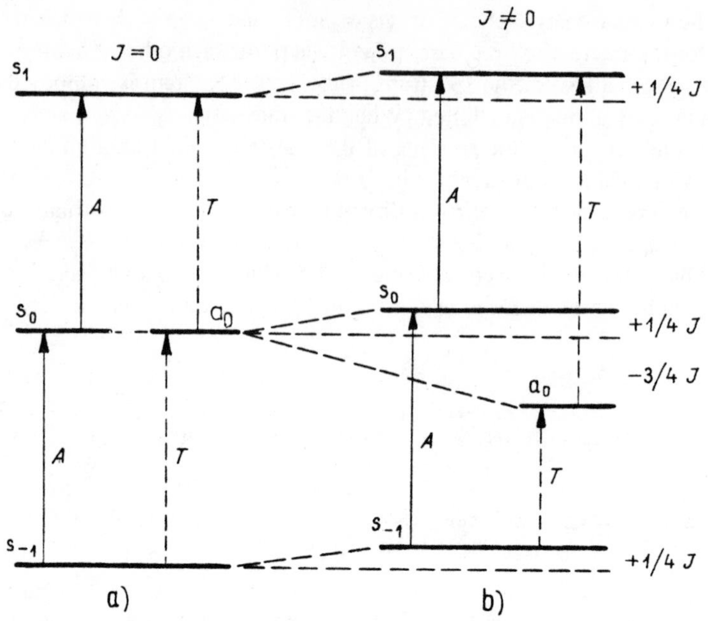

FIGURE 38. Energy levels and transitions of the A_2 spin system with (a) no interaction and (b) with spin-spin interaction. The broken arrows denote forbidden transitions (T).

isochrony. In such cases the nuclei can be regarded as approximate A_n spin system.) Since state a_0 cannot participate in any transition, nuclei on this level are presumably not able to enter resonance.

For an A_2 spin system, the H_2 molecule is a good example. From statistical reasons the protons are distributed so that three fourths of them are in *s* states, while the rest in state a_0. The latters (*para*-hydrogens) are not expected to participate in resonance. When it is true that only the molecules in *s* states change their magnetic quantum state (*ortho*-hydrogens), the total transition probability of the transitions must be unity, hence, must be identical to the observed value for, for example, an *AX* system. This has the following reasons. As there are three *s* levels, one third of the *ortho*-hydrogens, i.e., one fourth of all molecules, can take part in each of the transitions $s_{-1} \rightarrow s_0$ and $s_0 \rightarrow s_1$. The value 2 for the probability of the two allowed transitions means that the total transition probability of the transitions for hydrogen gas is indeed unity ($2 \cdot 1/4 + 2 \cdot 1/4 = 1$) independently of the fact that one fourth of the molecules (the *para*-hydrogens) are unable to enter resonance. Since in the spectrum of hydrogen (under identical conditions and at the same concentration) the line intensity is the same as the total intensity of the lines of hydrogen fluoride, where the total transition probability is also unity (the individual relative transition probabilities of the allowed four transitions are all 1/4,*) the inertness of *para*-hydrogens in resonance absorption can be taken as experimentally proved.

1.5.2. The *AX* Spin System

When $\Delta\nu AX$ is much greater than J_{AX}, two nuclei of spin 1/2 give rise to two symmetric doublets (two pairs of lines of 1:1 intensity). This is the simplest case for first-order coupling and requires (also for the general case A_nX_m) that:

$$J_{AX}/\Delta\nu AX \leqslant 0.1 \qquad (159)$$

* Compare p. 95.

Table 8
SPIN QUANTUM NUMBERS AND FREQUENCIES OF THE *AX* SPIN SYSTEM, WITHOUT AND WITH SPIN-SPIN INTERACTION

				Frequency ($\nu = E/h$)	
Level	I^A	I^X	I_{zT}^{AX}	$J^{AX} = 0$	$J^{AX} \neq 0$
1	1/2	1/2	1	$\nu A/2 + \nu X/2$	$\nu A/2 + \nu X/2 + J_{AX}/4$
2	1/2	−1/2	0	$\nu A/2 - \nu X/2$	$\nu A/2 - \nu X/2 - J_{AX}/4$
3	−1/2	1/2	0	$-\nu A/2 + \nu X/2$	$-\nu A/2 + \nu X/2 - J_{AX}/4$
4	−1/2	−1/2	−1	$-\nu A/2 - \nu X/2$	$-\nu A/2 - \nu X/2 + J_{AX}/4$

This condition* of first-order interaction permits, namely, the factorization of the K-th-order determinant ($K = 2^N$) into K first-order subdeterminants. The product functions are at the same time the basis functions, thus the energy of the levels can be calculated rather simply. For any level,

$$E_i^{AX}/h = \pm\frac{\nu A}{2} \pm \frac{\nu X}{2} \pm \frac{J_{AX}}{4} \tag{160}$$

where signs depend on the eigenvalues m corresponding to the factors of the product function (compare Equations 131b and 131d. The AX system has $2^N = 4$ different magnetic states, just like A_2 systems** which correspond to the possible number of combinations of the spin quantum numbers of nuclei A and X. These are listed in Table 8, where the frequencies of the levels are also given.

If the energy for transition A is greater than for X ($\Delta E_A > E_X$), then the order of the energy levels is the same as given in Table 8: the numbering of the levels increases in the order of decreasing energy values. According to the selection rule given by Equation 151, there are four transitions allowed. From Figure 39a it can also be seen that transitions **4** → **3** and **2** → **1**, further **4** → **2** and **3** → **1**, have the same energy. The resonance spectrum of the AX system is two lines: two-two lines of the transitions for A and X namely, coincide at higher and lower frequencies, respectively (see Figure 39b).

When spins A and X are coupled, a modification of the energy levels is expected. It has been shown earlier*** that coupled spins tend to align each other in the opposite sense. It is then clear that due to interaction, the energy of the less stable states **1** and **4** increases, while the energy of the levels **2** and **3** decreases. The energy change is of the same size, but of opposite sign. When $\Delta E/h = +J/4$, then the energy of the transitions **4** → **2** and **4** → **3** is lower by $J/2$, whereas the energy of the transitions **3** → **1** and **2** → **1** is higher by $J/2$ than in the interaction-free case (compare Table 8 and Figure 39c). Provided the probabilities of the two A and X transitions are identical, the resonance signal corresponding to both transitions A and X is composed of two lines of identical intensity at distance J from each other (see Figure 39d).

Transition probabilities can be calculated by Equation 150 and for the four allowed transitions the relative probabilities are the same, i.e., unity. For the allowed and forbidden transitions **4** → **2** and **4** → **1**, respectively, e.g.,

$$W^{4,2} \sim [\langle\alpha\beta|\ \mathcal{I}^+|\beta\beta\rangle]^2 = [\langle\alpha\beta|(\alpha\beta + \beta\alpha)\rangle]^2 = [\langle\alpha\beta|\alpha\beta\rangle + \langle\alpha\beta|\beta\alpha\rangle]^2 = [1 + 0]^2 = 1$$

$$W^{4,1} \sim [\langle\alpha\alpha|\ \mathcal{I}^+|\beta\beta\rangle]^2 = [\langle\alpha\alpha|(\alpha\beta + \beta\alpha)\rangle]^2 = [\langle\alpha\alpha|\alpha\beta\rangle + \langle\alpha\beta|\beta\alpha\rangle]^2 = [0 + 0]^2 = 0$$

* The condition given by Equation 159 illustrates well the significance of the measuring frequency in simplifying complex spectra; e.g., a value of $J/\Delta\nu = 0.17$ at 60 MHz entailing higher-order splitting changes to 0.1 at 100 MHz. It means, that instead of complex multiplets of the 60 MHz spectrum, at 100 MHz first-order spectrum is observable, from which δ and J values are obtainable easily, by direct observation, without calculations.

** Compare p. 89.

***Compare p. 46.

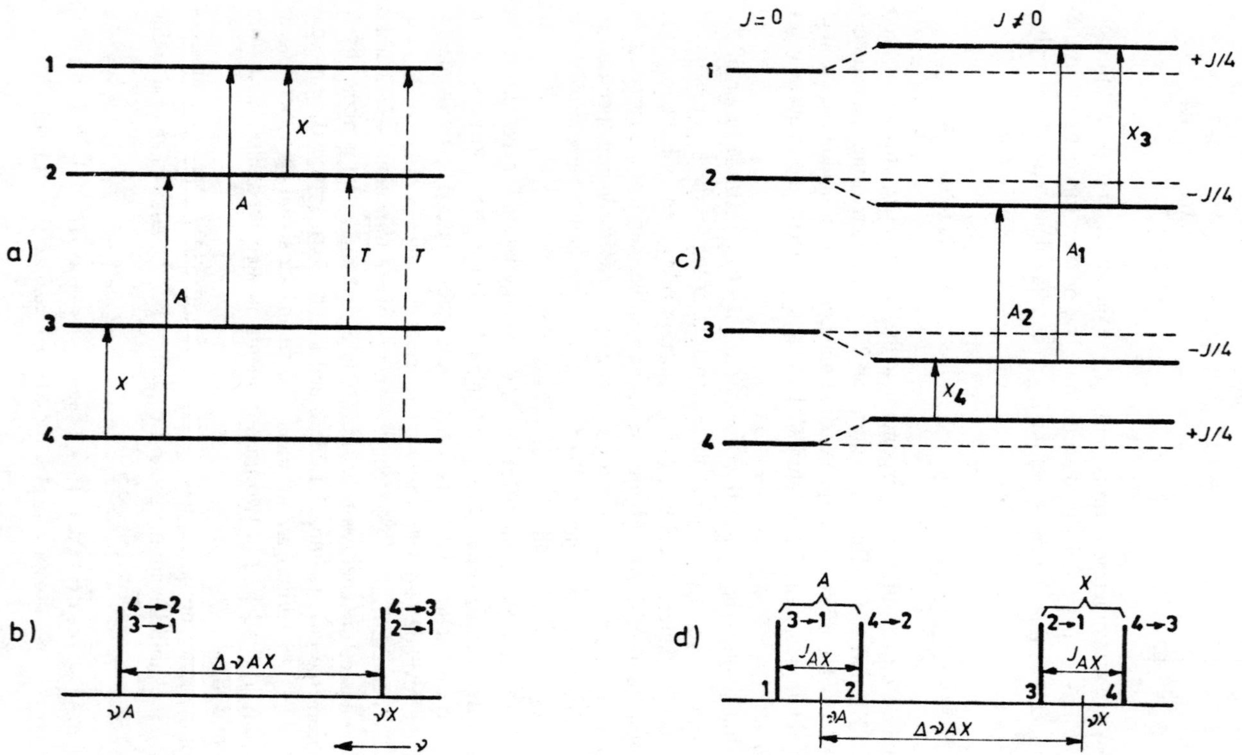

FIGURE 39 Energy levels and transitions (a) and (c) of the *AX* spin system and its schematic spectrum (b) and (d) without (a,b) and with (c,d) spin-spin interactions, respectively.

AX-type spectra are quite common and are characteristic, i.e., for the following classes of compounds: 1,2-disubstituted olefins, diastereotopic methylene protons, aromatic and heteroaromatic compounds containing two vicinal hydrogen atoms in the ring, 1,2-disubstituted oxyranes, etc. The oxyrane ring protons, for example, give *AX* spectrum in Compound **45** (see Figure 40). Coupling of two nuclei both of spin 1/2, but of different species, always gives *AX* spectra, since the heteronuclear coupling constant J_{AX} is much smaller than $\Delta\nu AX$, the shift difference.

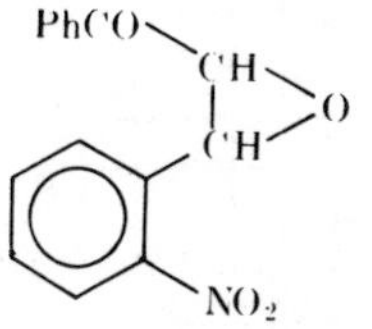

45

1.5.3. The *AB* Spin System

In the case of first-order interaction *AX*, the chemical shift diference ($\Delta\nu AX$) measured in hertz units is significantly greater than the coupling constant J_{AX}, consequently $J_{AX}/\Delta\nu AX$ is small (<0.1). For the A_2 system — in contrast — the chemical shifts of the two nuclei are identical: $\Delta\nu AA = 0$, i.e., $J_{AA}/\Delta\nu AA = \infty$. Therefore the *AB* system — where, according to definition J_{AB} and $\Delta\nu AB$ are comparable — is a transition case between the first-order *AX* and the zero-order A_2 systems.

The analysis of *AB* systems is analogous to that of A_2 or *AX* systems. First the energies of the four possible levels are to be determined. The calculation of the levels E_1 and E_4 does not present new problems, since the basis functions ($\alpha\alpha$ and $\beta\beta$) are the same. The only thing to be kept in mind is that in the energy expressions given in Table 7, the local field B_ℓ for the nuclei is not the same for *A* and *B* and therefore cannot be taken out of the brackets. Then

$$E_1^{AB} \equiv E(\alpha\alpha) = \frac{\gamma h B_0}{2}\left[(1-\sigma_A)+(1-\sigma_B)\right] + \frac{J_{AB}}{4}h = \gamma h B_0\left(1-\frac{\sigma_A+\sigma_B}{2}\right)+\frac{J_{AB}}{4}h \qquad (161)$$

Naturally $E(\alpha\alpha) \equiv E_1^{AB} = -E(\beta\beta) \equiv -E_4^{AB}$ is also valid for the coupling-free state ($J = 0$).

The energy of the levels $E(\alpha\beta) \equiv E_2^{AB}$ and $E(\beta\alpha) \equiv E_3^{AB}$ are obtained by finding the roots of the following secular determinant, analogous to Equation 139:

$$\begin{vmatrix} \mathbf{H}_{22}-E & \mathbf{H}_{23} \\ \mathbf{H}_{32} & \mathbf{H}_{33}-E \end{vmatrix} = 0 \qquad (162)$$

Upon expanding the determinant, we obtain

$$E^2 - E(\mathbf{H}_{22}+\mathbf{H}_{23}) + \mathbf{H}_{22}\cdot\mathbf{H}_{33} - \mathbf{H}_{23}^2 = 0$$

$$\text{as } \mathbf{H}_{23} \equiv \mathbf{H}_{32} \qquad (163)$$

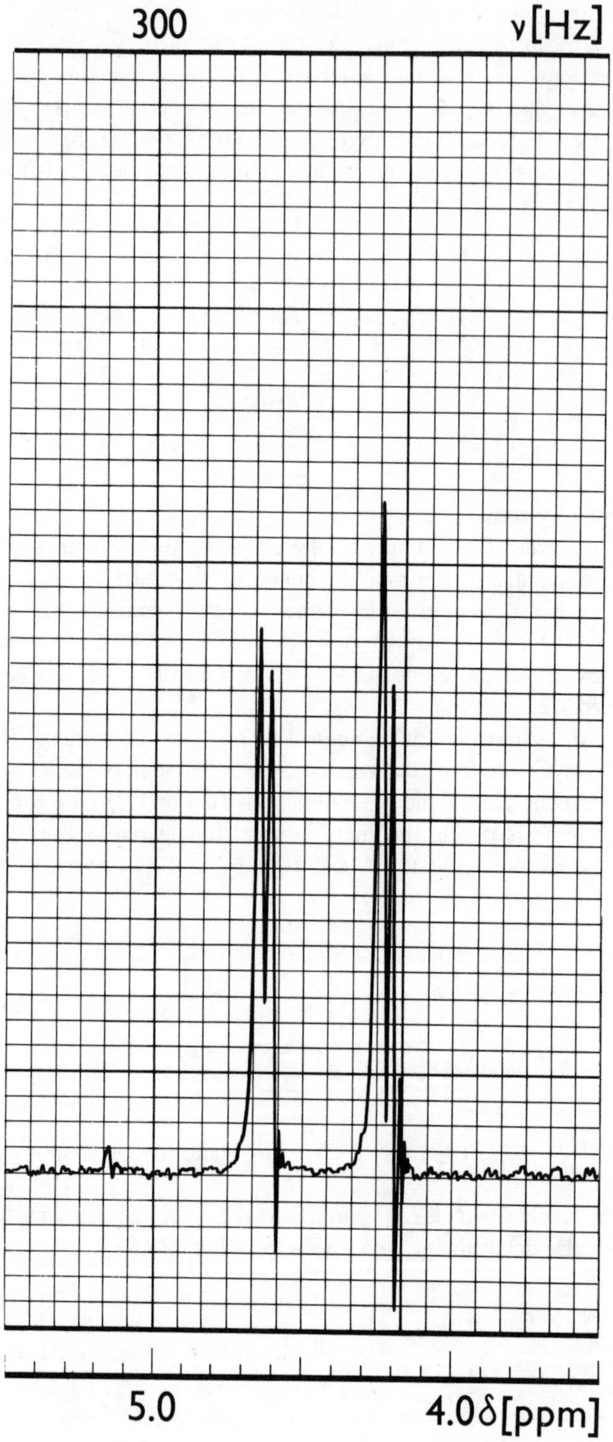

FIGURE 40. The *AX* signals of oxyrane protons in Compound **46.**

The matrix elements in Equation 163 are calculated according to Equation 157 and 134. Since $\varphi_2 = \alpha\beta$ and $\varphi_3 = \beta\alpha$, it follows that

$$\mathbf{H}_{22} = \langle\varphi_2|\mathcal{H}|\varphi_2\rangle = \langle\alpha\beta|\mathcal{H}|\alpha\beta\rangle = \frac{\gamma h B_0}{2}(-\sigma_A + \sigma_B) - \frac{J_{AB}}{4}h =$$

$$-\frac{\Delta\nu AB}{2} - \frac{J_{AB}}{4}h \tag{164}$$

The first term here was obtained in the same way as in the case of the A_2 system, i.e., by dividing the integral into two components, neglecting momentarily spin-spin coupling: $\langle\alpha\beta|\mathcal{H}|\alpha\beta\rangle = \langle\alpha|\mathcal{H}|\alpha\rangle + \langle\beta|\mathcal{H}|\beta\rangle$. Then the result, analogous to Equation 156, was written into this formula, but the different shielding factors of the two nuclei were distinguished by letters A and B in the index. If the difference in shielding is $\Delta\sigma AB = \sigma_A - \sigma_B$, the first term can be transformed in the following way: $h\gamma B_0\ \Delta\sigma AB/2$ or by denoting the local fields around nuclei A and B by B_A and B_B: $-\gamma\ h\,\Delta B_{AB}/2$. If now we express the chemical shift difference in frequency units, we obtain the result: $\Delta\nu AB/2$.

The second term is derived according to Equation 134 from the expression

$$\frac{J_{AB}}{4}[\langle\alpha\beta|(2\mathcal{P}_{AB} - 1)|\alpha\beta\rangle]$$

Upon performing the permutation, we get

$$\frac{J_{AB}}{4}[\langle\alpha\beta|(2\beta\alpha - \alpha\beta)\rangle] = \frac{J_{AB}}{4}[-1]$$

Similarly:

$$\mathbf{H}_{33} = \frac{\Delta\nu AB}{2} - \frac{J_{AB}}{4} \text{ and } \mathbf{H}_{23} = J_{AB}/2$$

Upon substituting these results into Equation 163, we get:

$$E^{AB}/h = -\frac{J_{AB}}{4} \pm \frac{1}{2}\sqrt{J_{AB}^2 + (\Delta\nu AB)^2} = -J_{AB}/4 \pm Q \tag{165}$$

where

$$Q = \frac{1}{2}\sqrt{J_{AB}^2 + (\Delta\nu AB)^2} \tag{166}$$

With appropriate conditions from Equation 165, we get identical results as for the A_2 and AX systems, respectively. Taking, namely, $\sigma_A = \sigma_B$ in Equation 165, $E_{2,3}^{AB} = E_{2,3}^{A2}$ will be $+J/4$ and $-3J/4$ (compare Table 7). Similarly, if $J \ll \Delta\nu$ hold, J^2 can be neglected in Equation 165 and $E_{2,3}^{AB}$ will be $-J/4 \pm \Delta\nu/2$, thus exactly the same as for AX case (compare Table 8).

The energy levels of the AB spin system are depicted in Figure 41 without and with spin-spin coupling. From the energy levels, the frequency of the transitions (spectral lines) can be calculated, too. Disregarding the term $h\gamma B_0(2 - \sigma_A + \sigma_B)/2$ which is found in each case for the transitions, we obtain the following expressions:

$$\begin{aligned}
&1 \quad \psi_3 \longrightarrow \psi_1 = Q + J/2\\
&2 \quad \psi_4 \longrightarrow \psi_2 = Q - J/2\\
&3 \quad \psi_2 \longrightarrow \psi_1 = -Q + J/2\\
&4 \quad \psi_4 \longrightarrow \psi_3 = -Q - J/2
\end{aligned} \tag{167}$$

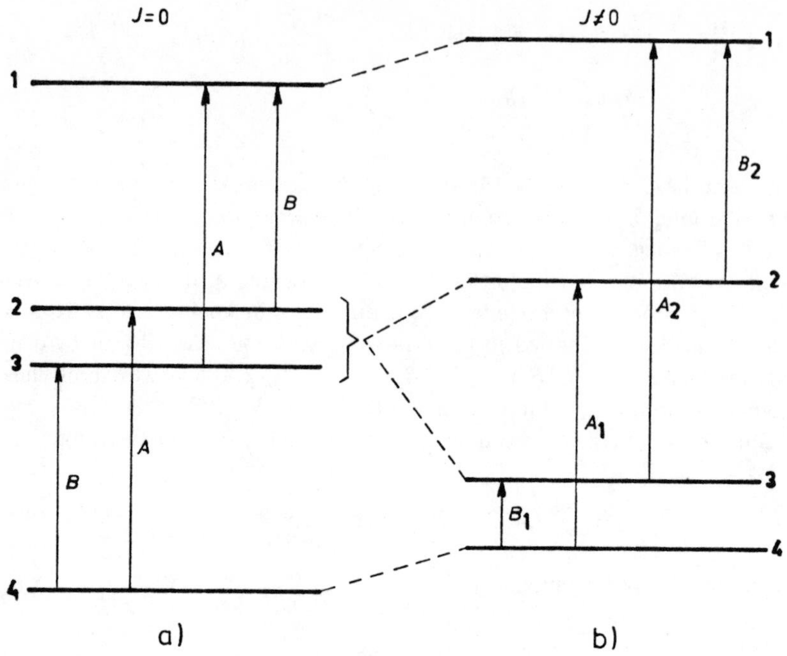

FIGURE 41. Energy levels and transitions of the AB spin system. (a) Without and (b) with spin-spin interaction.

It can be seen that although the value of the coupling constant J_{AB} can be directly extracted from the AB spectrum, as the difference of the transitions **1** and **2** or **3** and **4**, the chemical shift difference $\Delta\nu AB$ can only be obtained by calculation.

Upon expressing $\Delta\nu AB$ by means of Q from Equation 166:

$$\Delta\nu AB = \sqrt{4Q^2 - J^2} = \sqrt{(2Q+J)(2Q-J)} \qquad (168)$$

The quantities $(2Q + J)$ and $(2Q - J)$ are, however, identical to separation of the outer and inner spectral lines, i.e., **1** and **4** and **2** and **3**, respectively. Therefore,

$$\Delta\nu AB = \sqrt{(\mathbf{1}-\mathbf{4})(\mathbf{2}-\mathbf{3})} \qquad (169)$$

Transition probabilities are obtainable by the procedure described before when the basis functions of the AB system are substituted into Equation 150. From the four product functions, $\varphi_1 = \alpha\alpha$, $\varphi_2 = \alpha\beta$, $\varphi_3 = \beta\alpha$, and $\varphi_4 = \beta\beta$, the functions φ_2 and φ_3 should be, however, substituted by their linear combinations $\varphi^{2,3} = C_2\varphi_2 \pm C_3\varphi_3$.

The transition probabilities are then:

$$W^{3,1} \approx [\langle\psi_1|\mathcal{I}^+|\psi_3\rangle]^2 = [\langle\alpha\alpha|\mathcal{I}^+|(C_2\alpha\beta - C_3\beta\alpha)\rangle]^2 = (C_2 - C_3)^2$$

$$W^{4,2} \approx [\langle\psi_2|\mathcal{I}^+|\psi_4\rangle]^2 = [\langle(C_2\alpha\beta + C_3\beta\alpha)|\mathcal{I}^+|\beta\beta\rangle]^2 = (C_2 + C_3)^2$$

$$W^{2,1} \approx [\langle\psi_1|\mathcal{I}^+|\psi_2\rangle]^2 = [\langle\alpha\alpha|\mathcal{I}^+|(C_2\alpha\beta + C_3\beta\alpha)\rangle]^2 = (C_2 + C_3)^2$$

$$W^{4,3} \approx [\langle\psi_3|\mathcal{I}^+|\psi_4\rangle]^2 = [\langle(C_2\alpha\beta - C_3\beta\alpha)|\mathcal{I}^+|\beta\beta\rangle]^2 = (C_2 - C_3)^2$$

The above equations show that the intensity of the two inner and the two outer lines, respectively, are identical, and that for the calculation of the intensity ratio of the inner and outer lines, the coefficients C_2 and C_3 or at least their ratio should be known:

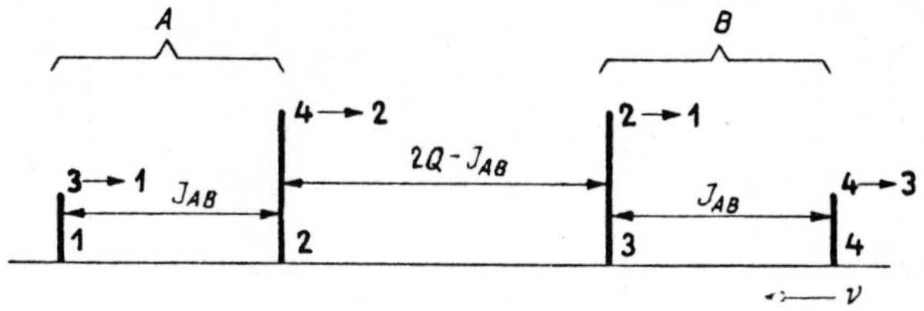

FIGURE 42. The schematic spectrum of the *AB* spin system.

$$A_{in}/A_{out} = \frac{(C_2 + C_3)^2}{(C_2 - C_3)^2} = \frac{(C_2/C_3 + 1)^2}{(C_2/C_3 - 1)^2} \tag{170}$$

and

$$\frac{C_2}{C_3} = \frac{E_2 - H_{22}}{H_{23}} = \frac{-J/4 + Q + J/4 + (\Delta\nu AB)/2}{J/2} =$$

$$\frac{2Q + \Delta\nu AB}{J} = \frac{2Q + \sqrt{4Q^2 - J^2}}{J} \tag{171}$$

Upon subsituting Equation 171 into Equation 170, simple calculations give:

$$A_{in}/A_{out} = (2Q + J) / (2Q - J) \tag{172}$$

From Equation 172 follows that the greater is the intensity difference between the inner and outer lines, the larger is the ratio $J/\Delta\nu$, i.e., the closer is the *AB* system to the limiting case A_2 and the farther it is removed from type *AX*.

The schematic spectrum of the *AB* spin system can now be constructed (see Figure 42) and the position and relative intensity of the lines can be calculated for different values of the ratio $J/\Delta\nu$ (Figure 43). Figure 43 demonstrates the change of the spectrum from *AX* to A_2, upon increasing of the ratio $J/\Delta\nu$.

If $J = 0$, the spectrum is two singlets of equal intensity associated with nuclei *A* and *X*, respectively. When *J* is small (type *AX*, first-order splitting), there are four lines of the same intensity, grouped to two doublets.

When *J* is further increased relative to $\Delta\nu$ the *AB*-type spectrum of four lines symmetric to the midpoint $(\nu A + \nu B)/2$ becomes apparent. The intensity of the outer lines is smaller (second-order splitting). On further increasing $J/\Delta\nu$, the inner lines are gradually approaching each other, while the outer ones diverge with respect to the midpoint. The intensity difference is increasing simultaneously up to the point when the outer lines disappear and the inner lines merge into a single one (A_2 zero-order spectrum). The value of J_{AB} can be directly read the *AB* spectrum, and by Equation 169, the chemical shift difference $\Delta\nu AB$ can also be calculated. From the spectrum being symmetrical, the sign of the two parameters cannot be determined. The question whether *A* or *B* has the greater chemical shift can only be answered by comparison with other compounds, i.e., empirically.

All types of organic compounds mentioned already in connection with *AX* systems can have *AB* spin systems, if the condition given by Equation 159 is not satisfied. The spectrum (see Figure 13) of 2-amino-thiazole(**2**) and its salt are examples for the *AB* system.

The spectrum of an *AB* spin system may simulate part *A* of an A_2X_3 system, i.e., a quartet of 1:3:3:1 intensity. When $A_{in}/A_{out} = 3$, then $Q = J$ and the *AB* spectrum is an equidistant quartet of 1:3:3:1 intensity, where the distance between the lines is just J_{AB} (compare Equation 172). It is possible when $\Delta\nu AB = J\sqrt{3}$. In the spectrum of Compound **46**, e.g., in $CDCl_3$,

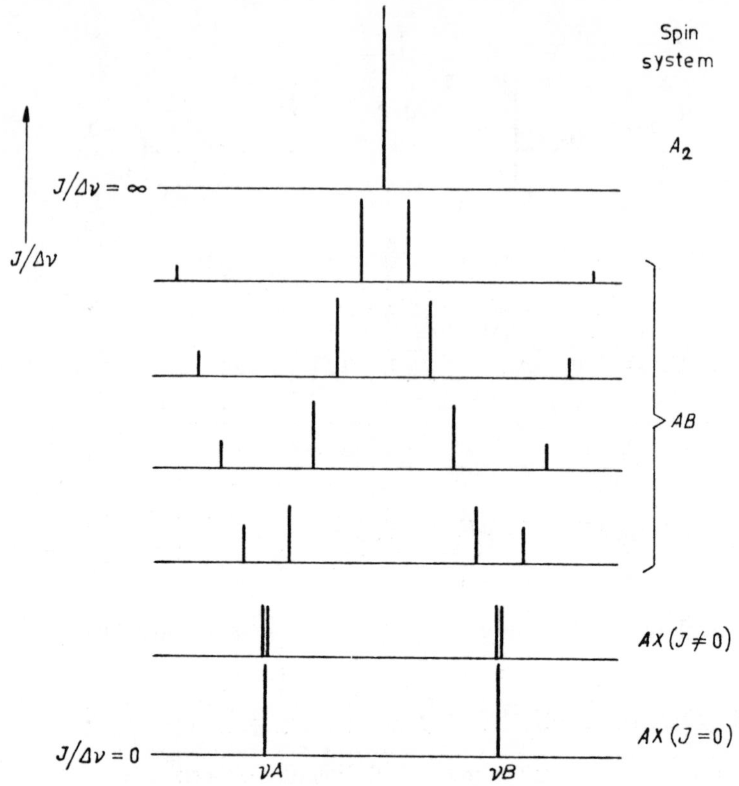

FIGURE 43. The dependence of the *AB* spectrum on $J/\Delta\nu$.

the signal of ring protons appear as a pseudo-quartet (see Figure 44a). In C_6D_6 solution, however, it becomes apparent that in fact it is an *AB* spectrum and not a quartet.*

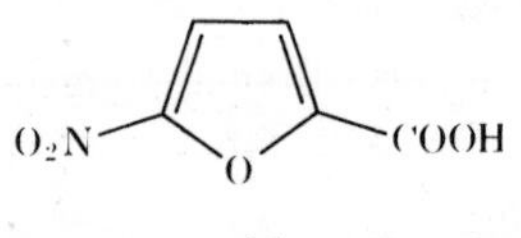

46

1.5.4. Further Spin Systems Characterized by a Single Coupling Constant

In the three-spin systems, AX_2 and AB_2, which can be described by two chemical shifts and by a single coupling constant, two chemically equivalent nuclei are coupled to a third one of different chemical environment. The behavior of AB_2 system is expected to be an intermediate between that of systems AX_2 and A_3.

1.5.4.1. The AX_2 Spin System

For a system of three spins, there are $2^3 = 8$ basis functions which are the following: ααα, ααβ, αβα βαα, αββ, βαβ, ββα, and βββ. Since the two nuclei *X* are equivalent — using the same arguments as for functions αβ and βα of A_2 system the wavefunctions ααβ and αβα, further βαβ and ββα we should be substituted by their respective linear combinations: $1/\sqrt{2}\alpha\ (\alpha\beta \pm \beta\alpha)$, and $1/\sqrt{2}\beta\ (\alpha\beta \pm \beta\alpha)$. The wavefunctions are contained in Table 9, together with their symmetry properties, the resultant spin quantum numbers, and the frequency values of the magnetic states belonging to them.

* Compare p. 40.

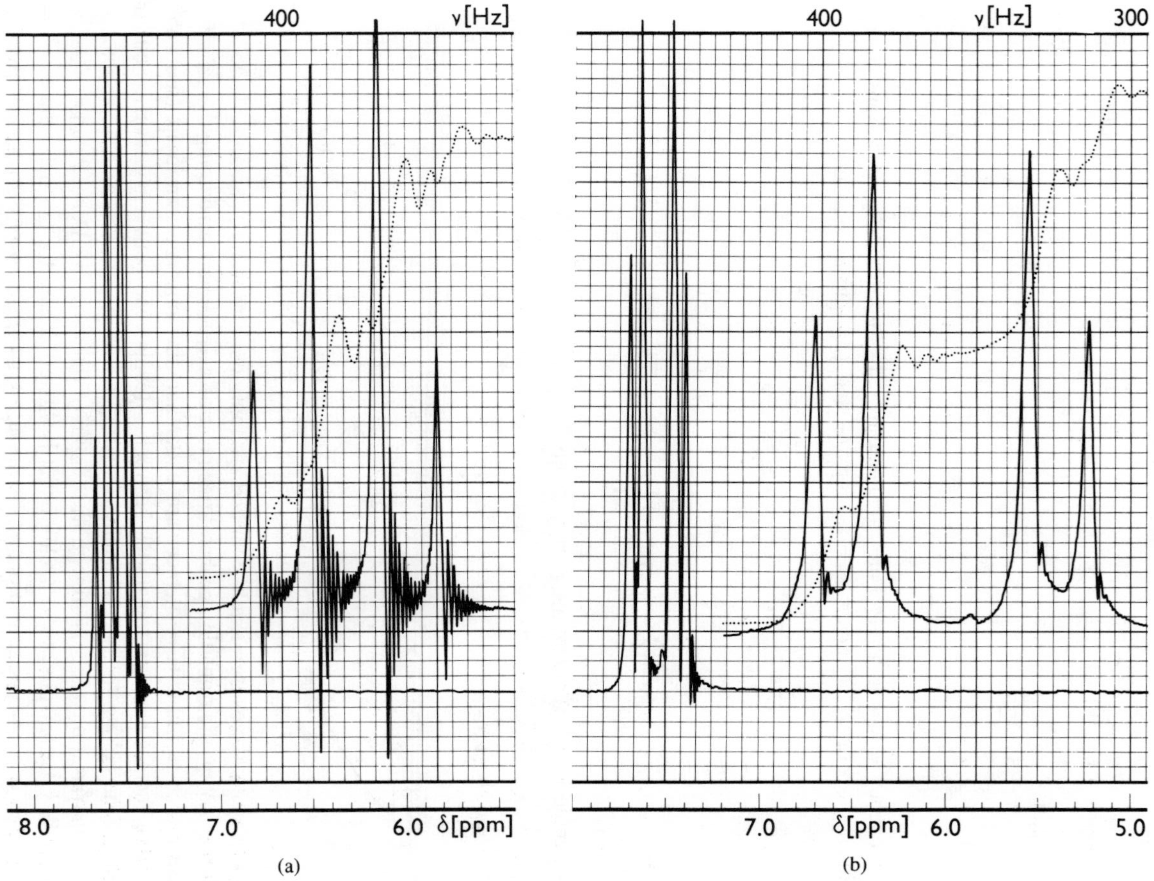

FIGURE 44. The signals for ring protons of Compound **46**. (a) In $CDCl_3$ and (b) in C_6D_6 solution.

Table 9
BASIS FUNCTIONS, RESULTANT SPIN QUANTUM NUMBERS, SYMMETRY, SYMBOLS, AND FREQUENCIES OF THE QUANTUM STATES OF THE AX_2 SPIN SYSTEM

Basis function				Quantum state	
Symbol	Form	I_{zT}	Symmetry	Symbol	Frequency
ψ_1	$\alpha\alpha\alpha$	3/2	s	$s_{3/2}$	$\nu A/2 + \nu X + J_{AX}/2 + J_{XX}/4$
ψ_2	$\alpha(\alpha\beta + \beta\alpha)/\sqrt{2}$	1/2	s	$1s_{1/2}$	$\nu A/2 + J_{XX}/4$
ψ_3	$\beta\alpha\alpha$	1/2	s	$2s_{1/2}$	$-\nu A/2 + \nu X - J_{AX}/2 + J_{XX}/4$
ψ_4	$\alpha(\alpha\beta - \beta\alpha)/\sqrt{2}$	1/2	a	$a_{1/2}$	$\nu A/2 - 3J_{XX}/4$
ψ_5	$\beta(\alpha\beta - \beta\alpha)/\sqrt{2}$	−1/2	a	$a_{-1/2}$	$-\nu A/2 - 3J_{XX}/4$
ψ_6	$\alpha\beta\beta$	−1/2	s	$1s_{-1/2}$	$\nu A/2 - \nu X - J_{AX}/2 + J_{XX}/4$
ψ_7	$\beta(\alpha\beta + \beta\alpha)/\sqrt{2}$	−1/2	s	$2s_{-1/2}$	$-\nu A/2 + J_{XX}/4$
ψ_8	$\beta\beta\beta$	−3/2	s	$s_{-3/2}$	$-\nu A/2 - \nu X + J_{AX}/2 + J_{XX}/4$

The frequency values in Table 9 are calculated by substituting the eigenfunctions ψ_n, and the frequency operator, $\mathcal{H}$, given by Equations 89 and 134 into the equation:

$$E/h = \langle\psi_n|\mathcal{H}|\psi_n\rangle \tag{173}$$

This gives:

$$E_n/h = \langle\psi_n|\sum_i \nu_i \mathcal{I}_{zi} + \frac{1}{4}\sum_{i<j}\sum J_{ij}(2\mathcal{P}_{ij} - 1)|\psi_n\rangle \tag{174}$$

As an example let us calculate the energy value E_7:

$$E_7 h = \frac{1}{2}\,[\langle(\beta\alpha\beta + \beta\beta\alpha)|\mathcal{H}|(\beta\alpha\beta + \beta\beta\alpha)\rangle] =$$

$$\frac{-\nu A}{2}(\langle\beta|\beta\rangle) + \frac{\nu X}{2}(\langle\alpha|\alpha\rangle) - \frac{\nu X}{2}(\langle\beta|\beta\rangle) + \frac{1}{8}\,\{J_{AX}[\langle(\beta\alpha\beta + \beta\beta\alpha)|4\alpha\beta\beta\rangle] +$$

$$J_{XX}[\langle(\beta\alpha\beta + \beta\beta\alpha)|(\beta\alpha\beta + \beta\beta\alpha)\rangle]\} = -\frac{\nu A}{2} + \frac{1}{8}\,[2J_{XX}] = -\frac{\nu A}{2} + \frac{J_{XX}}{4}$$

Figure 45 depicts the energy levels with and without coupling.

As the next step, the transition probabilities must be calculated. Considering the selection rules, $\Delta m = \pm 1$ and $s \leftarrow\!|\!\rightarrow a$, there are altogether nine possible transitions (see Table 10). Among them, eight are transitions between s levels, one between a levels. Four of these transitions belong to each of nuclei A and X, while the ninth is a s.c. *combination transition*. In transitions associated with A, the spin of only nucleus A does change, whereas in those of X, it is the spin of the nuclei X that is altered. Thus, for example, the $\psi_8 \rightarrow \psi_6$ ($\beta\beta\beta \rightarrow \alpha\beta\beta$) is an A-type transition, while the $\psi_8 \rightarrow \psi_7$ [$\beta\beta\beta \rightarrow 1/\sqrt{2}\,(\beta\alpha\beta + \beta\beta\alpha)$ is an X-type one. In the combination transition, all of the nuclei change spins simultaneously, e.g., $\psi_6 \rightarrow \psi_3$ ($\alpha\beta\beta \rightarrow \beta\alpha\alpha$). Transition probabilities (see Table 10) can be calculated as before* from Equation 150.

Let us calculate as examples the relative probabilities for a transition X ($\psi_8 \rightarrow \psi_7$) and for the combination transition ($\psi_6 \rightarrow \psi_3$):

* Compare p. 91, 93, and 98.

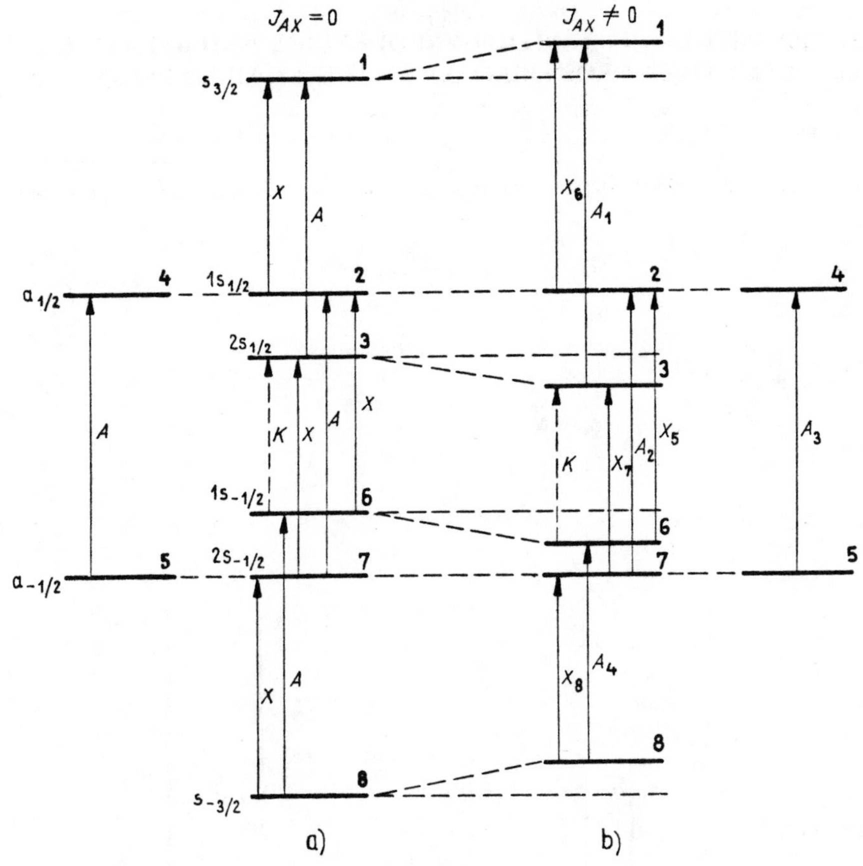

FIGURE 45. Energy levels and transitions of spin system AX_2. (a) Without and (b) with spin-spin interaction. Broken arrows denote combination (K) transitions.

$$W^{8,7} \sim \frac{1}{2}\,[\langle(\beta\alpha\beta + \beta\beta\alpha) \mid \mathcal{J}^{+} \mid \beta\beta\beta\rangle]^2 = \frac{1}{2}\,[\langle(\beta\alpha\beta + \beta\beta\alpha) \mid (\alpha\beta\beta + \beta\alpha\beta + \beta\beta\alpha)\rangle]^2 =$$

$$\frac{1}{2}\,[0 + 0 + 1 + 0 + 0 + 1]^2 = 2;$$

$$W^{6,3} \sim \frac{1}{2}\,[\langle\beta\alpha\alpha \mid \mathcal{J}^{+} \mid \alpha\beta\beta\rangle]^2 = [\langle\beta\alpha\alpha \mid (\alpha\alpha\beta + \alpha\beta\alpha)\rangle]^2 = [0 + 0]^2 = 0$$

From Table 10 the following conclusions can be drawn:

1. The probability of the allowed combination transition is zero, therefore only eight transitions lead to absorption of energy. The frequency of the transitions $\psi_7 \rightarrow \psi_2$ and $\psi_5 \rightarrow \psi_4$, furthermore of $\psi_6 \rightarrow \psi_2$ and $\psi_{c2} \rightarrow \psi_1$, as well as those of $\psi_7 \rightarrow \psi_3$ and $\psi_8 \rightarrow \psi_{c7}$, are pairwise equal. Thus, the spectrum AX_2 should contain five lines whose relative intensity is given by the respective transition probabilities (see Figure 46).
2. The coupling constant J_{XX} does not influence the spectrum, therefore its value cannot be determined from the latter.
3. For reasons of symmetry, the two coupling constants, J_{AX} are identical, thus the AX_2 system can be characterized by a single coupling constant whose magnitude can be directly read from the spectrum, whereas its sign cannot be obtained.
4. The chemical shift difference and its sign can also be determined from the spectrum,

Table 10
TRANSITIONS, TRANSITION FREQUENCIES, AND RELATIVE TRANSITION PROBABILITIES OF THE AX_2 SPIN SYSTEM

Transition				Transition	
Number	Type	Combining basis functions	Symbols	Frequency	Relative probability
1	*A*	$\psi_3 \rightarrow \psi_1$	$2s_{1/2} \rightarrow s_{3/2}$	$\nu A + J_{AX}$	1
2	*A*	$\psi_7 \rightarrow \psi_2$	$2s_{-1/2} \rightarrow 1s_{1/2}$	νA	1
3	*A*	$\psi_5 \rightarrow \psi_4$	$a_{-1/2} \rightarrow a_{1/2}$	νA	1
4	*A*	$\psi_8 \rightarrow \psi_6$	$s_{-3/2} \rightarrow 1s_{-1/2}$	$\nu A - J_{AX}$	1
5	*X*	$\psi_6 \rightarrow \psi_2$	$1s_{-1/2} \rightarrow 1s_{1/2}$	$\nu X + J_{AX}/2$	2
6	*X*	$\psi_2 \rightarrow \psi_1$	$1s_{1/2} \rightarrow s_{3/2}$	$\nu X + J_{AX}/2$	2
7	*X*	$\psi_7 \rightarrow \psi_3$	$2s_{-1/2} \rightarrow 2s_{1/2}$	$\nu X - J_{AX}/2$	2
8	*X*	$\psi_8 \rightarrow \psi_7$	$s_{-3/2} \rightarrow 2s_{-1/2}$	$\nu X - J_{AX}/2$	2
9	Comb.	$\psi_6 \rightarrow \psi_3$	$1s_{-1/2} \rightarrow 2s_{1/2}$	$2\nu X - \nu A$	0

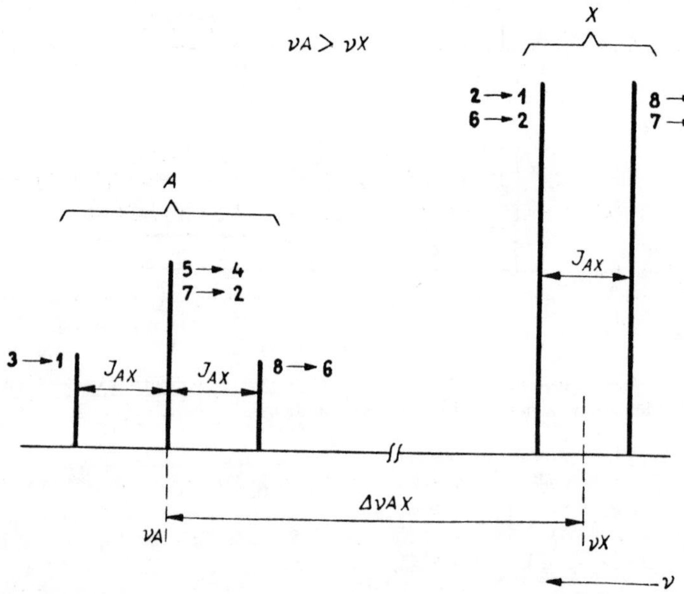

FIGURE 46. The schematic spectrum of the AX_2 spin system.

as the total intensity of the *X* part is twice that of the *A* part. From Figure 46 it can be seen that the spectrum derived before qualitatively* is identical to the calculated, as well to the measured (see Figure 28) ones.

1.5.4.2. The AB_2 Spin System

The combination of the different spin states in the AB_2 system leads qualitatively to the same changes as in the case of the system AX_2. The pair of states, $1s_{1/2} - 2s_{1/2}$ and $1s_{-1/2} - 2s_{-1/2}$, are combined. Because of the energy differences arising from the combination of the states, all *A*- and *B*-type transitions are of different energy. As compared to the AX_2 for the AB_2 system, the number of lines is increased, having nine lines at the most. The combination transition $\psi_6 \rightarrow \psi_3$ namely, though still of relatively small probability, becomes allowed, and the lines of the transition pairs, $\psi_6 \rightarrow \psi_2 - \psi_2 \rightarrow \psi_1$, $\psi_7 \rightarrow \psi_3 - \psi_8 \rightarrow \psi_7$,

* Compare p. 47.

Table 11
BASIS FUNCTIONS AND FREQUENCIES OF THE QUANTUM STATES OF THE AB_2 SPIN SYSTEM

Basis function		Frequency
Symbol	Form	
ψ_1	$\alpha\alpha\alpha$	$\nu A/2 + \nu B + J/2$
ψ_2	$C_1[\alpha(\alpha\beta + \beta\alpha)]/\sqrt{2} + C_2\beta\alpha\alpha$	$\nu B/2 - J/4 + Q_1$
ψ_3	$C_1\beta\alpha\alpha - C_2[\alpha(\alpha\beta + \beta\alpha)]/\sqrt{2}$	$\nu B/2 - J/4 - Q_1$
ψ_4	$\alpha(\alpha\beta - \beta\alpha)/\sqrt{2}$	$\nu A/2$
ψ_5	$\beta(\alpha\beta - \beta\alpha)/\sqrt{2}$	$-\nu A/2$
ψ_6	$C_3\alpha\beta\beta + C_4[\beta(\alpha\beta + \beta\alpha)]/\sqrt{2}$	$-\nu B/2 - J/4 + Q_2$
ψ_7	$C_3[\beta(\alpha\beta + \beta\alpha)]/\sqrt{2} - C_4\alpha\beta\beta$	$-\nu B/2 - J/4 - Q_2$
ψ_8	$\beta\beta\beta$	$-\nu A/2 - \nu B + J/2$

Note: $J \equiv J_{AB}$, $\Delta\nu \equiv \Delta\nu AB$;

$$Q_{1,2} = \sqrt{\Delta\nu^2 \pm \Delta\nu J + (3J/2)^2/2};$$
$$C_1 = J/(2\sqrt{Q_1^2 - \Delta\nu Q_1 - Q_1 J/4}); \quad C_2 = \sqrt{(Q_1 - \Delta\nu - J/4)/2Q_1};$$
$$C_3 = J/(2\sqrt{Q_2^2 - \Delta\nu Q_2 + Q_2 J/4}); \quad C_4 = \sqrt{(Q_2 - \Delta\nu + J/4)/2Q_2}.$$

Table 12
TRANSITIONS, TRANSITION FREQUENCIES, AND RELATIVE TRANSITION PROBABILITIES FOR THE AB_2 SPIN SYSTEM

Transition		Combining basis functions	Transition	
Number	Type		Frequency	Relative probability
1	A	$\psi_3 \rightarrow \psi_1$	$\nu A/2 + \nu B/2 + 3J/4 + Q_1$	$(C_2\sqrt{2} - C_1)^2$
2	A	$\psi_7 \rightarrow \psi_2$	$\nu B + Q_1 + Q_2$	$[\sqrt{2}(C_2C_3 - C_1C_4) + C_1C_3]^2$
3	A	$\psi_5 \rightarrow \psi_4$	νA	1
4	A	$\psi_8 \rightarrow \psi_6$	$\nu A/2 + \nu B/2 - 3J/4 + Q_2$	$(C_4\sqrt{2} + C_3)^2$
5	B	$\psi_6 \rightarrow \psi_2$	$\nu B + Q_1 - Q_2$	$[\sqrt{2}(C_1C_3 + C_2C_4) + C_1C_4]^2$
6	B	$\psi_2 \rightarrow \psi_1$	$\nu A/2 + \nu B/2 + 3J/4 - Q_1$	$(C_1\sqrt{2} + C_2)^2$
7	B	$\psi_7 \rightarrow \psi_3$	$\nu B - Q_1 + Q_2$	$[\sqrt{2}(C_1C_3 + C_2C_4) - C_2C_3]^2$
8	B	$\psi_8 \rightarrow \psi_7$	$\nu A/2 + \nu B/2 - 3J/4 - Q_2$	$(C_3\sqrt{2} - C_4)^2$
9	K	$\psi_6 \rightarrow \psi_3$	$\nu B - Q_1 - Q_2$	$[\sqrt{2}(C_2C_3 - C_1C_4) + C_2C_4]^2$

and $\psi_7 \rightarrow \psi_2 - \psi_5 \rightarrow \psi_4$, do not coincide anymore. The transition $\psi_5 \rightarrow \psi_4$ gives a line at a constant value (νA), and thus it is the reference point of the spectrum which remains fixed upon the alteration of ratio $J/\Delta\nu$.

The eigenfunctions, the frequency of levels as well as the frequency and relative intensity of the transitions, are listed in Tables 11 and 12. The energy levels of the AB_2 system are shown in Figure 47 in comparison of those of the AX_2 system. Using Table 12, the schematic spectra of the AB_2 system can be constructed as a function of the ratio $J/\Delta\nu$. Fixing first the value of J_{AB} and letting $\Delta\nu AB$ change, or vice versa, the spectra shown on Figures 48a and 48b are obtained, provided $\nu A > \nu B$. When $\nu A < \nu B$, the mirror images of the above spectra are obtained.

Figure 49 shows the signals of the aromatic protons of a few 2,6-symmetrically substituted benzene derivatives, for illustrating the transition $AX_2 \rightarrow AB_2 \rightarrow A_3$. Note that the change of the first-order interaction towards the higher-order couplings is manifested first in the distortion of the symmetry of multiplets. The signal groups come closer to each other, and inner lines become more intense, at the expense of the outer ones, exactly as it has been

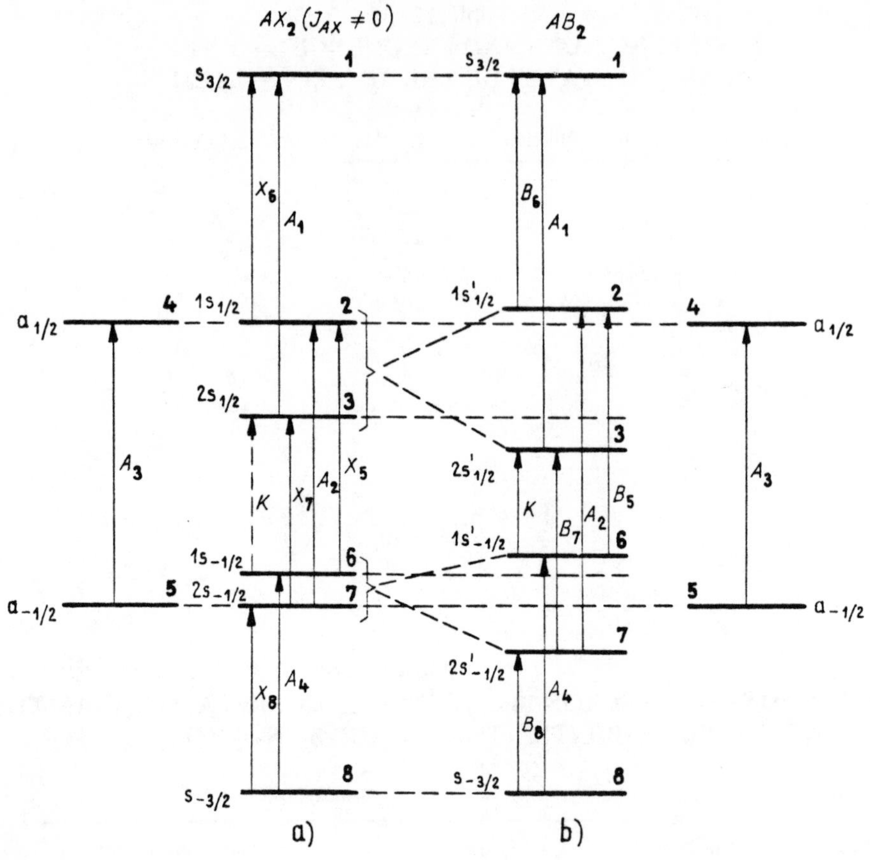

FIGURE 47. (b) Energy levels and transitions of the AB_2 spin system, as compared (a) to those of the AX_2 system.

already demonstrated for the transition $AX \rightarrow AB \rightarrow A_2$ (compare Figure 43). If through the tips of peaks straight lines are drawn (see Figure 50), these intersect to form an upside down $\wedge$; this is the s.c. "roof pattern". This pattern is characteristic of spin systems which are of higher order, but still very close to the first-order case. In the calculation of such spectra, the first-order approximation is allowed. The roof structure is very useful in the recognition of related multiplet-pairs. This may be a problem when there are more than two signal-pairs in the spectra, and it is doubtful which two belong together (compare Problem **18**).

The determination of the spectral parameters, i.e., J_{AB}, νA, and νB, for the AB_2 spin system is as follows. The chemical shift νA, as has already been mentioned, is identical with that of line **3**, corresponding to the antisymmetric transition $\psi_5 \rightarrow \psi_4$. νB is the arithmetic mean of lines **5** and **7** of transitions $\psi_6 \rightarrow \psi_2$ and $\psi_7 \rightarrow \psi_3$. When all lines of the transitions A and B are found in the spectrum, the value of $3|J_{AB}|$ can be directly obtained if the sum of the line frequencies **4** and **8** (transitions $\psi_8 \rightarrow \psi_6$ and $\psi_8 \rightarrow \psi_7$) are deducted from the frequency sum of lines **1** and **6** (transitions $\psi_3 \rightarrow \psi_1$ and $\psi_2 \rightarrow \psi_1$). When these eight lines are not all resolved, then the distance of the lines **5** and **7** from the line **3** yields the values $\Delta\nu - Q_1 + Q_2$ and $\Delta\nu + Q_1 - Q_2$ and J_{AB} can be derived from these quantities. Another possibility is the use of published tables.[301]* These enable to determine the ratio $J/\Delta\nu$ from the distance and relative intensity of the lines, thereafter J_{AB} can be calculated if $\Delta\nu$ is

* Today there are computer programs available that make possible determination of exact spectral parameters for different spin systems from observed line frequencies by iteration.

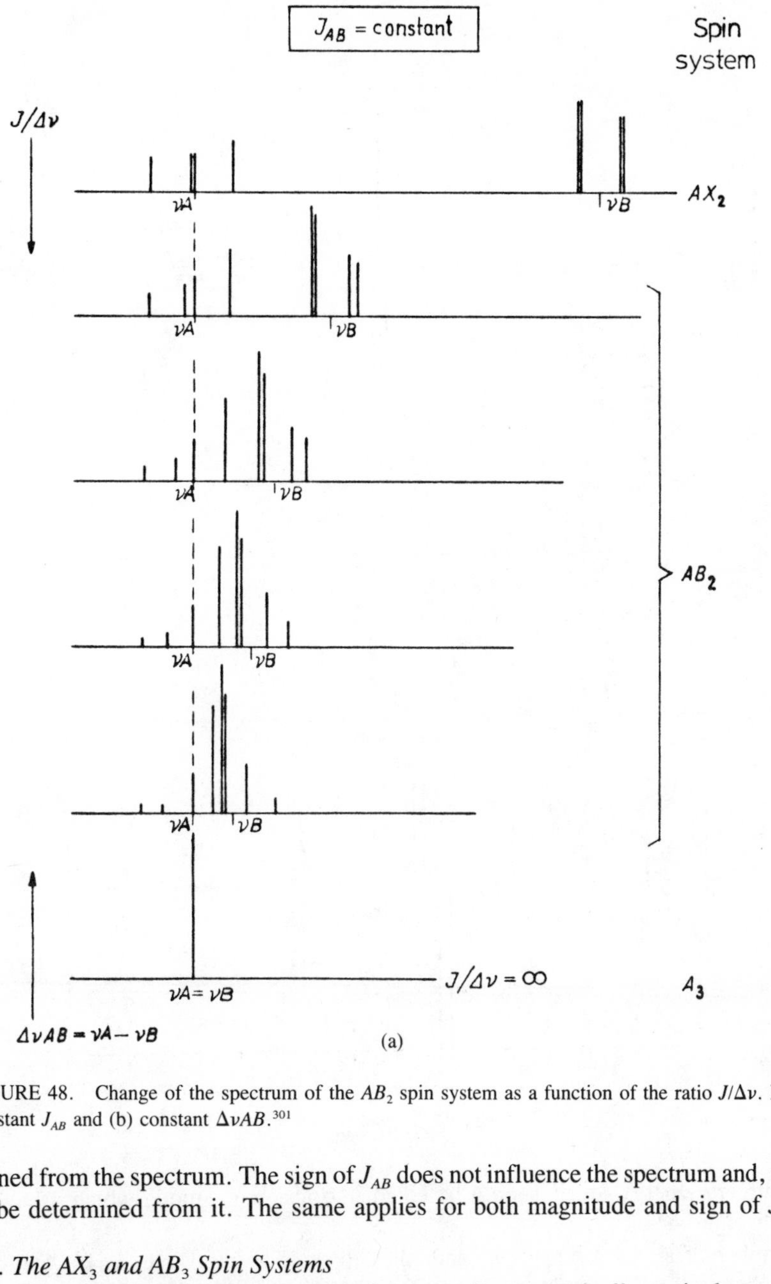

(a)

FIGURE 48. Change of the spectrum of the AB_2 spin system as a function of the ratio $J/\Delta\nu$. For (a) constant J_{AB} and (b) constant $\Delta\nu AB$.[301]

determined from the spectrum. The sign of J_{AB} does not influence the spectrum and, therefore, cannot be determined from it. The same applies for both magnitude and sign of J_{BB}.

1.5.4.3. The AX_3 and AB_3 Spin Systems

Systems comprising three chemically and necessarily magnetically equivalent nuclei coupled with one chemically nonequivalent nucleus of the same kind give rise to spectra of type AB_3, if $J/\Delta\nu > 0.1$. Otherwise, AX_3-type first-order splitting (compare Figure 27) is observed and the spectrum consists of an 1:1 doublet (X part) and an 1:3:3:1 quartet (A_3 part). J_{AX} is given by the distance of adjacent lines of either the doublet or the quartet. νA and νX are the midpoints of the quartet and the doublet, respectively.

An AB_3 spectrum is characterized by the values of νA, νB, and J_{AB}, since J_{BB} does not influence the spectrum. Such spectra have been observed, e.g., for methyl-mercaptan,methyl-acetylene, and many molecules containing the group >CH–CH_3. In all these compounds the methyl protons are equivalent because of fast free rotation about the carbon-carbon bond.

Four-spin systems have 16 energy levels and basis functions. As 4 pairs are degenerate,

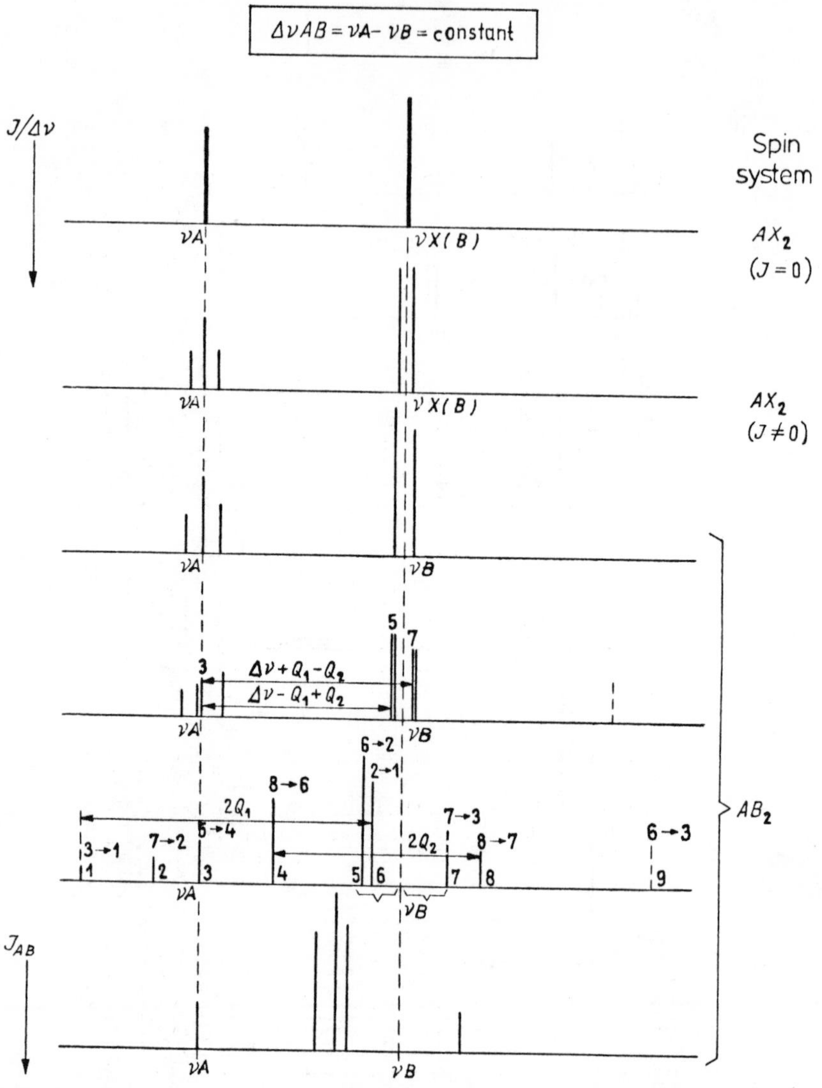

FIGURE 48b.

only 12 discrete energy levels should be taken into account, among which 4 have double statistical weight.

The number of allowed transitions and, therefore, of spectral lines is 16. Out of these, six and eight are of type *A* and *B*, respectively, while two are combinations. (For frequencies and relative intensities, see Table 13.) The structure of the spectrum is controlled again by the ratio, $J/\Delta\nu$ (compare Figure 51). By means of calculations similar to systems *AB* and AB_2 formulas can be derived for the position and intensity of the spectral lines. Their use is, however, rarely possible because some lines are usually very weak or are widely separated making their identification difficult. Thus, it is more practical to estimate the ratio $J/\Delta\nu$ using Figure 51 or to take recourse to tables available in the literature[301] for interpolation in order to determine the magnitude of J and $\Delta\nu$.*

If $J/\Delta\nu$ is not too large, the value of J_{AB} can be directly read from the spectrum (compare Figure 51). The value of J_{AB} is given by the distance of the lines **3** and **5** or **8** and **13**. νB

* See footnote p. 106.

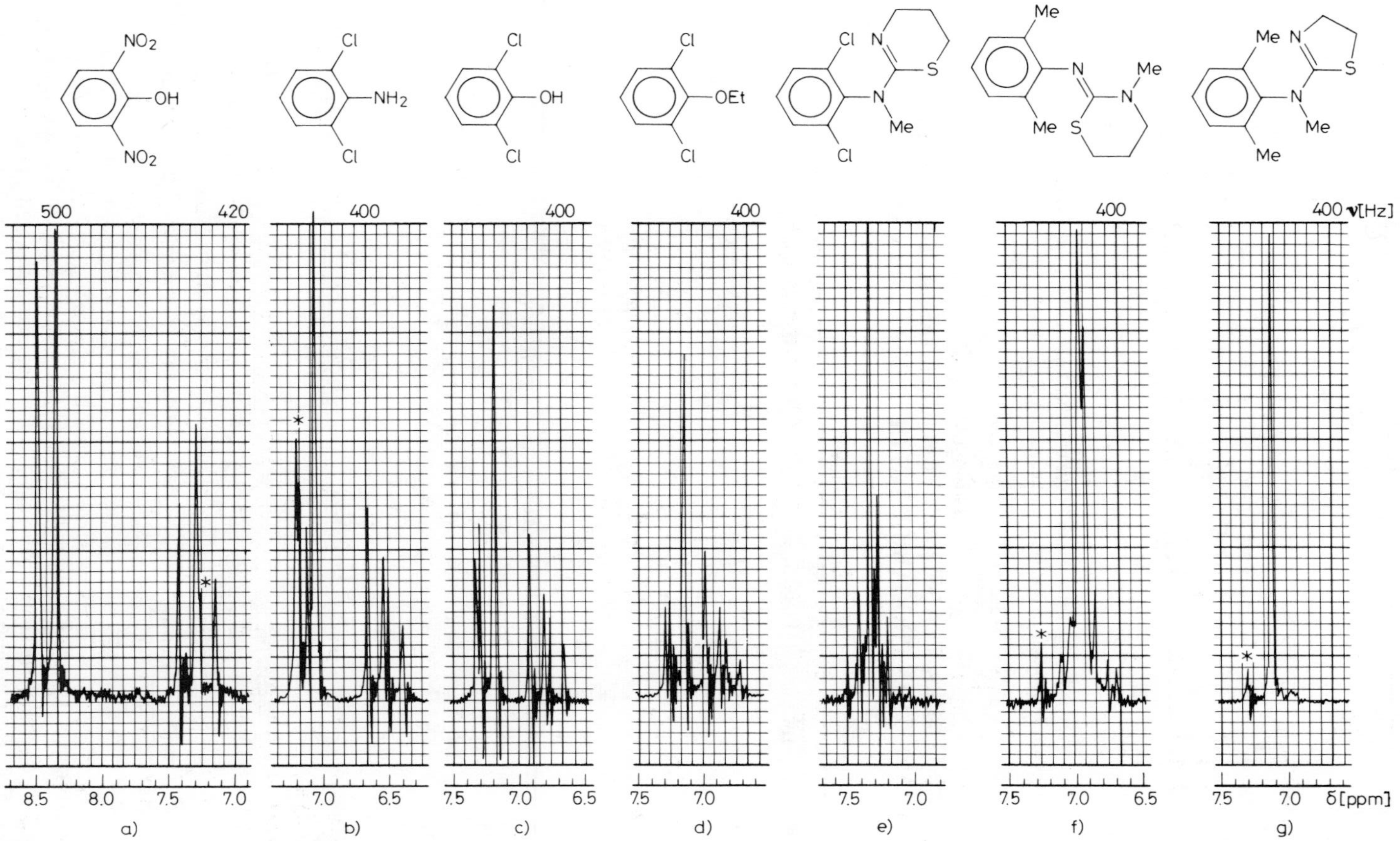

FIGURE 49. The multiplets of the ring protons of symmetrical 1,2,3-trisubstituted benzene derivatives, an illustration of the interconversion of the spin system AX_2–AB_2–A_3.

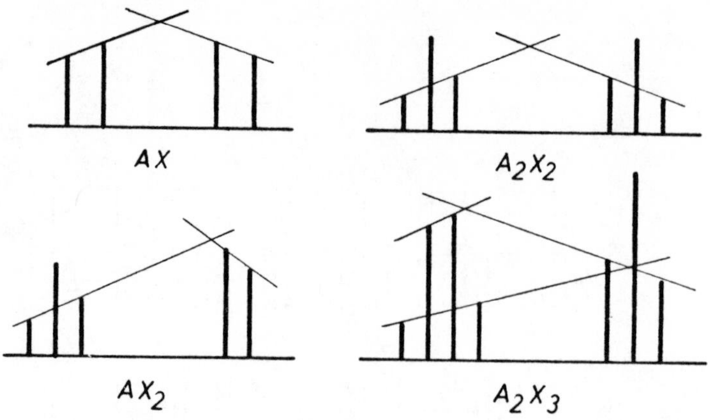

FIGURE 50. Roof patterns characteristic of higher-order splitting approximating first-order interaction for the systems AX, AX_2, A_2X_2, and A_2X_3.

Table 13
TRANSITIONS, TRANSITION FREQUENCIES, AND RELATIVE TRANSITION PROBABILITIES OF THE AB_3 SPIN SYSTEM

Transition			
Number	Type	Frequency	Relative intensity
1	A	$J + Q_1$	$(C_2\sqrt{3} - C_1)^2$
2	A	$-\Delta\nu/2 + Q_1 + Q_3$	$(C_1C_5 + C_2C_5\sqrt{3} - 2C_1C_6)^2$
3	A	$J/2 + Q_4$	$2(C_7 - C_8)^2$
4	A	$-\Delta\nu/2 + Q_2 + Q_3$	$(2C_3C_6 + C_3C_5 + C_4C_5\sqrt{3})^2$
5	A	$-J/2 + Q_4$	$2(C_7 + C_8)^2$
6	A	$-J + Q_2$	$(C_3 + C_4\sqrt{3})^2$
7	B	$-\Delta\nu/2 - Q_2 + Q_3$	$(C_3C_5\sqrt{3} + C_4C_5 + 2C_4C_6)^2$
8	B	$J/2 - Q_4$	$2(C_7 + C_8)^2$
9	B	$-\Delta\nu/2 + Q_1 - Q_3$	$(2C_1C_5 + C_1C_6 + C_2C_6\sqrt{3})^2$
10	B	$J - Q_1$	$(C_1\sqrt{3} + C_2)^2$
11	B	$-\Delta\nu/2 - Q_1 + Q_3$	$(C_1C_5\sqrt{3} - C_2C_5 + 2C_2C_6)^2$
12	B	$-\Delta\nu/2 + Q_2 - Q_3$	$(2C_3C_5 - C_3C_6 + C_4C_6\sqrt{3})^2$
13	B	$-J/2 - Q_4$	$2(C_7 - C_8)^2$
14	B	$-J - Q_2$	$(C_3\sqrt{3} - C_4)^2$
15	K	$-\Delta\nu/2 - Q_2 - Q_3$	$(2C_4C_5 - C_3C_6\sqrt{3} - C_4C_6)^2$
16	K	$-\Delta\nu/2 - Q_1 - Q_3$	$(C_1C_6\sqrt{3} - 2C_2C_5 - C_2C_6)^2$

Note: $J \equiv J_{AB}$, $\Delta\nu \equiv \Delta\nu AB$;
$Q_{1,2} = \sqrt{\Delta\nu^2 \pm 2J\Delta\nu + 4J^2/2}$; $Q_3 = \sqrt{\Delta\nu^2 + 4J^2/2}$; $Q_4 = \sqrt{\Delta\nu^2 + J^2/2}$;
$C_1 = \sqrt{(2Q_1 - \Delta\nu - J)/2Q_1}$; $C_2 = \sqrt{3J^2/[2Q_1(2Q_1 - \Delta\nu - J)]}$;
$C_3 = \sqrt{(2Q_2 - \Delta\nu + J)/2Q_2}$; $C_4 = \sqrt{3J^2/[2Q_2(2Q_2 - \Delta\nu + J)]}$;
$C_5 = \sqrt{(2Q_3 - \Delta\nu)/2Q_3}$; $C_6 = \sqrt{2J^2/[Q_3(2Q_3 - \Delta\nu)]}$;
$C_7 = \sqrt{(2Q_4 - \Delta\nu)/2Q_4}$; $C_8 = \sqrt{J^2/[2Q_4(2Q_4 - \Delta\nu)]}$.

can also be estimated, since its value falls close to the center of the two groups of lines constituting part B of the spectrum, whereas νA can only be calculated. The sign of J_{AB} does not have any effect upon the spectrum, analogously to systems AB and AB_2, therefore it is not available from spectrum analysis. When $\nu A < \nu B$ the mirror images of the spectra in Figure 51 are obtained.

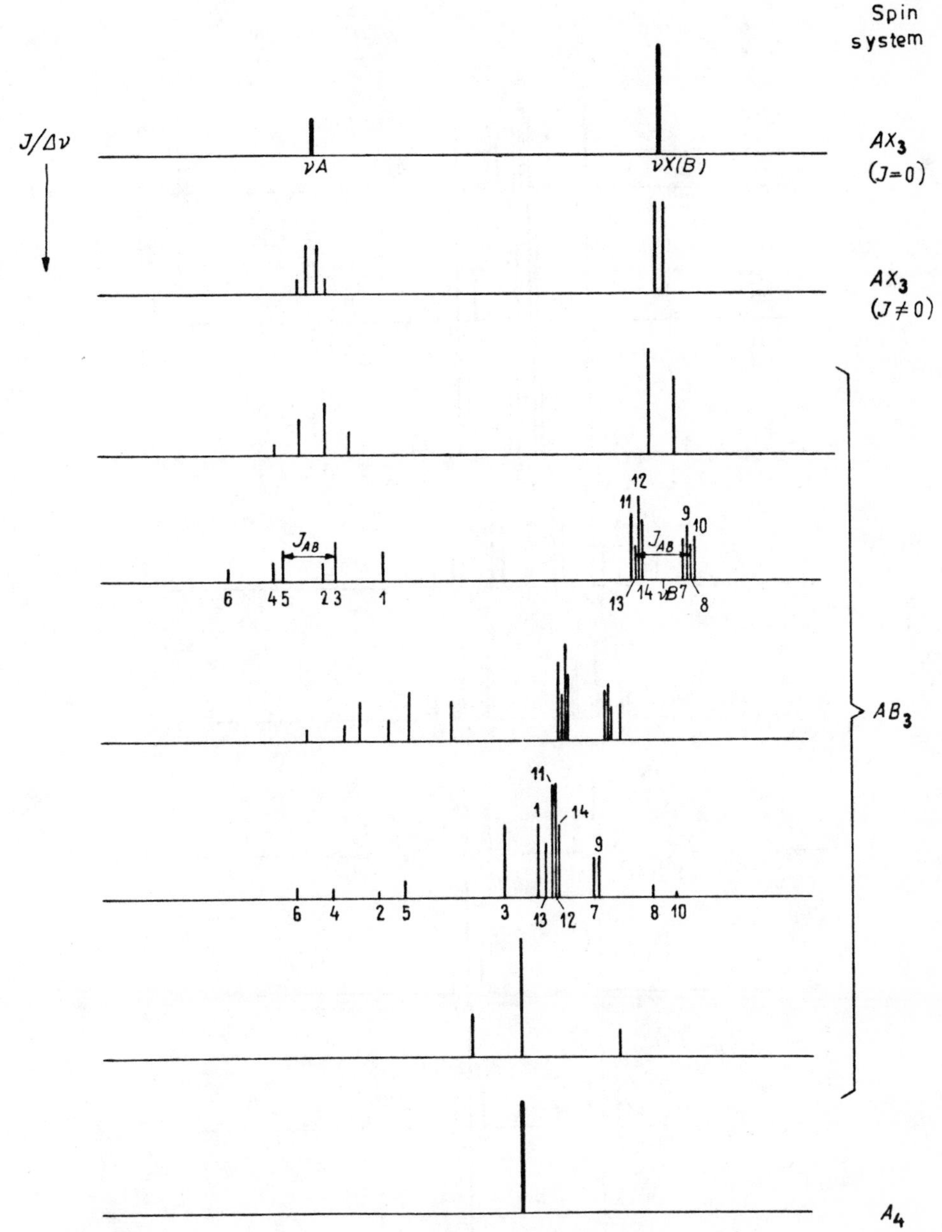

FIGURE 51. Change of the spectra of AB_3 spin systems as a function of the ratio $J/\Delta\nu$.[301]

1.5.4.4. Spin Systems A_2X_2 and A_2B_2

Molecules in which two pairs of nuclei are coupled and the nuclei composing the two pairs are chemically and magnetically equivalent give rise to either A_2X_2 or A_2B_2 spectra, depending on the ratio $J/\Delta\nu$. As types discussed so far, this system is also determined by two chemical shifts and a single coupling constant.

Difluoromethane and 1,1-difluoro-allene and many asymmetrically substituted open chain-type compounds of structure R–CH_2–CH_2–R′ give A_2X_2 spectra. However, many 1,2-di-substituted ethanes have A_2B_2 spectra, and when the condition of magnetic equivalence is not fulfilled, instead of A_2X_2- or A_2B_2-type spectra, $AA'XX'$ or $AA'BB'$ multiplets appear (see below).

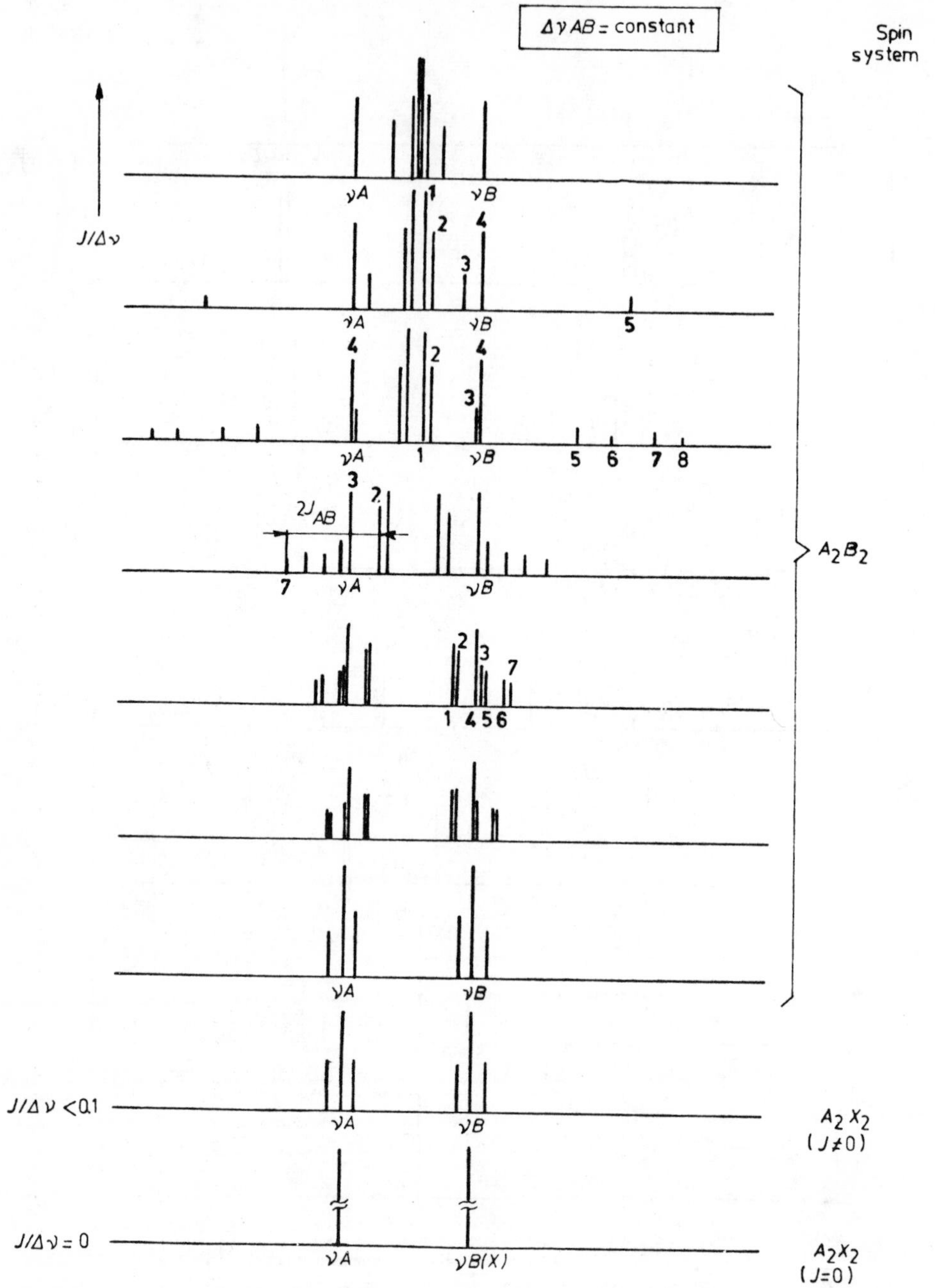

FIGURE 52. Change of the spectra of A_2B_2 spin systems as a function of the ratio $J/\Delta\nu$, where $\Delta\nu AB$ is constant.[301]

First-order, i.e., A_2X_2, spectra consist of two 1:2:1 triplets.* νA and νX are given by the central line of the triplets, and the distance of two neighboring lines within any of the triplets is equal to J_{AX}.

The A_2B_2 spectrum is a pattern of 18 lines symmetrically distributed about the center of the spectrum at $(\nu A + \nu B)/2$. Four lines arising from combinations are weak and lie far from the center; therefore, they are not significant. The dependence of the spectrum on the ratio $J/\Delta\nu$ is illustrated by Figures 52 and 53. The former depicts the theoretical spectra

* Compare p. 49.

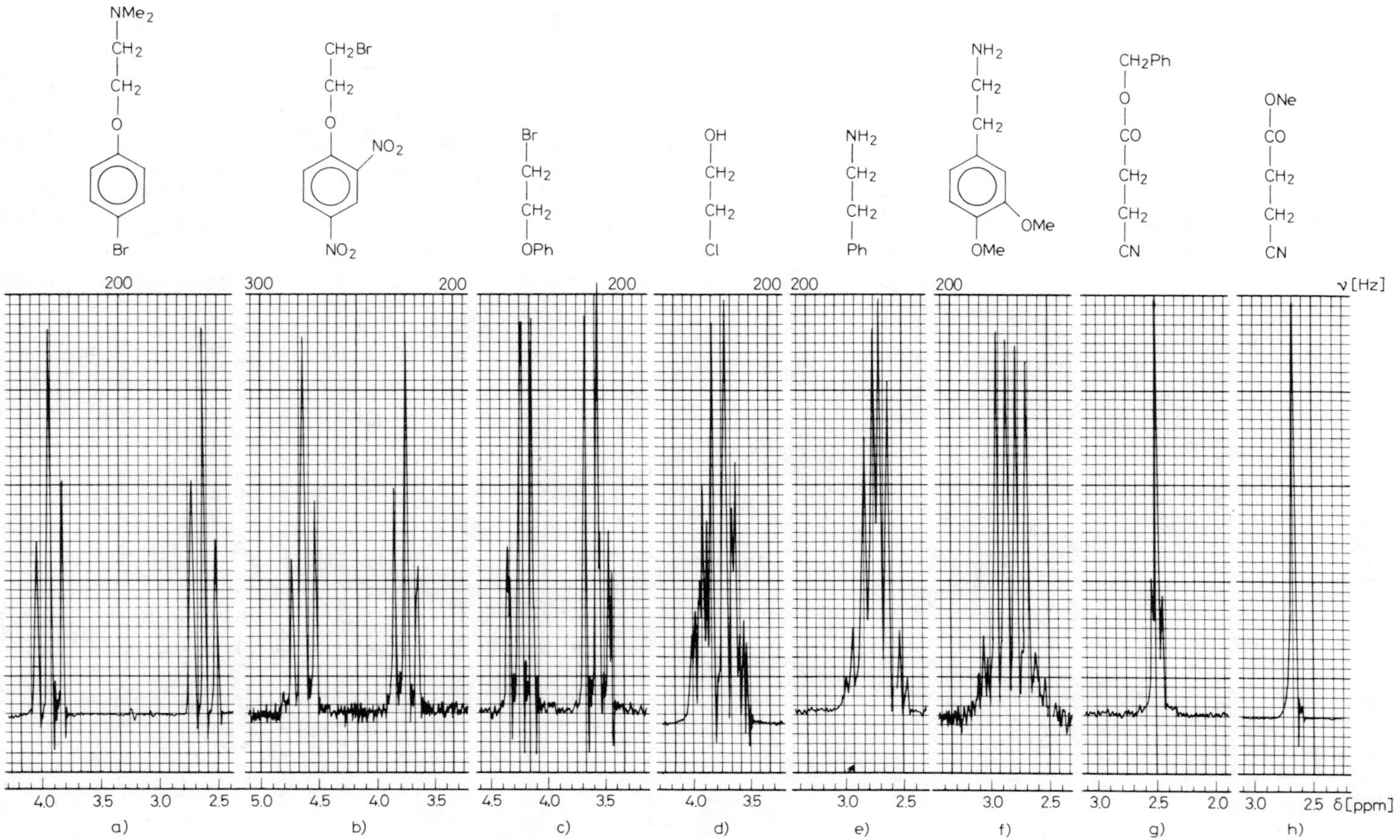

FIGURE 53. Methylene multiplets for type R–CH_2–CH_2–R′, illustrating the interconversion of the spin systems A_2X_2–A_2B_2–A_4.

belonging to different values of $J/\Delta\nu$, while the latter shows parts of spectra assigned to protons of group $-CH_2-CH_2-$ in compound $R-CH_2-CH_2-R'$. It can be seen from Figure 53 that the greater the difference between the shielding effects of R and R', the closer the spectrum resembles the limiting cases A_2X_2. *N,N*-dimethyl-2(*p*-bromophenoxy)-ethylamine and 2-(2,4-dinitrophenoxy)-ethylbromide are examples of the roof pattern. When all the lines are separated, chemical shifts and the coupling constant can be directly extracted in a simple way from the spectrum (compare Problem **29**). The chemical shifts are given by the frequency, relative to the center of lines **3**, or, if $J/\Delta\nu > 0.5$, of lines, **4**, both of outstanding intensity. The distance of the line **2** and the outermost line **7** from the center equals $2J_{AB}$ (compare Figure 52). With less completely resoluted spectra, computer methods or the tables[301] mentioned earlier have to be used for calculating J/Δ and therefrom J_{AB}.

1.5.4.5. Spin Systems A_2X_3 and A_2B_3

Two different groups of nuclei, consisting of three and two chemically and magnetically equivalent nuclei, respectively, give rise, when there is higher-order coupling between them, to the A_2B_3 spin system. Ethyl groups frequently belong to this spin system. The corresponding first-order case (see Figure 22) has already been discussed briefly* and consists of a symmetric triplet and a symmetric quartet.

When ratio $J/\Delta\nu$ is greater than 0.1, a higher-order splitting is observed. For cases approximating first-order splitting, the triplet and the quartet are still recognizable, but all lines are further split; the inner lines of the quartet are split into four components, while all other lines are split into doublets (compare Figures 30c and 54). The spectrum contains 34 lines at the most and its structure is again a function of the ratio $J/\Delta\nu$. Lines **12** and **13** belong to transitions A and B, respectively, and **9** belongs to combinations. The latter are, as usual, weak and often cannot be identified. Having assigned lines B_6, A_4, and A_5, the chemical shift difference is given by $\Delta\nu = (A_4 + A_5)/2 - B_6$. Line B_6, which is a fixed point of the spectrum, can be easily identified by its outstanding intensity. Until νB remains constant, line B_6 does not change its position as a function of $J/\Delta\nu$. Lines A_4 and A_5 can be found with reference to Figure 54. When we already know $\Delta\nu$, the ratio $J/\Delta\nu$ and, therefore, J_{AB} can be obtained from the reference tables[301] or one can make use of computer programs.

1.5.5. Spin Systems Characterized by Three Coupling Constants

1.5.5.1. The AMX Spin System

The simplest spin system possessing three coupling constants is the *AMX* type, where $\nu A \neq \nu M \neq \nu X$ and $J_{AM} \neq J_{AX} \neq J_{MX} \neq 0$, furthermore, there is only first-order coupling between any pairs of nuclei. Consequently, the product-functions are at the same time the basis functions of the spin system. The latter and the energy values of the corresponding levels are found in Table 14. Frequencies are obtained by substituting the wavefunctions, ψ_n, in Equation 174.

Selection rule given by Equation 151 permits 15 transitions; the probability of the combination transitions $\psi_7 \rightarrow \psi_2$, $\psi_6 \rightarrow \psi_3$, and $\psi_5 \rightarrow \psi_4$ are, however, zero. Frequencies and the transition probabilities as calculated according to Equation 150 are given in Table 15.

The *AMX* spectrum is, therefore, composed of twelve lines of the same intensity. Four-four lines belong to each of the transitions A, M, and X. For the conditions $\nu A > \nu M > \nu X$, and $J_{AM} > J_{AX} > J_{MX}$, the schematic spectrum is shown in Figure 55 (e.g., see the 220-MHz spectrum of acrylonitrile in Figure 31c and Problem **24**).

The chemical shifts and the coupling constants can be read directly from the spectra. The center of the double doublets A, M and X gives the corresponding shifts; J_{AX} is the distance of the outermost lines (lines **1** and **2** or **3** and **4**) in the A part of the spectrum. J_{MX} is determined

* Compare p. 48.

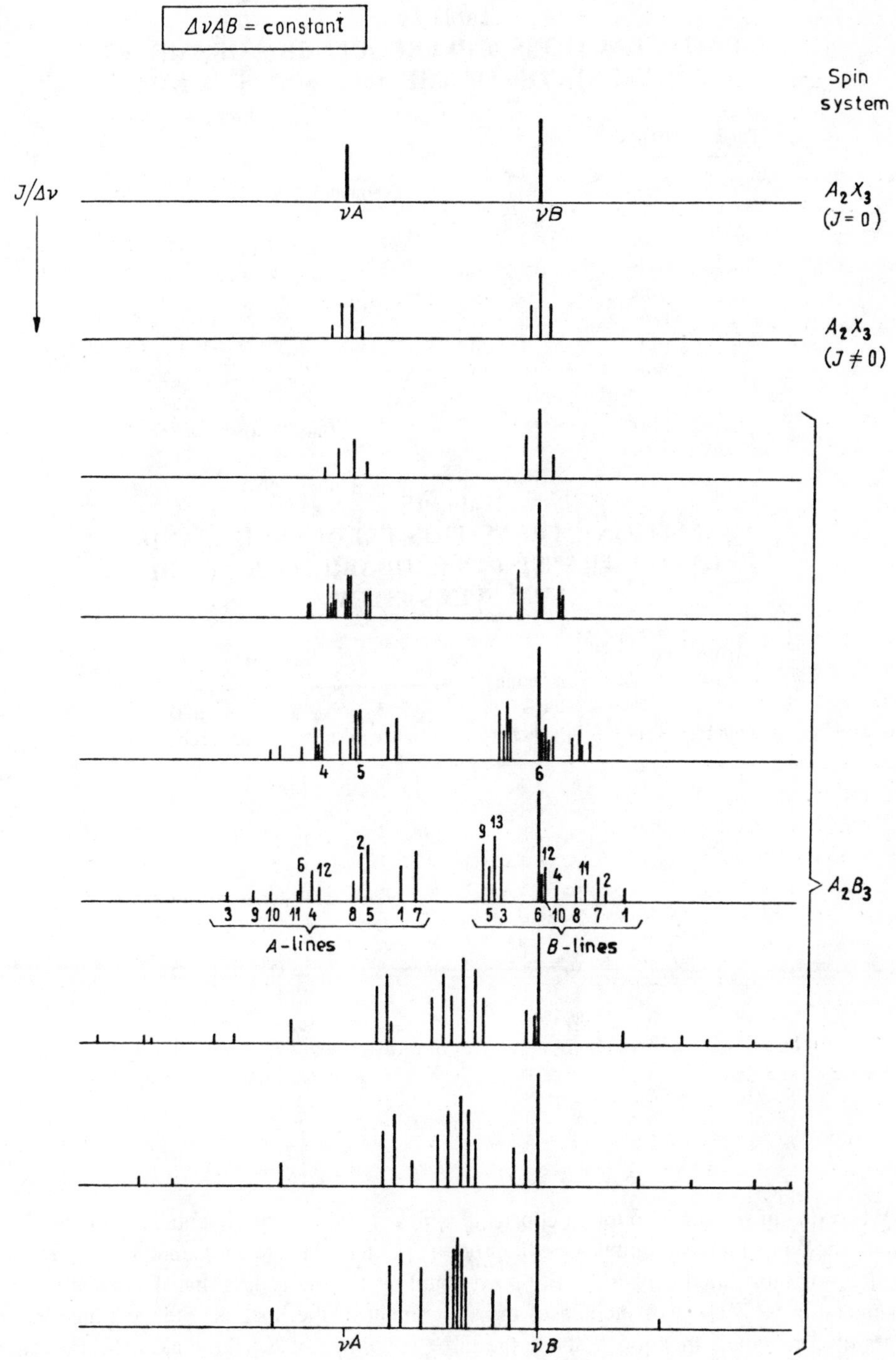

FIGURE 54. Change of the spectra of A_2B_3 spin systems as a function of the ratio $J/\Delta\nu$, with $\Delta\nu AB$ kept constant.[434]

in the same way from the *M* or *X* part. Finally J_{AM} is obtained from lines *A* or *M*, either as the distance of the centers of the two outer doublets or as the distance of the lines **1** and **3**, **2** and **4**, **5** and **7**, or **6** and **8**, respectively. The sign of neither the coupling constants nor of the chemical shifts can be derived from the spectrum.

Table 14
BASIS FUNCTIONS AND FREQUENCIES OF THE QUANTUM STATES OF THE *AMX* SPIN SYSTEM

Basis function		
Symbol	**Form**	**Frequency**
ψ_1	ααα	$(\nu A + \nu M + \nu X)/2 + (J_{AM} + J_{AX} + J_{MX})/4$
ψ_2	ααβ	$(\nu A + \nu M - \nu X)/2 + (J_{AM} - J_{AX} - J_{MX})/4$
ψ_3	αβα	$(\nu A - \nu M + \nu X)/2 + (-J_{AM} + J_{AX} - J_{MX})/4$
ψ_4	βαα	$(-\nu A + \nu M + \nu X)/2 + (-J_{AM} - J_{AX} + J_{MX})/4$
ψ_5	αββ	$-(-\nu A + \nu M + \nu X)/2 + (-J_{AM} - J_{AX} + J_{MX})/4$
ψ_6	βαβ	$-(\nu A - \nu M + \nu X)/ 2 + (-J_{AM} + J_{AX} - J_{MX})/4$
ψ_7	ββα	$-(\nu A + \nu M - \nu X)/2 + (J_{AM} - J_{AX} - J_{MX})/4$
ψ_8	βββ	$-(\nu A + \nu M + \nu X)/2 + (J_{AM} + J_{AX} + J_{MX})/4$

Table 15
TRANSITIONS, TRANSITION FREQUENCIES, AND RELATIVE TRANSITION PROBABILITIES OF THE *AMX* SPIN SYSTEM

Transition		Combining basis functions	Transition	
Number	**Type**		**Frequency**	**Relative probability**
1	*A*	$\psi_4 \rightarrow \psi_1$	$\nu A + J_{AM}/2 + J_{AX}/2$	1
2	*A*	$\psi_6 \rightarrow \psi_2$	$\nu A + J_{AM}/2 - J_{AX}/2$	1
3	*A*	$\psi_7 \rightarrow \psi_3$	$\nu A - J_{AM}/2 + J_{AX}/2$	1
4	*A*	$\psi_8 \rightarrow \psi_5$	$\nu A - J_{AM}/2 - J_{AX}/2$	1
5	*M*	$\psi_3 \rightarrow \psi_1$	$\nu M + J_{AM}/2 + J_{MX}/2$	1
6	*M*	$\psi_5 \rightarrow \psi_2$	$\nu M + J_{AM}/2 - J_{MX}/2$	1
7	*M*	$\psi_7 \rightarrow \psi_4$	$\nu M - J_{AM}/2 + J_{MX}/2$	1
8	*M*	$\psi_8 \rightarrow \psi_6$	$\nu M - J_{AM}/2 - J_{MX}/2$	1
9	*X*	$\psi_2 \rightarrow \psi_1$	$\nu X + J_{AX}/2 + J_{MX}/2$	1
10	*X*	$\psi_5 \rightarrow \psi_3$	$\nu X + J_{AX}/2 - J_{MX}/2$	1
11	*X*	$\psi_6 \rightarrow \psi_4$	$\nu X - J_{AX}/2 + J_{MX}/2$	1
12	*X*	$\psi_8 \rightarrow \psi_7$	$\nu X - J_{AX}/2 - J_{MX}/2$	1
13	*K*	$\psi_7 \rightarrow \psi_2$	$\nu A + \nu M - \nu X$	0
14	*K*	$\psi_6 \rightarrow \psi_3$	$\nu A - \nu M + \nu X$	0
15	*K*	$\psi_5 \rightarrow \psi_4$	$-\nu A + \nu M + \nu X$	0

When one of the three coupling constants is near to zero (experimentally unobservable), two of the three double doublets are collapsed to doublets. This phenomenon is very frequent for 1,2,4-trisubstituted benzenes. The *para* coupling is usually less than 1 Hz and remains unnoticed at 60 MHz.* In such cases the assignment of the lines is evident. Since for the types of compounds mentioned above the three couplings of the three aromatic hydrogens (*A*, *M*, and *X*) correspond to interactions between protons in *ortho*-, *meta*-, and *para*-positions, respectively, further $J^o > J^m$,** the double doublet belongs to the proton having both an *ortho* and *meta* partner. The doublet belonging to the *ortho* partner is split to a larger extent.

* Compare p. 65 and Volume II, p. 69.
** Compare p. 65 and Volume II, p. 68.

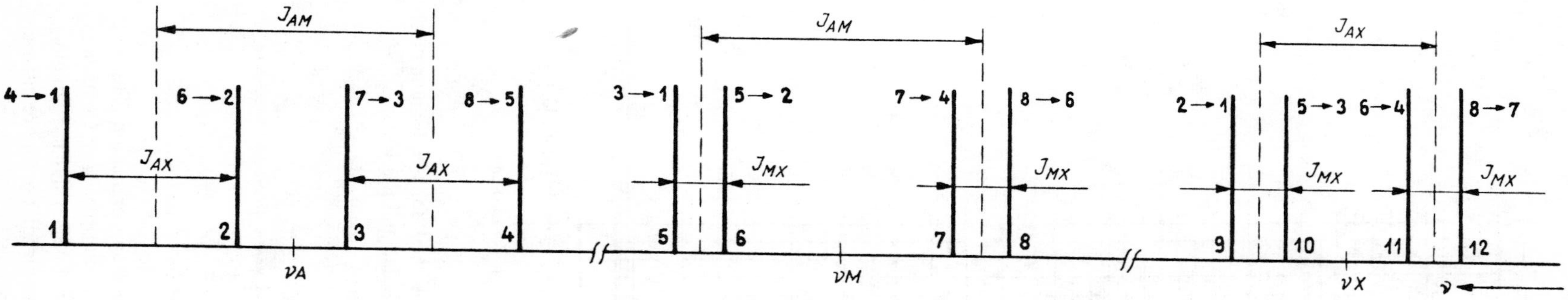

FIGURE 55. The schematic spectrum of the *AMX* spin system.

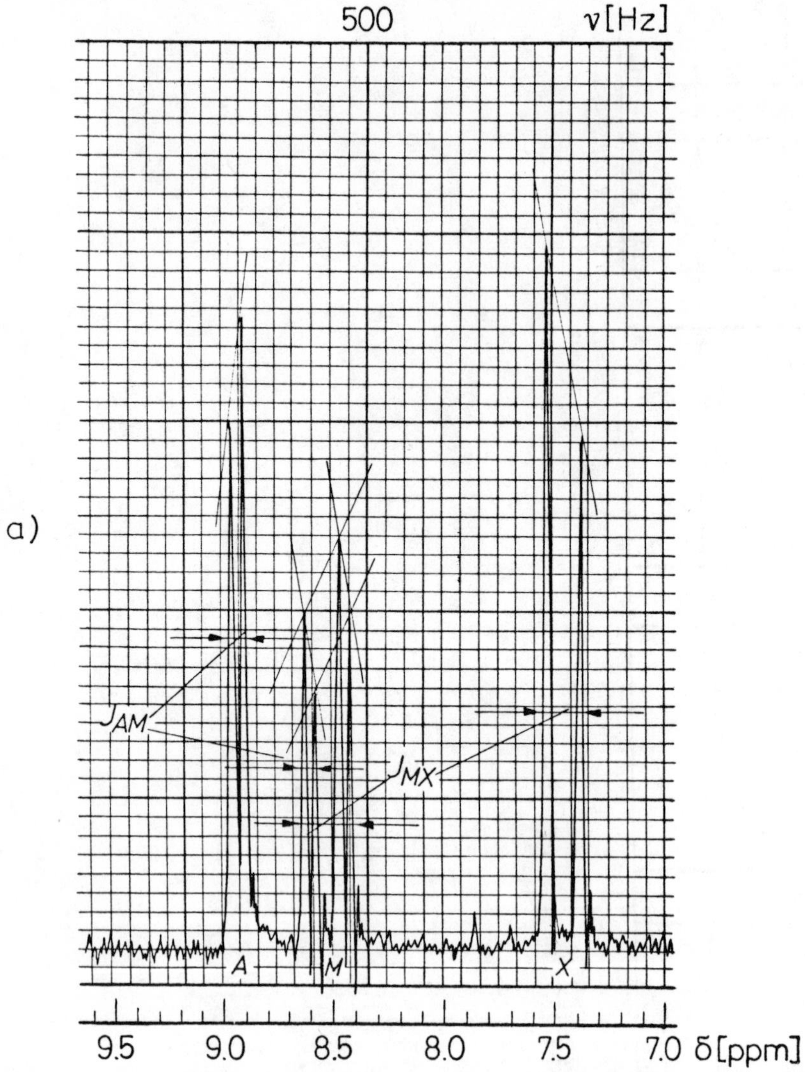

FIGURE 56. (a)The multiplets of the aromatic protons in 2,4-dinitrophenol (**46**) and (b) 2-nitro-5-hydroxy-benzylalcohol (**47**) in acetone-d_6.

This is illustrated by spectra (see Figure 56) of Compounds **47** and **48**. The assignments following the convention $\nu A > \nu M > \nu X$ are given in the formulas and the Figure.

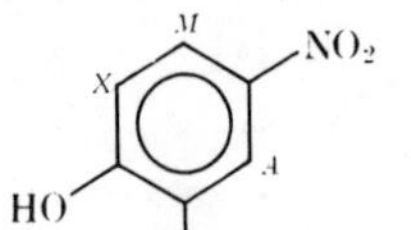

47

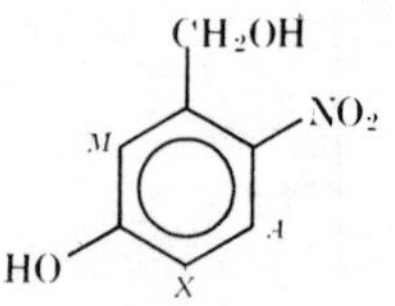

48

The treatment of the two spectra as systems *AMX* is only a good approximation. The roof pattern indicates clearly the effects of higher-order coupling. The multiplet *M* of Compound **47** is especially instructive; for pairs split by J_{AM}, the lines lying closer to signal *A* are more intense, whereas in those split according to J_{MX}, the ones closer to signal *X* are more intense.

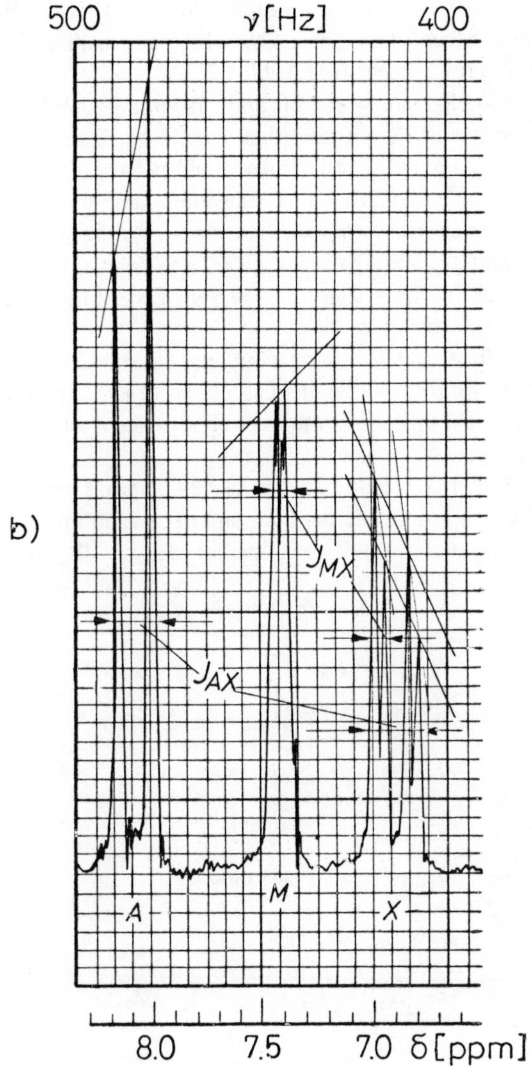

FIGURE 56b.

1.5.5.2. The ABX Spin System[1051]

ABX systems contain three, nonequivalent nuclei, with higher-order coupling between two nuclei (J_{AB} and $\Delta\nu AB$ are of the same order of magnitude). Many asymmetrically trisubstituted benzene derivatives, disubstituted pyridines and monosubstituted furans, thiophenes, and pyrrols, furthermore, compounds containing >CH–CH_2– group, which have chemically nonequivalent methylene protons, e.g., monosubstituted oxyranes, azirydines, and thiiranes and larger saturated rings, show *ABX*-type spectra. The determination of chemical shifts and coupling constants is rather complicated and ignoring the pertinent theory may lead to false conclusions.

Out of the eight basis functions and the corresponding energy levels, respectively, there are four that are the same as for the *AMX* system (see eigenfunctions ψ_1, ψ_2, ψ_7, and ψ_8 in Table 13), only the index "*M* " has to be substituted by "*B* ". The other four spin states become, however, combined — since $\nu A \approx \nu B$. As a result of this interaction, the energy levels are changed as a function of the ratio $J_{AB}/\Delta\nu AB$ (see Figure 57).

The eigenfunctions and frequency values of the combined states can be found in Table 16. These data and the values of systems *AMX* still valid here allow us to calculate the frequency and transition probabilities (see Table 17).

From Table 17 the following can be established. As with the *AMX* system, there are again eight transitions of type *A* and *B*, but now they cannot be distinguished from each other. The originally forbidden combination transitions **14** and **15** have (exceptionally) a rather large relative probability and can be regarded as *X* transitions, due to the not too high values of Q_1 and Q_2. (This approximation is more true the closer are the values of νA and νB.) The values of $\Delta\nu AB$, J_{AX}, and J_{BX} are small enough relative to J_{AB}, consequently, $Q_1 \approx Q_2 \approx J_{AB}/2$. Thus, the *X* part of the spectrum consists of six lines. The probability of the combination transition **13** is still zero.

The three chemical shifts and the three coupling constants can be obtained from the experimental spectrum as follows. Since the magnitudes of νX and νA or νB are significantly different, the *X* and *AB* parts of the spectrum are well separated, and therefore it is convenient to study them separately.

The six lines of the *X* part (see Table 17) are placed pairwise symmetrically to the center of the *X* multiplet (νX). νX is therefore the midpoint of this multiplet. From Table 17 it also becomes evident that the following relationships hold for the distance of the line-pairs:

$$\mathbf{15} - \mathbf{14} = |2(Q_1 + Q_2)|$$

$$\mathbf{10} - \mathbf{11} = |2(Q_1 - Q_2)|$$

$$\mathbf{9} - \mathbf{12} = |J_{AX} + J_{BX}|$$

$$\mathbf{10} - \mathbf{14} = \mathbf{15} - \mathbf{11} = 2Q_1$$

$$\mathbf{15} - \mathbf{10} = \mathbf{11} - \mathbf{14} = 2Q_2$$

The intensities of the lines (*A*) are related as follows:

$$A_{10} = A_{11} \text{ and } A_{14} = A_{15}$$

$$A_9 = A_{12} = A_{10} + A_{14} = A_{11} + A_{15} = 1$$

$$A_{10} = A_{11} < A_9 + A_{12} > A_{14} = A_{15}$$

Since both Q_1 and Q_2 are positive the distance of the lines **14** and **15** from the center is always greater than that of lines **10** and **11**. Accordingly the line-pairs can be distributed in three ways around the center

15	**10**	**9**		**12**	**11**	**14**
15	**9**	**10**		**11**	**12**	**14**
9	**15**	**10**		**11**	**14**	**12**
			νX			

The schematic structure of the *X* part of an *ABX* spectrum according to the arrangement corresponding of the second line is illustrated by Figure 58. The procedure for the assignment of the lines is the following. The two most intense lines belong to transitions **9** and **12**. The members of the pairs cannot be distinguished from one another, therefore although the distance of the lines **9** and **12** does give the absolute value of the sum $J_{AX} + J_{BX}$ it does not yield the sign. Having made the assignment for the lines **9** and **12** the assignment for the other lines follows automatically (according to the three schemes given above).

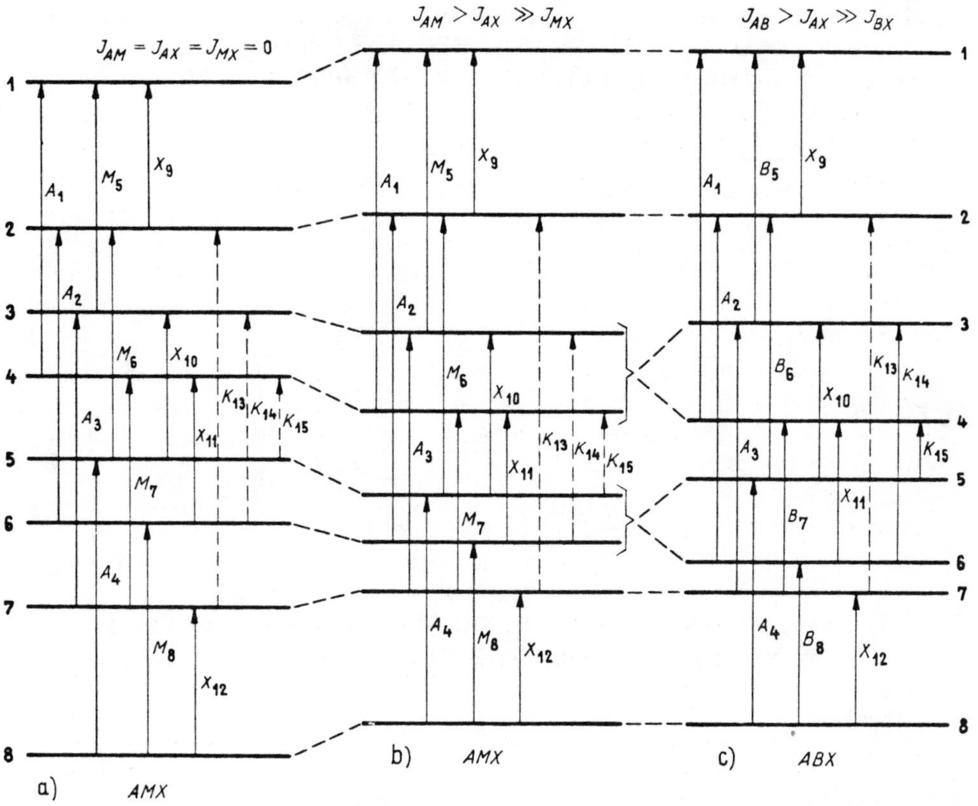

FIGURE 57. Energy levels and transitions of the spin systems (c) *ABX* and *AMX*, (a) without and (b) with spin-spin interactions, assuming that $J_{AM} > J_{AX} > J_{MX}$.

The values of νX, $|J_{AX} + J_{BX}|$, and also of Q_1 and Q_2 can therefore be calculated from the *X* multiplet, if all six lines appear. It happens, however, quite often that the intensity of the line-pair **14** and **15** is very small or zero; then the four remaining lines have the same intensity, just like for the *AMX* spectra. This happens because, although the relative probability of the transitions **9** and **12** is, independently of the spectral parameters, always unity, the relative intensity of the other two pairs of transitions can vary from one to zero, while the sum of their intensity remains unity. In such a case, the only information which can be obtained from the *X* multiplet is νX.

The *AB* part of the spectrum (see Figure 59) is made up of eight lines: from two overlapping *AB* quartets, corresponding to the two spin orientations of the nucleus *X*. For the distance of the lines — utilizing the data of Table 17 — the following relationships are obtained:

$$\mathbf{1} - \mathbf{3} = \mathbf{2} - \mathbf{4} = \mathbf{6} - \mathbf{8} = \mathbf{5} - \mathbf{7} = J_{AB},$$

$$\mathbf{1} - \mathbf{5} = \mathbf{3} - \mathbf{7} = 2Q_1,$$

$$\mathbf{2} - \mathbf{6} = \mathbf{4} - \mathbf{8} = 2Q_2$$

$$\mathbf{3} + \mathbf{5} - \mathbf{4} - \mathbf{6} = \mathbf{1} + \mathbf{7} - \mathbf{2} - \mathbf{8} = \mathbf{3} + \mathbf{5} - \mathbf{2} - \mathbf{8} = \mathbf{1} + \mathbf{7} - \mathbf{4} - \mathbf{6} = |J_{AX} + J_{BX}|$$

The absolute value of J_{AB} can therefore be obtained from the *AB* multiplet, but not its sign. The shifts νA and νB can also be calculated if the formula (Equation 169) derived for the *AB* systems is applied to the two *AB* quartets and the two values for each of *A* and *B* are averaged or the average frequencies of the corresponding line-pairs of the two quartets are

Table 16
BASIS FUNCTIONS AND FREQUENCIES FOR THE COMBINED STATES OF THE *ABX* SPIN SYSTEM

Basis function		
Symbol	Form	Frequency of the combined state
ψ_3	$C_1\varphi_3 + C_2\varphi_4$	$\nu X/2 - J_{AB}/4 + Q_1$
ψ_4	$C_1\varphi_4 - C_2\varphi_3$	$\nu X/2 - J_{AB}/4 - Q_1$
ψ_5	$C_3\varphi_5 + C_4\varphi_6$	$-\nu X/2 - J_{AB}/4 + Q_2$
ψ_6	$C_3\varphi_6 - C_4\varphi_5$	$-\nu X/2 - J_{AB}/4 - Q_2$

Note: $Q_{1,2} = \sqrt{[\Delta\nu AB/2 + (J_{AX} - J_{BX}/4]^2 + J_{AB}^2/4}$;
$C_1 = \sqrt{J_{AB}^2/2Q_1(4Q_1 - 2\Delta\nu AB - J_{AX} + J_{BX})} = J/4Q_1C_2$;
$C_2 = \sqrt{(4Q_1 - 2\Delta\nu AB - J_{AX} + J_{BX})/8Q_1}$;
$C_3 = \sqrt{J_{AB}^2/2Q_2\,(4Q_2 - 2\Delta\nu AB + J_{AX} - J_{BX})} = J/4Q_2C_4$;
$C_4 = \sqrt{(4Q_2 - 2\Delta\nu AB + J_{AX} - J_{BX})/8Q_2}$.

Table 17
TRANSITIONS, TRANSITION FREQUENCIES, AND RELATIVE TRANSITION PROBABILITIES OF THE *ABX* SPIN SYSTEM

Transition		Combining basis function	Transition	
Number	Type		Frequency	Relative probability
1	*AB*	$\psi_4 \rightarrow \psi_1$	$(\nu A + \nu B)/2 + (2J_{AB} + J_{AX} + J_{BX})/4 + Q_1$	$(C_3 - C_4)^2$
2	*AB*	$\psi_6 \rightarrow \psi_2$	$(\nu A + \nu B)/2 + (2J_{AB} - J_{AX} - J_{BX})/4 + Q_2$	$(C_1 - C_2)^2$
3	*AB*	$\psi_7 \rightarrow \psi_3$	$(\nu A + \nu B)/2 + (-2J_{AB} + J_{AX} + J_{BX})/4 + Q_1$	$(C_3 + C_4)^2$
4	*AB*	$\psi_8 \rightarrow \psi_5$	$(\nu A + \nu B)/2 + (-2J_{AB} - J_{AX} - J_{BX})/4 + Q_2$	$(C_1 + C_2)^2$
5	*AB*	$\psi_3 \rightarrow \psi_1$	$(\nu A + \nu B)/2 + (2J_{AB} + J_{AX} + J_{BX})/4 - Q_1$	$(C_3 + C_4)^2$
6	*AB*	$\psi_5 \rightarrow \psi_2$	$(\nu A + \nu B)/2 + (2J_{AB} - J_{AX} - J_{BX})/4 - Q_2$	$(C_1 + C_2)^2$
7	*AB*	$\psi_7 \rightarrow \psi_4$	$(\nu A + \nu B)/2 + (-2J_{AB} + J_{AX} + J_{BX})/4 - Q_1$	$(C_3 - C_4)^2$
8	*AB*	$\psi_8 \rightarrow \psi_6$	$(\nu A + \nu B/2 + (-2J_{AB} - J_{AX} - J_{BX})/4 - Q_2$	$(C_1 - C_2)^2$
9	*X*	$\psi_2 \rightarrow \psi_1$	$\nu X + (J_{AX} + J_{BX})/2$	1
10	*X*	$\psi_5 \rightarrow \psi_3$	$\nu X + Q_1 - Q_2$	$(C_1C_3 + C_2C_4)^2$
11	*X*	$\psi_6 \rightarrow \psi_4$	$\nu X - Q_1 + Q_2$	$(C_1C_3 + C_2C_4)^2$
12	*X*	$\psi_8 \rightarrow \psi_7$	$\nu X - (J_{AX} + J_{BX})/2$	1
13	*K*	$\psi_7 \rightarrow \psi_2$	$\nu A + \nu B - \nu X$	0
14	*K*	$\psi_6 \rightarrow \psi_3$	$\nu X - Q_1 - Q_2$	$(C_2C_3 - C_1C_4)^2$
15	*K*	$\psi_5 \rightarrow \psi_4$	$\nu X + Q_1 + Q_2$	$(C_2C_3 - C_1C_4)^2$

substituted into Equation 169. This procedure can only be followed, however, when the lines of the two quartets can be properly assigned. This is often not difficult — owing to the different relative intensities of the lines in the two quartets (e.g., see Figure 59 or Problem **58**). When, in the other hand, there is no essential difference between the intensity of the line-pairs **1** and **2**, **3** and **4**, **5** and **6**, or **7** and **8**, the lines cannot be assigned to the appropriate quartets. Figure 60a shows two alternative assignments of the lines of the *AB* multiplet. Other theoretically possible arrangements can be disregarded on the ground of line distances and intensities.

Since the intensity difference of the corresponding lines of the quartets is larger (for the same value of J_{AB}) when the quantity $\Delta\nu AB$ is more different for the two quartets, in doubtful cases the assignment given in Figure 60b is *a priori* more plausible (if the inner lines of the

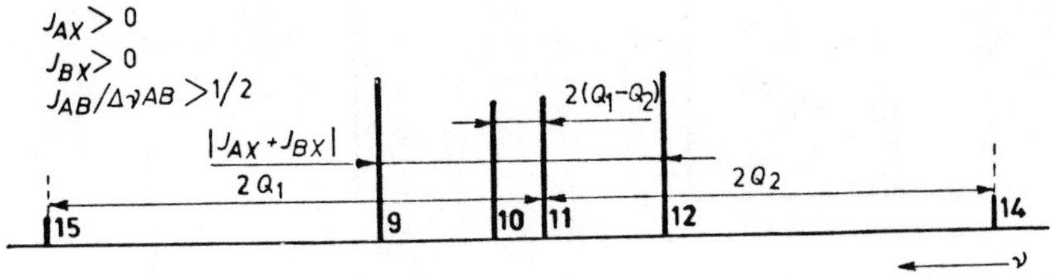

FIGURE 58. The X part of the schematic spectrum of the ABX spin system.

quartet fall closer to one another, their intensity further increases relatively to the intensity of the outer ones*).

The grouping of lines to quartets can be carried out without ambiguity when the X part could properly be assigned. The distance of the centers of the two quartets should be namely $(|J_{AX} + J_{BX}|)/2$, compare Table 17, and this quantity can also be read from the X part of the spectrum. The reverse of this statement is also true; if the assignment of the lines AB is possible, the assignment of the X lines becomes feasible even when only four of the lines are observable.

Having completed the assignment of the spectral lines, it is easy to calculate J_{AX} and J_{BX}. One can, namely, construct, using the formula given in Table 16 for Q_1 and Q_2, the following equations:

$$\sqrt{(2Q_1 + J_{AB})(2Q_1 - J_{AB})} = \sqrt{(\mathbf{1} - \mathbf{7})(\mathbf{3} - \mathbf{5})} = \Delta\nu AB + \frac{1}{2}(J_{AX} - J_{BX}) \tag{175a}$$

$$\sqrt{(2Q_2 + J_{AB})(2Q_2 - J_{AB})} = \sqrt{(\mathbf{2} - \mathbf{8})(\mathbf{4} - \mathbf{6})} = \Delta\nu AB - \frac{1}{2}(J_{AX} - J_{BX}) \tag{175b}$$

From the above relationships we obtain:

$$\sqrt{(\mathbf{1} - \mathbf{7})(\mathbf{3} - \mathbf{5})} - \sqrt{(\mathbf{2} - \mathbf{8})(\mathbf{4} - \mathbf{6})} = |J_{AX} - J_{BX}| \tag{176}$$

Since $|J_{AX} + J_{BX}|$ is known already, a comparison of the two yields the absolute value and the relative sign of the two coupling constants. The symmetry of the problem prevents, however, to determine the absolute signs the pairs $J_{AX} + J_{BX}$ and $-J_{AX} - J_{BX}$ as well as $J_{AX} - J_{BX}$ and $-J_{AX} + J_{BX}$, yield identical spectra.

Addition of Equations 175a and 175b gives:

$$\frac{1}{2}\left[\sqrt{(2Q_1 + J_{AB})(2Q_1 - J_{AB})} + \sqrt{(2Q_2 + J_{AB})(2Q_2 - J_{AB})}\right] = \Delta\nu AB \tag{177}$$

Equation 177 is nothing else but the mathematical restatement of the fact that $\Delta\nu AB$ is obtained by calculating separately the quantities $\Delta\nu AB(Q_1)$ and $\Delta\nu AB(Q_2)$ for the two AB quartets corresponding to the two orientations of the spin X, by means of Equation 169 and then taking their averages. As it has been mentioned, we may also proceed simply by substituting the average frequency values of the line-pairs **1** and **2**, **3** and **4**, **5** and **6**, and **7** and **8** into Equation 169.

* Compare p. 99.

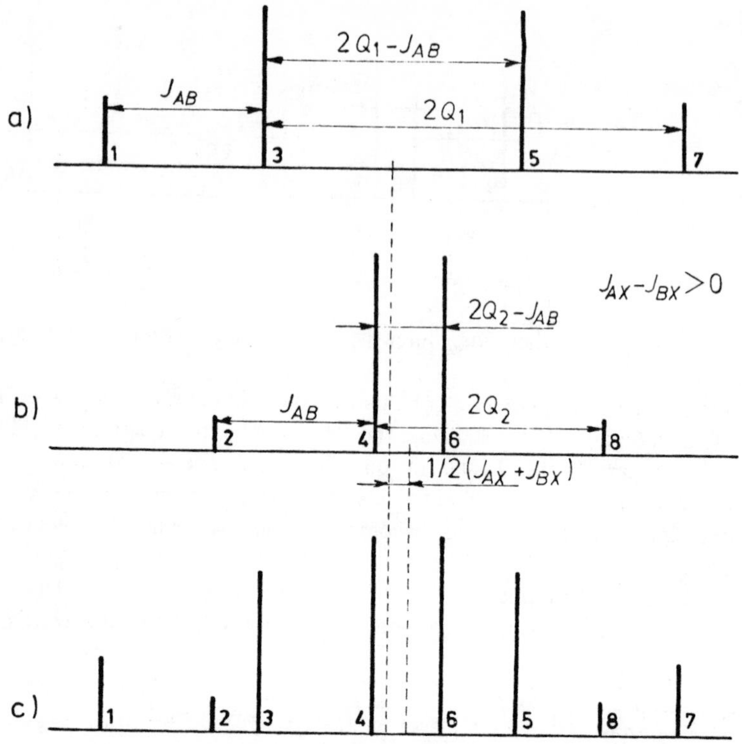

FIGURE 59. The *AB* part of the schematic spectrum of the *ABX* spin system.

$$\Delta\nu AB = \sqrt{\left(\frac{1+2}{2} - \frac{7+8}{2}\right)\left(\frac{3+4}{2} - \frac{5+6}{2}\right)} \quad (178)$$

The structure of the spectrum depends on relative sign of J_{AX} and J_{BX}. In view of its practical importance, we will discuss this question in more detail. When J_{AX} and J_{BX} have the same sign, $Q_1 - Q_2$ is smaller, whereas $|J_{AX} + J_{BX}|$ is greater. Therefore in fortunate cases the lines of transitions **10** and **11** in the *X* part are closer to the center *X*, whereas lines **9** and **12** are located farther away (see Figure 58). When J_{AX} and J_{BX} are of opposite sign, the situation is reversed. This, of course, requires that the relation $Q_1 - Q_2 < J_{AX} + J_{BX}$ should be hold. At the same time, if the signs are identical, the centers of the *AB* quartets lie farther away (see Figure 60b), while if the signs J_{AX} and J_{BX} are opposite, they are closer spaced (see Figure 60a), and this statement is unconditional.

If $J_{AB}/\Delta\nu AB \geq 1/2$, the relative sign of J_{AX} and J_{BX} can be inferred from the intensity ratio of the line-pairs: **10/14** = **11/15**. A calculation based on data in Tables 16 and 17 shows that in case of the same sign for J_{AX} and J_{BX}, intensity of the lines **10** and **11** is greater than that of lines **14** and **15**, whereas if the signs are opposite, this intensity relationship is reversed. This can be understood considering the following arguments. When the signs of J_{AX} and J_{BX} are the same, the expressions $\pm J_{AX} \mp J_{BX}$ can be neglected, since the two coupling constants are found always with opposite signs in the coefficients C_1 to C_4, determining the intensity, and because their values are not significantly different. Consequently, $Q_1 \approx Q_2$, furthermore $C_1 \approx C_3$ and $C_2 \approx C_4$. The intensity ratio **11/15** = **10/14** is then given by:

$$\frac{(C_1C_3 + C_2C_4)^2}{(C_2C_3 - C_1C_4)^2} \approx \frac{(C_1^2 + C_2^2)^2}{(C_1C_2 - C_1C_2)^2}$$

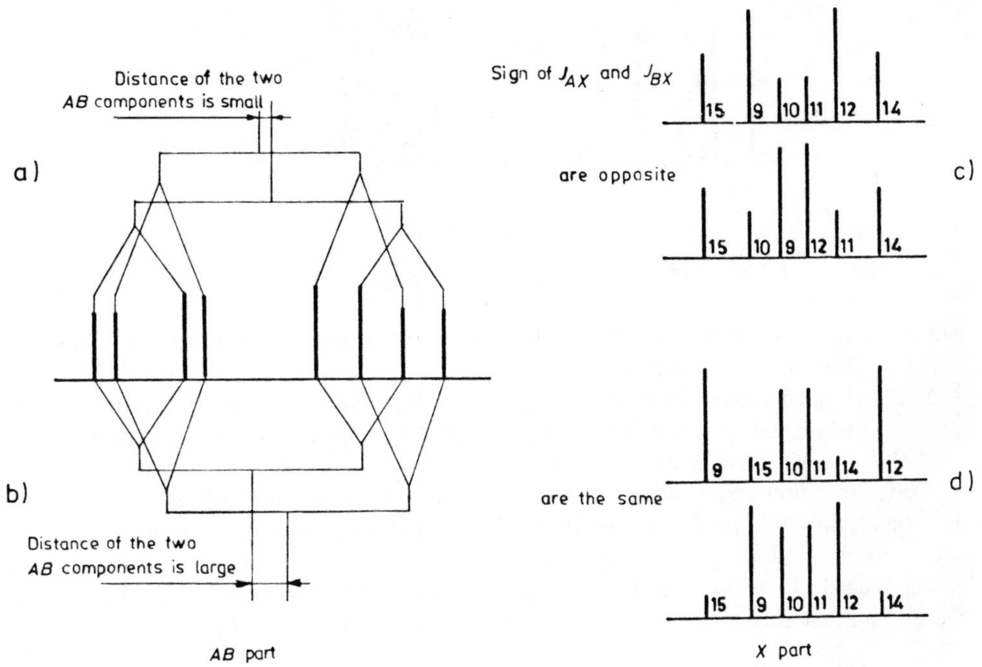

FIGURE 60. Assignment (a) and (b) of the *AB* and (c) and (d) *X* lines of an *ABX* spectrum, when the signs of J_{AX} and J_{BX} are identical (b and d) or opposite (a and c).

and is large because $C_2C_3 \approx C_1C_4$ and therefore the denominator is small. When the signs of J_{AX} and J_{BX} are different, it can be shown by using the expressions formerly given for Q_1, Q_2 and C_1 to C_4 that $(C_2C_3 - C_1C_4)^2 > (C_1C_3 + C_2C_4)^2$. For illustration see Figure 60.

Under certain conditions, some of the 14 lines of the *ABX* spectrum cannot be observed, while others may coincide; thereby the spectrum appears to be simpler (s.c. rudimentary spectra). The analysis of such spectra is, however, by no means simpler. In many instances, such simplified multiplets cannot be used for the determination of spectral parameters. In addition, rudimentary spectra may create the impression that one deals with simpler spin systems.

Most often the *X* part is reduced to only four lines, due to the small relative probabilities of combination transitions. The larger is $\Delta\nu AB$, i.e., the more we approach the type *AMX*, the weaker are the combination lines.

Simplification is sometimes more extensive. When, for example, either $\Delta\nu AB = 0$* or $J_{AX} = J_{BX}$, then $Q_1 = Q_2$ and $C_1 = C_3$, and $C_2 = C_4$ (see Table 16). It then follows (compare Table 17) that the *X* part of the *ABX* spectrum collapses to a symmetrical triplet, whereas the *AB* part still consists of eight lines. Under such circumstances, the lines **10** and **11** coincide at the value of νX, the intensity of the lines **14** and **15** is in turn zero. The question as to which of the simplifying conditions is operative can be decided upon comparing theoretically calculated and experimentally measured line intensities.

If $J_{AB} \gg \Delta\nu AB + J_{AX} \mp J_{BX}$, then $Q_1 \approx Q_2 \approx J_{AB}/2$ and $C_1 \approx C_2 \approx C_3 \approx C_4 = 1/\sqrt{2}$. In such a case, transitions **10** and **11** overlap and the probability of the transitions **14** and **15** become zero. The *X* part of the *ABX* spectrum is again simplified into a symmetric 1:2:1 triplet.

Under the same conditions, the *AB* part of the spectrum is an 1:1 doublet, the distance of the components is $(J_{AX} + J_{BX})/2$. The intensity of the lines **1**, **2**, **7**, and **8** is, namely, zero

* This is the s.c. *AA'X* spin system ($\nu A = \nu A'$, but $J_{AX} \neq J_{A'X}$).

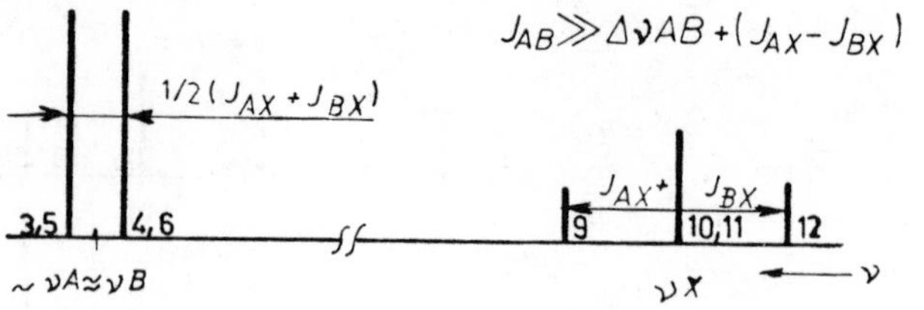

FIGURE 61. The schematic spectrum of the *ABX* spin system, when $J_{AB} \gg \Delta\nu AB + |J_{AX} - J_{BX}|$.

and the pairs **3** and **5**, as well as **4** and **6**, composed of lines of unit intensity, coincide (compare Problem **23** and Figure 61).

As a consequence under the above conditions, the *ABX* spectrum simplifies in a misleading way to a rudimental spectrum, of the AX_2 type having apparent coupling constant $J'_{AX} = (J_{AX} + J_{BX})/2$ which is the average of the real values.

Such a spectrum may lead to the false conclusion that $J_{AX} = J_{BX}$. The value of J_{AX} and J_{BX} cannot, however, be determined at all, since the spectrum is invariant as long as $J_{AX} + J_{BX}$ is constant. If a "pseudo" AX_2 spectrum is suspected on chemical evidence, it is recommended to measure the spectra in another solvent, or, after adding shift reagents to the solution, take recourse to partial deuteration or measuring at another frequency.

1.5.5.3. The ABC Spin System

The spin system consisting of three nuclei gives an *ABC*-type spectrum when all chemical shift differences are small but nonzero and of the same order of magnitude as the coupling constants. Many vinyl derivatives, compounds containing a terminal oxyrane ring, asymmetrically trisubstituted benzenes, disubstituted pyridines, monosubstituted furans, thiophenes, and pyrrols may give spectra of this type.

The parameters νA, νB and, νC, J_{AB}, J_{AC}, and J_{BC} can be calculated theoretically from *ABC* spectrum, although only the relative signs of the coupling constants are determinable. The calculation is, however, rather tiresome, and it may happen that the experimental spectrum is reproduced by more than one set of parameters. The expansion of the 8th order determinant of the three-spin system leads, namely, to expressions containing third-order subdeterminants as well from the particular solutions of which it is difficult to select the appropriate one.[474] Therefore mainly computer methods are used to approximate iteratively the observed spectrum.[474] As a starting point for the iteration, a set of approximate values, calculated for the *ABX* system, can be used ("*ABK*" approximation). There are many examples in the literature for the analysis of the *ABC* system by computer methods.[258,434,474,725,1042,1187,1518]

1.5.6. Spin Systems Characterized by Four Coupling Constants

1.5.6.1. The AA'XX' Spin System

In principle, six coupling constants could be assigned to the A_2X_2 spin systems. Denoting chemically equivalent nuclei by *A* and *A'* and by *X* and *X'*, respectively, they are $J_{AA'}$, $J_{XX'}$, J_{AX}, $J_{AX'}$, $J_{A'X}$, and $J_{A'X'}$.

If, the nuclear pairs *A* and *X* are magnetically nonequivalent, e.g., when no free rotation is possible around the carbon-carbon bond in an RCH_2–CH_2R' type compound, the system can be characterized by four coupling constants. In 1,1-difluoro-ethylene (**38**), the protons are coupled differently to the *cis* and *trans* fluorine nuclei, but as a consequence of symmetry, it still holds that $J_{AX} = J_{A'X'}$ and $J_{AX'} = J_{A'X}$. Hence for the *AA'XX'* system, it is necessary to take into consideration four coupling constants, $J_{AA'}$, $J_{XX'}$, $J^{cis}_{AX} \equiv J_{A'X'}$, and $J^{trans}_{A'X} \equiv J_{AX'}$, and two chemical shift values. The 16 (2^4) basis functions and the frequency of the levels are listed in Table 18.

Figure 62 shows the energy level schemes corresponding to the conditions $J_{AA'} = J_{XX'} = J_{AX} = J_{AX'} = 0$ (limiting case A_2X_2, $J_{AX} = 0$), also for $J_{AA'} = J_{XX'} = 0$ and $J_{AX} = J_{AX'}$ (limiting case A_2X_2, $J_{AX} \neq 0$), and finally $J_{AA'} \neq J_{XX'} \neq J_{AX} \neq J_{AX'} \neq 0$ (the general case $AA'XX'$). The corresponding spectra are composed of two lines, two symmetric triplets and two multiplets of ten lines each, respectively.

In Table 18 the following abbreviations are used:

$$Q_1 = J_{AA'} + J_{XX'}$$

$$Q_2 = J_{AA'} - J_{XX'}$$

$$Q_3 = J_{AX} - J_{AX'}$$

$$Q_4 = J_{AX} + J_{AX'} \qquad (179)$$

The energy values can be obtained in the same way as before by the use of the Equation 174. As an example, let us calculate the energy of the level $2a_o$, E_{14}

$$E_{14} = \langle\psi_{14}\Big|\sum_i \nu_i \mathcal{I}_{zi} + \frac{1}{4}\sum_{i<j}\sum J_{ij}(2\mathcal{P}_{ij} - 1)\Big|\psi_{14}\rangle =$$

$$\frac{1}{4}\,[\langle(\alpha\beta\alpha\beta - \beta\alpha\alpha\beta + \alpha\beta\beta\alpha - \beta\alpha\beta\alpha)\Big|\left(\frac{\nu A}{2} - \frac{\nu A}{2} + \frac{\nu X}{2} - \frac{\nu X}{2}\right) +$$

$$\frac{1}{4}\sum_{i<j}\sum J_{ij}(2\mathcal{P}_{ij} - 1)\Big|(\alpha\beta\alpha\beta - \beta\alpha\alpha\beta + \alpha\beta\beta\alpha - \beta\alpha\beta\alpha)\rangle] =$$

$$\frac{J_{AA'}}{16}\,[\langle(\alpha\beta\alpha\beta - \beta\alpha\alpha\beta + \alpha\beta\beta\alpha - \beta\alpha\beta\alpha)\,|\,(-3)\,(\alpha\beta\alpha\beta - \beta\alpha\alpha\beta + \alpha\beta\beta\alpha - \beta\alpha\beta\alpha)\rangle] +$$

$$\frac{J_{XX'}}{16}\,[\langle(\alpha\beta\alpha\beta - \beta\alpha\alpha\beta + \alpha\beta\beta\alpha - \beta\alpha\beta\alpha)\,|\,(\alpha\beta\alpha\beta - \beta\alpha\alpha\beta + \alpha\beta\beta\alpha - \beta\alpha\beta\alpha)\rangle] =$$

$$-\frac{3}{4}J_{AA'} + \frac{1}{4}J_{XX'} = -\frac{Q_2}{2} - \frac{Q_1}{4}$$

From the data in Table 18, the energy of the allowed transitions and the relative intensity of the corresponding lines can be determined. As, however, the eigenfunctions ψ_{13} and ψ_{14} are combined, the data in Table 18 relating to these eigenfunctions must be modified (the data in that table refer to coupling-free cases).

The energy of the coupled levels is obtained from the following determinant:

$$\begin{vmatrix} \mathbf{H}_{13,13} - E & \mathbf{H}_{13,14} \\ \mathbf{H}_{14,13} & \mathbf{H}_{14,14} - E \end{vmatrix} = 0 \qquad (180)$$

The value of $\mathbf{H}_{13,13}$ and $\mathbf{H}_{14,14}$ is found in Table 18 and equals $Q_2/2 - Q_1/4$ and $-Q_2/2 - Q_1/4$, respectively.

$$\mathbf{H}_{13,14} = \frac{1}{4}\,[\langle(\alpha\beta\alpha\beta + \beta\alpha\alpha\beta - \alpha\beta\beta\alpha - \beta\alpha\beta\alpha)\Big|\sum_i \nu_i \mathcal{I}_{zi} +$$

$$\frac{1}{4}\sum_{i<j}\sum J_{ij}(2\mathcal{P}_{ij} - 1)\Big|(\alpha\beta\alpha\beta - \beta\alpha\alpha\beta + \alpha\beta\beta\alpha - \beta\alpha\beta\alpha)\rangle] = \frac{1}{4}\left[\frac{\nu A}{2} - \frac{\nu A}{2} - \frac{\nu X}{2} + \frac{\nu X}{2}\right] +$$

$$\frac{1}{16}\,[J_{AA'}(2-2) + J_{XX'}(2-2) + J_{AX'}(8) + J_{AX}(-8)] = (J_{AX'} - J_{AX})/2 = -Q_3/2$$

Table 18
BASIS FUNCTIONS, SYMBOLS, AND FREQUENCIES OF THE QUANTUM STATES OF THE $AA'XX'$ SPIN SYSTEM

Basis function		Quantum state	
Symbol	Form	Symbol	Frequency $(-Q_1/4)$
ψ_1	$\alpha\alpha\alpha$	s_2	$\nu A + \nu X + Q_4/2$
ψ_2	$(\alpha\beta + \beta\alpha)\alpha\alpha/\sqrt{2}$	$1s_1$	νX
ψ_3	$\alpha\alpha(\alpha\beta + \beta\alpha)/\sqrt{2}$	$2s_1$	νA
ψ_4	$\beta\beta\alpha\alpha$	$1s_0$	$\nu X - \nu A - Q_4/2$
ψ_5	$\alpha\alpha\beta\beta$	$2s_0$	$\nu A - \nu X - Q_4/2$
φ_6[a]	$(\alpha\beta - \beta\alpha)(\alpha\beta - \beta\alpha)/2$	$3s_0$	$-Q_1$
φ_7[a]	$(\alpha\beta + \beta\alpha)(\alpha\beta + \beta\alpha)/2$	$4s_0$	0
ψ_8	$(\alpha\beta + \beta\alpha)\beta\beta/\sqrt{2}$	$1s_{-1}$	$-\nu X$
ψ_9	$\beta\beta(\alpha\beta + \beta\alpha)/\sqrt{2}$	$2s_{-1}$	$-\nu A$
ψ_{10}	$\beta\beta\beta\beta$	s_{-2}	$-\nu A - \nu X + Q_4/2$
ψ_{11}	$(\alpha\beta - \beta\alpha)\alpha\alpha/\sqrt{2}$	$1a_1$	$\nu X - (Q_1 + Q_2)/2$
ψ_{12}	$\alpha\alpha(\alpha\beta - \beta\alpha)/\sqrt{2}$	$2a_1$	$\nu A + (Q_2 - Q_1)/2$
φ_{13}[a]	$(\alpha\beta + \beta\alpha)(\alpha\beta - \beta\alpha)/2$	$1a_0$	$(Q_2 - Q_1)/2$
φ_{14}[a]	$(\alpha\beta - \beta\alpha)(\alpha\beta + \beta\alpha)/2$	$2a_0$	$-(Q_1 + Q_2)/2$
ψ_{15}	$(\alpha\beta - \beta\alpha)\beta\beta/\sqrt{2}$	$1a_{-1}$	$-\nu X - (Q_1 + Q_2)/2$
ψ_{16}	$\beta\beta(\alpha\beta - \beta\alpha)/\sqrt{2}$	$2a_{-1}$	$-\nu A + (Q_2 - Q_1)/2$

The states $3s_0$ and $4s_0$, furthermore $1a_0$ and $2a_0$, are combined; this is not indicated in the table. Thus, φ_6, φ_7, φ_{13}, and φ_{14} are the noncombined product functions; the frequencies of the combined states $1a'_0$, $2a'_0$, $3s'_0$, and $4s'_0$ are not identical to the values given in the table.

Thus, the determinant can be rewritten in the following form:

$$\begin{vmatrix} Q_2/2 - Q_1/4 - E & -Q_3/2 \\ -Q_3/2 & -Q_2/2 - Q_1/4 - E \end{vmatrix} = 0 \tag{181}$$

and from this

$$E_{13,14} = -Q_1/4 \pm \sqrt{Q_2^2 + Q_3^2}/2 = -Q_1/4 \pm Q_6/2 \tag{182a}$$

E_8 and E_7 can be calculated in the same way:

$$E_{6,7} = -Q_1/4 \pm \sqrt{Q_1^2 + Q_3^2}/2 = -Q_1/4 \pm Q_5/2 \tag{182b}$$

The energy of the transitions and the relative transition probabilities are then given in Table 19, where,

$$(2C_1^2 - 1):(2C_1\sqrt{1 - C_1^2}):1 = Q_1:Q_3:Q_5 \tag{183a}$$

$$(2C_2^2 - 1):(2C_2\sqrt{1 - C_2^2}):1 = Q_2:Q_3:Q_6 \tag{183b}$$

As the A and X parts of the spectrum are identical, it is enough to study one of them. Therefore in Table 19 only transitions A are given.

For the sake of clarity, let us separate transitions between antisymmetric and from those

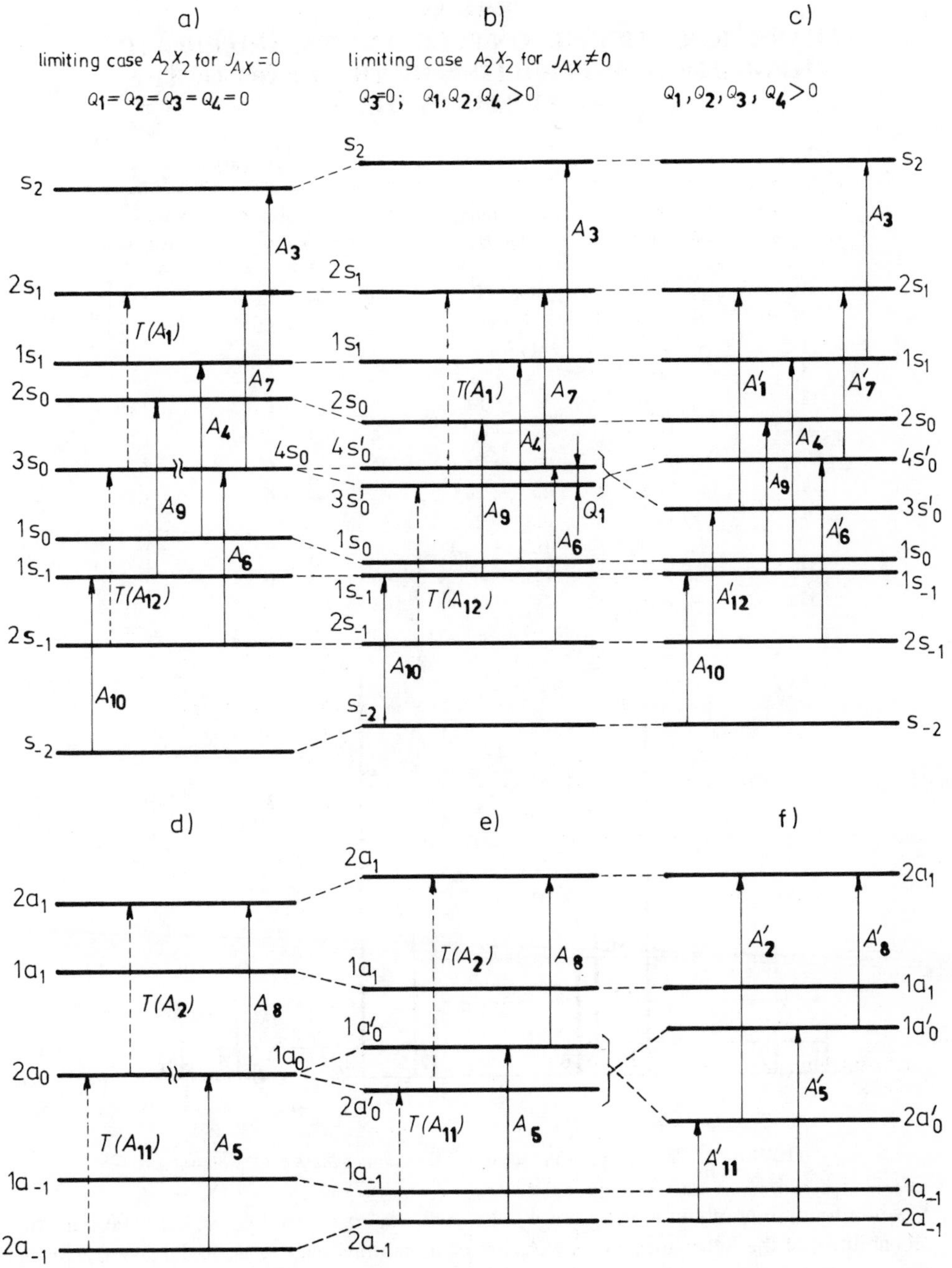

FIGURE 62. The (a) to (c) symmetric and (d) to (f) antisymmetric energy levels and transitions of the $AA'XX'$ spin system for (a) and (d) $Q_1 = Q_2 = Q_3 = Q_4 = 0$, (b) and (e) $Q_3 = 0$ and $Q_1,Q_2,Q_4 > 0$, and (c) and (f) $Q_1,Q_2,Q_3,Q_4 > 0$.

between symmetric levels. This can safely be done since any transition of type $a \leftrightarrow s$ is forbidden.

According to the selection rule given by Equation 151, there are eight allowed antisymmetric transitions, four belonging to each of A and X. The four A transitions are $\psi_{16} \rightarrow \psi_{14}$, $\psi_{16} \rightarrow \psi_{13}$, $\psi_{14} \rightarrow \psi_{12}$, and $\psi_{13} \rightarrow \psi_{12}$.

There are 20 allowed symmetric transitions. Out of these, eight belong to each of A and

Table 19
TRANSITIONS, TRANSITION FREQUENCIES, AND RELATIVE TRANSITION PROBABILITIES OF THE *A* PART OF THE *AA'XX'* SPIN SYSTEMS

Number of transitions	Combining basis function	Combining levels	Transition: Frequency	Transition: Relative probability
1	$\psi_6 \rightarrow \psi_3$	$3s'_0 \rightarrow 2s_1$	$\nu A + (Q_1 + Q_5)/2$	$1 - C_1^2$
2	$\psi_{14} \rightarrow \psi_{12}$	$2a'_0 \rightarrow 2a_1$	$\nu A + (Q_2 + Q_6)/2$	$1 - C_2^2$
3	$\psi_2 \rightarrow \psi_1$	$1s_1 \rightarrow s_2$	$\nu A + Q_4/2$	2
4	$\psi_4 \rightarrow \psi_2$	$1s_0 \rightarrow 1s_1$		
5	$\psi_{16} \rightarrow \psi_{13}$	$2a_{-1} \rightarrow 1a'_0$	$\nu A + (Q_6 - Q_2)/2$	C_2^2
6	$\psi_9 \rightarrow \psi_7$	$2s_{-1} \rightarrow 4s'_0$	$\nu A + (Q_5 - Q_1)/2$	C_1^2
7	$\psi_7 \rightarrow \psi_3$	$4s'_0 \rightarrow 2s_1$	$\nu A + (Q_1 - Q_5)/2$	C_1^2
8	$\psi_{13} \rightarrow \psi_{12}$	$1a'_0 \rightarrow 2a_1$	$\nu A + (Q_2 - Q_6)/2$	C_2^2
9	$\psi_8 \rightarrow \psi_5$	$1s_{-1} \rightarrow 2s_0$	$\nu A - Q_4/2$	2
10	$\psi_{10} \rightarrow \psi_8$	$1s_{-2} \rightarrow 1s_{-1}$		
11	$\psi_{16} \rightarrow \psi_{14}$	$2a_{-1} \rightarrow 2a'_0$	$\nu A - (Q_2 + Q_6)/2$	$1 - C_2^2$
12	$\psi_9 \rightarrow \psi_6$	$2s_{-1} \rightarrow 3s'_0$	$\nu A - (Q_1 + Q_5)/2$	$1 - C_1^2$

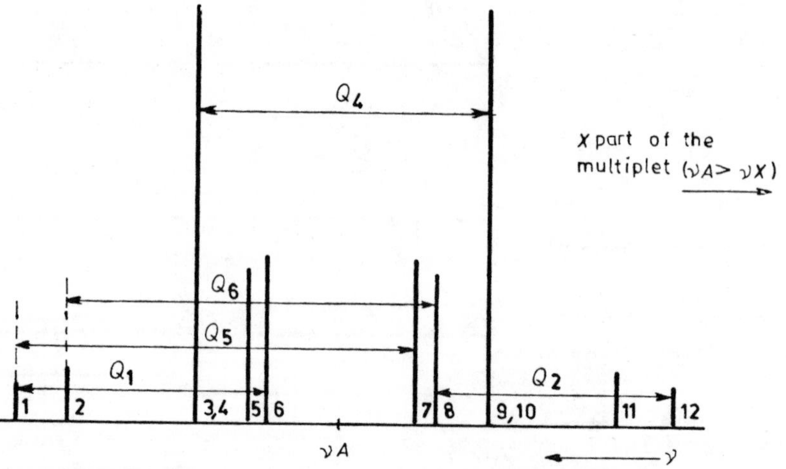

FIGURE 63. The *A* part of the schematic spectrum of the *AA'XX'* spin system.

X, while four are combinations ($\psi_9 \rightarrow \psi_5$, $\psi_8 \rightarrow \psi_4$, $\psi_5 \rightarrow \psi_2$, and $\psi_4 \rightarrow \psi_3$). The transition probabilities of the latter are zero. Two pairs of transitions ($\psi_{10} \rightarrow \psi_8$ and $\psi_8 \rightarrow \psi_5$, further $\psi_4 \rightarrow \psi_2$ and $\psi_2 \rightarrow \psi_1$) are degenerate; therefore six *A* lines due to symmetric transitions are expected. With the 4 lines of antisymmetric transitions, the *AA'XX'* spectrum consists of 10 + 10 = 20 lines.

The schematic structure of part *A* of the spectrum is shown in Figure 63. The quantities Q_1, Q_2, Q_4, Q_5, and Q_6 can be obtained directly from the spectrum, while Q_3 can be calculated from the former using Equations 182a or 182b. The coupling constants can easily be calculated from the former quantities using Equations 179. The intensity ratios can be calculated by Equations 183a and 183b. The chemical shifts νA and νX are, of course, read directly as the centers of the *A* and *X* parts of the spectrum.

Upon examining Table 19 and Figure 63 one observes that the spectrum has two symmetric

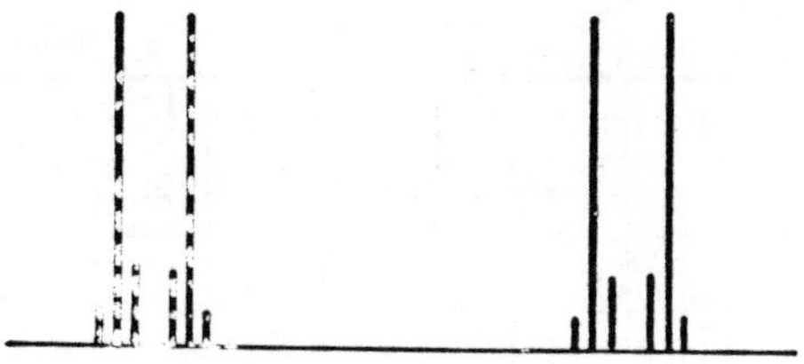

FIGURE 64. The schematic *AA'XX'*-type spectrum of *para*-disubstituted benzenes, resembling the *AX* doublet pair.

AB components (lines **2, 5, 8,** and **11** and **1, 6, 7,** and **12,** respectively) whose center is found at value νA. The total intensity of the degenerate line-pairs **3, 4** and **9, 10** is identical to the total intensity of the rest of the lines (compare Table 19). The entire spectrum is symmetric to the center, and this is also true for parts *A* and *X* separately.

The olefinic signals of the *Z,Z*-muconic acid dimethyl ester provide a nice example for the *AA'XX'* system (compare Problem **62**). A rudimentary form of it is encountered in Problem **66**.

Simplified *AA'XX'*-type spectra are found often with *para*-disubstituted benzene or γ-substituted pyridines. For these molecules applied that: $J_{AX} \equiv J^o \gg J_{AA'} \approx J_{XX'} \equiv J^m$ and $J_{AX'} \equiv J^p \approx 0$ (compare Table 5). In this case (compare definition Equation 179), $Q_2 = 0$, so $Q_3 \approx Q_4 \approx Q_6$. Furthermore, as $Q_1 < 1/2\ Q_3 \approx 1/2\ Q_4$, although to a much worse approximation, the approximate equality $Q_3 \approx Q_4 \approx Q_5 \approx Q_6$, also holds. According to Table 19 with the above conditions, *A* and *X* parts of the *AA'XX'* spectrum are both simplified, as a rough approximation, into doublets, where $J = J_{AX}$. This is so because among the ten lines, four-four, namely, **2** to **5**, as well as **8** to **11**, coincide (the intensity factors are 3 - 3) and only the lines **1, 6, 7,** and **12** are separated. Even the latter ones are close to the two intense maxima and are found on the right- and left-hand sides of them (with unit intensity for each pair). After all a pair of *AX*-like doublets can be observed, each accompanied by two minor satellites, one lying closer to the center, being more intense (see Figure 64). As an example, the spectrum of *p*-nitrophenol is reproduced (see Figure 65). The simplified *AA'XX'* multiplet, much resembling an *AX* spectrum, is slightly degenerated (roof structure) towards type *AB* (*AA'BB'*).

1.5.6.2. The AA'BB' Spin System

Spectra of type *AA'BB'* are expected for spin systems consisting of four nuclei which, due to their symmetry, are pairwise chemically equivalent, but magnetically nonequivalent, and the differences in the chemical shifts are of the same magnitude as that of the coupling constants. Many organic compounds belong to this category (see Table 6).

As for the *AA'XX'* system, the number of spin wave-functions is 16, but since all of them are combined, their form is, apart from the states s_2 and s_{-2} ($\psi_1 = \alpha\alpha\alpha\alpha$, $\psi_{16} = \beta\beta\beta\beta$, different. No explicit expression can be given for the levels $1s_{o'}$, $2s_{o'}$, $3s_{o'}$, and $4s_{o'}$ because in order to calculate the combination coefficients, one would have to solve a fourth-order determinant which is impossible. Therefore the energy of these levels cannot be calculated directly. As in the case of the *AA'XX'* system, there are altogether 28 allowed transitions. Four of these are combinations, having (unlike those of the *AA'XX'* system) finite, but nevertheless very low, probabilities. These lines are, therefore, insignificant in experimental spectra. Among the remaining 24 lines, 12 belong to each of the *A* and *B* transitions, and parts *A* and *B* of the spectrum are as in the case of *AA'XX'* mirror images of each other. Explicit frequency and intensity formulas can only be given for six lines of them, as the other six lines belong to transitions whose initial or final state coincides with one of the levels $1s_{o'}$, $2s_{o'}$, $3s_{o'}$, or $4s_{o'}$. It is noteworthy that the transition pairs **3** and **4**, as well as **9**

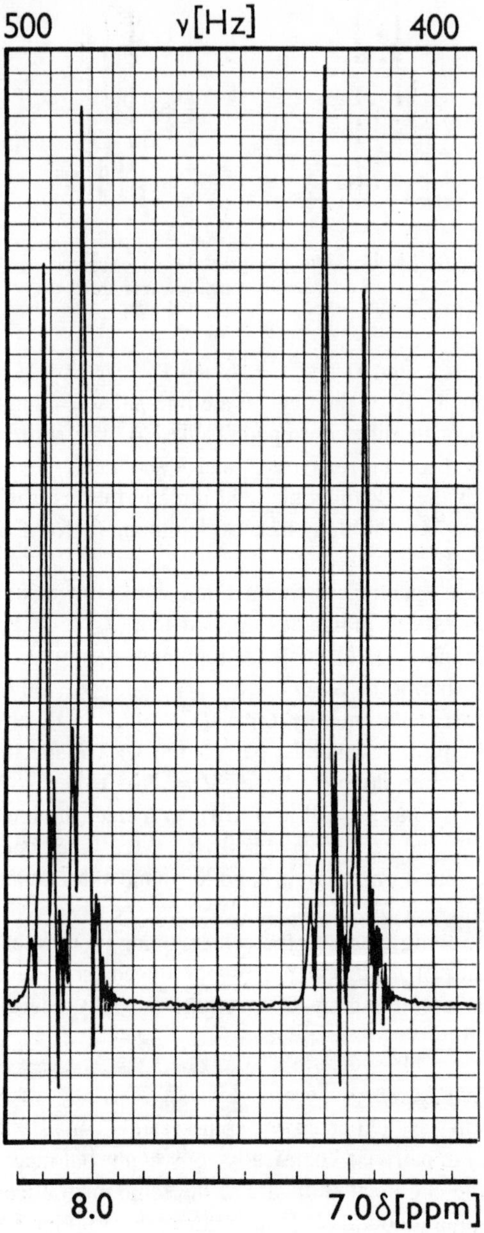

FIGURE 65. The *AA'XX'* multiplet of the aromatic protons of *p*-nitrophenol in $CDCl_3$.

and **10**, respectively, which are degenerate for the system *AA'XX'*, are split; thus, the *A* part of the spectrum is composed of 12 lines.

Due to the previously outlined problems, the direct determination of the chemical shift and coupling constant values for *AA'BB'* spectra is difficult usually and various indirect methods are applied for this purpose.

One of the possibilities is to put in trial values for coupling constants, and the theoretical *AA'XX'* spectrum is calculated by these data. Thereafter the $\Delta\nu AX(B)$ is gradually decreased and the trend of the frequency and intensity changes is noted. With a correct set of coupling constants, the calculation will properly reproduce the experimental spectrum when the correct $\Delta\nu$ value is reached, and an assignment of the individual lines will become possible. When

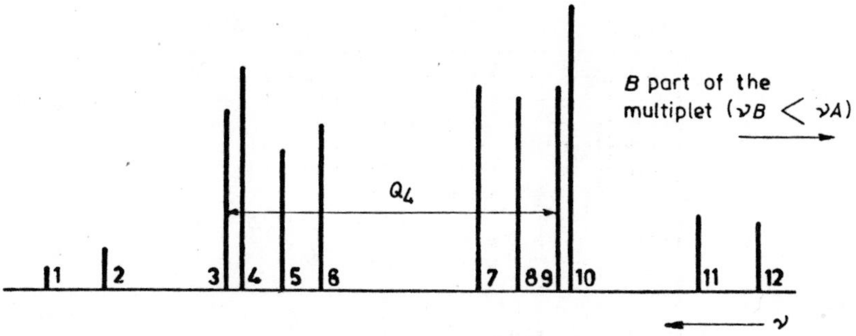

FIGURE 66. The *A* part of the schematic spectrum of the *AA'BB'* spin system.

the agreement is unsatisfactory, an iterative approach involving the correction of coupling constants has to be used.

Let us take 1,2-dichlorobenzene as an example. It is known that for benzene derivatives $J^o > J^m > J^p$ (about 8.0, 2.0, and 0.5 Hz), i.e., $J_{BB'} > J_{AA'}$ and $J_{AB} > J_{AB'}$. For the conditions that $\nu A > \nu B$ and that couplings constants are all positive, the form of the *AA'BB'* spectrum is shown in Figure 66 (compare the spectrum of 1,2-dichlorobenzene in Figure 69).

Upon comparing Figure 66 to the schematic spectrum of the system *AA'XX'* (see Figure 63), it can be seen that:

1. The intensity of lines closer to the *B* multiplet in the *A* part of the spectrum increases.
2. The originally degenerate line pairs **3** and **4** and **9** and **10** are split.
3. The distance of the lines **5** and **6** and of **7** and **8** changes. As the chemical shift difference decreases with respect to case *AA'XX'*, the *A* and *B* parts of the spectrum approach each other so that the lines **11** and **12** may even overlap with the *B* lines. Since although the complete *AA'BB'* spectrum is still symmetrical, while in contrast to case *AA'XX'*, the *A* and *B* multiplets are not.

The chemical shift difference $\Delta\nu AB$, as well as Q_2, Q_3, and Q_4, can be obtained from the experimental spectra, $Q_4 = \nu_3 - \nu_9$ and $\sqrt{\Delta\nu AB^2 + Q_4^2} = \nu_3 + \nu_9$, further $\sqrt{Q_2^2 + Q_3^2} = \nu_5 - \nu_{11}\sqrt{(\Delta\nu\, AB + Q_2)^2 + Q_3^2} = \nu_5 + \nu_{11}$, Q_1, however, cannot be directly derived from the spectrum, but can be calculated in an iterative way from the frequencies of lines **6** and **7**. This calculation of Q_1 can be avoided assuming $J_{AA'} \approx 0$ and thus taking $Q_1 \approx -Q_2$. This approximation is justified with many models, e.g., also for 1,2-dichlorobenzene. When the values of Q_1, Q_2, Q_3, and Q_4 are known, the absolute values of all coupling constants can be calculated.

The relative sign of the coupling constants can be, in principle, determined, but in a rather complicated way. It is also impossible to decide which of the pair of constants represents $J_{AA'}$ and which represents $J_{BB'}$ and the situation between J_{AB} and $J_{AB'}$ is similar. This problem may be, however, often resolved empirically. For 1,2-dichlorobenzene, for example, it is known that $J_{AB} > J_{AB'}$ and $J_{BB'} > J_{AA'}$.

In asymmetrically *para*-disubstituted benzene derivatives when $J_{AB} \gg J_{AB'} \approx 0$ and $J_{AB'} < J_{AA'} \approx J_{BB'} < J_{AB}$, the *AA'BB'*-type spectrum is simplified to an *AB* quartet containing similar satellite lines as *AA'XX'* multiplets characterized by $J_{AX} \gg J_{AX'} \approx 0$ and $J_{AX'} < J_{AA'} \approx J_{XX'} \ll J_{AX}$. For examples see Problem **6.**

The literature contains the discussion of a number of other spin systems, too. For a more detailed discussion of spin-spin coupling, textbooks[99,300,434,1051,1132,1211,1375,1379] have to be consulted. Practical applications of the theory of spin-spin coupling are illustrated in the Problems (Volume III).

Chapter 2

NMR SPECTROMETERS, RECORDING TECHNIQUES, MEASURING METHODS

INTRODUCTION

The operation principles and construction of NMR spectrometers will be discussed here only briefly, as for the chemist the equipment is rather a tool than the subject of investigations. The basic principles of measurement, however, provide many important viewpoints for selecting the best conditions for obtaining the best possible spectra. Hence we may not neglect these problems completely.

At the very beginning of the development of NMR spectroscopy, the main problem was just the construction of instruments having sufficient sensitivity.

In this section a very schematic description of special measurement techniques and their applications in structural research will be given which are possible by the modern NMR spectrometers and which yield additional information to routine spectra. We cannot, however, give here the theoretical discussion of them or even a detailed description of the experimental procedure. The main objective has been to discuss briefly the basic principles to point out the fields of application and to give a few typical examples.

2.1. CW SPECTROMETERS AND THEIR MAIN PARTS

Operation principles of FT spectrometers becoming very widespread in the 1970s are completely different from the classical CW (continuous wave) spectrometers. Although in the last decade the FT spectrometers have become more current, still several CW instruments are working. Therefore, and since some measuring methods are identical in the CW and FT mode, we review the operating principles of CW spectrometers briefly.

In this section we will mention some of measuring techniques (as integration of spectra, measuring at variable temperature, double resonance), which may be performed on CW instruments. The measuring of relaxation time is discussed in connection with the FT technique, because its routine use has become possible by the FT mode.

2.1.1. Magnetic Field Stability and Field Homogeneity

It has been shown* that the energy** and population difference of the levels*** is proportional to the magnitude of B_0, therefore the intensity of the resonance signal and the signal-to-noise ratio, ζ increase by the 3/2 power of B_0. The application of a higher field is all the more required, since the chemical shift differences also increase proportionally to B_0† so the signals are separated better. Three types of magnets are used to establish the field $\mathbf{B}_0$: permanent magnets, electromagnets, and superconducting magnets.

Permanent magnets are inexpensive, can be easily handled, and have good field stability. The highest field, however, that can be produced by permanent magnets is only about 1.4 T, so that they can only be used in 60-MHz instruments (for protons).

By the use of electromagnets, posing greater problems from the viewpoint of field stability, fields around 2.5 T can easily be achieved, therefore in 100-MHz spectrometers these magnets are used.

* Compare p. 24.
** Compare p. 9.
***Compare p. 11.
† Compare p. 26.

In spectrometers working at still higher measuring frequencies (200 to 600 MHz for protons), superconducting electromagnets are applied. By their use fields between 5 and 15 T can be reached. The cooling of the superconducting magnet is achieved by liquid helium.

Conditions necessary for high resolution NMR spectra are extremely stringent, not only for the magnitude, but also with respect to the stability and homogeneity of the polarizing field. The change ΔB_0 in the effective B_0 value may not exceed $10^{-9}B_0$. Such a high stability can only be achieved directly if well-thermostated permanent magnets are applied.

For the automatic stabilization of the field of the electromagnets, the technique of the proton-stabilization (lock-in) is applied. Without this technique, B_0 is drifted in time, due to temperature fluctuations. Usually a reference substance is placed in the vicinity of the sample to be measured (e.g., water) which possesses a sharp resonance signal. When B_0 is changed, the reference signal alters its frequency. This frequency change is amplified after transformation into a voltage difference, and this voltage is used to induce a secondary coil. The magnetic field produced by the latter compensates for ΔB_0. Reproducibility is especially important as the basic requirement for spectrum accumulation.* Thermostating of magnet and sample is achieved by applying a cooling jacket around the magnet. Conditioning of the laboratory also significantly improves the stability of magnet.

The homogeneity of B_0 is another fundamental condition for high resolution. Fluctuations of 10^{-8} T can entail significant line broadening. This means that the homogeneity should be better than 10^{-8}, even in the case of spectrometers operating at 60-MHz frequency.

Maximum resolution is limited by the natural width of the resonance lines which is 1.0 to 0.1 Hz for protons in liquids. In practice, lines are always broader due to field inhomogeneity. In order to reduce the inhomogeneity, the pole caps are designed to have a maximum area and the smallest possible gap between them. The gap, however, has to accomodate the sample tube and the detection devices (transmitting and receiving coils, etc.). The inhomogeneities across the sample are smaller when the volume of the sample is reduced; this is, however, limited by sensitivity. The reduction of inhomogeneities can also be achieved by regulating the current fed into the shim coils by means of which the profile of the magnetic field can be corrected. In most spectrometers, there are several such regulating systems — among them shim coils placed in the three main directions — which are set by hand. In modern spectrometers most of the regulation systems are operated automatically.

Homogeneity can be improved by as much as one and a half orders of magnitude when the sample tube is evenly spinned at a constant frequency of about 25 Hz, perpendicularly to $\mathbf{B}_0$. (This is usually accomplished by using plastic turbines fixed onto the sample tubes and rotated by compressed air.) In such a case the inhomogeneities are averaged out in a plane perpendicular to the rotation axis, and only those inhomogeneities are left unchanged which appear along the axis of rotation. It is important to achieve an even and smooth rotation; furthermore, it is essential that a maximum homogeneity should be achieved before sample spinning starts, otherwise there appear disturbing spinning sidebands on both sides of the signals at distances corresponding to the spinning frequency and its multiples (see Figure 67). Their intensity can be reduced when homogeneity is better or when the rotational speed is increased. When the spinning speed is too high, however, vortexes may develop in the sample which again lead to inhomogeneities. To ensure even spinning, one should use balanced sample tubes, and upon eventually sealing the tubes, one must be careful not to upset this balance. When applying a sample tube with thicker walls, the spinning side bands are reduced; this way, however, the sample volume is also decreased.

Note that the spinning side bands are easy to distinguish from spectral lines, since their position depends on the speed of rotation. The spinning side bands of the reference signal

* Compare p. 141-142.

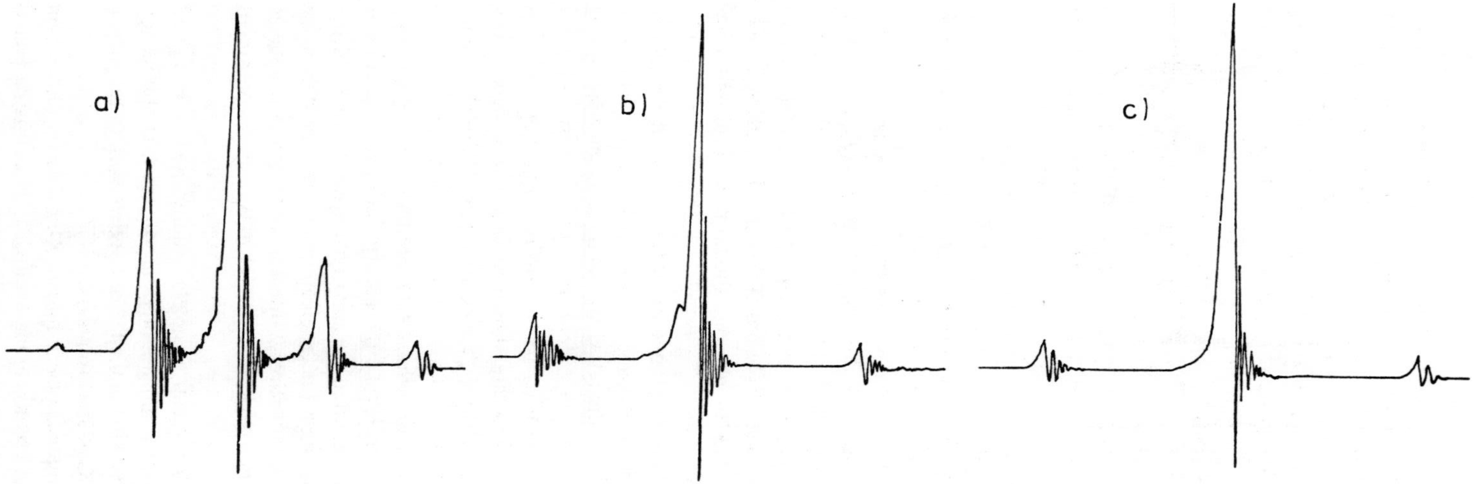

FIGURE 67. Spinning sidebands on the chloroform signal for spinning frequencies. (a) 12, (b) 22, and (c) 26 Hz.

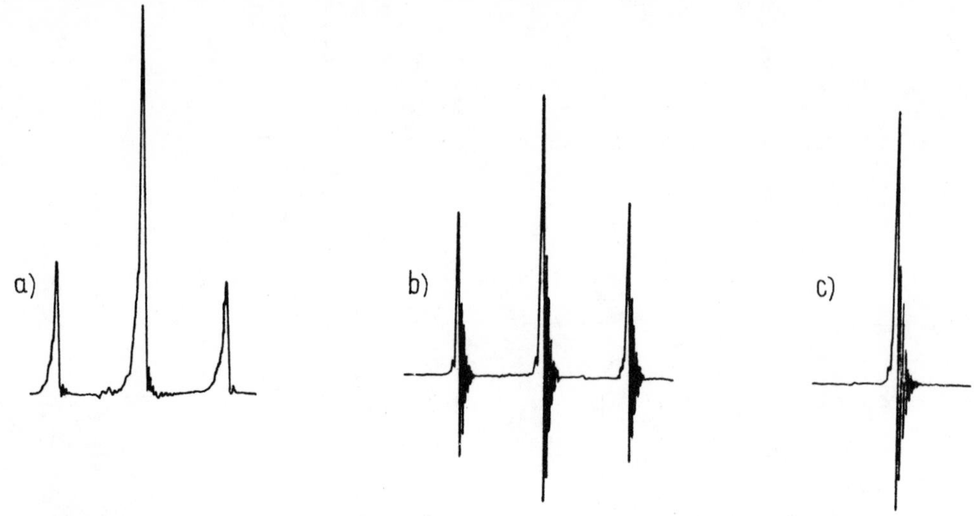

FIGURE 68. Signal forms in (a) and (b) inhomogeneous and (c) homogeneous polarizing magnetic field.

may be confused with the signals of the sample, when amplification is large (compare Figures 84 and 85).

By a careful preparation of the sample, one can achieve a significant reduction of the inhomogeneities. Contamination of the outer wall of the sample tube may also distort homogeneity.

Solid particles dispersed in the sample, especially ferromagnetic particles (''fish''), lead to a decrease of homogeneity and instability of the resolution. Filtering, repeatedly if necessary, of the solutions is therefore recommended (though one should be careful with filter paper since its fibers also impair homogeneity!).

Any sharp signal can be used to get information of the homogeneity, e.g., the signal of the reference substance. In inhomogeneous field, the signals are broadened and have an asymmetric shape, ''hump'' (see Figure 68a), or although sharp enough, their spinning side bands are too intense (see Figure 68b).

The spectrum of *ortho*-dichlorobenzene can be used for setting resolution. On Figures 69a, b, and c, the spectra of *ortho*-dichlorobenzene are shown recorded with these spectrometers: VARIAN® A 60-D (CW), JEOL® PS-100 (FT), and BRUKER® WH-500 (FT).

2.1.2. The RF Transmitter. The Detector and the Phase Correction

To induce energy transitions between magnetic levels, the absorption of RF radiation is necessary. Therefore an RF radiation source is required for creating the excitation field $\mathbf{B}_1$.

According to the different chemical shifts of the same nuclear species, either RF, or the field $\mathbf{B}_0$, should be altered continuously in a certain narrow interval to effect the resonance of differently shielded nuclei. (This is the origin of the terminology ''CW'', i.e., continuous wave.)

The current — CW spectrometers that utilize electromagnets usually modify $\mathbf{B}_0$, since the even variation of this can be realized easier, and for this purpose an audiofrequency generator working at frequency 2 to 5 kHz is usually applied. For the detection of resonance absorption, there are two methods applied in CW spectrometers.

In the regenerative (Purcell) method, the sample is placed into an RF oscillator of constant voltage of high stability, and the change of the voltage in the oscillating circuit is measured as a function of $\mathbf{B}_0$. For this purpose, the usual compensation method (e.g., balancing bridges) is applied. In the case of resonance, the balance is upset and on the meter of the

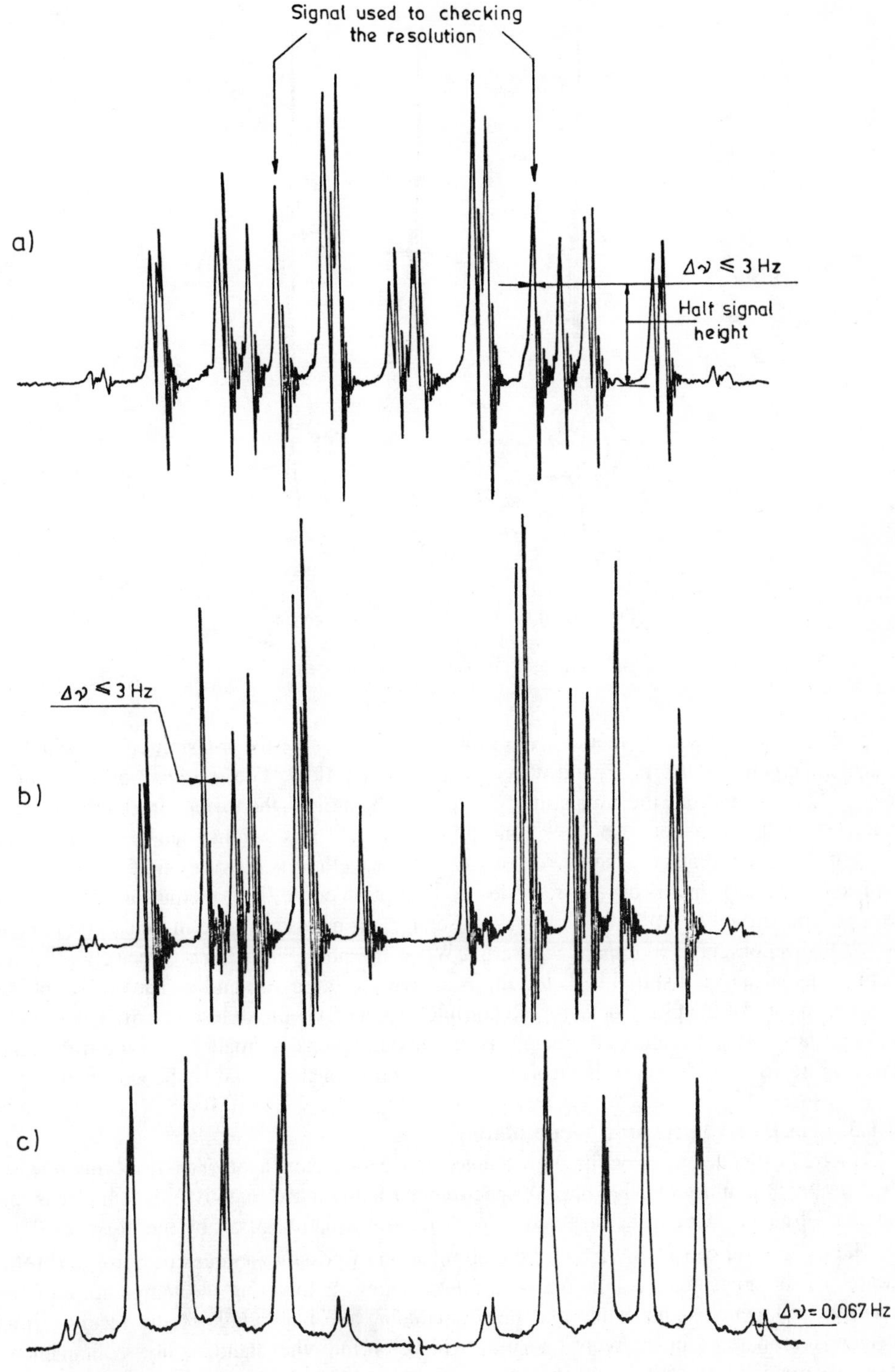

FIGURE 69. The *AA'BB'* multiplet of *ortho*-dichlorobenzene for checking the resolving power recorded on (a) VARIAN® A-60D,(b) JEOL® PS-100, and (c) BRUKER® WM-500 spectrometers. (Reproduced by courtesy of VARIAN, JEOL, and BRUKER, respectively.)

compensation circuit, a deflection proportional to the amount of absorbed energy can be observed or recorded.

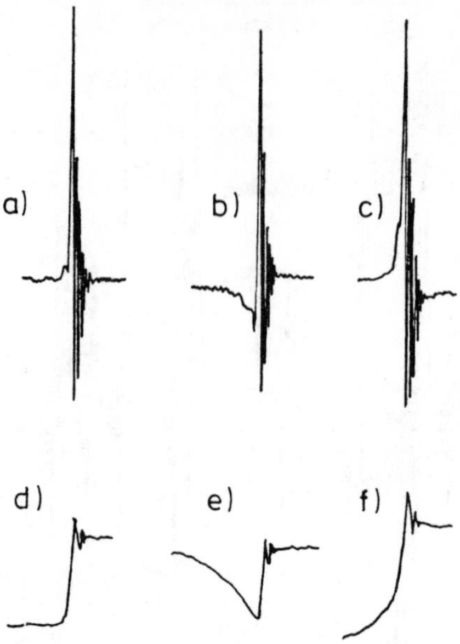

FIGURE 70. Absorption signal (a) to (c) and integral curve (d) to (f) in the case of proper phase setting (a, d) and for shifted phase (b, c, e, f).

In the inductive (Bloch) method, the probe is placed inside the transmitter coil which is oriented perpendicularly to $\mathbf{B}_0$ and which creates the RF field. The receiver coil is oriented perpendicularly to both the transmitter coil and to $\mathbf{B}_0$ around the probe. In the receiver, a voltage is induced under resonance conditions only, and this voltage can be recorded. In current CW spectrometers, the more sensitive Bloch method is most often applied.

The resonance signal is usually recorded as absorption curve, i.e., the out-of-phase signal component, shifted by 90° relative to the phase of B_1, is measured. The in-phase component is the dispersion curve (compare Figure 8). When the phase shift is not precisely 90°, the form of the absorption signal is distorted, as shown in Figure 70b and c. The correct phase is set by means of the phase detector. Recordings of incorrect phase lead to errors especially through false integral values: the height of the integral steps is smaller than the true value (see Figure 70e and f), and this error is also a function of the signal shape and intensity.

2.1.3. Sensitivity, Spectrum Accumulation

Among factors determining the performance of spectrometers apart from resolving power, sensitivity is the most decisive one. A spectrometer is the more sensitive, the smaller is the number of nuclei that can be detected. This is determined first of all by the noise level.

The noise level can be lowered by the elimination of possible sources of noise, a careful design of the amplifiers, oscillators, and power supplies, lowering the temperature of the receiver coil which reduces thermal noise, reducing the bandwidth of the receiver (this demands an increase in the recording time which, on the other hand, is not recommended over a certain limit), and finally by improving the so-called filling factor of the receiver coil which can be realized by reducing the distance between the probe and the receiver coil.

For characterizing sensitivity, the signal-to-noise ratio, ζ, is used. This can be determined for ^{1}H NMR spectra using a 1% solution of ethylbenzene in CCl_4 by measuring the height of the most intense line of the methylene quartet and dividing this value by four tenths times the noise as measured on the base line (see Figure 71a).

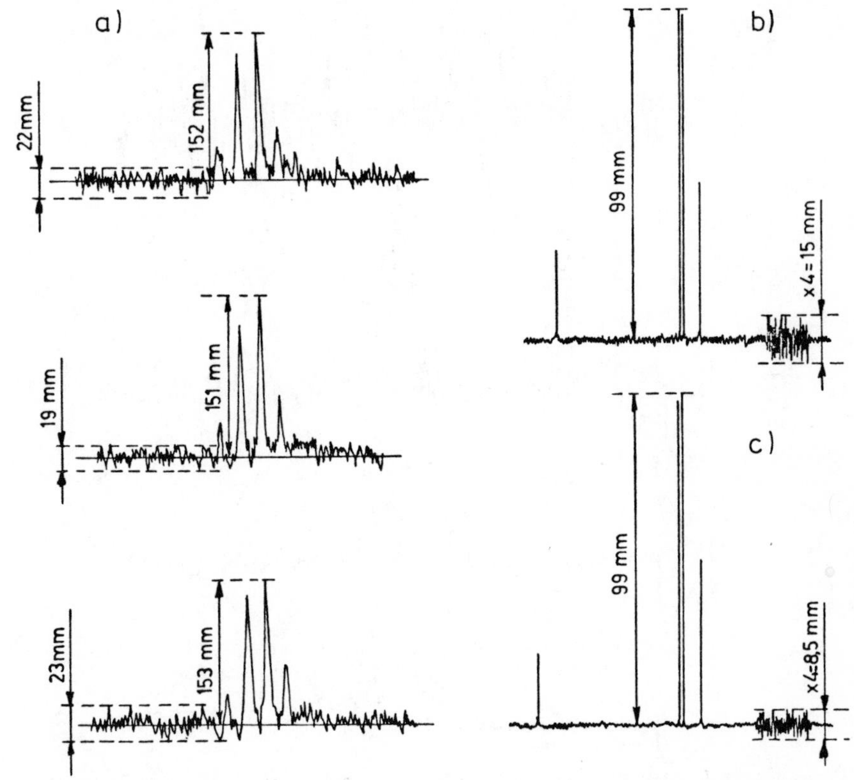

FIGURE 71. ^{1}H NMR spectrum part (a) recorded on VARIAN® A-60D spectrometer of 1% ethylbenzene in CCl_4 with the methylene quartet, for the determination of the signal-to-noise ratio (ζ) from three parallel measurements: $(151 + 152 + 153)/3 = 152$; $(22 + 23 + 19)/3 = 21.3$; $21.3/2.5 = 8.5$; $\zeta = 152/8.5 = 18{:}1$. (Courtesy of VARIAN.) ^{13}C NMR spectrum part (b) and (c) of 10% ethyl benzene in $CDCl_3$ recorded on BRUKER® WM-250 and WM-500 spectrometers (1 scan) to determine ζ, 66:1 (b) and 115:1 (c). $\zeta = 10A/4B$, where A is the intensity of line at 127.9 ppm and B is the noise.[218]

With 60-MHz CW spectrometers, this ratio is $\sim$ 18:1, whereas at 100 MHz, it is 50:1. To get a sufficiently high ζ-value for ^{1}H NMR spectra at 60 MHz, the necessary conditions are 0.1 to 0.4 $m\ell$-volume of sample solution and a minimum concentration of 10^{-2}, optimally 10^{-1} to 10 mol/ℓ. To determine ζ for ^{13}C NMR spectra, the most intense line (signal of the *ortho*-carbon of ethylbenzene in 1% CCl_4 solution is generally used (see Figure 71b). As in more concentrated solutions molecular interactions cause signal broadening, dilute solutions are preferable.

A significant increase in ζ is made possible by means of spectrum accumulation techniques (CAT, computer average of transients). The spectra are recorded many times in succession; correspondingly, the frequency and intensity data are added up and stored in the memory of a computer connected to the spectrometer. This way signal intensity builds up, while random noise is cancelled. With pulsed technique, the computer also decomposes the spectrum obtained as the superposition of the n spectra according to the principles of Fourier transformation (FT), and having averaged individual frequency, signal shape, and intensity data, the final result is fed into the recording unit of the spectrometer. Thus, after n scans, signal intensities increase by a factor of n, whereas noise increases only by a factor of $\sqrt{n}$. By this method it is possible to study the spectra of very dilute solutions or even gases.

CAT is an integral part of the operation principles of the FT spectrometers, but it can also be applied for CW equipment when the reproducibility and stability are satisfactory for

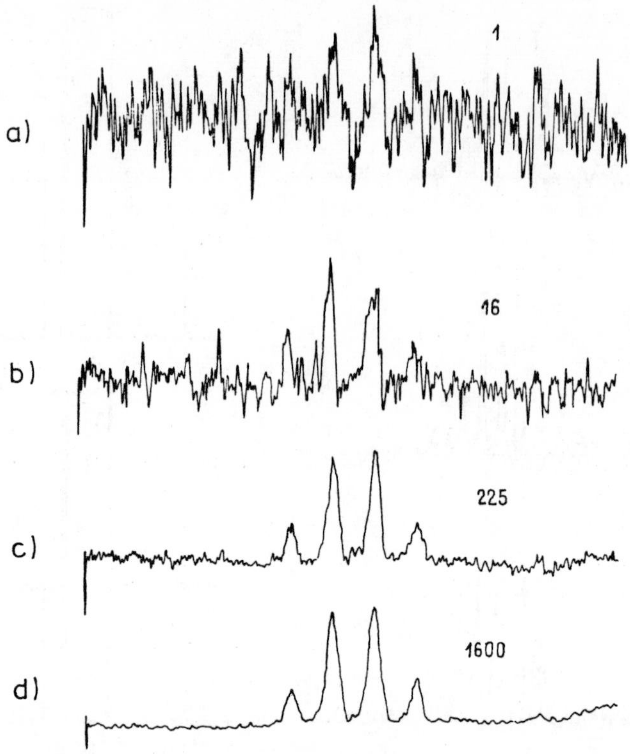

FIGURE 72. Spectrum part of 0.1% ethylbenzene in CCl_4 with the methylene quartet. (a) After a single scan and (b) to (d) after 16, 225, and 1500 scans, respectively. (Courtesy of VARIAN.)

this purpose. On Figure 72 the methylene quartet of ethylbenzene taken in 0.1% CCl_4 solution is shown obtained in a simple run (a) and after 16, 225, and 1600 accumulated runs (b to d). The CAT procedure is very time consuming. Taking scanning time of CW spectrometers (about 4 to 5 min), an increase of ζ by a factor of 10 requires 100 repetitive scans, i.e., 7.5 hr. The FT spectrometers (see below) solve this problem by introducing new principles of measurement.

2.1.4. Spectrum Integration; Signal Intensity Measurement, Quantitative Analysis

It has become clear at the very beginning of NMR studies on organic compounds that ^{1}H NMR signal intensities are directly proportional to the number of nuclei that give rise to these signals and are independent of the chemical environment of these nuclei.* Thus, as opposed to optical spectroscopy, absolute intensities can be measured to a very good approximation and the intensities can be directly compared, since even for different protons the extinction coefficients are in most cases practically the same.

In fact the proportionality is only a good approximation, as the signal intensities depend also on the relaxation times T_1 and T_2 and on the magnitude of the excitation field B_1. T_1 and T_2 depend also on the chemical environment of the nuclei, whereas the value of B_1 giving maximum intensities is determined by the former parameters.** Therefore if high precision is endeavored, calibration is necessary. Apart from these considerations, only intensities from spectra run under exactly the same experimental conditions can be compared.

* See Reference 1051, compare p. 24, and see Figure 10.

** Compare p. 22 and Equation 60.

Signal heights depend to a large extent on the signal shape; therefore they give the actual ratios only exceptionally. Therefore it is unavoidable to determine the area under the signals.

Modern spectrometers contain built-in integrators. The integral curve is a horizontal line in an absorption-free spectral region, and concurrently with signals, steps appear whose height is proportional to the absorbed amount of energy (compare Figure 10). By measuring the step heights, the intensity of the signal in question is obtained directly.

The integrator is basically an amplifier placed between a phase-sensitive rectifier and the recording unit which amplified a voltage difference, rectified and fed back through a condenser. Upon recording the integrated spectrum repeatedly, averaged step heights can be obtained and used in calculations.

For obtaining precise integral values, the correct choice of B_1 is essential; when the latter is too high, saturation sets in and the intensities will be lower fluctuating. Correct phase setting is very important (compare Figure 70). The stability and high resolution also helps to obtain accurate integrals. For closely lying signals, namely, the horizontal parts of the integral curve between the signals, become inflections and, therefore, no accurate individual step heights can be obtained, only the summarized height of two or more signals.

The precise measurements of the NMR signal intensities are very important, both in structure elucidation work (determination of the relative number of the chemically different nuclei) and in analytical work (determination of component ratios). Symmetrical dimers and polymers naturally cannot be distinguished from monomers.

The weight percentage of hydrogen in compounds of unknown composition can be determined with high accuracy (0.1%) when a known amount of an inert substance is given to the solution of the sample in known concentration. This inert (e.g., water, chloroform, etc.) substance should exhibit a sharp signal well separated from the signals due to the unknown compound. The amount of solvent is immaterial. The relative intensities are then determined at different relative concentrations. When the total intensity of the signals of the unknown is equal to the intensity of the reference signal, the amount of corresponding nuclei is equal in both samples.

By this method, for example, the H/D ratio in a sample of isotopically impure heavy water can be easily determined; proton signal intensity is determined in the original sample, then a known amount of water is added, and the signal intensity is determined once again. When the signal intensity is plotted against the amount of water added, a straight line is obtained, the intercept of which directly gives the amount of water originally present.

An example is the determination of the acetone content of the bioactive penicillin derivative **49**.[977] The final step of synthesis of **49** is precipitation of the product by the addition of acetone. The end product, despite drying, contains some amount of acetone, the maximum allowed content of which is 1%. Acetone content can be quickly and accurately determined by taking the integrated ^{1}H NMR spectrum.

Acetone gives a singlet at 2.22 ppm. This is cointegrated with the C-4′ methyl signal of **49**, giving a height of *A;* the two C-2 methyl signals are also cointegrated at about 1.6 ppm to give a step of height *B*. Since signal *B* is for six protons, the intensity of the acetone signal is $A - B/2$.

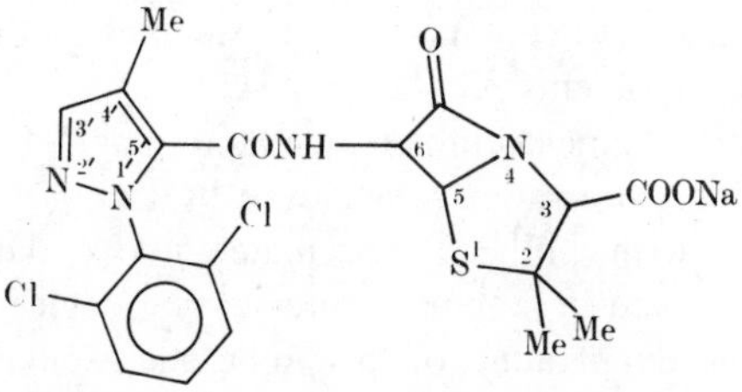

49

Knowing the molecular weight of the acetone and Compound **49** (58 and 491.3), the following formula can be deduced for the weight percent of acetone:

$$\frac{58(A-B/2)/B}{58(A-B/2)/B + 491.3} \cdot 100\% \qquad (184)$$

On Figure 73, a part of the spectra of three drug samples of **49** can be seen with the integral steps. The first sample (a) did not contain acetone, in the second (b), there is a smaller amount, whereas in the third (c), there is a greater amount of impurity. From Figures 73b and c, A (b) = 60, B (b) = 110, A (c) = 80, and B (c) = 106 mm. Using Formula 184, the acetone content of the two samples are 0.53 and 3.00 wt%.

In a similar fashion, the determination of the composition of simple two-component mixtures, and in many cases of multicomponent mixtures, is simple. It is enough when each component possesses at least one signal or group of signals, well separated from the rest of the signals for which the corresponding atom number is known. From the integral ratio, the composition of mobile equilibria, e.g., that of tautomers, isomers, and conformers, is directly obtainable.

As an example, the reduction of ethyl nicotinate (**50**) to 3-pyridyl-carbinol (**51**) is mentioned. Reduction is fast, and it is of economic importance to follow the process. In Figure 74, a part of the spectra of two mixtures (a,b) is shown with the quartet (4.55 ppm) and singlet (4.70 ppm) methylene signals of the starting material and the reduced product, respectively, and with the methyl triplet of the former.

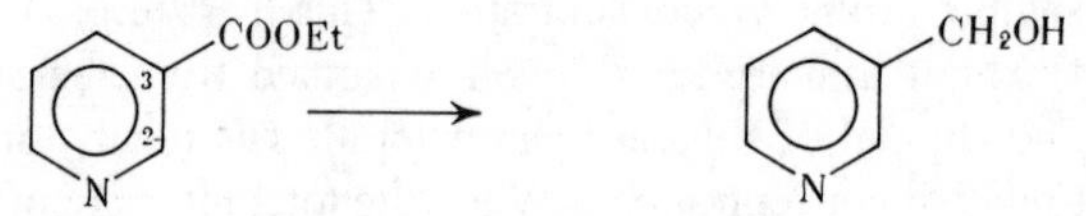

50 **51**

In order to calculate component ratio, it is not recommended to compare these signals, since they are too close to one another. Their summarized step heights can, however, be used in comparison, e.g., with the methyl signal. The product ratio in the mixture can be obtained when two thirds of the step height of the methyl signal is deduced from the total intensity of the methylene signals, and then the former is divided by the latter. The calculation using these data (see Figure 74a and b) gives 70.5 and 23% as the product content of the first and the second sample, respectively.

For the use of the integrated spectra in structural work, there are some examples in the Problem section, and a few examples for determining the composition are also given there (Problems **29, 37, 38, 39**, and **60**).

2.1.5. Variable Temperature Measurements

Investigation of the temperature dependence of NMR spectra is a very powerful method for the study of the internal molecular motions and association.* In this section only, the technical problems of spectral recording will be dealt with, and applications will be discussed in detail later. The Problem section contains further examples for the temperature-dependent studies of spectra (compare Problems **33, 42**, and **51**).

Spectrometers permit measurement in the temperature range of −120 and +200°C. The automatic control of sample temperature is achieved by letting a gas stream of controlled temperature (nitrogen, air) into the probe. The accurate setting of the temperature and keeping it at a constant level is achieved by a temperature-sensing device placed in the vicinity of the probe. This detects the temperature of the probe and regulates at the same time the

* Compare p. 71 and Section 3.7.1.1 (Volume II, p. 92-97).

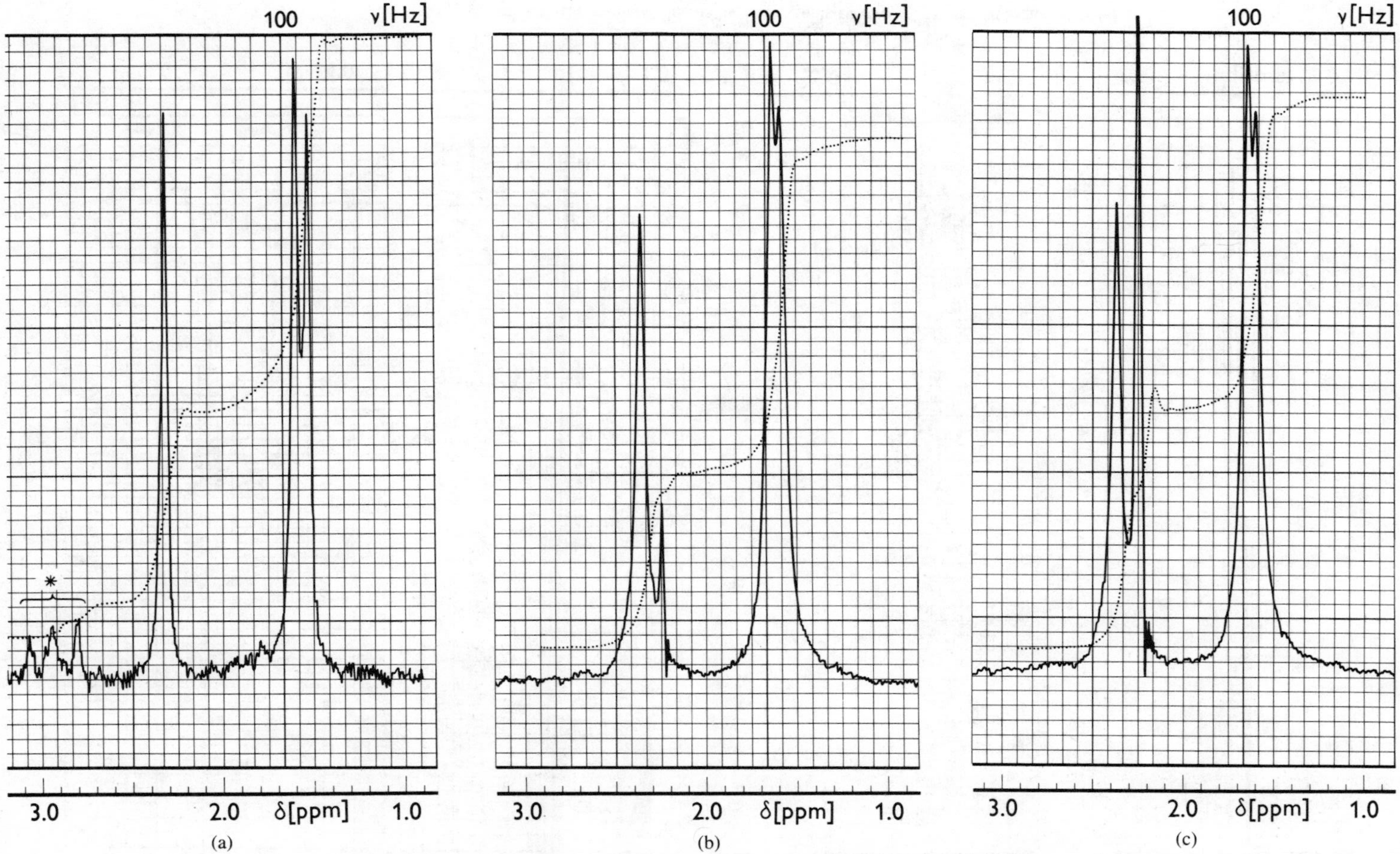

FIGURE 73. The spectrum part (1 to 3 ppm) of Compound **49** in D_2O, containing various amounts (a-c) of acetone.

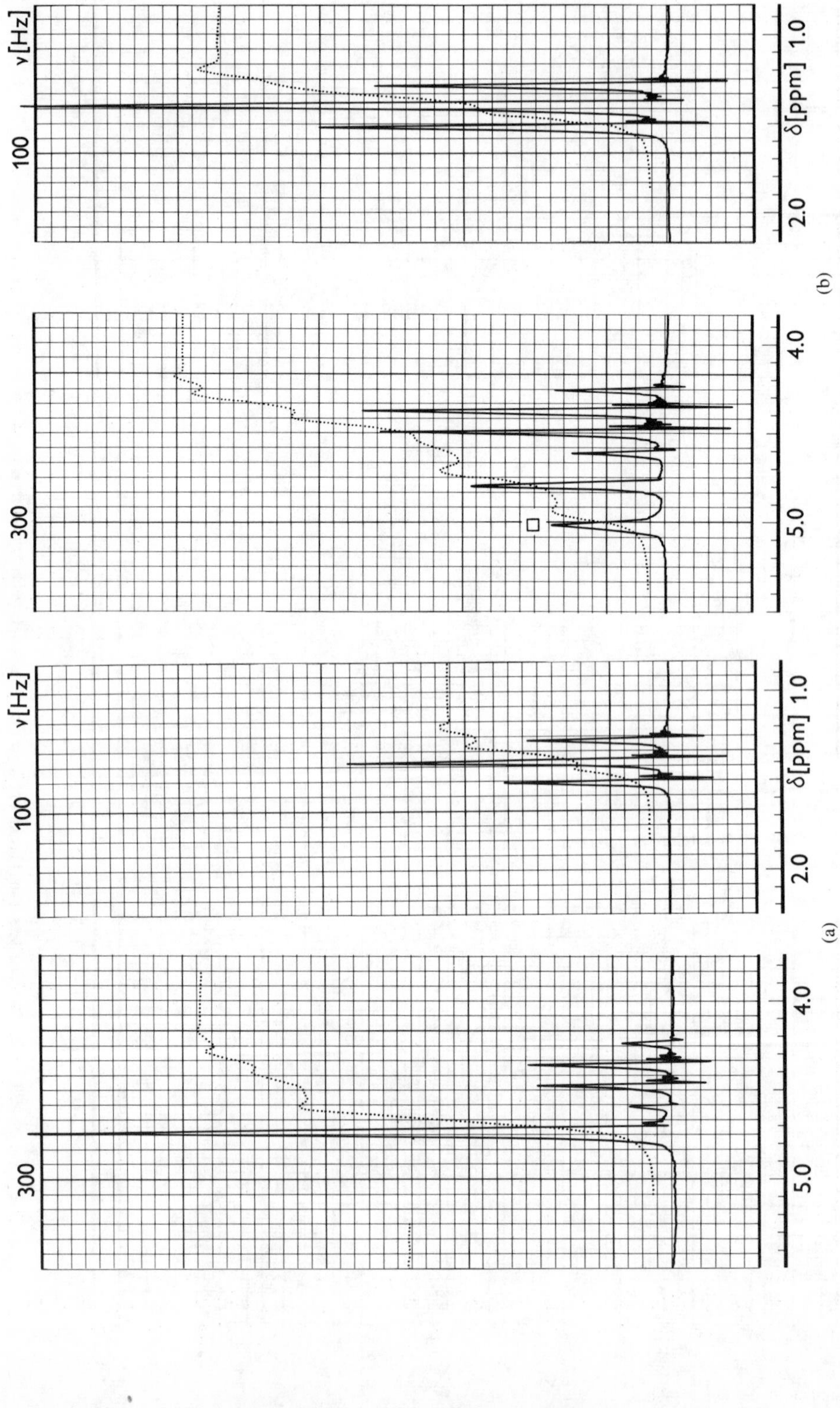

FIGURE 74. The spectrum part with the methyl and methylene signals of two mixtures of Compounds **50** and **51**.

working of the heating unit. The sensor is the element of a bridge circuitry balanced at the temperature desired.

When the temperature is changed, the circuit is off-balance and the resulting current regulates the heating power. Corresponding to the direction of the current, the heating power transferred to the nitrogen gas is increased or decreased until the former equilibrium is achieved again.

For measurements above room temperature, a gas stream at room temperature is heated to the required temperature. For measurements below room temperature, dry nitrogen gas is used which is first let through a coil immersed into liquid nitrogen and then reheated to the necessary temperature. Cooling may also be achieved by means of expanding the gas through an orifice. Drying of both the nitrogen gas and the pipes is very important in order to prevent icing. The gas used for spinning the sample must also be thoroughly dried.

The thermoelement used to control temperature is calibrated for higher temperatures by using the temperature dependence of the spectrum of ethylene glycol, whereas the same effect on methanol is utilized for lower temperatures. The signal-separation of ethylene-glycol and methanol is directly proportional to temperature and is a sensitive measure of the latter.

These signal shifts are related to the formation and reopening of the hydrogen bonds and are, therefore, influenced markedly by the water content of the samples. Therefore, only calibration sample in a sealed sample tube may be used. For a new sample, a new calibration curve must be made.

Measurements above room temperature are limited by the boiling point of the sample and those at low temperature are limited by the solubility of the solute and by the freezing of the solvent. The measurements at subambient temperatures are often hampered by line broadening* due to increased viscosity. This may be partly compensated for by lowering the concentration. Since a change in temperature causes significant drift in the magnetic field, the homogeneity should always be reset before measurement.

Finally it is noted that high-temperature measurements may be useful, e.g., to attain the desired concentration of poorly soluble substances, study thermally catalyzed processes (e.g., rearrangement reactions), improve spectral quality when the lines are broad at room temperature due to high viscosity, and to identify signals (of acidic protons) that change their location upon changing temperature, etc.

2.1.6. Double Resonance in CW-Mode

Though in routine frequency- or field-sweep NMR experiments only one set of isochronous nuclei is excited at a time, Bloch pointed out as early as in 1954[149] that by the simultaneous application of two or more RF oscillators, it is possible to excite two or more sets of nuclei with different chemical shifts or even types (e.g., proton and fluorine). When a sufficiently broad-band RF oscillator is used which emits polychromatic radiation in a frequency interval, $\Delta\nu$, then in the case of a sufficiently large bandwidth (500 to 1000 Hz for protons), all anisochronous sets of a given type of magnetic nuclei can be simultaneously excited.[450]

The recording of NMR spectra in the presence of more than one excitation field is called multiple resonance (double — DR, triple or broad-band — BBDR). Depending on whether nuclei of the same kind or different kind are excited simultaneously, one talks about homo- or heteronuclear double resonance. There are various types of multiple resonance methods — which provide different sorts of information — depending on the number of excitation frequencies, the scanning mode (field or frequency sweep), and the range and intensity of the excitation frequencies. The multiple resonance methods may be very useful both in structural research and theory as well as for solving various technical problems of recording spectra. There is no space here for even a schematic description of all the aspects of multiple

* Compare p. 19 and p. 160.

resonance or to deal with the theoretical background of these phenomena. A short summary of the features important for structural research is given here; for details we refer to monographs[51,300,434,1147] and reviews.[73,1110]

2.1.6.1. Spin Decoupling

The complexity of multiplets frequently makes the interpretation of the NMR spectra difficult or even impossible. The main complicating factor leading to spectral complexity, the spin-spin splittings, cannot be removed by a change in the recording conditions. Multiple resonance is, however, a way to achieve this. By spin decoupling, the simplest kind of DR technique, the signals of mutually coupled spin groups can be recognized and their splitting due to mutual spin-spin coupling can be eliminated.

In the simplest case, when two nonequivalent nuclei of spin 1/2 interact (*AX* spin system), the doublet of nucleus *A* corresponding to the two spin states of nucleus *X* by irradiating the sample in a constant field B_0 with a second and fixed excitation frequency $\nu_2 = \nu X$, corresponding to the resonance frequency of *X*, collapses to a singlet at frequency $\nu_1 = \nu A$. Spin-spin coupling requires that both interacting nuclei are for a sufficient time in a given spin state. The necessary minimum average lifetime of the interacting spin states is of the order $1/J$. If the lifetime is shorter than that, the spin orientation also changes during the resonance and the spectrometer will record an unsplit signal corresponding to the average spin state.* The role of the second RF field B_2 is to accelerate transitions between the spin states of the selected group of nuclei in order that nuclei coupled with them should "feel" only the average of these spin states. Decoupling can be expected when the amplitude γB_2 of the excitation field is much greater than the distance of the outer lines of the multiplet to be simplified (in A_nX_m spin systems $\gamma B_2 >> mJ_{AX}$). For complete decoupling, $\Delta\nu AX$ must be sufficiently great, otherwise disturbing phenomena as signals shifts,** intensity changes, and further splitting occur. Complete decoupling is therefore easiest to realize in the case of first-order spin systems. The first spin decoupling experiments were carried out on $^{13}CH_3I$, simplfying its methyl doublet in the ^1H NMR spectrum into a singlet when the sample was irradiated at the resonance frequency of the ^{13}C nucleus.[1227]

DR experiments can be characterized by three experimental parameters: the magnitudes of the polarizing field B_0 and the excitation fields B_1 and B_2 (or the corresponding frequencies ν_1 and ν_2). Recording of the DR spectrum is achieved by varying one of the three parameters, keeping the other two constant. Simple NMR spectra obtained in field or frequency modulation mode are identical. The same does not hold for DR spectra. When B_0 and ν_2 are unchanged and ν_1 is varied, the signals of nuclei of different chemical shifts appear successively, and the fine structure of the signals will be simplified for nuclei in spin-spin interaction with the irradiated nucleus *X*. The *X* signal will be missing from the spectrum, since under the strong irradiation of $\nu_2 = \nu X$ nucleus will be saturated. When in turn both $\nu_1 = \nu A$ and $\nu_2 = \nu X$ are constant, only the signals appearing at the field $B_0' = B_0(A)$ or $B_0(X)$, corresponding to the resonance frequency νA or νX, will be simplified. For other $\mathbf{B}_0'$ fields, the resonance of nucleus *X* and *A* does not occur. For this reason, when nucleus *X* is coupled to *i* sets of nuclei, *i* experiments are necessary for the decoupling of all interacting nuclei. Therefore, only the frequency modulation method is applied for DR experiments. The third version, when ν_0 and ν_1 are constant and ν_2 is varied, is the s.c. INDOR, internuclear double resonance (see Section 2.1.6.4).

Spin decoupling is illustrated by a simple example. In the ^1H NMR spectrum of 2-formyl-3-bromo thiophene (**52**), the H-4, H-5, and the aldehydic signals at 7.18, 7.75, and 10.03 ppm are doublet, double doublet, and doublet, respectively. J(H-4, H-5) = 6 Hz, J(H-4, CHO) $\approx$ 0 Hz, and J(H-5,CHO) = 2 Hz (see Figure 75a). Upon irradiating the H-4 signal

* Compare p. 50 and Volume II, p. 97, 238, and 251.

** Signal shifts under the effect of DR (Bloch-Siegert effect)[152] may also be observed on spin-decoupled spectra. The shift is paramagnetic and the greater, the closer the signal is to ν_2.

(see Figure 75b), the double doublet is simplified into doublet ($J = 2$ Hz), whereas the aldehydic signal remains unchanged (elimination of interaction between H-4 and H-5). When H-5 is saturated, the other two lines become singlets (see Figure 75c); couplings to both are eliminated. Finally, when the aldehydic signal is decoupled, an *AB* spectrum ($J_{AB} = 6$ Hz) can be observed for the ring protons (see Figure 75d).

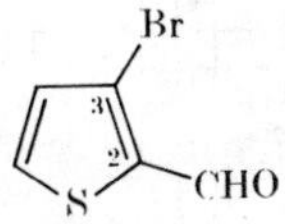

52

A further example of homonuclear decoupling will be shown in connection with the spectrum of PVC (compare Figure 121). Heteronuclear decoupling can be realized easier since the condition prescribing that the chemical shift difference between the nuclei to be decoupled is large.

Line broadening caused by quadrupole moment of the nuclei of spin $I \geqslant 1$ can also be eliminated using DR. Thus, the spectral lines of nitrogen-containing compounds become sharper on irradiation of ^{14}N resonance frequency.*

In the routine ^1H NMR spectrum of formamide, a strongly broadened formyl doublet and a broad absorption of the amide protons appear (see Figure 76a). Eliminating the ^{14}N,H couplings by heteronuclear DR, the spectrum can be described in the *AMX* approximation. The coupling constant of the interaction between the formyl hydrogen and the amide proton in *cis*-position and between the two *geminal* amide hydrogens is 2.1 Hz for both, while that of the hydrogens in *trans*-position is 13.3 Hz.[179] Accordingly the formyl proton signal and the signal of the amide proton in *trans*-position to the former are double doublets, while the signal of the amide hydrogen in *cis*-position is a triplet (see Figure 76b).

2.1.6.2. Selective Decoupling

The individual lines of more complex multiplets may also be decoupled separately. By means of such experiments the relative sign of the coupling constants can be determined, for example. The simplest case is that of an *AMX* spin system. When the amplitude of the RF field B_2 is reduced to a level that only transitions corresponding to two lines of a double doublet (e.g., that of the *M* spin) are saturated ($\gamma B_2 \approx J$), only one of the doublets of the other signal (e.g., *A*) collapses to a singlet. Depending on whether the relative sign of the coupling constants J_{AM} and J_{AX} is the same or opposite, the lower or the higher field doublet is simplified, respectively. This can be explained as follows.

The two doublets of the *M* double doublet correspond to molecules differing in the orientation of spin *X*. The same applies to the *A* signal (see Figure 77). The two-two lines of the *M* doublets differ in the orientation of spin *A*, whereas those of the *A* doublets in the spins of *M* (see Table 14). Taking $J_{AM} > 0$ as shown in Figure 77, spin orientations corresponding to the different lines are different for cases $J_{AX} > 0$ and $J_{AX} < 0$. The change in the sign of J_{AX} is reflected by the opposite spin orientations of nucleus *X* in part *A* of the spectrum and by that of *A* in the *X* part. Upon irradiating one of the doublets, e.g., the right-hand side one, the *M* part by frequency ν_2, there are two possibilities. When J_{AM} and J_{AX} have identical signs, only the lower-frequency doublet in the *A* part is reduced into a singlet; if in turn the signs are opposite, it is the higher-frequency doublet of the *A* lines that is simplified into a singlet. The change in the *M* part upon irradiating the *A* multiplet is naturally analogous.

* Compare Volume II, p. 78.

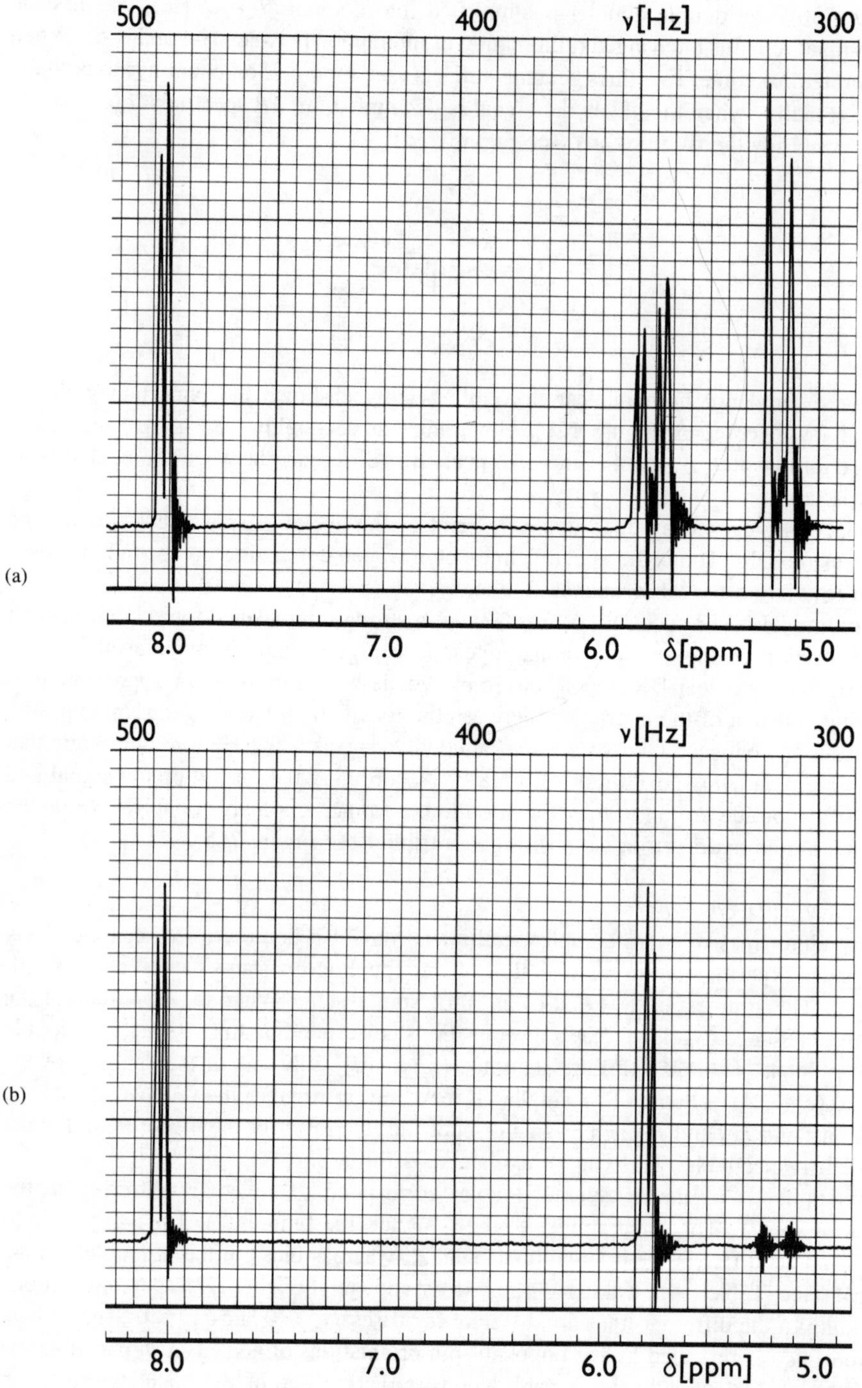

FIGURE 75. The spectrum of (a) Compound **52** and (b), (c), and (d) the DR spectra observed by saturation of H-4, H-5, and the aldehydic proton. (Reproduced by permission of Professor H. Wamhoff, Friedrich Wilhelm University, Institute of Organic and Biochemistry, Bonn.)

As an example,[525,1110] furan-2-carboxylic acid **53** is mentioned, in which H-5, 3, and 4 represent an *AMX* spin system (see Figure 78). Irradiating the lower-frequency doublet of

(c)

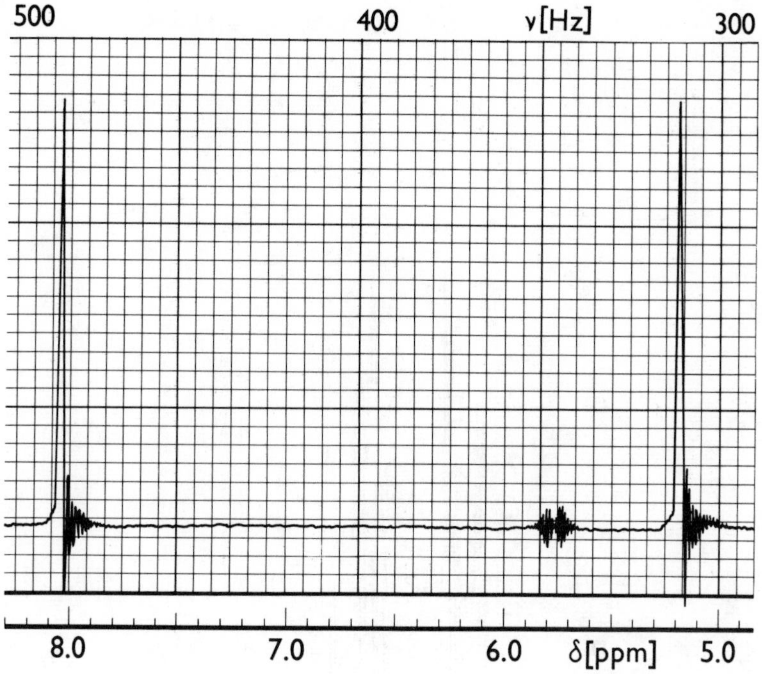

(d)

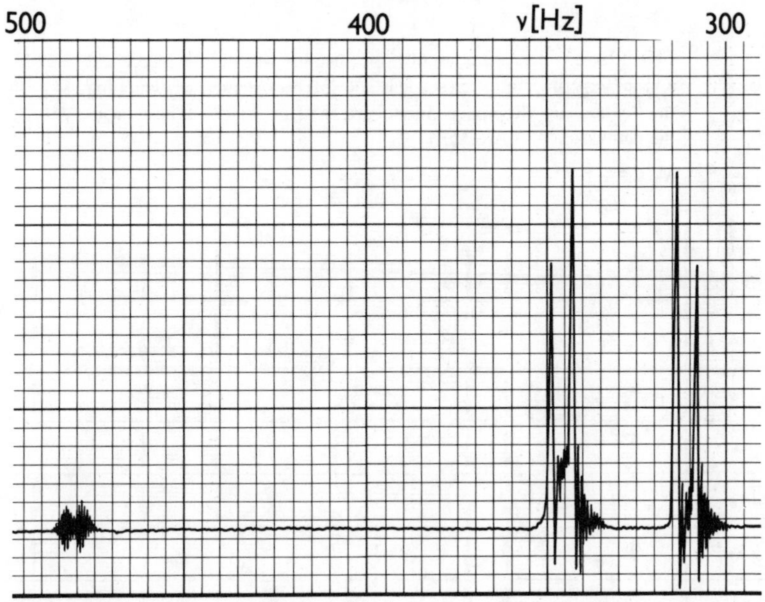

FIGURE 75c and d.

the signal *M*, the lower-frequency doublet in part *A* is simplified into singlet (see Figure 78b), proving the identical signs of J_{AM} and J_{AX}.

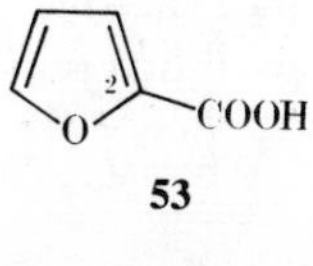

53

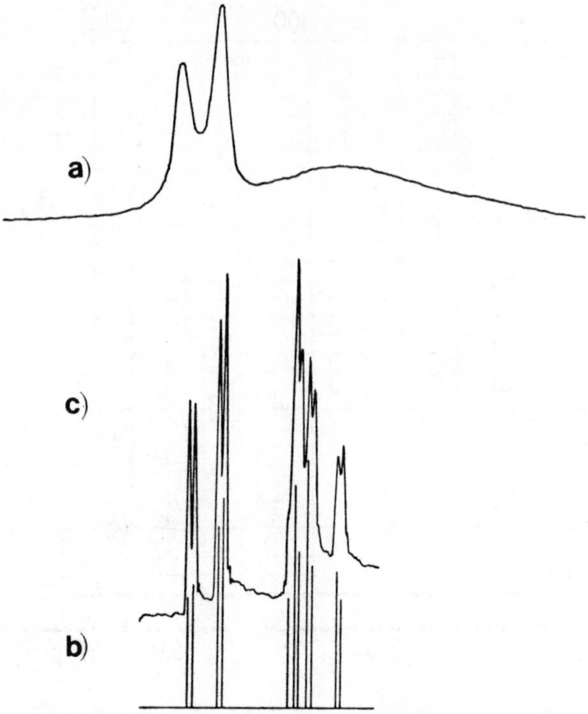

FIGURE 76. The spectrum of (a) formamide with (b) the theoretical and (c) experimental heteronuclear DR spectra obtained by saturating the ^{14}N nucleus. (Reproduced by courtesy of Dr. H. W. Wehrli, with the permission of VARIAN.)

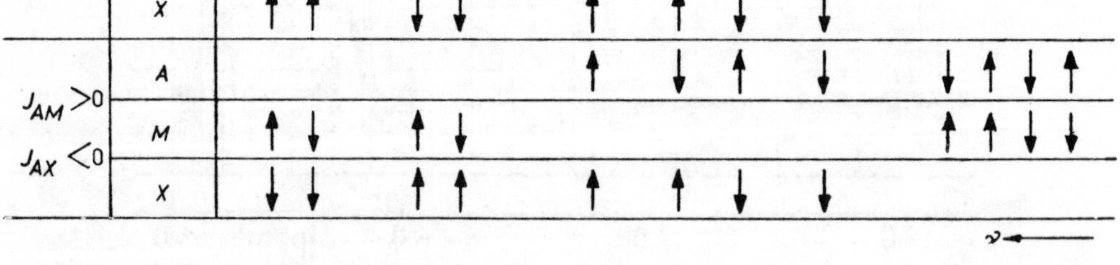

FIGURE 77. Spectral lines and spin orientations of the *AMX* spin system for cases (a) $J_{AM} > 0$, $J_{AX} > 0$, and (b) $J_{AM} > 0$, $J_{AX} < 0$ for the illustration of the principle of selective spin decoupling.

2.1.6.3. *Tickling*[514]

When the amplitude of the RF is decreased further ($\gamma B_2 \approx \Delta\nu^*_{1/2} \approx 1/T^*_2$, where $\Delta\nu^*_{1/2}$ is the half-width of the irradiated line and T^*_2 is the spin-spin relaxation time, shortened by inhomogeneities in B_0),* it can be achieved that only one transition of the spin system will

* Compare p. 177.

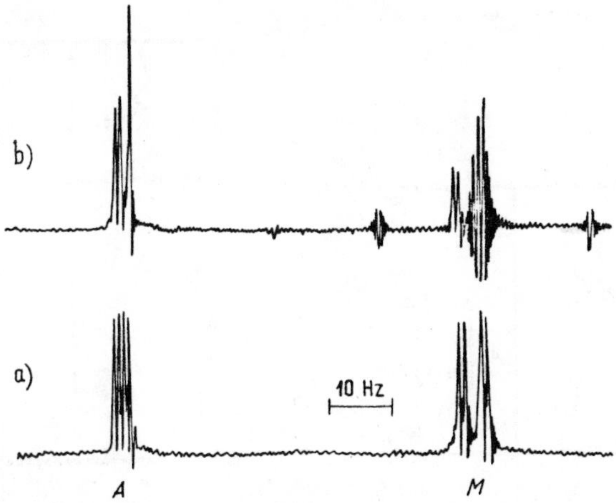

FIGURE 78. (a) The A and M part (H-5 and H-3 multiplets), respectively, of **53** and (b) the change of the A multiplet upon saturating the two upfield lines of the M multiplet.[525,1110]

be saturated. In this case there will be no simplification of the signals of the coupled partners, on the contrary, all lines belonging to transitions whose ground or excited state are common with the transition excited by DR will be split. This is called spin tickling. Three cases are possible: either the ground or excited states are common (regressive transitions) or the ground state of one of the transitions is common with the excited state of the other (progressive transitions). These three cases are exemplified by the transition pairs A_4 and M_8, A_1 and M_5, or M_8 and A_2 of the *AMX* system in Figure 57. The signal arising from regressive transitions is a doublet of well-separated lines, whereas progressive transitions give rise to less resolved broad doublets. (In the case of degenerate levels, multiple splitting is possible.)

This can be understood in the following way. Take the *AX* spin system. Its four allowed transitions are depicted on Figure 79a. These give rise to four lines in the usual spectrum (see Figure 79b). On irradiation at the frequency of one of the lines, say of X_1, upper and lower energy levels of transition X_1 are perturbed, and the originally forbidden transitions **4** → **1** and **3** → **2** (A''_2 and A''_1) become allowed (see Figure 79c). Simultaneously, the energy of the transitions A_1 and A_2 is also changed (A'_1 and A'_2). As a result, both transitions are split (A_1 and A_2) and the doublet of the A_1 regressive transition arising this way is sharper than the signal of the more perturbed (progressive) transition, (see Figure 79d). The transition X_2 is not influenced.

For illustration the *AMX* proton system of 2,3-dibromo-propionic* acid was chosen. In the spectrum, the double doublets of the methylene protons (M, X) and of the methine proton (A) are well separated (see Figure 80a). Irradiating the highest-frequency A_1 line (compare Figure 55) in the order of decreasing frequencies in part M the second and the fourth lines and in part X the third and fourth lines are split (see Figure 80b). Provided that both J_{AM} and J_{AX} are positive, the first line corresponds to transition **1**(**4** → **1**) (compare Figure 55), therefore the four perturbed lines must belong to transitions from or to level **4** or **1**. These transitions are **5** and **7** (M) and **9** and **11** (X). Related to transition **1** transitions **5** and **9** are regressive (common upper level), and **7** and **11** are progressive (the ground state of the irradiated transition coincides with the excited states of the observed ones). As expected, the splitting of the lines assigned to transitions **5** and **9** is greater.

When now line A_2 (transition **6** → **2**) is excited, the third M and the third X lines correspond

* CH_2Br-CHBr-COOH.

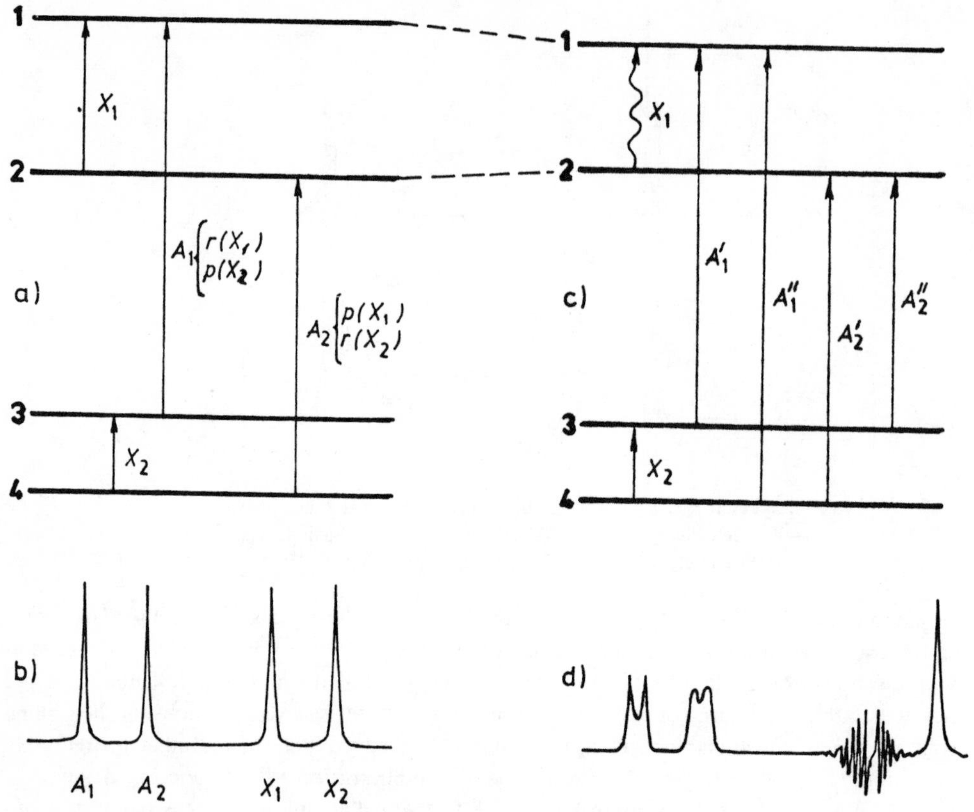

FIGURE 79. (a) The energy levels and transitions of the *AX* spin system, with (b) the corresponding spectral lines, (c) the level diagram, and (d) the tickling spectrum: r(X_1) and r (X_2), regressive; p(X_1) and p(X_2), progressive *A* transitions saturating X_1 and X_2 transitions, respectively.

to progressive transitions, while the first *M* and the fourth *X* are regressive. In fact (see Figure 80c) the former line pairs are less separated, while the latter doublets are better resolved. Thus (compare Figure 55), the first and third *M* lines correspond to transitions **6** and **8**, respectively, and the third and fourth *X* lines correspond to transitions **9** and **11**, respectively. The assignment of the line to transitions shown in Figure 80a (different from the scheme in Figure 55) follows from these considerations.

From this assignment it follows (compare Table 15) that $J_{MX} < 0$, for example, is opposite in sign to the other two coupling constants (in the case of identical sign, the energy level sequence in Table 15 and Figure 55 would remain unaltered). This is in agreement with the fact that *geminal* coupling constants ^2J of the methylene hydrogens is generally negative.* The absolute signs, of course, cannot be determined. It would also be in harmony with the mirror image relationship: $J_{AX} < 0$, $J_{AM} < 0$, $J_{MX} > 0$. All other combinations (three more mirror-image pairs) can, however, be excluded. The starting assumption, $J_{AX} > 0$ and $J_{AM} > 0$, is based on the convention of defining the sign of $^1J(^{13}C,H)$ couplings as positive; couplings ^3J(H,H) proved to be almost without exception to be positive.** The method for the determination of the absolute sign of the coupling constants is heteronuclear tickling of ^{13}CH multiplets.

Tickling experiments can be of use in many other fields of theoretical and practical NMR

* Compare p. 55.
** Compare p. 55.

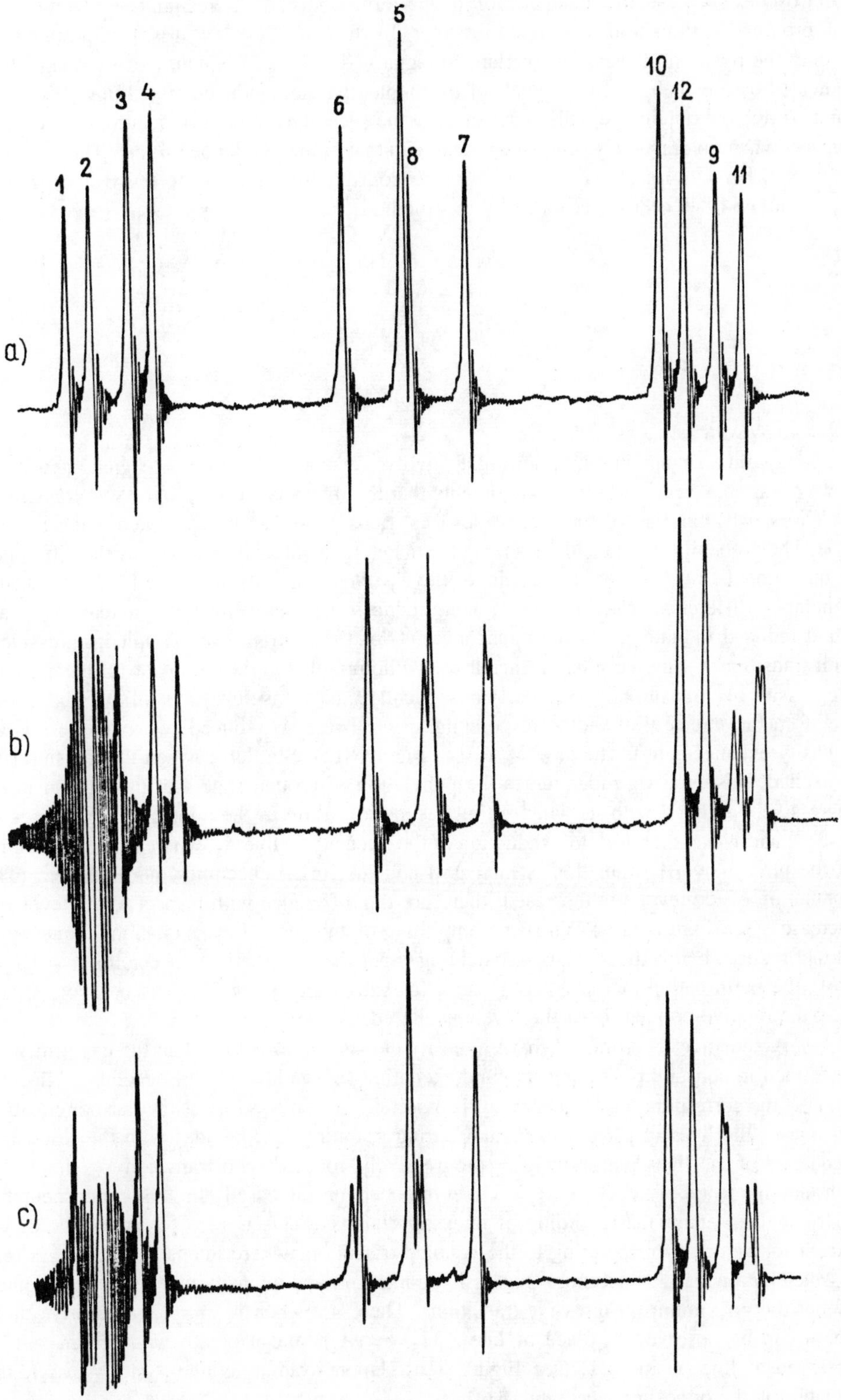

FIGURE 80. The *AMX* multiplet of 2,3-dibromo propionic acid (a) recorded in benzene-d_6 on JEOL® PS-100 spectrometer and the tickling spectra, by saturating (b) the outside downfield *A* line (leftmost one) or (c) the neighboring line. (Reproduced by courtesy of JEOL .)

spectroscopy, e.g., in the identification of the term levels by the recognition of regressive and progressive transitions belonging together. In structural research it is of importance for tracing the position of weak or overlapped signals. The exact frequency of such signals is indicated by the change of the signals of its coupled partners induced by tickling. When the range of the overlapping signals is scanned by the second excitation frequency ν_2, a change appears when the frequency of ν_2 is identical with that of the overlapped signal. The chemical shift of the α-vinyl protons as well as its *cis*-configuration relative to the β-hydrogen of Compound **54** has been determined by this method.[727]

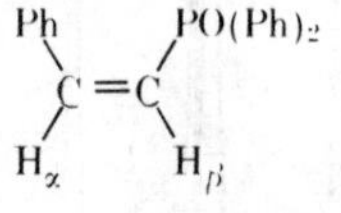

54

2.1.6.4. INDOR

Along with a change in multiplicity, DR involves also a modification of signal intensities. It was first observed in tickling experiments that the intensity of the progressive transitions increases, whereas that of the regressive ones drops. This can be explained using the *AX* case. The probability of transitions among the magnetic levels is determined by the difference in their population.* When, for example, the X_1 transition is excited (see Figure 81a), the population difference between levels **1** and **2** reduces: number of nuclei is increased in state **1** and reduced in state **2**. On the upper level (**1**) of the A_1 transition, coupled regressively with transition X_1, the population increases, while population on the upper (**2**) level of the progressive A_2 transition decreases. Consequently, the transition probability for A_1, and therefore the intensity of the corresponding spectral line, is reduced, whereas there is an intensity enhancement in the case of A_2 (see Figure 81b). Similarly, when the transition X_2 is excited, this entails a reduction in the population of ground state **4** of the A_2 transition, regressively coupled with it, which in turn leads to a drop in the population difference of levels **2** and **4**, and therefore to a reduction of the intensity of line A_2, whereas the probability of the progressive A_1 transition is increased and the line A_1 becomes more intense. (The population of the level **3** is increased, therefore the difference with respect to the level **1** is increased, see Figure 81c.) When the amplitude of the secondary excitation frequency is further reduced below the level used in tickling experiments, to satisfy the condition $\gamma^2 B_2^2 T_1 T_2 \approx 1$, the perturbation of the levels **1** and **2** (as shown in Figure 79), can be avoided and only the relative population of the levels is altered.

A very sensitive recording of these intensity changes is possible when the experiment is performed in constant polarizing (B_0) and excitation (B_1) fields, and the secondary RF field B_2, i.e., the corresponding frequency ν_2, is modified. When B_1 satisfies the Larmor equation for one of the lines (e.g., X_1) in B_0 and during scanning ν_2 coincides with the resonance frequency of the lines regressively or progressively coupled with transition X_1, the so far constant intensity of the line X_1 is either reduced or increased. In this experiment the instrument is set for the recording of intensity changes. This is zero for any value of ν_2, except for values corresponding to the coupled transition, where intensity decrease is represented by a negative maximum (regressive transition) and intensity increase is represented by positive maximum (progressive transition). Therefore when $\nu_1 = \nu X_1$, a negative maximum can be observed in place of line $A_1(\nu_2 = \nu A_1)$, and a positive maximum can be observed at line position A_2 (see Figure 81d). Upon irradiating line $X_2(\nu_1 = \nu X_2)$, the situation is the opposite (see Figure 81e).

* Compare p. 14.

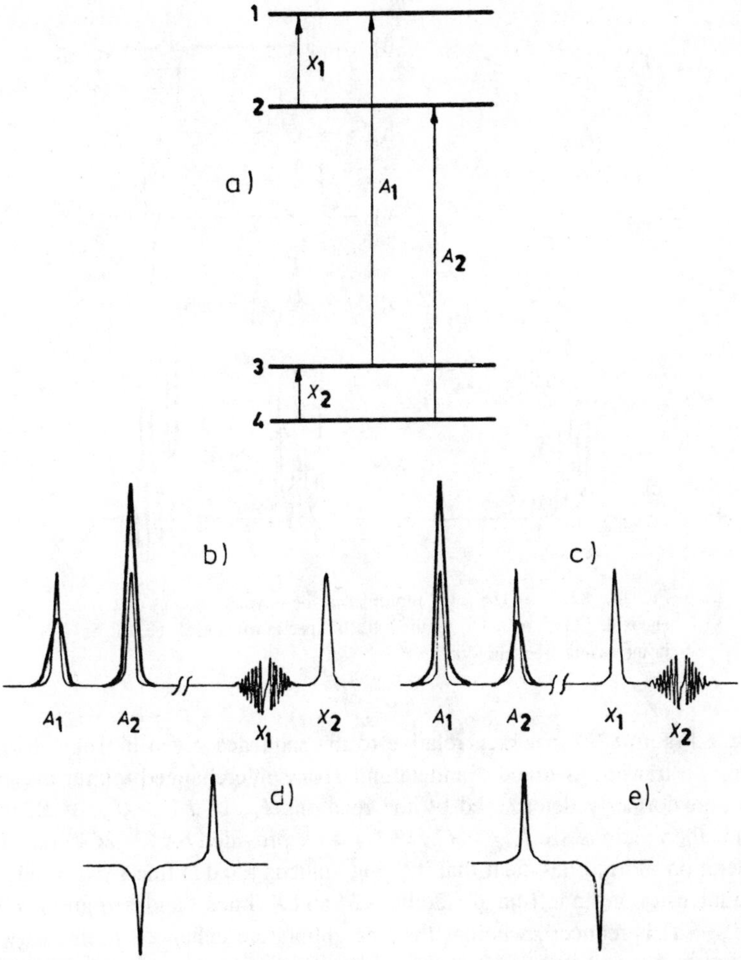

FIGURE 81. Level diagram and transitions (a), the tickling (b) and (c), and the INDOR spectra (d) and (e) obtained upon irradiating the lines X_1 and X_2 for the *AX* system.

For a successful INDOR experiment, the amplitude of B_1 may not exceed 0.1 to 0.2 Hz, otherwise saturation would reduce the intensity of the signal. The amplitude of ν_2 must be around 0.5 Hz to meet the condition $\gamma^2 B_2^2 T_1 T_2 \sim 1$.

In early years of NMR spectroscopy, heteronuclear INDOR has proved especially useful for the accurate measurement of chemical shifts of less-sensitive nuclei (^{13}C, ^{15}N, etc.) from available ^1H or ^{19}F NMR spectra. In the first study of this kind,[75] the ^{13}C satellite lines were chosen as fixed frequency ν_1, and frequency ν_2 varied in the range characteristic of ^{13}C until the intensity of the satellite line has changed. This is the origin of the name of the method,[75] INDOR.

Similarly to tickling, INDOR may also be used for the determination of the shifts of overlapped signals, in the determination of the absolute sign of coupling constants, in selecting the energy levels belonging together and thereby in constructing energy level diagrams, and in many other mainly theoretical aspects. As an illustration to the INDOR spectra, the signals of the methylene protons of styrene oxide (**73**) are shown here[1104] which are obtained by choosing successively one by one of the four lines of the methine ring proton as the fixed frequency ν_1 (see Figure 82a to d). The ring protons give rise to an *AMX*

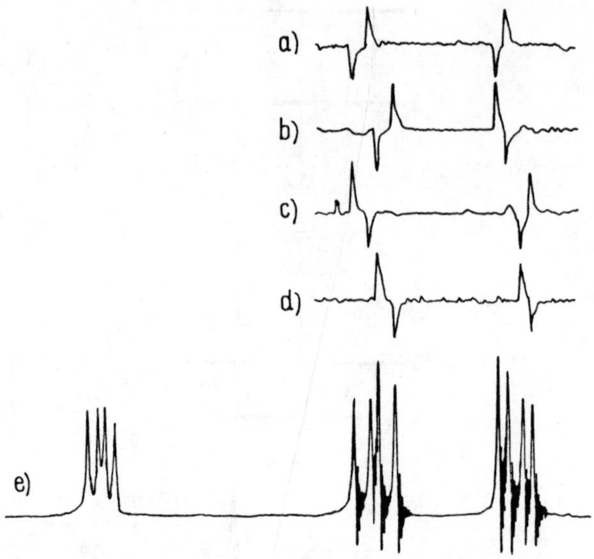

FIGURE 82. (a) The *AMX* multiplet of the oxyrane protons of styreneoxide (**73a**) and (b) to (e) the INDOR spectra irradiating the *A* lines in increasing frequency order.[1104]

spectrum (see Figure 82e) in which relative to the sequence given in Table 15 and Figure 55, the two-two transitions **6** and **7** and **10** and **11** are interchanged within the pairs, since transitions were formerly determined by the relations $J_{AM} > J_{AX} > J_{MX} > 0$, whereas for styrene oxide the relations are $J_{MX} > J_{AM} > J_{AX} > 0$, provided $\delta A > \delta M > \delta X$. Taking this into consideration, it becomes clear that if the ν_1 value is fixed at line **1** ($\nu_1 = \nu A_1$, transition **4** → **1**), the intensity of the left-hand side lines *M* and *X* (lines **5** and **9**: regressive transitions **3** → **1** and **2** → **1**) is reduced, whereas their neighbors are enhanced in intensity (these are the transitions **7** → **4** and **6** → **4**, corresponding to lines **7** and **11**; therefore with respect to A_1 the latter are progressive transitions, see Figure 82a). If $\nu_1 = \nu A_2$, there is no intensity change for the left-hand side line pair *M* (they possess no common levels with those of the transition **6** → **2**), whereas the lines *X* change their roles, namely, the transition **2** → **1** is progressive relative to transition **6** → **2**, whereas the transition **6** → **4** is now regressive. Intensity reduction is observed at line **7** (*M* transition **5** → **2**), and an intensity increase occurs at line **8** (extreme right-hand *M* line) arising from the now progressive **8** → **6** transition (see Figure 82b). A completely analogous interpretation can be given for the INDOR spectra of Figure 82c and d also. Overhauser effect which also increases intensities by DR will be considered later (Section 2.2.5.3).

2.1.7. NMR Solvents and Reference Substances; Calibration

In order to obtain high resolution NMR spectra, it is advisable to measure dilute solutions. Pure liquids often have high viscosity and therefore give broad lines. For CW studies, concentrations in the range 0.05 to 5 mol/ℓ are necessary. In FT mode, concentrations of one order smaller are sufficient. The investigation of gases, due to low molar concentration, and the high resolution measurement of solids, due to line broadening arising from dipolar interactions and inhomogeneities, respectively, needs special measuring conditions.

The solvents must satisfy the following requirements: the absence of the type of magnetic nuclei to be measured (e.g., protons in ^{1}H NMR), low polarity, no interaction with the sample, no decomposition at the temperature of the measurements, low viscosity, moderate

Table 20
DATA FOR THE COMMONLY USED NMR SOLVENTS[281]

Solvent	The absorption of the solvent (ppm)	Impurity	Light isotope: Melting point (°C)	Light isotope: Boiling point (°C)	Remarks
CCl_4	—	—	−22	76	
CS_2	—	—	−108	46	
$Cl_2C = CCl_2$	—	—	−22	121	
$CDCl_3$ (Chloroform-d)	7,27*s* ~1.5[a]	$CHCl_3$ H_2O (rarely)	−63	61	
$CDBr_3$	6.80*s*	$CHBr_3$	7	149	
CD_3CN	2.00*qi*	CD_2HCN	−42	82	
CD_3COCD_3 (Acetone-d_6)	2,17*qi* ~2,8[a]	CD_3COCD_2H H_2O (often)	−95	56	
CD_3SOCD_3 (DMSO-d_6)	2.62*qi* ~3.3[a]	CD_3SOCD_2H H_2O (mostly)	19	189	Hygroscopic internal standard TMS or DSS
C_6D_6(benzene-d_6)	7.37*s*	C_6D_6H	6	80	
CD_3OD (Methanol-d_4)	3.47*qi* ~3.0[a]	CD_2HOD CD_3OH,H_2O (always)	−97	64	
CD_3NO_2	4.33*qi*	CD_2HNO_2	−25	101	
D_2O	~5.2[a]	HDO,H_2O	0	100	Hygroscopic internal standard DSS
CF_3COOD (TFA)	>10[b]	CF_3COOH	−15	72	
$O{=}CD{-}N(CD_3)_2$ (DMF-d_7)	{2.88*qi* 2.97*qi*} 8.02*s*	$CDON(CD_3)CD_2H$ $CHON(CD_3)_2$	−58	152	Internal standard DSS
C_5D_5N (Pyridine-d_5)	7.1—7.8*m* 8.6*m*	C_5D_4HN	−42	115	

Note: *s*, singlet; *qi*, quintet; *m*, multiplet.

[a] Signal position and shape are variable, especially for solutions containing acidic protons.
[b] Usually an intensive signal at about 11.5 ppm; its position and shape are variable.

vapor pressure, high dissolving power (compare Table 20). The solvents most often used are, therefore, perhalogenated hydrocarbons, CS_2, and perdeuterated organic solvents.

When only a part of the ^1H NMR spectrum is necessary, nondeuterated solvents may also be used. In such cases, however, because of intense spinning side bands, a much wider range than that actually covered by solvent signals is useless.

Perdeuterated solvents also contain partially deuterated isotopic isomers. In the spectra of deuterated acetonitrile, acetone, DMSO, and methanol, for example, the 1:2:3:2:1 quintet of the CD_2H group of the isotopic impurity can always be found. The deuterium atoms ($I = 1$) lead to a $(2nI + 1)$-fold splitting of the proton signal (see Figure 83).

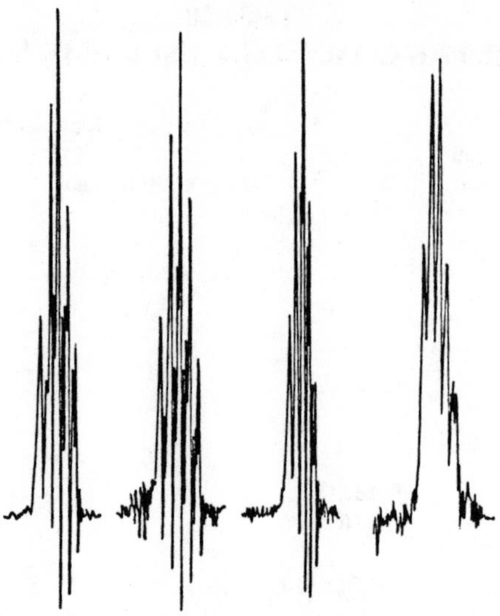

FIGURE 83. Quintets of CD_2H groups in the spectrum of (a) acetone-d_6, (b) acetonitrile-d_3, (c) methanol-d_4, and (d) DMSO-d_6.

Since the chemical shifts are significantly influenced by anisotropic solvent – solute interactions, solvents (CCl_4, $CDCl_3$, CD_3COCD_3, CD_3SOCD_3) in which these effects are smaller (0.1 to 0.3 ppm) are preferred. When comparing chemical shifts, they should be measured in the same solvent, at approximately the same temperature and concentration. Strongly anisotropic solvents (CS_2, liquid SO_2, CF_3COOD, C_6D_6, C_5D_5N) may alter ^1H NMR chemical shifts even by 1 to 2 ppm.*

The solubility of organic substances may change in deuterated solvents, primarily in heavy water, as compared to nondeuterated solvents. Solubility in commercial chloroform, stabilized by small amounts of ethanol, is often misleading; in $CDCl_3$ there is no ethanol, and mixed solvents dissolve many substances better. Hence, in case of samples sensitive to acids, before solving it is necessary to neutralize the solvent (e.g., adding K_2CO_3 to the solvent). For the same reason $CDCl_3$ is frequently somewhat acidic due to the presence of phosgene formed by decomposition.

A relatively small increase in viscosity can already result in a significant line broadening due to slower relaxations,** so viscous solvents (CD_3OD, CD_3SOCD_3) should possibly be avoided. The effect of viscosity may be compensated by dilution and by using elevated temperatures. Field homogeneity should be reset prior to the measurements on solutions in TFA, acetic acid-d_4, DMSO-d_6, and DMFA-d_7, since these solvents have a strong field-distorting effect.

Certain solvents always contain some water, e.g., methanol-d_4, acetone-d_6 (sometimes $CDCl_3$, too), and especially heavy water and DMSO-d_6 which are very hygroscopic. In the study of compounds containing acidic protons, the signal of these protons is merged with the signal of the water protons as a result of fast exchange processes.*** For the same reason, the signal of the acidic protons also disappears upon the addition of heavy water to the solution in solvents nonmiscible with water. Exchange is relatively slow in DMSO-d_6 and very acidic solvents, e.g., in TFA, therefore, for the study of the spin-spin coupling of acidic protons these solutions can be used.†

* Compare p. 38-40.
** Compare p. 15 and p. 147.
***Compare p. 71 and Volume II, p. 97 and 105.
† Compare Volume II, p. 100, 105, and 106.

In acidic solvents there is always a possibility for protonation; this always takes place when there is a basic group in the solute. The protonation process always leads to significant spectral changes.*

For variable temperature studies, the boiling and melting point, respectively, of the solvent is also important. For higher temperatures, DMSO-d_6, C_2Cl_4, and $CDBr_3$ solvents are the best; for low temperatures, DMSO-d_6, $CDCl_3$, $CFCl_3$, CS_2, CD_3OD, CCl_2F_2 or a 1:1 mixture of $CDCl_3$ and CCl_4 (melting point approximately $-90°C$) are best.

Upon using chiral solvents, the optical purity of the sample may be determined (compare Section 3.7.2). For this purpose, CH_3–$CHPhNH_2$, 1-α-naphtyl-ethylamine, or trifluoro-methyl-phenyl-carbinol ($CF_3CHPhOH$) can be used. The use of pyridine-d_5 and DMFA-d_7 should be avoided, if possible since they are usually strongly contaminated, among others by light isotopes.

The signals of the most often used solvents are found in Table 20. The δ-values are given for neat liquids; strong solvent-solute interactions may alter these shifts by 0.1 to 0.2 ppm.

With the first series of NMR spectrometers, each spectrum had to be calibrated separately. For this purpose the s.c. side band technique was applied: the RF field was modulated by an audiofrequency generator and side bands were observed in the spectra on both sides of the spectral lines at distances equal to the modulation frequency. In the case of coincidence between a side band and another signal, the shift difference of the latter and the modulated signal is equal to the modulation frequency.

In current CW and FT spectrometers equipped with proton-stabilizers field stability is sufficient to permit the use of precalibrated recording charts. To check the calibration, the signal of the solvent is quite satisfactory.

Naturally the zero point of the spectrum should be set prior to all measurements by adjusting the signal of the reference substance on the preprinted recording chart. As pointed out earlier, for ^{1}H and ^{13}C NMR measurements, almost invariably TMS is used as reference, or in aqueous solutions DSS is used,** while in the resonance study of other nuclei, various other reference substances are used.***

On Figure 84 the spectrum of acetone as a dilute solution in chloroform can be seen, taken at high amplification. In such a case the signal of TMS should be recorded by lower amplification. When the spinning side band of the latter is sufficiently intense, it can have an intensity commensurable to the sample signals (in this case those of acetone) and therefore may give rise to misunderstanding.† In doubtful cases the origin of the signal can be decided by altering the spinning frequencies, which causes the shift of a spinning side band (see Figure 84b and compare Figure 67, too).

At high amplification any sharp line of DSS can be source of misassignments, too, since they can be mixed up with the sample signals, and, what is more often the case, they can temper the measurements of intensity. It is therefore useful if the positions and shapes of the reference signals are known (see Figure 85a and b). The spinning side band at 0.83 ppm in Figure 85a, for example, seems to have the same intensity as the signal of $\delta SiMe_3$ — due to the higher gain. (This side band is superimposed upon the multiplet of $\delta SiCH_2$.)

The chemical shift differences arising from volume susceptibility when applying external standards‡ must be corrected for. The accuracy of the chemical shift values measured and recorded on precalibrated paper is ± 0.02 ppm, which is completely satisfactory in the study of organic chemical problems.

* Compare Volume II, p. 17, 22, 44, 62, 66-67, 81, 82-83, 90, 106-107, and 108-109.

** Compare p. 43 and Volume II, p. 89.

*** Compare Volume II, p. 239, 240, 260, 261-262, and 269.

† Compare p. 136.

‡ Compare with the second footnote p. 43.

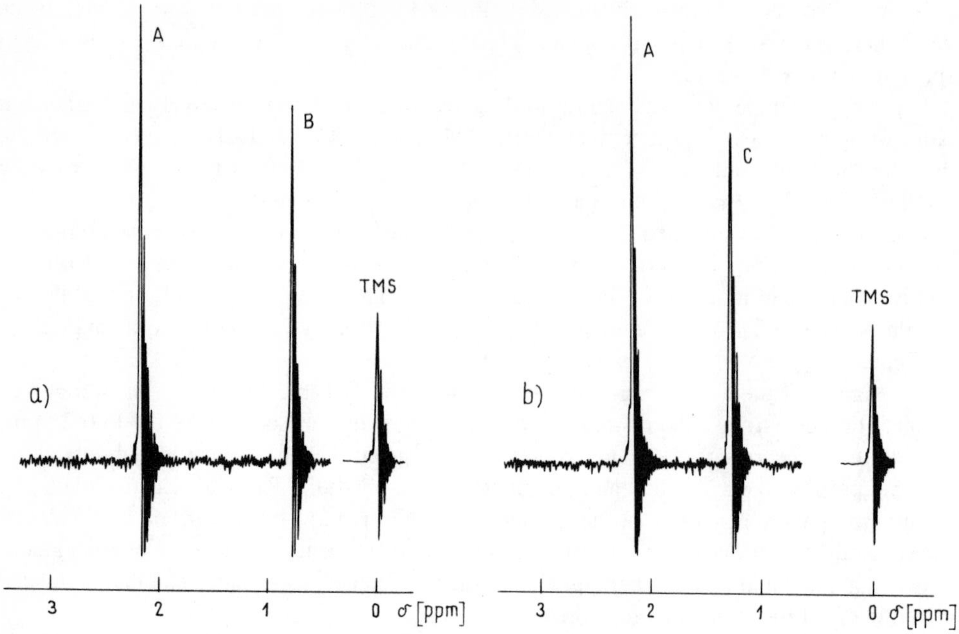

FIGURE 84. The spectrum of a dilute $CDCl_3$ solution of acetone with the singlet of acetone (A) and the TMS spinning sidebands (B,C) of commensurable intensity, (a) and (b) at two different spinning frequencies.

2.2. FOURIER TRANSFORM NMR SPECTROSCOPY[464,556,1284-1286]

The first two decades of NMR spectroscopy were the period of CW instruments. In this technique the excitation of spins is carried out by RF magnetic field of slowly changing frequency, which is present during the entire period of spectrum recording (i.e., the RF field interacts continuously with the spin system, hence the name CW).

The FT spectrometers became widespread at the beginning of 1970s. Here the magnetic nuclei are excited by short RF pulses of discrete frequency, and the spectroscopic information is recorded after switching off the RF field. In this case an interferogram is obtained (see Figure 86a) which is transformed by the computer of the spectrometer, using the principle of Fourier analysis, into the usual frequency-dependent absorption spectrum (see Figure 86b).

The interferogram is a superposition of the decay patterns (wiggles) of the absorption signals. If the spectrum consisted of a single signal, this decay would be a sine wave of exponentially decreasing amplitude. The interferogram is produced by the difference between the ν_1 frequency of the B_1 field and the Larmor frequencies of the magnetic moment components (the resonance frequencies of the various nuclei) induced by B_1.* The two waves are in some instances in-phase, in other instances more or less out-of-phase, causing a typical interference phenomenon. In the meantime, the magnetic moment components $\mathbf{M}_{x,y}$ are reduced exponentially to zero by means of relaxation processes. This decay (FID, free induction decay),** governed by T_2, is usually fast (below 1 s for protons). The interferogram may be produced by means of RF-pulses some microseconds in length. If pulse amplitude B_1 is high enough, the interferogram will embrace all absorption lines (excitation of all nuclei with different chemical shifts) with amplitudes identical to those measureable in the CW spectrum. Consequently, the interferogram contains all the information present in the

* Compare p. 178.

** Compare p. 178.

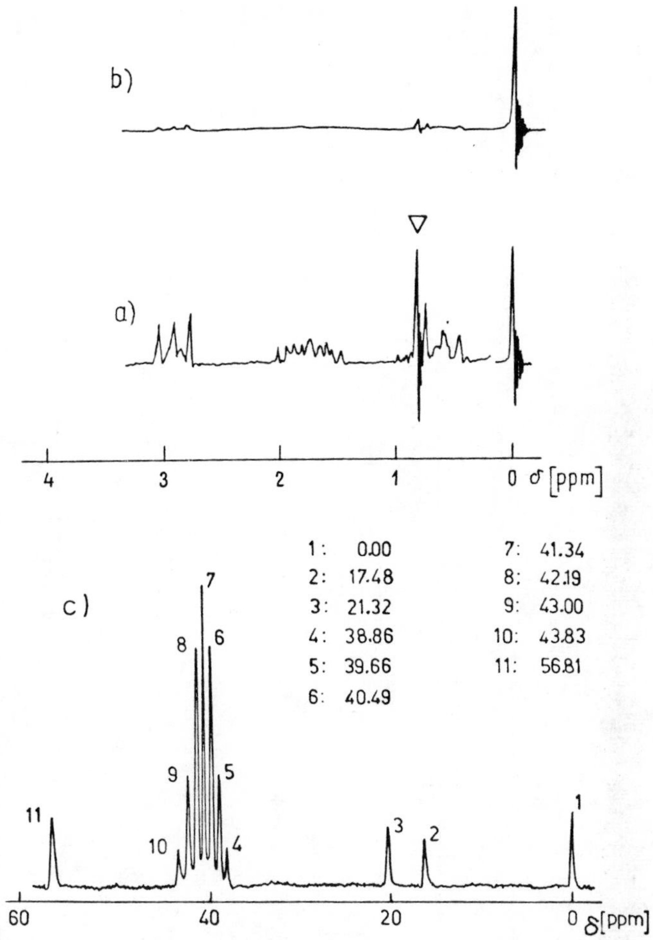

FIGURE 85. The multiplets of DSS in D_2O. At higher and lower amplifications in proton resonance (a) and (b) and the ^{13}C NMR lines (c) in DMSO-d_6 of DSS with the septet of solvent (due to ^{13}C-^2H interaction in groups CD_3).

absorption spectrum. The development and measurement of FID, data acquisition in computer memory, and transformation of interferogram into absorption spectrum takes less time than the time required to measure a CW spectrum, including plotting. The gain in time may be invested into spectrum averaging, which improves signal-to-noise ratio (ζ), even under the invariance of reproducibility conditions. Averaging the interferograms produced by repeated RF pulses and then transforming them requires much less time than to measure a spectrum of the same quality by CW method and CAT (computer of average transients) procedure. (The condition of adiabatic sweep, i.e., that the spin system should always be in equilibrium, restricts the sweep rate to a value permitted by the relaxation times T_1, determining the rate of equilibration, which is approximately 0.1 to 1 Hz/s for protons.)

The possibility of the excitation of spins by means of short and intense RF pulses was suggested by Bloch[148,151] simultaneously with the CW method, but this technique was used for the recording of absorption spectra only two decades later. Pulsed techniques had been used by that time only for the investigation of time-dependent phenomena (molecular motions, exchange processes, and, first of all, relaxations). The theoretical and experimental pioneers of this field were Torrey[1422] and Hahn.[617] Ernst[454] and Anderson were the first who directed the attention, in 1966, to the fact that an absorption NMR spectrum may also be

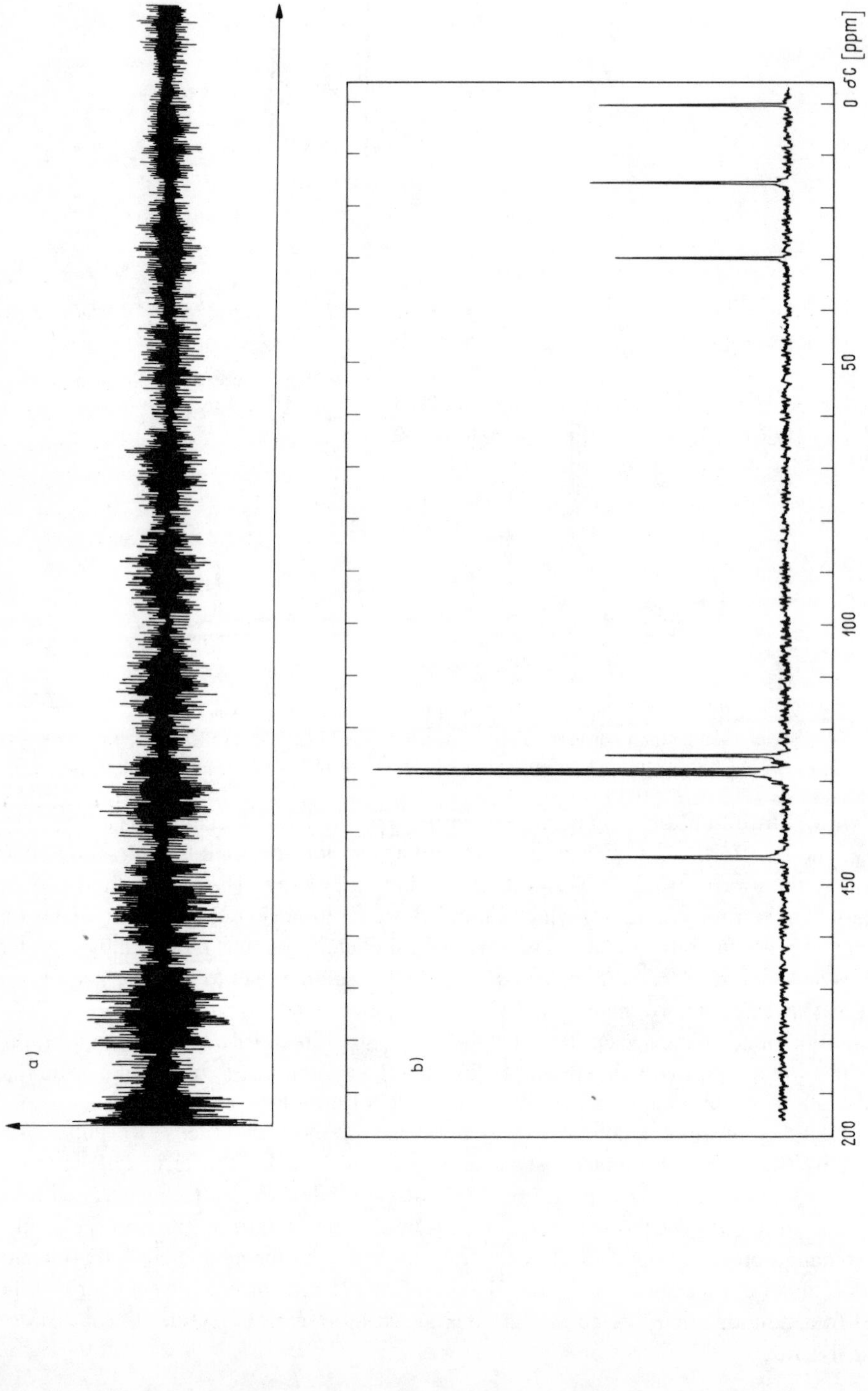

FIGURE 86. Interferogram (FID) of an 8.5 *M* solution of ethylbenzene (a) in acetone-d_6 and (b) proton-decoupled ^{13}C NMR spectrum recorded by a VARIAN® XL-100 FT spectrometer: 0.8-s measurement time, 90° pulse. (Reproduced by courtesy of VARIAN.)

produced by means of RF pulse excitation and subsequent FT, and that in the resulting PFT (pulsed Fourier Transform) spectra, ζ is much better than in the spectra obtained by the CW technique.

2.2.1. The Fourier Transformation[184]

FT is a completely general, classical[503] mathematical method, applied to set up a relationship between two variables. If the magnitude of a dependent variable $f(x)$ is known as a function of a given physical parameter x (independent variable), the FT of this function yields the same dependent variable as a function of another physical parameter (y) as independent variable:

$$f(y) \equiv \int_{-\infty}^{\infty} f(x)e^{-ixy}\mathrm{d}x \equiv \mathcal{F}[f(x)] \tag{185}$$

It can be proved that the inverse operation also exists:[184]

$$f(x) = \frac{1}{2\pi}\int_{-\infty}^{\infty} f(y)e^{ixy}\mathrm{d}y \equiv \mathcal{F}^{-1}[f(y)] \tag{186}$$

Originally, Fourier was concerned with the theory of thermal conduction and developed the mathematical transformation named after him for this theory.

In NMR spectroscopy the two independent variables are time and frequency, and their relationship must be known in order to derive the absorption spectrum of a spin system excited by pulses, since it is the macroscopic magnetization of excited spins as a function of the excitation frequency, whereas the interferogram, the recorded response of the spin system to pulsed RF excitation, is the variation of magnetization in time induced by the RF pulses in the spin system. Thus, by means of FT the interferogram may be converted into absorption spectrum:

$$M(\nu) = \int_{-\infty}^{\infty} M(t)e^{-i2\pi\nu t}\mathrm{d}t \equiv \mathcal{F}[M(t)] \tag{187}$$

where M is the magnetization of the spin system.

Any complicated function which changes periodically as a function of time may be expanded into an infinite series of sine and cosine functions, its harmonic components, i.e., it may be given as a sum of the sine and cosine components of a given ground frequency and its overtones, multiplied by appropriate coefficients:

$$f(t) = f(t+T) = \sum_{n=-\infty}^{\infty}\left[A_n\cos 2\pi\frac{nt}{T} + B_n\sin 2\pi\frac{nt}{T}\right] \tag{188}$$

where T is the length of one period and

$$A_n = A_{-n} = T^{-1}\int_0^T f(t)\cos 2\pi\, nt\cdot T^{-1}\,\mathrm{d}t \tag{189}$$

and

$$B_n = B_{-n} = T^{-1}\int_0^T f(t)\sin 2\pi\, nt\cdot T^{-1}\,\mathrm{d}t \tag{190}$$

are the Fourier coefficients. Equation 188 expresses the Fourier series (infinite trigonometrical series).

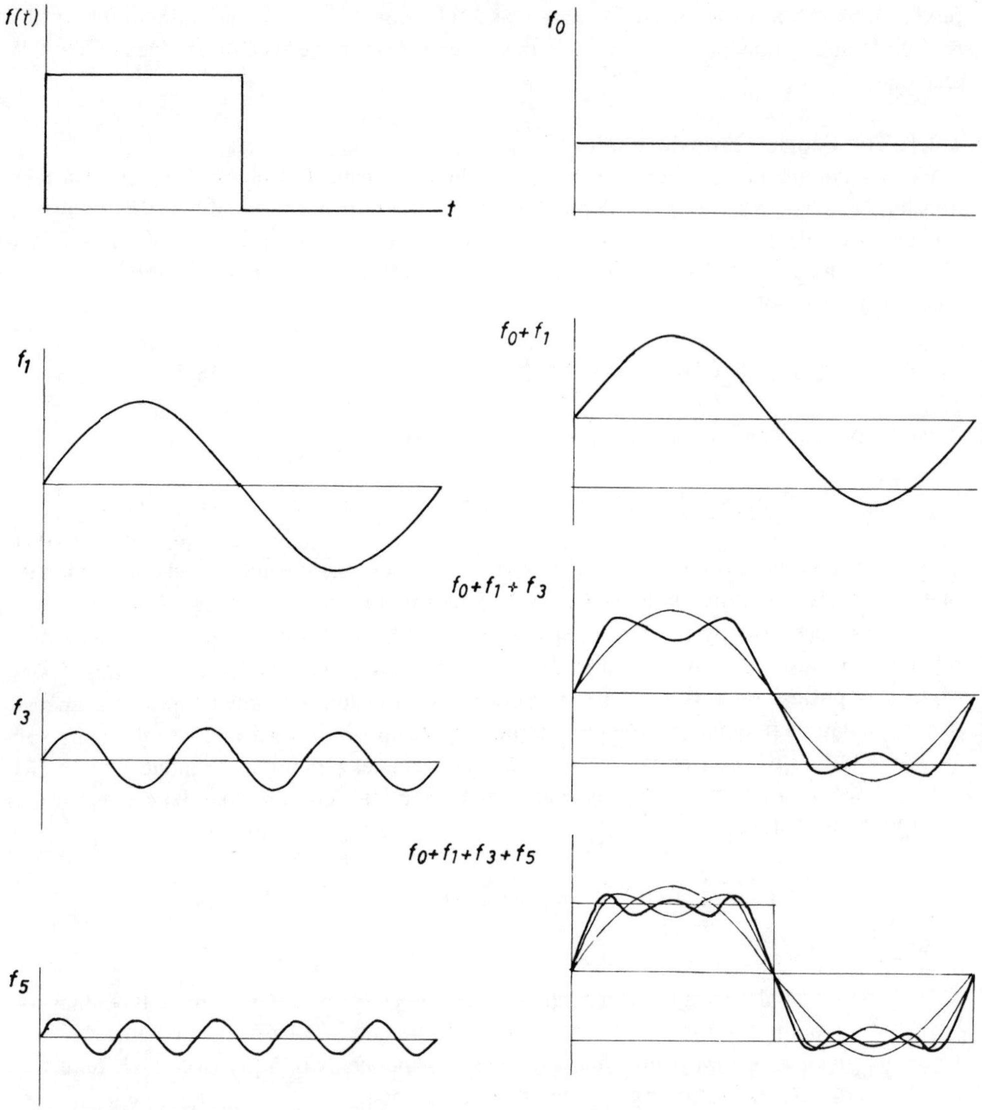

FIGURE 87. Square function $f(t)$ representing an RF pulse, and its approximation with a series of sine Fourier components. Zero-, first-, third-, and fifth-order components (f_0, f_1, f_3, f_5), and functions $f_0 + f_1$, $f_0 + f_1 + f_3$, and $f_0 + f_1 + f_3 + f_5$ synthesized from them.

If function $f(t)$ is even, i.e., $f(t) = f(-t)$, $B_n = 0$ for all n, and if it is odd, i.e., $f(t) = -f(-t)$, $A_n = 0$ for all n. Thus, a rectangular function representing an RF pulse may be expanded into an infinite series of Fourier sine-components, the more accurately, the more components are taken into account (see Figure 87).

On analogy of Fourier series, nonperiodical functions may also be approximated by means of Fourier integrals. If a periodic function $f'(t)$ is identical in a sufficiently long interval with a nonperiodic function, $f(t)$ may be expressed by $f'(t)$. Expressing the Fourier coefficients in a form C_n/T and using the Euler formula ($\cos\alpha + i\sin\alpha = e^{i\alpha}$), the sine and cosine terms may be collected in a single term:

$$f'(t) = \sum_{-\infty}^{\infty} \frac{C_n}{T} e^{2\pi i n t/T} \qquad (191)$$

where

$$C_n = \int_{-T/2}^{+T/2} f'(t)e^{-2\pi int/T} dt \qquad (192)$$

Introducing the new variable n/T, i.e., frequency ν, and regarding it as a continuous variable if $T \rightarrow \infty$, one obtains

$$\lim_{T\rightarrow\infty} f'(t) = f(t) = \int_{-\infty}^{\infty} f(\nu)e^{2\pi i\nu t}d\nu \qquad (193)$$

and

$$f(\nu) = \lim_{T\rightarrow\infty} f'(\nu) = \int_{-\infty}^{\infty} f(t)e^{-2\pi i\nu t}dt \qquad (194)$$

The last two Fourier integrals are identical to Equations 185 and 186 defining the FT, now in terms of actual physical quantities, time, and frequency, instead of general variables x and y. It is clear that the Fourier integrals mean the mutual transformation of two functions necessarily demanding the existence of one another. This mathematical operation is called FT.

Figure 88 shows some simple functions of importance in NMR spectroscopy, together with their Fourier transforms. The first "Fourier twins" are the so-called Dirac-delta function and its transform (see Figure 88a). Its definition is

$$\delta(t) = \begin{cases} \infty & \text{if } t = 0 \\ 0 & \text{if } t \neq 0 \end{cases} \qquad (195)$$

with

$$\int_{-\infty}^{\infty} \delta(t)dt \equiv 1 \qquad (196)$$

The main properties of δ are

$$\int_{-\infty}^{\infty} \delta(t)f(t)dt = f(0) \qquad (197)$$

and

$$\int_{-\infty}^{\infty} \delta(t - \tau)f(t)dt = f(\tau) \qquad (198)$$

Function δ is the mathematical definition of an impulse. Its FT is $f(\nu) \equiv \mathcal{F}[\delta(t)] \equiv 1$, since

$$\mathcal{F}[\delta(t)] = \int_{-\infty}^{\infty} \delta(t)e^{-i2\pi\nu t}dt = e^0 = 1 \qquad (199)$$

i.e., an impulse of infinitely short duration corresponds to a continuous frequency spectrum with a constant amplitude from $-\infty$ to $+\infty$.

If the origin in the time domain is shifted by τ, a periodic complex Fourier transform is obtained in the frequency domain, with a period of $\nu = 1/\tau$ (see Figure 88b). The continuous

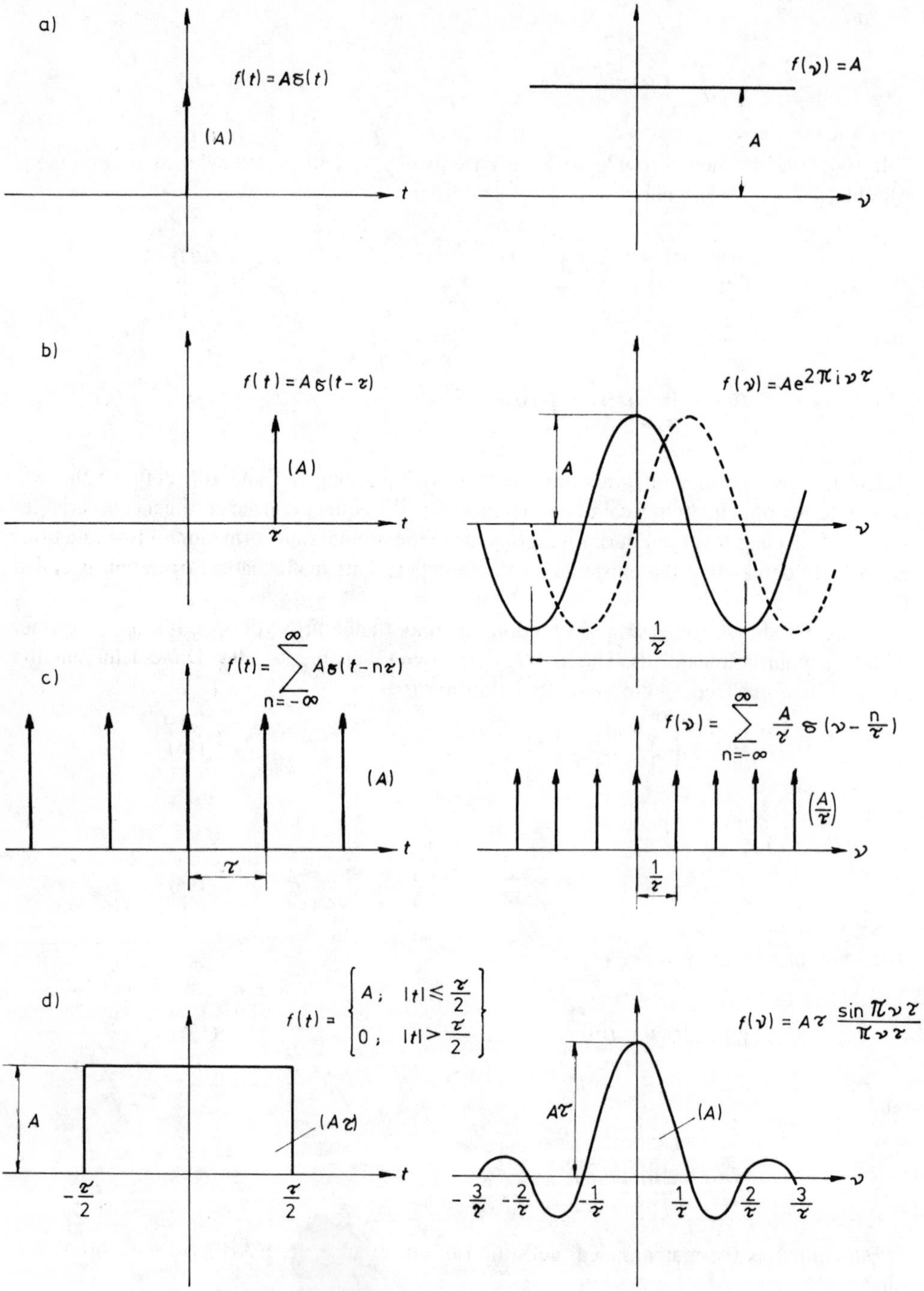

FIGURE 88. Some functions of importance in FT NMR spectroscopy and their Fourier transforms. Amplitudes and, in parentheses, integrals (areas under the curves) are given for (a) to (c), also taking into account Equation 198 defining Dirac's delta.

RF excitation of the spin system in the CW may be described by a complex, periodic function.

When in the NMR experiment nucleus i is off-resonance, i.e., $\nu_i \neq \nu_1$ (where ν_i is the Larmor frequency of nucleus i and ν_1 is the frequency of RF-field), the effective field acting

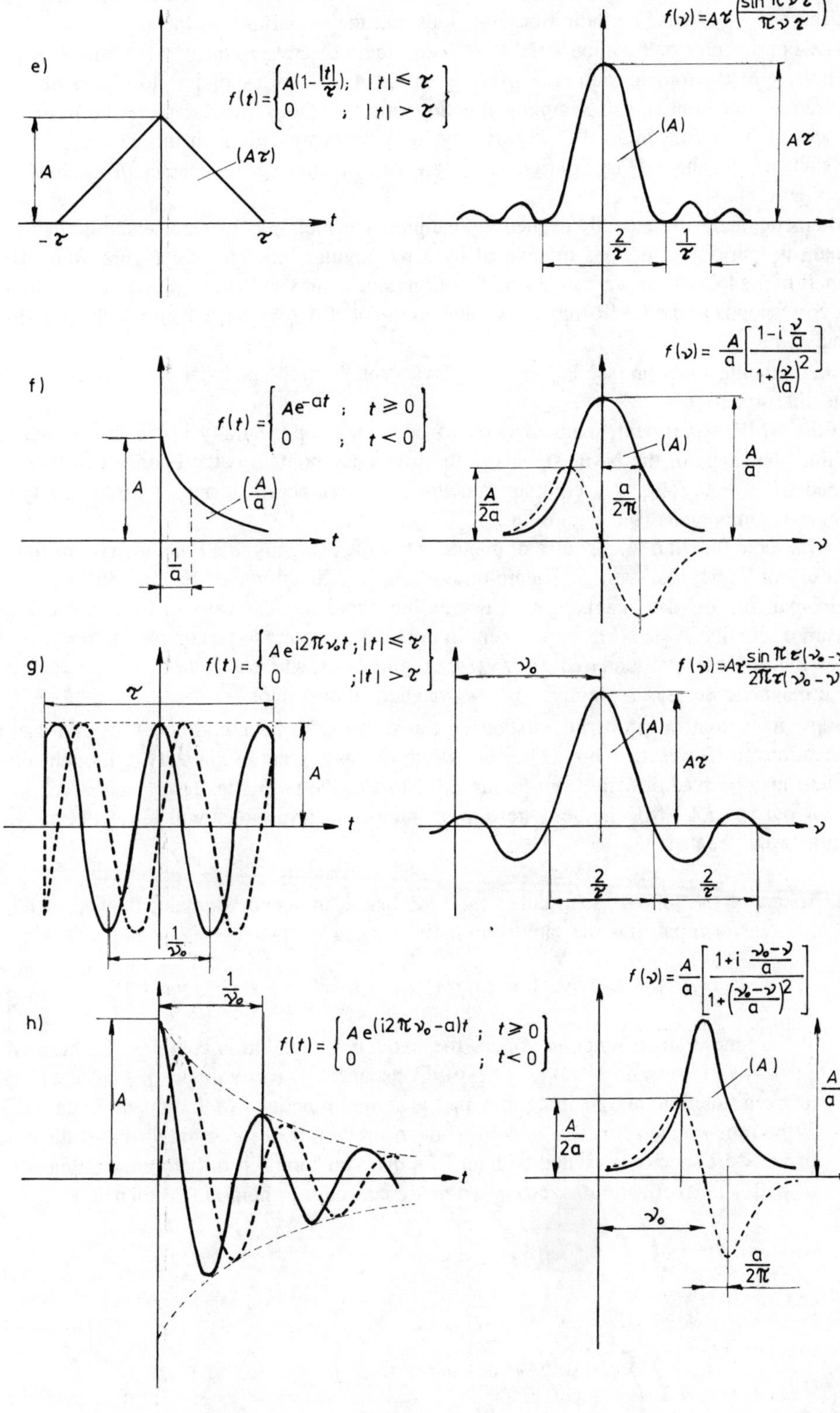
e)
$f(t)=\begin{cases} A(1-\frac{|t|}{\tau}); & |t| \leqslant \tau \\ 0 & ; |t| > \tau \end{cases}$
A
(Aτ)
−τ
τ
t
$f(\nu)=A\tau\left(\frac{\sin \pi\nu\tau}{\pi\nu\tau}\right)^2$
(A)
Aτ
2/τ
1/τ
ν
f)
$f(t)=\begin{cases} Ae^{-at} & ; t \geqslant 0 \\ 0 & ; t<0 \end{cases}$
A
(A/a)
1/a
t
$f(\nu)=\frac{A}{a}\left[\frac{1-i\frac{\nu}{a}}{1+\left(\frac{\nu}{a}\right)^2}\right]$
(A)
A/a
A/2a
a/2π
ν
g)
$f(t)=\begin{cases} Ae^{i2\pi\nu_0 t}; & |t| \leqslant \tau \\ 0 & ; |t| > \tau \end{cases}$
τ
A
t
1/ν₀
$f(\nu)=A\tau\frac{\sin \pi\tau(\nu_0-\nu)}{2\pi\tau(\nu_0-\nu)}$
ν₀
(A)
Aτ
2/τ
2/τ
ν
h)
1/ν₀
$f(t)=\begin{cases} Ae^{(i2\pi\nu_0-a)t} & ; t \geqslant 0 \\ 0 & ; t<0 \end{cases}$
A
t
$f(\nu)=\frac{A}{a}\left[\frac{1+i\frac{\nu_0-\nu}{a}}{1+\left(\frac{\nu_0-\nu}{a}\right)^2}\right]$
(A)
A/a
A/2a
ν₀
a/2π
ν

FIGURE 88e-h.

on magnetization **M** and thus the magnitude of **M** as well changes periodically in time. For this reason, we discuss periodic time functions and their Fourier transforms.

The Fourier pair of function $e^{i2\pi\nu t} \equiv e^{i\omega t}$ is a single discrete frequency (ν), since this is the inverse of the transformation $\mathcal{F}\,[A\delta(t - \tau)] = Ae^{i2\pi\nu t}$ corresponding to Figure 88b.

Dirac's comb function or sampling function, most suitable for describing the pulse sequences of basic role in NMR spectroscopy, may be expressed as an infinite series of δ-functions of spacing τ. The Fourier pair is a series of discrete frequencies of spacing $1/\tau$ (see Figure 88c).

Magnetic nuclei are actually excited by real pulses, which may be represented in the time domain by periodic functions modulated by a rectangular function (see Figure 88d). The FT-pair of the latter is the so-called sinc-function (sinc ≡ sin x/x). Its central, large-amplitude part corresponds to the NMR frequency bandwidth $\Delta\nu_i$ if $\Delta\nu_i = 2/\tau$, where τ is the duration of pulse.

The triangular function (see Figure 88e) plays a role in the ''apodization'',* its transform is the sinc^2 function.

In the NMR experiment, magnetization **M** decreases exponentially in the time domain, and thus the shape of the NMR signals in the frequency domain is the Fourier pair of e^{-at} defined for $t \geqslant 0$. This is a complex function of Lorentzian real and of dispersion-type imaginary components (see Figure 88f).

If a periodic function, e.g., pulse or magnetization, exists only in a finite interval of time, the result of FT is, like with the square function, a sinc function (see Figure 88g). Finally, the FT-pair of complex magnetization decreasing exponentially in time, i.e., of the FID measured actually in the NMR experiment in the time domain,** is a complex Lorentz- and dispersion-type curve around frequency ν_0 (see Figure 88h), where ν_0 is the Larmor frequency of the magnetic nucleus represented by the magnetization vector.

Some mathematical properties of Fourier transform pairs have consequences and useful applications in FT spectroscopy. They are worth discussing here separately, although some of them may be read directly from Figure 88. Most of the properties have general validity, but for our special NMR studies, the role of independent variables will always be played by time t and frequency ν.

1. If one of the functions (in either the time or the frequency domain) is even or odd, its transform pair has the same symmetry, i.e.,

$$\mathcal{F}\,[f(t)] = \pm\mathcal{F}\,[f(-t)] = \mathcal{F}\,[\pm f(-t)] \tag{200}$$

2. If the time domain function is real, the frequency function is complex, furthermore, $f(\nu) = f^*(-\nu)$ and $f^*(\nu) = f(-\nu)$. Consequently, the real part of $f(\nu)$ is always an even function, and its imaginary part is an odd function of ν (compare Figure 88f).
3. If the time domain function is defined over a region $0 < t < \infty$ only (this is the case of an NMR spectrum excited with an RF pulse; for time $t < 0$, before the application of pulse, magnetization is zero — compare causality principle),*** then

$$f(t) = \int_0^\infty\int_0^\infty f(\tau)\cos 2\pi\nu\tau \cos 2\pi\nu t \, d\tau \, dt \tag{201}$$

and

$$f(t) = \int_0^\infty\int_0^\infty f(\tau)\sin 2\pi\nu\tau \sin 2\pi\nu t \, d\tau \, dt \tag{202}$$

* Compare p. 183.
** Compare Section 2.2.2.
***Compare p. 185.

where the first integrals are the sine and cosine transforms of $f(t)$, corresponding to the absorption (real) and dispersion (imaginary) signals (see Figure 88f).

4. The transform of a sum of two (time or frequency) functions is equal to the sum of the corresponding transforms:

$$\mathcal{F}[f(t) \pm g(t)] = \mathcal{F}[f(t)] \pm \mathcal{F}[g(t)] \tag{203}$$

This property makes it possible to average interferograms (in the time domain) by measuring an arbitrary number of FIDs (with the application of a series of RF pulses or pulse sequences) and transform them into absorption spectra only after averaging (application of CAT technique with FT spectrometers), since the coaddition of interferograms has the same effect as that of spectra.

5. A scaling of variable in the time domain by factor a involves the following change in the frequency domain:

$$\mathcal{F}[f(at)] = (1/a)f(2\pi\nu/a) \tag{204}$$

As can be seen, scaling of function $f(t)$ does not affect the properties of its $f(\nu)$ pair, changing only its scale.

6. The transform of a function multiplied by constant a in any domain (expansion or contraction of coordinate system) is equivalent to multiplying the transform of the function by a:

$$\mathcal{F}[af(t)] = a\mathcal{F}[f(t)] \tag{205}$$

This is the theoretical basis of the scaling of FID* which is necessary, e.g., in the case of too strong signals, in order to prevent the computer word or ADC from overflowing (compare Section 2.2.3).

7. The area below the curve (the integral of the function) in one of the domains is equal to the initial values of the transform (compare Figure 88):

$$\int_{-\infty}^{\infty} f(\nu)\,d\nu = f(t)_{t=0} \tag{206}$$

Consequently, the initial value of the interferogram (at $t = 0$) determines the intensity of the absorption signals, and the further run of the interferogram in the time domain is immaterial. The integrals are not affected by further operations or parameters, such as data acquisition time and weighting for improving resolution or sensitivity,** unlike line shape (and thus height). However, it is important to start data collection without delay at $t = 0$, otherwise the intensities will be distorted in the absorption spectrum. Namely, a change in the origin $t = 0$ appears as a phase shift (mixing of absorption and dispersion line shapes) in the frequency domain (compare Figures 88b and 89, as well as point 9 of this Section).*** Since the integral of a dispersion signal is zero, it is evident that a mixed line shape leads to false intensities. Phase shifts may occur for other reasons, too: thus, in order to measure accurate intensities, phase errors must be corrected.

* Compare p. 183.
** Compare p. 173-174.
*** Compare p. 172-173 and 181-182.

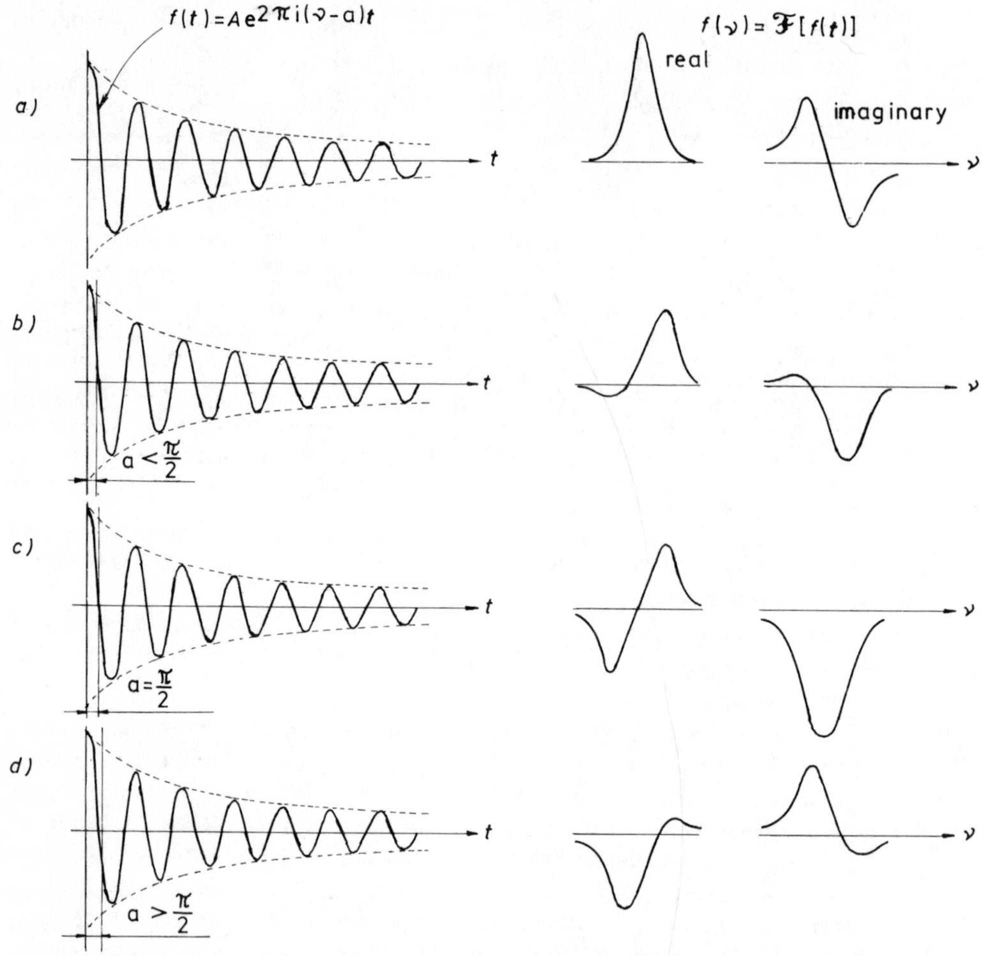

FIGURE 89. Effect of phase shift in the time domain on the Fourier transform pair in the frequency domain (i.e., the absorption spectrum). (a) The real and imaginary components of the function $f(t) = A \cdot e^{(i2\pi\nu-a)t}$ representing FID and function $f(\nu) = f(t)$ for $a = 0$, (b) for $0 < a < \pi/2$, (c) for $a = \pi/2$, and (d) for $\pi/2 < a < \pi$.

8. If the time domain function is a series of functions with spacing τ, the transform is a series of functions with spacing $1/\tau$:

$$\mathcal{F}\left[\sum_{n=-\infty}^{\infty} \delta(t - n\tau)f(t)\right] = \frac{1}{\tau}\sum_{n=-\infty}^{\infty} f\left(\nu - \frac{n}{\tau}\right) \quad (207)$$

This relationship (compare Dirac's sampling function, Figure 88c) is the basis of the computer processing in FT spectroscopy; if the time domain function is digitized, the transformation of the resulting discrete set of values yields the digitized form of the absorption spectrum in the frequency domain.

9. If the origin of one of the functions (i.e., in the time domain) is displaced by τ (coordinate shift, see Figure 88b), the Fourier pair is multiplied by the factor $e^{\pm i2\pi\nu t}$.

$$\mathcal{F}[f(t \pm \tau)] = e^{\pm i2\pi\nu t} f(\nu) \quad (208)$$

It follows from this property that a phase shift in the time domain distorts the absorption spectrum (frequency domain), leading to a mixture of absorption and dispersion signals (see Figure 89).

It is worth noting that in infrared (FT-IR) spectroscopy[104,292,1385,1437] the radiation source is not phase sensitive, and thus only the magnitudes (absolute values) of the intensity belonging to the individual frequency components are obtained as the square root of the sum of squares of real (absorption) and imaginary (dispersion) components. The advantage of magnitude spectrum is that distortions due to phase errors do not occur, but in NMR spectroscopy it has major disadvantages, owing to the "hump" of signals (they are not Lorentzian, and the dispersion signal drops more slowly to zero away from the resonance frequency). The intensities of overlapping signals are not additive; since the intensity of negative dispersion signal is subtracted from that of the coinciding other signal, the intensities partly cancel one another, and the spectrum is distorted. Intensities may not be determined theoretically, either, since the integral is divergent. Two similar transforms of the real time domain function are the energy and power (mean energy) spectrum (or spectral density function), defined as

$$E(\nu) = f(\nu)f^*(\nu) \tag{209}$$

and

$$P(\nu) = \lim_{T\to\infty} \frac{1}{2T} f(\nu)f^*(\nu) \tag{210}$$

where $f(t) = 0$ if $|t| > T$. $E(\nu)$ gives the energy corresponding to frequency ν. If $f(t)$ [the Fourier transform of which is $f(\nu)$] exists only for finite time, the time average of energy, represented by $P(\nu)$, is more useful. It should be noted that although $E(\nu)$ is a Lorentzian for a given signal, its intensity is proportional to the square of the number of nuclei producing the signal.

10. Convolution is a special transformation: convolution $K_{fg}(t)$ of functions $f(t)$ and $g(t)$ is the moving average of one function weighted with the other one. It is, therefore, customary to call $g(t)$ the weight function. All points of $K_{fg}(t)$ are obtained as the weighted average of $f(t)$ about point τ, where

$$K_{fg}(t) = \int_{-\infty}^{\infty} g(\tau)f(t-\tau)\mathrm{d}\tau \tag{211}$$

and

$$K_{fg}(t) = f(t)\circledast g(t) \equiv g(t)\circledast f(t) \tag{212}$$

The operation of convolution is denoted by $\circledast$. Mathematically, convolution is a multiplication of one of the functions (in this case g) with the shifted value $f(t-\tau)$ of the other function and the integration of the product over τ. It may be shown that the Fourier transform of the product of two functions is identical with the convolution of the transformed functions, i.e.,

$$\mathcal{F}[f(t)\,g(t)] = f(\nu)\circledast g(\nu) \tag{213}$$

and, conversely, the Fourier transform of the convolution of two functions is identical with the product of the transformed functions:

$$\mathcal{F}[f(t)\circledast g(t)] = f(\nu)\,g(\nu) \tag{214}$$

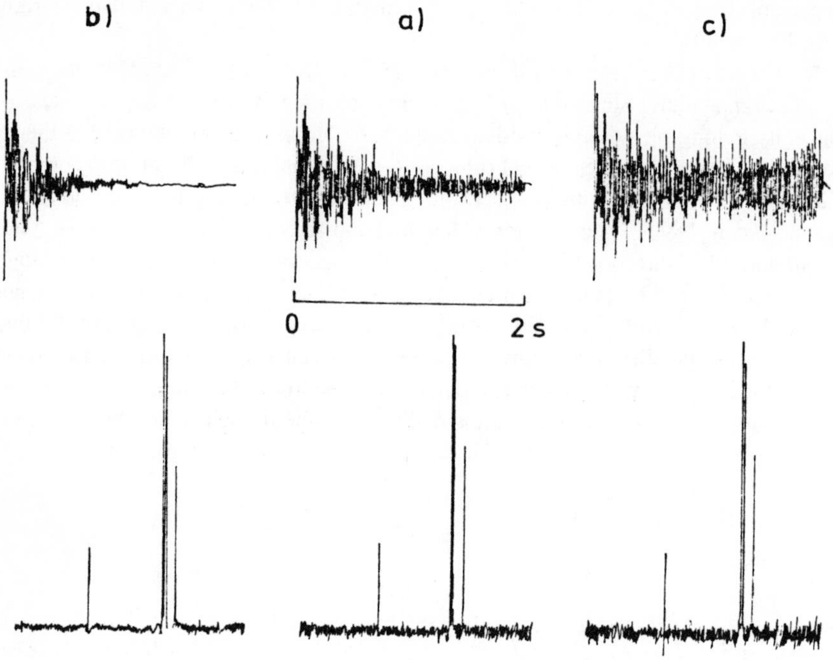

FIGURE 90. The effect of weight functions on the ^{13}C NMR spectrum. Spectrum of ethylbenzene without weight function (a) and after multiplying FID by (b) a positive or (c) negative weight function. (From Shaw, D., *Fourier Transform NMR Spectroscopy*, Elsevier, Amsterdam, 1976, 191. With permission.)

The convolution of the NMR spectrum in the frequency domain with an appropriate weight function makes it possible to increase either resolution or sensitivity (ζ) at the expense of the other parameter. This shows one of the major advantages of the mathematical properties of convolution and FT. Instead of the lengthy operation of convolution in the frequency domain, one has to multiply the time domain function by the Fourier transform of the weight function which has the very simple form of $e^{\pm a/t}$. This "exponential multiplication" of FID with positive exponent yields lower ζ, but better resolution; a negative exponent yields the opposite result. The Fourier transform of the FID multiplied by the window function is, therefore, either a higher resolution spectrum or a lower resolution one with a better ζ. Since PFT spectrometers measure the spectrum in the time domain (as FID), the enhancement of resolution or ζ requires only a simple operation, the multiplication of FID with the appropriate weight function (see Figure 90).

11. Correlation function, a concept in the theory of probability, is the expectation value

$$R_{fg}(\tau) = \overline{f^*(t)g(t+\tau)} \tag{215}$$

where $f(t)$ and $g(t)$ are probability distributions. For any given function

$$R_{fg}(\tau) = \lim_{T\to\infty} \frac{1}{T} \int_{-T/2}^{T/2} f^*(t)g(t+\tau)\mathrm{d}\tau = f(t) \otimes g(t) \tag{216}$$

where $\otimes$ denotes the mathematical operation of correlation. Correlation, mathematically similar to convolution, is the measure of connection between two functions (e.g., in the time domain), and similarly to convolution,

$$R_{fg}(\nu) = f^*(\nu)\, g(\nu) \tag{217}$$

where

$$\mathcal{F}^{-1}[R_{fg}(\nu)] = R_{fg}(t) \quad (218)$$

However, unlike with convolution, the two correlated functions do not commute, since $R_{fg}(t) = R^*_{fg}(-t)$. For two, completely independent functions, $R_{fg}(t)$ is zero for any value of the common variable t, whereas for two functions connected by some causal relationship, correlation is finite for all or at least certain values of t. There is a correlation, e.g., between the frequency of excitation and the magnetization induced by it. This concept may be utilized, e.g., in noise modulation technique (see Section 2.2.5.5), where the excitation function is known, but not periodical. However, by multiplying the resulting FID by the complex conjugate of this function and Fourier transforming the product, the absorption spectrum may be obtained.

12. Autocorrelation is the correlation of a function with itself:

$$R_{ff}(\tau) = \overline{f^*(t)\, f(t+\tau)} = \lim_{T\to\infty} \frac{1}{T} \int_{-T/2}^{T/2} f^*(t) f(t+\tau) \mathrm{d}t = f^*(t) \otimes f(t) \quad (219)$$

Like with cross-correlation, if a function is coherent, its averaged value (integral) is $R_{ff}(t) \neq 0$, otherwise it approaches zero: at time $t = 0$, all functions are coherent. If an $f(t)$ function has a coherent and an incoherent (noise) component, autocorrelation may be used to filter out the coherent component, which can be transformed into absorption signal by FT. The resulting spectrum is the $P(\nu)$ power spectrum defined by Equation 210, and its square root is the (magnitude) spectrum.* Autocorrelation is utilized in the theories of stochastic excitation** and relaxations: in the former, the distribution of excitation energy in the time domain (i.e., the noise); in the latter, the motions causing relaxation are random.

2.2.2. Pulsed Excitation and the Free Induction Decay (FID): Pulsed Fourier Transform (PFT) Spectroscopy

In CW spectroscopy, the spin systems are excited by RF field $\mathbf{B}_1$ perpendicular to the static magnetic field $\mathbf{B}_0$ of direction z and rotating with the Larmor frequency. In PFT spectroscopy, this is replaced by a square RF pulse or pulse sequence of discrete frequency ν_1, duration t_p, and direction x. (In contrast with impulse, of infinitely short duration, by pulse we mean an RF radiation applied for a finite time, but not continuously.)

The Fourier transform of square pulse is the sinc $\nu \equiv \sin \nu/\nu$ function (see Figure 88d), whereas the frequency-domain function of an "impulse", i.e., a δ-function is a "smooth" function, having unit value everywhere (see Figure 88a). However, if t_p is small enough (in the order of μs), there is no significant relaxation during this time (corresponding to the s order of relaxation processes), and thus no difference is observable between the response of a spin system excited by a pulse or a δ-function (the reaction of the spin system is the absorption spectrum in both cases). The pulse tips the macroscopic magnetization **M** from direction z in the same way as a rotating, continuous excitation field $\mathbf{B}_1$, causing the same resonance.

For a better understanding of the phenomenon, let us regard the variation of magnetization as a function of time in a coordinate system rotating with frequency ν_0 (rotating frame):

$$\dot{M} = \dot{M}_x\mathbf{i} + M_x \frac{\partial i}{\partial t} + \dot{M}_y\mathbf{j} + M_y \frac{\partial j}{\partial t} + \dot{M}_z\mathbf{k} + M_z \frac{\partial k}{\partial t} \quad (220)$$

* Compare p. 173.

** Compare Section 2.2.5.5.

where $\dot{M}_{x,y,z}$ are the partial derivatives of the components of magnetization with respect to time, and **i, j,** and **k** are the unit vectors. The magnitude of the latter are always unity; only their directions may change when they rotate with angular velocity $\boldsymbol{\omega}$:

$$\partial i, j, k/\partial t = \omega \times \mathbf{i}, \mathbf{j}, \mathbf{k} \tag{221}$$

Thus,

$$\dot{M} = \dot{M}_{rot} + \omega \times \mathbf{M} \tag{222}$$

since

$$\dot{M}_x\mathbf{i} + \dot{M}_y\mathbf{j} + \dot{M}_z\mathbf{k} = \dot{M}_{rot} \tag{223}$$

and

$$\omega \times (M_x\mathbf{i} + M_y\mathbf{j} + M_z\mathbf{k}) = \omega \times \mathbf{M} \tag{224}$$

where $\dot{M}$ and $\dot{M}_{rot}$ are the changes of magnetization in time in the fixed and rotating frames, respectively. Regarding Equation 47, we obtain:

$$\dot{M}_{rot} = \gamma\mathbf{M} \times \mathbf{B} - \omega \times \mathbf{M} = \gamma\mathbf{M} \times \mathbf{B} + \gamma\mathbf{M} \times \frac{\omega}{\gamma} \tag{225}$$

since, from vector algebra, $\boldsymbol{\omega} \times \mathbf{M} = -\mathbf{M} \times \boldsymbol{\omega}$, and thus

$$\dot{M}_{rot} = \gamma\mathbf{M} \times \left(\mathbf{B} + \frac{\omega}{\gamma}\right) = \gamma\mathbf{M} \times \mathbf{B}_{eff} \tag{226}$$

i.e., in the rotating frame, $\mathbf{B}_{eff}$ acts on macroscopic magnetization, where the magnitude of the z component differs by ω/γ from $\mathbf{B}_0$. This ratio has the dimension of magnetic field, and it may be regarded as a fictive field, arising from the rotation of the coordinate system and opposite to the direction of $\mathbf{B}_0$ (see Figure 91). If only $\mathbf{B}_0$ is present ($\mathbf{B} = \mathbf{B}_0$), $\mathbf{B}_{eff} = 0$, since $\boldsymbol{\omega} = -\gamma\mathbf{B}_0$ and $\dot{M}_{rot} = 0$. Thus, the magnetization is constant in time, if the frame rotates with the ν_0 Larmor frequency, i.e., $\boldsymbol{\omega} = \boldsymbol{\omega}_0$. In the presence of $\mathbf{B}_1$ which rotates with frequency ω_0 and is perpendicular to $\mathbf{B}_0$,

$$B_{eff} = B_0 + \omega/\gamma + B_1 = B_1 \tag{227}$$

therefore, in the $\omega = \omega_0$ case, $\mathbf{M}$ precesses around $\mathbf{B}_1$ (if $\mathbf{B}_1$ is of direction x, then around the x axis) with an angular velocity of $\omega_1 = \gamma B_1$. By a given t_p time, $\mathbf{M}$ has rotated by an angle of

$$\theta = \gamma B_1 t_p \quad \text{[rad]} \tag{228}$$

around the x axis (about $\mathbf{B}_1$). If the resonance condition is not satisfied ($\omega \neq \omega_0$), the higher $\Delta\omega = \boldsymbol{\omega} - \boldsymbol{\omega}_0$, the more $\mathbf{B}_{eff}$ tilts toward $\mathbf{B}_0$:

$$|B_{eff}| = \sqrt{(B_0 - \omega/\gamma)^2 + B_1^2} \tag{229}$$

If the change in $\mathbf{B}_{eff}$ is slow enough (in the case of CW), $\mathbf{M}$ follows the direction of $\mathbf{B}_{eff}$.[1,1309a] The condition of "slow", adiabatic sweep is the inequality $\dot{B}_0 << \gamma B_1^2$. However, significant relaxation may not occur, also during the sweep, i.e., the relation $T_2^{-1} << \dot{B}_0/B_1 << \gamma B_1$

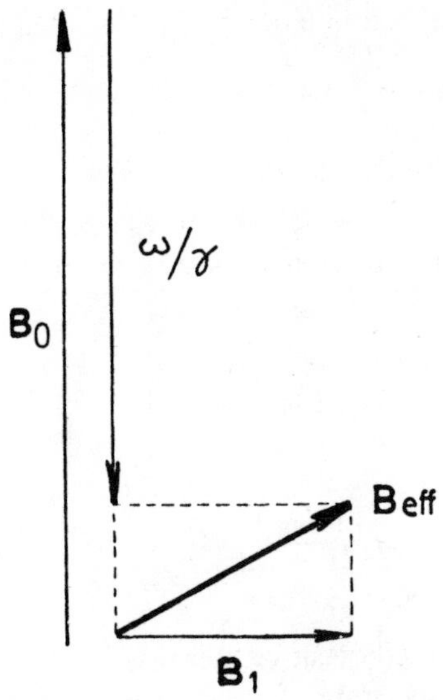

FIGURE 91. Field $\mathbf{B}_{eff}$ acting on the magnetic moment in the rotating frame and its composition from polarizing field $\mathbf{B}_0$ and fictive field ω/γ of direction z and excitation field $\mathbf{B}_1$ of direction x.

must hold. Let us see now the response of the spin system to an excitation by rectangular RF pulses of width t_p and amplitude $2B_1$ at intervals t_{pi}. The response is, again, the variation of the components of induced magnetization in time in a frame rotating at an RF of $\nu_1 \equiv \nu_c(\nu_c$ is the carrier frequency).

In the PFT method, the strong and suddenly applied field $\mathbf{B}_1$ flips $\mathbf{B}_{\text{eff}}$ from direction z into direction x. Magnetization is unable to follow it, and thus it precesses in the $[z,y]$ plane, about the x axis (i.e., around $\mathbf{B}_1$). Since $\Delta\omega$ is small with respect to $\boldsymbol{\omega}_0$, $\mathbf{B}_{eff} \approx \mathbf{B}_1$, and if $\mathbf{B}_1$ is strong enough, it may be able to excite, despite of its fixed frequency, the nuclei of slightly different ν_i resonance frequencies simultaneously, supposing that $\gamma B_1 \gg 2\pi\Delta\nu_i$, and for a 90° pulse, $t_p \ll 1/4\ \Delta\nu_i$ [s], where $\Delta\nu_i$ is the frequency range embracing the resonance frequencies of all N_i nuclei. Accordingly, t_p must be in the μs range.

Regard now the spin system left alone and precessing freely after switching off the short RF pulse of duration t_p. Assuming that the spin system was excited by a 90° pulse parallel to the x axis of the rotating frame, in the instant of switch-off macroscopic magnetization is in the y direction (it has no x or z components), i.e., a signal of maximum intensity may be observed, since the intensities of measured signals are determined by the $M_{x,y}$ components.* This is the free induction signal where "free" indicates that the spin system studied precesses freely after the disappearance of the RF field, i.e., $M_{x,y}$ was induced by an already absent RF field.

During the interval t_{pi}, the individual spins are dephased by spin-spin relaxations, and signal intensity gradually decreases, dropping finally to zero. The time constant of the process is T_2 in a homogeneous magnetic field; in reality it is less, T_2^*, since owing to the inhomogenity ΔB_0 the frequencies of the individual nuclei are slightly different which also causes dephasing. (In ^{13}C NMR spectroscopy, the BB-decoupling of protons further decreases T_2 (compare Section 2.2.5.5) by creating a further possibility for the spin-spin relaxation of carbon nuclei.) Owing to the inhomogeneities

$$T_2^{*-1} = T_2^{-1} + \gamma\Delta B_0 \tag{230}$$

** Compare p. 21-22.

Time constant T_2^* may be defined in terms of the measured average line width

$$\Delta\nu_{1/2} = (\pi T_2^*)^{-1} \qquad (231)$$

where $T_2^* \leqslant T_2 \leqslant T_1$ holds.

If, for the moment, inhomogeneities are disregarded, it can be shown[520] that M_y (the component which determines signal intensity) is the highest, in steady state, under the condition $t_{pi} > T_2$, when

$$\cos\theta = e^{-t_{pi}/T_1} \qquad (232)$$

Then

$$M_y = \frac{\sin\theta}{1 + e^{-t_{pi}/T_1}} M_0 = \frac{\sin\theta}{1 + \cos\theta} M_0 \qquad (233)$$

When $t_{pi} > 3T_1$, the effect of consecutive pulses is independent, and when $M_y = M_0$, sin θ is constant. This yields a spectrum of steady phase in which the maximum of signals is at a pulse angle of $\theta = 90°$, independently of T_1. Thus, the optimum signal intensity depends on two factors, t_{pi} and t_p (or θ equivalent to the latter). Since t_{pi} is determined by the resolution,* an optimum should be determined for θ.

If the condition $t_{pi} > T_2$ is not fulfilled, θ_{opt} is different for the nuclei with different ν_i. In this case, $\overline{\cos\theta_{opt}} = \theta_E$, the Ernst angle,** i.e., an average for all nuclei is applied to determine the optimum of pulse angle, t_p and of t_{pi}.

For the extreme case, $t_{pi} \gg T_1, T_2$, θ_{opt} is, of course, 90°. If, on the other hand, $t_{pi} \approx T_1, T_2$, the conditions of CW measurements are realized in the PFT measurement, and from the average amplitude of the pulses following rapidly one another, defined as

$$(\nu_1)_{opt}\, t_p/t_{pi} = \theta_{opt}/t_{pi} \qquad (234)$$

it may be obtained[454] that a maximum signal intensity may be expected when the saturation factor is unity, [i.e., $(\nu_i t_p)_{opt}/t_{pi}]^2 T_1 T_2 = 1$, like in the CW technique (compare Equation 60).

The FID following the RF pulse represents an exponential decrease of the $M_{x,y}$ components. This is measured by the detector of the spectrometer, the phase being related to that of the RF pulse as reference. (The reference is continuously monitored by the detector, even when the RF field is switched off.) If a single line of frequency ν_i is excited, ($\nu_i = \nu_1$), the FID is a single exponential (see Figure 92a), and its transform is a Lorentzian of width $1/\pi T_2$ Hz. If the RF is slightly off-resonance ($\nu_0 \neq \nu_i$), the M_{xy} component rotates at a different speed with respect to the frame, which has an exactly constant frequency ν_0. Consequently, magnetization and ν_0 will alternately be in- and out-of-phase, respectively. This leads to the interference phenomenon and makes the FID to be a series of alternating minima and maxima with exponentially decreasing amplitudes (see Figure 92c).

The real situation is more complex. The magnetization vector precesses around the z axis with frequency ν_i, while its magnitude decreases exponentially with a time constant T_2. The point of the vector passes therefore a cork-screw-curve in the x-y-t coordinate system (see Figure 92b), the (y,t) projection of which is shown in Figure 92c. The (x,y) projection is a spiral: **M** approaches the z axis along this line. Since, however, the spectrometer detects the variation of the M_y component only, the FID appears as the simpler (y,t) projection.

FID is analogous to the wiggles observed in the CW spectra. The FID corresponding to

* Compare p. 185.

** Compare Section 2.2.4.4.

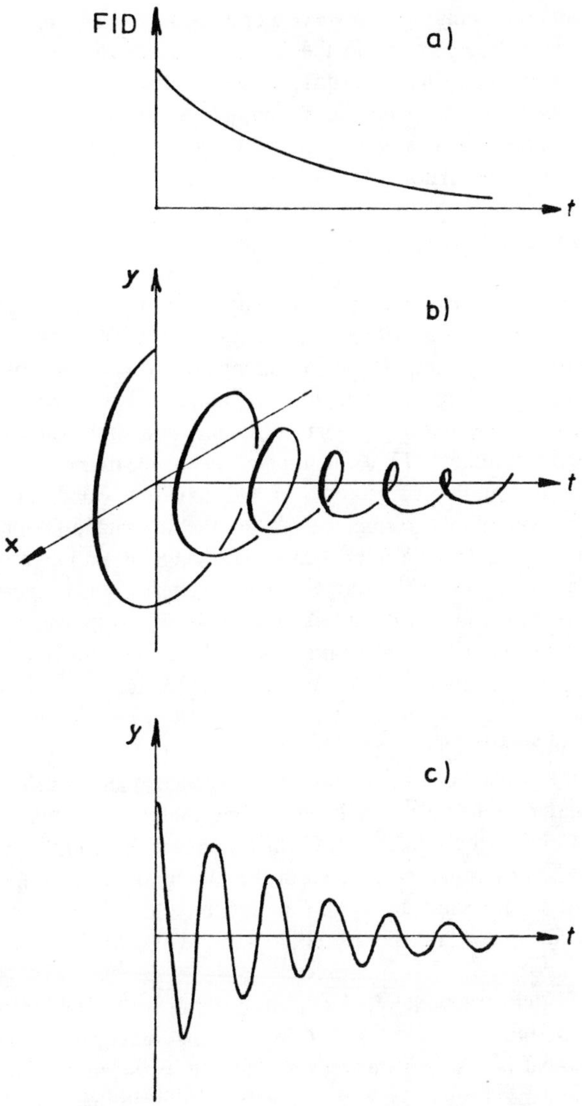

FIGURE 92. FID: the exponential decay of precessing magnetic moment. The precession of the magnetic moment about z (a) in the $\nu_i \neq \nu_0$ case, i.e., off-resonance, (b) the projection of precessing momentum vector in the $[y,t]$ plane, i.e., the y component of the point of vector as a function of time, and (c) the interferogram.

a multiline spectrum is a superposition of the interferograms of individual nuclei with ν_i frequencies similar to the one shown in Figure 92.

In the CW experiment, the frequency (or the field) may be changed only at a rate limited by the relaxations (T_1), since the spin system must be in equilibrium with its environment during the whole period of experiment (corresponding to typical T_1 of some seconds, the sweep rate is limited to some Hz/s or some tenth of Hz/s), and the nuclei with various chemical shifts are excited only one after the other. In contrast, the RF pulse excites all nuclei of the same type (e.g., all protons) simultaneously, i.e., complete information may be obtained during a time $\leqslant T_1$.

This gain in time may be invested into the accumulation of spectra or FIDs. Then, instead of a single pulse, a series or sequence of pulses is applied, and the FID is remeasured several times. The coaddition or averaging of n spectra or FIDs improves ζ by a factor of $\sqrt{n}$. This advantage of recording the spectrum in the time domain (advantage of FT spectroscopy) was first pointed out by Fellgett[473] in 1951. However, the number of pulses may not be increased beyond any limit. Its plausible optimum is equal to the number of measurable lines in the measured spectrum region $\Delta\nu_i$, i.e., the increase in ζ depends on the ratio of $\Delta\nu_i$ and a typical line width, $\Delta\nu_{1/2}$:

$$\zeta_{PFT}/\zeta_{CW} \sim \sqrt{\Delta\nu_i/\Delta\nu_{1/2}} \tag{235}$$

The relationship between B_1 and signal intensity is defined, as we have seen, by angle θ (compare Equations 228 and 233). However, in contrast with CW, signal width is independent of the magnitude of B_1 in pulsed NMR. The increase in ζ is approximately 10- to 20-fold for protons, and corresponding to the broader $\Delta\nu_i$ range, it is in principle 60- to 80-fold for carbons.[454] The real improvement is less, since ζ increases with the ratio T_2^*/T_1 (it is maximum when $T_2^* = T_1$), and the value of T_2^*, as mentioned, is also decreased by the inhomogeneities and the broad-band decoupling of protons. In addition, low values of T_2^* make it possible to speed up CW measurements, further decreasing the advantage factor of FT.

In PFT spectroscopy, ζ may be improved by applying, instead of a series of equidistant (t_{pi} = constant) and equally long (t_p = constant) pulses, periodically repeated pulse sequences in which the distance and length of pulses are varied. These pulse sequences are applied mainly for the measurement of relaxation times, and their use for improving ζ is relatively small (SEFT and DEFT methods, compare Section 2.2.9.1).

2.2.3. FT NMR Spectrometers

PFT spectrometers consist of two main units: the spectrometer itself, which contains the magnet, RF transmitter, and detector as the most important components, and the data system, comprised of a computer, the interfaces of which connect the computer to the units of the spectrometer, and the programs which control the measurement and data processing (software). The magnet is the same as with CW instruments. Corresponding to the higher frequencies required, up-to-date spectrometers (above 100 MHz) may contain supraconducting magnets.

Of the methods of field-frequency lock ("proton stabilization"),* internal deuterium lock (heterolocking) is the most suitable for PFT instruments, since in the case of homonuclear locking the lock signal is also off during the intervals between the pulses, i.e., it is not maintained continuously. Proton lock may not be applied either, for ^{13}C measurements, since the RF field used for BB-proton decoupling (compare Section 2.2.5.5) would also saturate the lock signal. (Although the deuterium nucleus is insensitive, this is compensated by the high concentration in which deuterium occurs in deuterated solvents.)

The *RF transmitter* must fulfill three tasks; it has to provide the excitation (ν_1), decoupling (ν_2), and lock frequencies. In CW instruments the RF transmitter is a crystal oscillator of basis frequency ν_i corresponding to the nucleus investigated, and the exciting and homonuclear decoupling frequencies are obtained by audiofrequency modulation. Heteronuclear decoupling and locking require, however, separate oscillators. IN PFT instruments, all required frequencies are synthesized from one basis frequency, which is produced by a quartz master oscillator (MO). One of the methods for producing frequency ν_1 is the use of a voltage-controlled RF oscillator (VCO) with a mean frequency of ν_1. A fraction (e.g., n-th part) of this frequency produced by frequency dividers is fed into a phase-sensitive detector. A different fraction (e.g., m-th part) of the MO frequency (ν_M) is fed into the same detector, and m and n are chosen so as to satisfy the relationship $\nu_M/m = \nu_1/n$. The output signal of the detector (error signal), which is proportional to the frequency difference, is fed back

* Compare p. 136.

into the VCO, correcting its ν_1 frequency to the required value, $(\nu_M/m)\ n$. In such a manner the VCO is phase-locked to the MO. By applying different *m, n* pairs, arbitrary phase-stabilized frequencies may be synthesized[1286b] and stabilized with respect to ν_M. If continuously varying frequency is required, the above method is applied again, using, however, AF-VCO instead of RF-VCO. For the production of pulses with appropriate phases (required, for instance, in measurement of relaxation times), electronic delay units may be used, which pass the pulses at different rates. The controlled RF obtained from the MO-VCO system must be amplified in order to attain the approximate 20-μs duration of the 90° pulses (this corresponds to a power of approximately 10^2 to 10^4 W, depending on the way of detection). Since in multinuclear instruments the amplification of the most diverse frequencies (excitation, decoupling, lock, etc.) is required, BB pulse amplification must be used, unlike with CW systems. It is fortunate, however, that the amplification need not be continuous, the pulsed mode enables the power to be increased by about a factor of ten. It is also advantageous that the greater bandwidth corresponds to a faster development and disappearance of pulse, i.e., the time distribution of the pulse is closer to the ideal rectangular function.

The transmitter system that passes the RF pulse into the sample also includes, in addition to the amplifiers, a gate which switches on and off the power to be transmitted and the transmitter coil which in the case of regenerative detection* also plays the role of receiver coil.

The gate is opened and closed by means of the pulse programmer, i.e., this gate controls the duration and phase of the RF transmitted to the sample. The duration of pulse t_p must be controlled in the 1 μs to 10 ms range, the spacing of pulses t_{pi} in the 10 μs to 10^3 s range, and the duration of pulse sequences in the 0.2 ms to 10^3 s range. The accuracy of t_p must be some percent. The actual magnitude of t_{pi} is less important, but it must be reproduced extremely accurately, within 1 ppm. The pulse programmer must work, therefore, on two time scales. A 10-MHz high-frequency "clock oscillator" gives t_p accurate to some hundredth of μs, whereas t_{pi} is controlled by a "slower" clock (at an accuracy of some ms).

The gate, which also controls the decoupling field in addition to the timing of pulses, must operate very efficiently; the field ratio of on and off states must be $>10^9$. (At a power of 10 kW, the power remaining after switch-off may not exceed 1 μW![464a])

Of the two types** regenerative and inductive *detection systems* used in CW spectrometers, the former is most often applied in PFT instruments. This has two reasons. First, the single-solenoid arrangement is simpler and electrically more efficient, which is very important in PFT systems, applying several kilowatts (10^{-3} to $4 \cdot 10^{-2}$ T) RF power[464b] (in CW spectrometers the corresponding parameters are approximately 1W and 10^{-8}T). The main advantage of two-coil induction detection is that in the CW technique there is less leakage (interference of excitation field into the receiver coil increasing the electronic noise) between the crossed coils. This problem is insignificant in PFT systems, since the receiver and transmitter is operated alternately; the coils are never active simultaneously. A weak interference may be filtered out more easily (by compensation using "paddles").

The *detector (receiver coil)* must meet very high, in part conflicting, requirements. Whereas the CW method requires the application of only weak fields, the very strong pulses of PFT systems, although for short times (<10 μs), represent extremely high loads (100 to 1000 V) on the receiver, which must recover its original state from the "oversaturation" present during the duration of pulse very quickly (within 2 to 3 μs). At the same time, the detector must be sensitive enough to weak NMR signals. Finally, since the technique develops toward multinuclear systems, the frequency range of detectors is also relevant.

The task of the detector is to demodulate from the carrier frequency, ν_1 the AF-components arising from the precession of measured nuclei (frequencies $\nu_1 - \nu_i$ in the $\Delta\nu_i$ frequency

* Compare p. 138-140.

** Compare p. 140.

range), and to amplify the original voltages of some μV to about 10 V, corresponding to the range of ADC (see below). The receiver must be linear, otherwise the overtones of very strong signals (e.g., solvent peaks) will also appear in the spectrum (with alternating amplitudes). The great bandwidth of detector and amplifier is a requirement raised by the simultaneous presence of all ν_i frequencies and the multinuclear mode of operation. This was not required in CW technique where only one ν_i frequency of one type of nucleus was measured at a time. The receiver must detect all frequency components with the same sensitivity and the same response time. If response time is frequency dependent, phase shifts appear in the spectrum that cannot be eliminated. Finally, the elasticity of receiver is also important; after encountering a pulse, it must return to equilibrium very quickly, in order to start the detection of FID after the pulse, at $t = 0$. If this is not the case, again a phase shift occurs, and in the transformed spectrum the absorption signals become distorted, owing to the dispersion component.

An integral part of FT spectrometers is the small, usually digital, computer built into the instrument. The capacity of the computer also determines the applicability of the instrument as a whole, and thus some requirements raised against the computer are also mentioned here. In this respect, the most important parameters are as follows: the speed of the computer (cycle time: time required to load or store a word, usually in the ns range), the "vocabulary", rapid-access memory (size of data set directly stored and accessible), and finally the word length, i.e., the number of bits in a computer word that determines the accuracy of the measurement (nowadays 16 to 24 bits).

The most important functions of the computer are data acquisition and averaging (coaddition of FIDs induced by repeated pulses or pulse sequences) in order to increase ζ, scaling and FT of the acquired data, and additional data processing (e.g., apodization, weighting, phase correction, shifting, rescaling, etc.), either preceding or following the transformation. A very important component of the system is "software", i.e., the program modules that coordinate and control the basic functions of the computer and the spectrometer, transform the spectra according to the various requirements, and assist in presenting the spectra or spectroscopic information, in general, in various forms.

The computers of FT spectrometers provide, however, a number of other possibilities for the presentation of spectra, enhancing the applicability of spectroscopic information, by processing the same FID. Thus, the frequencies with respect to arbitrary points (TMS signal or carrier frequency), chemical shifts and heights and intensities of signals may be listed automatically, signals below a given threshold may be omitted, the frequency and intensity scales may be chosen arbitrarily, etc. The availability of spectroscopic data in digital form makes, in fact, the plot of the spectrum superfluous; the plot may be used for illustrative purposes, and the list of spectrum parameters or derived data is used in structure elucidation.

The most important elements of the system interfacing the spectrometer to the computer are the digitizer (ADC, analogue to digital converter) and DAC. The main task of ADC is the digital representation of FID, i.e., the equidistant sampling of the detector signal. Here the sampling rate of the ADC is important. As the FID may be conceived as consisting of sine wave components, and one sine function may be defined by two points in one period,[142] the sampling rate must be at least twice the measured frequency range $\Delta\nu_i$ (the s.c. Nyquist frequency). The signals of frequency ν_k outside the range $\Delta\nu_i$ are "folded back" by the ADC into the range measured, where they occur as spurious signals at frequency $\nu = 2\Delta\nu - \nu_k$. Thus, it is important to filter out such ν_k frequencies. The accuracy of digitization is determined by the word length of ADC; if this is, for example, 10 bits, ν may be given with an accuracy of 0.1%, since $1/2^{10} \approx 10^{-3}$. Similarly, for intensities with a 10-bit converter, the weakest signal which may still be distinguished from noise is $1/2^8 = 0.004$ times the strongest signal (1 bit is reserved for the sign, and if the noise level is reduced to $1/2^9$ by means of accumulation, the weakest signal must be at least the double of this if it emerge from the noise). The communication between the system and the operator is maintained by means of console, teletype, oscilloscope, numerical display, etc.

The following sections of this chapter deal with a number of experimental techniques for obtaining the spectra in forms providing diverse types of information. These methods analyze the response of the same spin system to different forms of excitation.

2.2.4. The PFT Spectrum

2.2.4.1. Accuracy, Dynamic Range, Filtering, Apodization, and Scaling

The quality of the spectrum obtained from digitized FIDs depends on several factors, and it may be improved by various arithmetic methods. It is important that during data acquisition the word length of the computer should not be exceeded, since otherwise the information corresponding to the "most intense" part of FID is lost and the transformed absorption spectrum is distorted. If the word lengths of the ADC and the computer are n and N bits, respectively, the computer words may be filled after 2^{N-n} FID accumulations, and the addition of further FIDs may lead to "overflow", and thus to the distortion of transformed spectra. If, for example, a FID fills the 10-bit ADC completely, and the word length of the computer is $N = 12$ bits, only 4 FIDs may be coadded. If, however, $N = 20$, 1024 spectra may be averaged under the same conditions, causing a 32-fold improvement in ζ.

If the computer word is full, all data must be scaled down, (e.g., shifted one bit to the right: division by two) and averaging may be continued. However, if this scaling is to be carried out many times, the weakest signals are gradually lost because of "underflow". The maximum improvement of ζ may be achieved by at most 2^N accumulations. Of course, the overflow of ADC must also be avoided; this is accomplished by adjusting the gain of receiver, which may be done by observing some FIDs after reaching the steady state.

If the FID data do not fill the ADC completely, the accuracy of measurement is unnecessarily reduced. However, the overflow of ADC leads to the distortion of the spectrum. Overflow cuts off the beginning of FID which has the same effect as a multiplication of FID by a window function. This is equivalent in the frequency domain to a convolution by the Fourier transform of the window function (sinc), causing side lobes to appear on either side of the absorption bands, which cannot be eliminated (see Figure 93).

The distortions around the zero point of the spectrum due to the leakage of systematic noise and to the breakthrough of the excitation (carrier) frequency may be eliminated ("filtered") in part, e.g., by producing the FID by means of alternating pulses, changing in phase by 180°. In this case, consecutive FIDs have alternating signs, whereas noise is unchanged. By subtracting the average of even and odd FIDs, signal intensities are added with a simultaneous noise reduction.

FID lasts in principle for infinite time, but the spectrometers measure it for only a finite t_{aq} acquisition time. This is equivalent to a multiplication of the real FID by a window function, which is 1 in the $0 - t_{aq}$ period and 0 elsewhere. In the frequency domain, this corresponds to a convolution of the Fourier transform of the window function by the real spectrum, i.e., the appearance of side lobes (see Figure 88d), as in the case of ADC overflow (compare Figure 93).

This distortion of the spectrum may be compensated by an operation called apodization (cutting off of the "feet", i.e., the side lobes of bands). This is performed by multiplying the end of FID with an appropriate function, for example, a triangle or $1/2(\cos\theta + 1)$, where $0° < \theta < 90°$. This is necessary in the case of particularly low ζ values or when the amplitude is increased in the domain of longer t_{aq} values for resolution enhancement (multiplication of FID with weight function).* Figure 94 shows how the absorption spectrum changes upon apodizing an increasingly longer part of FID. It can be seen that apodization may even eliminate some weaker signals. Side lobes increasingly disappear, but the Lorentzian line shapes and the intensities are gradually distorted.

* Compare p. 174.

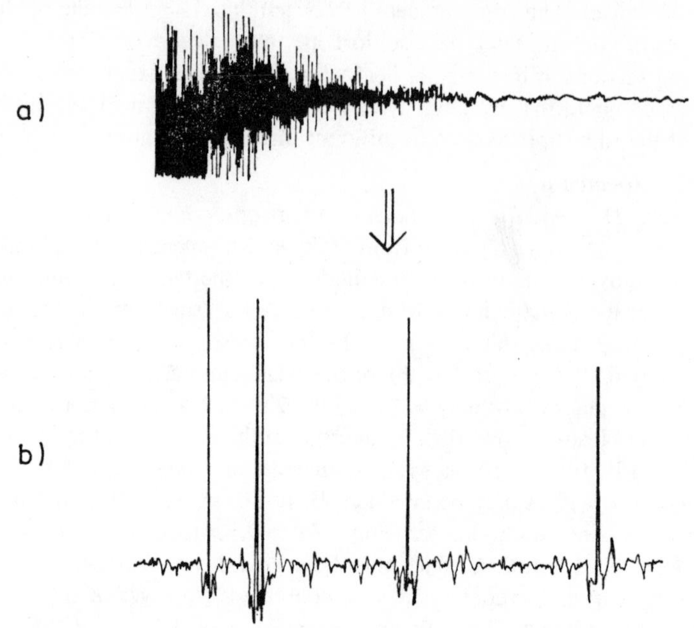

FIGURE 93. The effect of the overflow of ADC on the spectrum. The multiplication of FID by a window function (a) and the consequence (b): side bands in the spectrum. (From Shaw, D., *Fourier Transform NMR Spectroscopy,* Elsevier, Amsterdam, 1976, 189. With permission.)

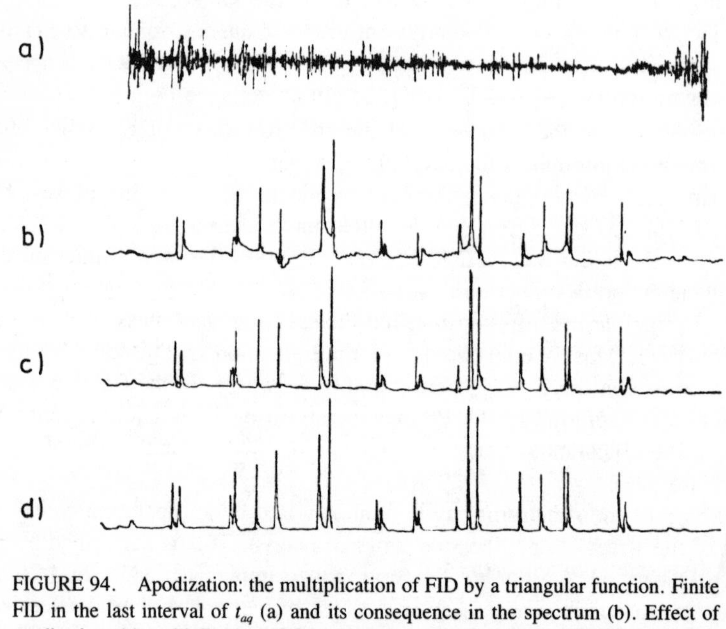

FIGURE 94. Apodization: the multiplication of FID by a triangular function. Finite FID in the last interval of t_{aq} (a) and its consequence in the spectrum (b). Effect of apodization: (c) multiplication of the last tenth or (d) nine tenth of FID by triangular function, i.e., the effect of 10 or 90% apodization. (From Shaw, D., *Fourier Transform NMR Spectroscopy*, Elsevier, Amsterdam, 1976, 195. With permission.)

2.2.4.2. The FT of FID and the Resolution

Signal averaging is followed by FT. As we have seen, a digitized FID may be transformed into a digitized absorption spectrum, which is then converted into the usual frequency spectrum by the DAC. It is noted here that the frequency-dependent spectrum is obtained after FT in the CW technique as well. Upon the slow adiabatic sweep, a simple phase detector successively analyzes the amplitudes and phases of the individual frequency components, i.e., it carries out a consecutive, sequential Fourier analysis for the components. In the CW spectra the varying excitation frequency represents time, playing the role of independent variable according to the relationship $\nu = v \cdot t$, where v is the sweep rate, and frequency occupies its place as real independent variable only in the absorption spectrum recorded after the phase detector.[451]

Digital FT means that instead of the integration prescribed in the definition of transformation $\mathcal{F}$, a summation of finite (N) data points must be performed. The FID function must be redefined according to Equation 207, as a series of δ functions with spacing τ multiplied with the actual amplitudes of FID as coefficient:

$$A(t) = \sum_{n=0}^{N-1} A_n \delta(t - n\tau) \tag{236}$$

The discrete FT of this function is

$$A(\nu) = \frac{1}{N} \sum_{k=0}^{N-1} A(t) e^{-i2\pi\nu k/N} = \frac{1}{N} \sum_{k=0}^{N-1} A(t)(\cos \omega k/N - i \sin \omega k/N) \tag{237}$$

Here $\tau = 1/(2\Delta\nu_i)$,* and thus the resolution is

$$1/t_{aq} = 1/N\tau = 2\Delta\nu_i/N \tag{238}$$

and the data acquisition rate is $1/\tau$ data points per second; t_{aq} is the acquisition time of FID.

Consequently, the longer t_{aq} the higher the resolution is. However, t_{aq} is restricted by the computer memory, and it is not worth, either, increasing t_{aq} beyond a certain value. Namely, FID decreases exponentially in time (according to the function e^{-T/T_2^*}), and after a time of T_2^* it drops to approximately 5% of its maximum. If $t_{aq} \approx \pi T_2^*$, the resolution corresponds to the typical line width, $1/\pi T_2^*$. Thus in ^{1}H NMR, where typical line width is about 0.2 Hz, t_{aq} should be in the range of seconds in order to reproduce these line widths satisfactorily. For ^{13}C nuclei, line widths are in the hertz range, and thus t_{aq} may be chosen as some tenths of second. Consequently, if the available computer memory is 4, 8, or 32 K, respectively, the ^{1}H and ^{13}C resonance spectra may be obtained by acquisition times defined by the line widths, under the resolution given in Table 21 (at 100 MHz). It can be seen that the resolution attainable at 32 K is already unnecessary. In this case the number of data points may be decreased, e.g., to $N/2$, and by filling up the remaining places with zeros, ζ may be improved. The effect of zero filling may be seen from the following consideration.

Since $A(\nu)$ is complex, half of the data points are lost. The reason for this apparent "loss" of information is that on the transformation of real FID data the imaginary part of the time domain function remains undefined. Consequently, the sine and cosine components of the transformed frequency domain function will not be independent. This may be overcome by using the principle of causality.[88] The real and imaginary components are independent if the time domain function, too, is transformed in complex form, i.e., another interval, $-N\tau$ to $t = 0$ (time — t_{aq}) is added to the time interval $t_{aq} = N\tau$, where $t > 0$. Since in the time preceding the pulse FID was continuously zero, this may be done by adding N

* Compare p. 182 on the Nyquist frequency.

Table 21
DEPENDENCE OF THE RESOLUTION ON COMPUTER MEMORY FOR ^1H AND ^{13}C NMR SPECTRA

Memory size [words]	$\Delta\nu_i$ [Hz]	t_{aq} [s]	$1/t_{aq}$ (resolution) [Hz]
2^{12}	1000	2.05	0.5
	5000	0.4	2.4
2^{13}	1000	4.1	0.2
	5000	0.8	1.2
2^{15}	1000	16.4	0.06
	5000	3.3	0.3

zeros to the actually real data. Consequently, the number of data point are doubled, leading to an twofold increase in ζ under unchanged resolution.

It is noted that the loss of information may be recovered by another method, too. If two detectors are applied for the detection of FID (quadratic detection), all data points may be characterized by two components, which can be regarded as the real and imaginary parts of FID. Upon the transformation of these signals, the information gain occurs as the sign of signals with respect to carrier frequency ν_1. Under normal condition this is a superfluous information, but it enables ν_1 to be placed into the center of interval $\Delta\nu_i$ (normally it must be outside the frequency range to be measured, since otherwise the signals will appear on both sides of ν_1). Of course, the information on the signs excludes the possibility of the enhancement of ζ by means of zero filling.

The process of discrete FT is performed by means of the Cooley-Tukey FFT (fast Fourier transformation) algorithm.[295] The determination of $A(\nu)$ coefficients by direct summation requires N^2 multiplications and N additions for N data points, whereas in the FFT algorithm, the number of multiplications is only $2N \log_2 N$ (for $N = 2^{12} = 4096$ data points, the gain in time is 170-fold; for more data points it is even greater!). The FFT algorithm requires a repeated pairing of data points, and thus the number of data points must be a power of two: $N = 2^x$. Another advantage of the FFT algorithm is that it requires relatively little computer memory, the $A(\nu)$ coefficients are produced successively from the initial FID data (by a combination of progressively increasing weighted sums of data points), and thus the results of the new step of calculation usually replace the data of the previous step in the memory. Thus, the transformation of N data points requires just slightly more than N words. However, this involves the loss of the original FID data, too. If the transformed, frequency domain spectrum is for some reason unsatisfactory (e.g., other weight function or apodization should have been used, or more FIDs should have been coadded for better ζ), the experiment has to be repeated. Of course, in modern instruments, this is no longer a problem, since a copy of the FID may be stored in other parts of the memory (preferably on disk or magnetic tape) until it is decided on the basis of the frequency domain spectrum whether to continue the measurement or apply other ways of processing. Thus, further FIDs may be added to the already collected ones, or they may be transformed under different conditions, without remeasuring already stored experimental data.

2.2.4.3. *Tailored Excitation*[1404]

It is sometimes necessary to excite the spin system by frequencies with a nonuniform power spectrum, i.e., the energy should either increase (a) or decrease (b) in a given frequency range (see Figure 95). Therefore, the excitation function should be "tailored" to a given shape: in the appropriate frequency domain, it should have a spike or a notch. The term "window function" may also be used, and the window is "open" or "closed" for the

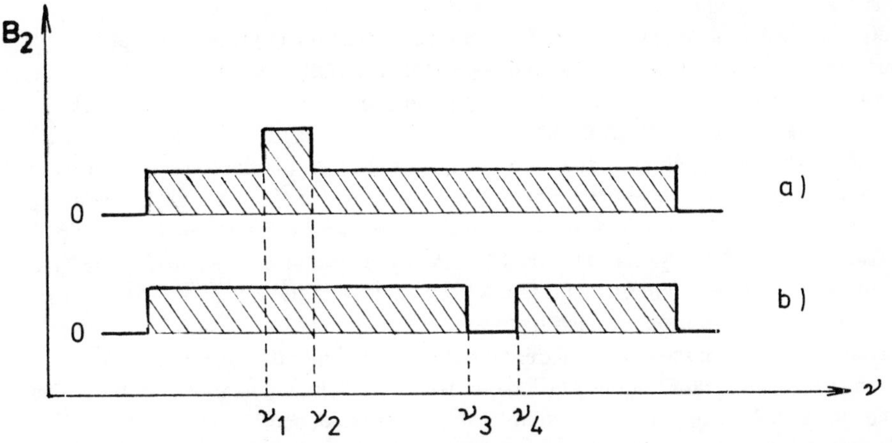

FIGURE 95. Window functions of uneven energy distribution. With a notch (a): in the frequency region $\nu_1 - \nu_2$ the amplitude of excitation function is higher, and the nuclei of $\nu_1 < \nu_i < \nu_2$ resonance frequencies saturate. With a spike (b): in the frequency region $\nu_3 - \nu_4$, the excitation energy is zero, and the nuclei with $\nu_3 < \nu_i < \nu_4$ resonance frequencies are not excited.

above corner or dip. The former may be used, e.g., in DR when the frequency to be saturated is in the $\nu_1 - \nu_2$ range. The latter is useful in cases when an intense signal covering the range $\nu_3 - \nu_4$ of the spectrum is to be eliminated or suppressed (e.g., the strong absorption of the DHO content of the solvent in the case of biological systems measured in D_2O solution, compare Section 2.2.9.3). Solvent signal should be suppressed, since the signals of the solute which are very weak in comparison with the solvent peak fill only a proportionally smaller portion of the ADC register (otherwise it would overflow because of the solvent signal, and the spectrum would be distorted).

FT technique is advantageous in this respect, too, since it enables arbitrary energy distribution functions to be produced for excitation by the application of a series of modulated RF pulses, in the following way. First, the energy function $f(\nu)$ with the required energy distribution in the frequency domain should be transformed. If this is done by a digital computer, the function in the time domain is represented by a series of N discrete coefficients. An RF pulse sequence with a spacing t_{pi}, will be modulated by these coefficients producing a periodic pulse sequence of N members, where the length t_{si} of the pulse sequence is determined by the required resolution: $t_{si} = 1/\Delta\nu_i$, and $\Delta\nu_i$ is the frequency range to be measured. The excitation function is thus a series of modulated delta-functions:

$$g(t) = \sum_{n=1}^{N} B_1 f(t) \cdot \delta(t - n\,\tau) \tag{239}$$

where $\tau = t_{si}/N$ and, of course, $\mathcal{F}[f(t)] = f(\nu)$. Modulation may be accomplished equally well by changing the amplitudes, lengths, or phases of the pulses; the magnitude of $f(t)$ is determined usually by the length of the pulse; while its sign by the phase of the pulse $(\pm\,\pi)$.

By means of window functions, it is also possible to carry out double and multiple resonance experiments if the two (or more) types of nuclei may be excited by a single RF transmitter. Thus, solvent peak elimination is possible not only by "closed" window functions (containing a notch, excluding thereby the solvent proton excitation), but also be an "open" window function (which selectively saturates the solvent signal). Solvent peaks may also be eliminated by utilizing its longer T_1 relaxation (compare Section 2.2.9.3) by means of appropriate pulse sequences.

2.2.4.4. *Pulse Width*

Of the measurement parameters, the most difficult to select properly is pulse width (which may be expressed as angle θ according to Equation 228. Signal intensities depend in a complex way on θ; in this respect, the optimum angle is higher the larger the ratio t_{pi}/T_1 for a given line of the spectrum is, where t_{pi} is pulse spacing. If the latter is equal to t_{aq} (this is the most efficient choice, enabling the fastest data acquisition), θ_E (the Ernst angle),[454] corresponding to the optimum pulse width, is smaller for the more slowly relaxing nuclei: $\cos \theta_E = e^{-t_{pi}/T_1}$. If, therefore, θ is chosen to be about 90°, some signals, e.g., in ^{13}C NMR those of the quaternary carbon atoms, including carbonyl carbons with longer relaxation times, may become saturated and thus unobservable. With too low θ (about 0°), the intensities decrease and their ratios are also distorted. After all, smaller θ values are generally more advantageous, since the corresponding optimum signal intensity is higher and the distortions owing to the nonideality of pulses (departures from rectangular) are smaller for longer pulses. The success of the application of smaller θ angles also depends on instrumental parameters. If digitization contributes to noise, a greater number of FIDs decreases the value of ζ. Moreover, the requirements concerning the exact reproduction of shorter pulses are also stricter. The conversion of angle θ to pulse width, i.e., to duration t_p of pulses, may be performed on the basis of Equation 228 by using the effective field B_1 characteristic of the given instrument. This is usually available in tabular form or in the form of a calibration curve supplied with the instrument. The calibration of the spectrometer is easy; one has to choose an arbitrary singlet signal and record the spectra for a series of pulse angles. In the 0 to 90° region, the signal gradually increases, reaches a maximum at 90°, and then decreases to 0° when $\theta = 180°$. In the 180 to 360° region, the signal should be negative, passing a minimum at 270°. By noting the t_p values corresponding to the extremal and zero signals, the pulse widths corresponding to $\theta = 90$, 180, 270, and 360° may be determined in s units (see Figure 96).

2.2.5. Double Resonance Methods in PFT Spectroscopy

Most of the spectrum types obtained by DR experiments are independent of the way of excitation, and thus the principles discussed at CW spectroscopy apply almost without change in PFT spectroscopy. Therefore, we will discuss here only those DR methods which are different from CW technique or which are applied exclusively in PFT, mostly in ^{13}C NMR spectroscopy.

2.2.5.1. *Off-Resonance*

The off-resonance decoupling of protons is one of the routine methods of ^{13}C NMR. It does not eliminate the splittings due to $^1J(^{13}\text{C},^1\text{H})$ couplings, however, it decreases the distance between the lines of multiplets.[450] Off-resonance technique may be regarded as a special case of spin decoupling, since the decoupling field is strong, but the frequency is not exactly the resonance frequency of protons, differing from it by some hundred hertz.

The method may be understood most easily from a vector diagram,[155,662] given in Figure 97, for the simplest AX case. If nucleus X is placed into an excitation field of frequency $\nu_2 = \gamma B_2 \neq \nu X$, the z component of the static field in the rotating frame will be $\Delta\nu \pm J/2$ in frequency units, where $\Delta\nu = \nu X - \nu_2$, and the sign of J depends on the spin state of nucleus A.* Thus, nucleus X feels an effective field of frequency:

$$\nu_{\text{eff}}^{\pm} = \sqrt{(\Delta\nu \pm J/2)^2 + \nu_2^2} \qquad (240)$$

* In a field rotating with frequency ν_2, the polarizing field $B_o = \nu X/\gamma$ is on the one hand decreased by the fictive field of magnitude ν_2/γ (cf. p. 176 and Figure 91) and on the other hand it varies with the quantum state of nucleus A (for parallel A and X, it increases by $J/2$; for antiparallel it decreases by $J/2$, cf. p. 93 and Figure 39c.)

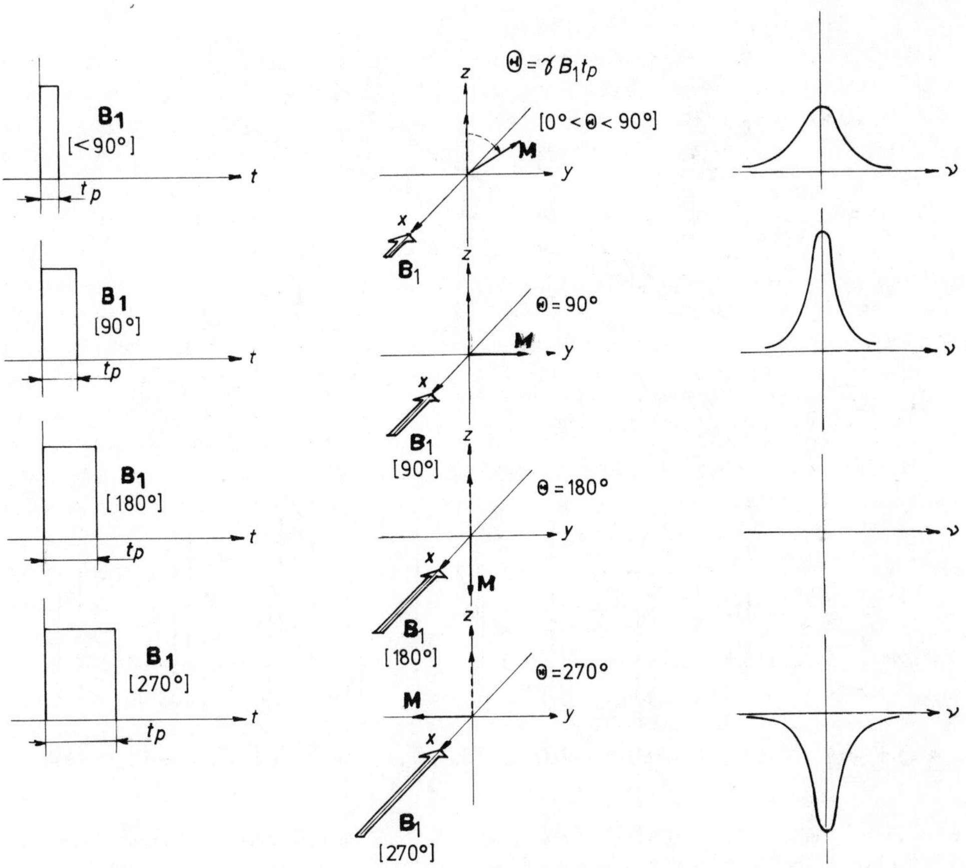

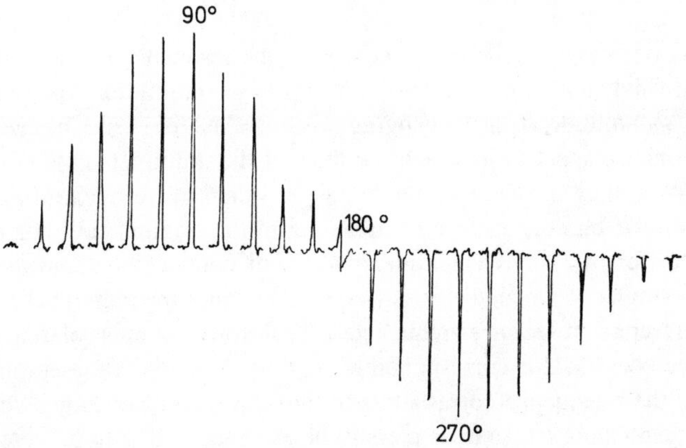

FIGURE 96. The effect of changing pulse width (pulse angle θ or pulse length t_p) on intensity. Calibration of pulse angle in the ^{13}C NMR spectrum of benzene (a): change in intensity upon changing t_p from 0 to 66 μs in steps of 3 μs. (From Shaw, D., *Fourier Transform NMR Spectroscopy,* Elsevier, Amsterdam, 1976, 189. With permission.)

arising from nuclei A. It can be shown[662] that the reduced coupling constant J^*, corresponding to the remaining, smaller coupling, is $J^* = \nu_{eff}^+ - \nu_{eff}^-$, and under the conditions $\nu_2 \gg \mathcal{F}$ and $\nu_2 \gg \Delta\nu_1$

$$J^* = J\Delta\nu/\nu_2 \tag{241}$$

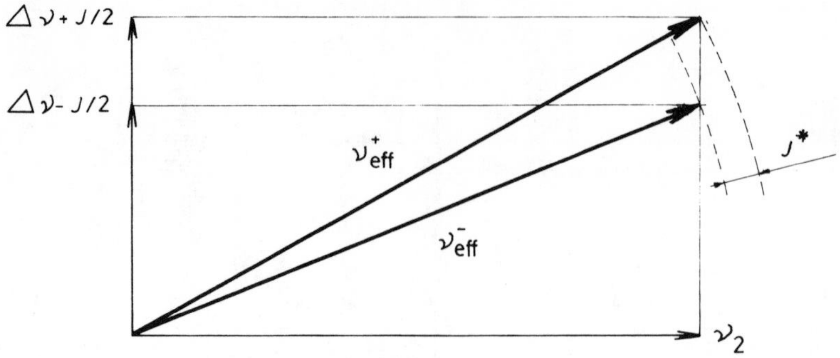

FIGURE 97. Vector diagram for the illustration of off-resonance, with the magnitude of reduced coupling constant J^* determined by νX, ν_2, and J_{AX}.[155]

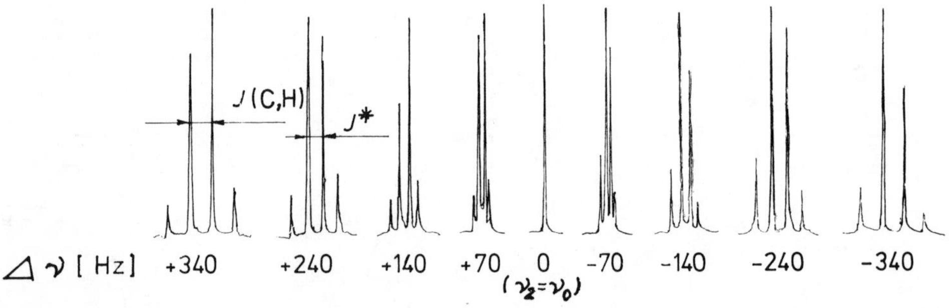

FIGURE 98. The effect of decoupling frequency on the off-resonance multiplets: the methyl quartet of methanol as a function of offset $\Delta\nu = \nu_0 - \nu_2$. (From Wehrli, F. W. and Wirthlin, T., *Interpretation of Carbon-13 NMR Spectra*, Heydon, London, 1976, 68. With permission.)

i.e., line spacing decreases (J^* decreases) closer to the resonance (small $\Delta\nu$) and for higher decoupling field (higher ν_2, i.e., for higher $B_2 = \nu_2/\gamma$ amplitude). Applying an excitation field of constant amplitude B_2 and changing $\Delta\nu$ in a $\Delta\nu_F$ to $-\Delta\nu_F$ range, one obtains a series of off-resonance spectra, in which the lines of the multiplets gradually approach one another; at $\Delta\nu_i = 0$ they coalesce into a singlet and then move outward again (see Figure 98) until reaching the line distances corresponding to the original coupling constant J. The changes in the different signals (multiplets) have, of course, the same direction, but the coalescence into singlet occurs at different frequencies, since the individual protons or proton groups have different ν_i resonance frequencies. Therefore, by interpolating the J values of the off-resonance spectra, $\Delta\nu_i$ corresponding to $J^* = 0$ for the various multiplets may be determined, and their sequence corresponds to the sequence of chemical shifts of protons. This allows corresponding ^{13}C and ^1H pairs to be assigned.* It is to be noted that although the multiplicities of the signals do not change in the off-resonance spectra, the relative intensities are distorted as in the case of roof structure corresponding to higher-order interactions. As an example, let us consider the case of 1,3-butanediol: Me–CHOH–CH_2–CH_2OH. The chemical shifts in its ^1H NMR spectrum (see Figure 99a) are δH-1, 3.80; δH-2, 1.68; δH-3, 4.03, and δH-4, 1.23 ppm.[1044] In the ^{13}C NMR spectrum (see Figure 99b), four signals appear at 26.9, 44.8, 63.2, and 69.3 ppm[190] of quartet, triplet, triplet, and doublet multiplicity, according to the off-resonance spectra (see Figure 99c to e). The assignment of the triplets at 44.8 and 63.2 to C-1 and C-2 is ambiguous. By recording a few off-resonance spectra with different offsets in sign and magnitude and interpolating the multiplets

* Compare p. 209.

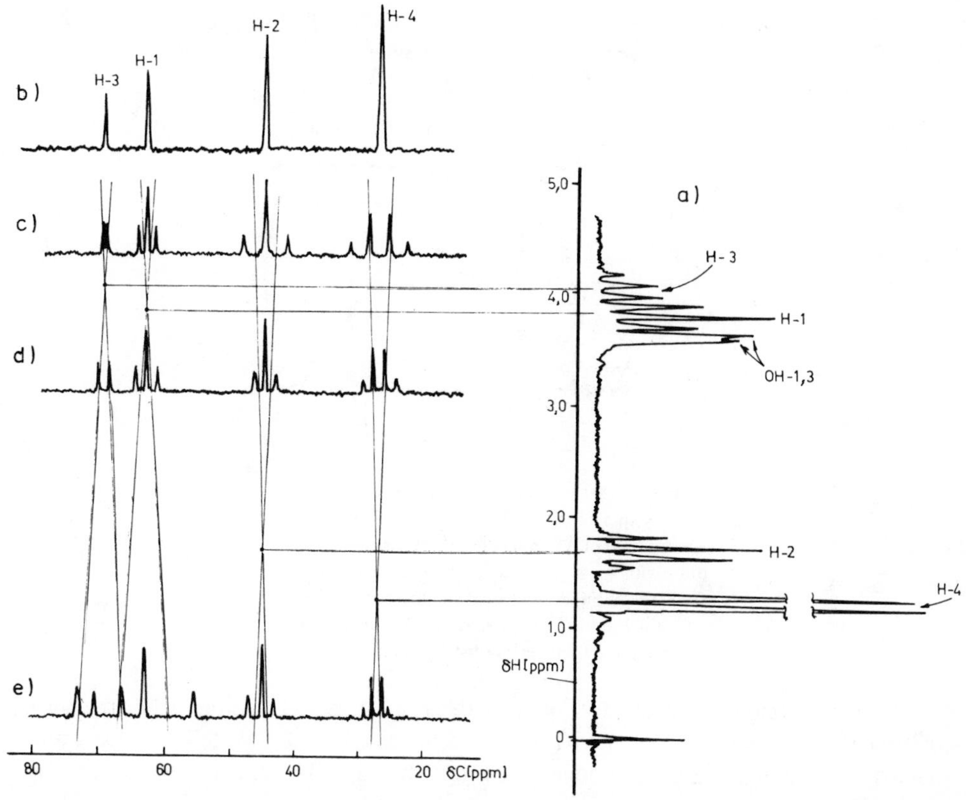

FIGURE 99. Graphical method for cross-correlation of ^{1}H and ^{13}C NMR spectra. The ^{1}H (a) and (b) the ^{13}C BBDR spectrum; (c), (d), and (e) off-resonance spectra with offsets different in magnitude and sign, respectively.

to find the offsets corresponding to $J^* = 0$ for the individual multiplet, one can get a graphical scheme (see Figure 99), in which the sequence of intersection points of zero residual splittings is the same as the sequence of the ^{1}H NMR signals. From this graphical correlation of the ^{1}H and ^{13}C NMR spectra, the assignment given in Figure 99b is straightforward.

In the off-resonance spectra, usually first-order multiplets occur, but higher-order splitting is also possible if $J^*(C,H)$ and $J(Y,H)$ couplings between the decoupled hydrogens and their Y neighbors (usually protons, too) in higher-order interaction with the former are similar in magnitude to one another and to the chemical shift difference, $\Delta\nu YH$. In this case, the carbon signal is split via virtual coupling,* i.e., instead of the A multiplet of the AX_n spin system corresponding to the CH_n group, the A multiplet of an AX_nY_m spin system corresponding to a CH_nH_m group may be observed.

2.2.5.2. Double Resonance Difference Spectroscopy (DRDS) and Selective Nuclear Population Inversion (SPI)

Of the DR experiments feasible by means of CW technique, the one directly corresponding to INDOR is missing in PFT spectroscopy. However, DRDS and SPI methods may be regarded as analogues of INDOR.

DRDS[468] is the subtraction of two spectra, one normal spectrum and one spectrum taken upon excitation of line ν_2 (monitored in the CW-INDOR experiment) by a weak B_2 field. In the latter spectrum, the signal intensities of nuclei coupled progressively or regressively

* Compare p. 63.

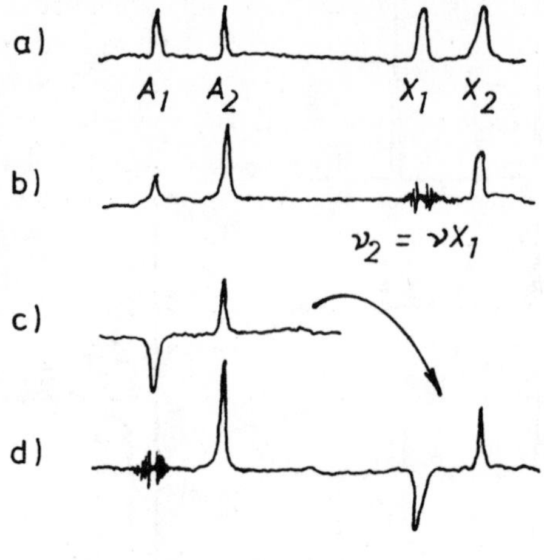

FIGURE 100. Selective inversion double resonance experiment (SPI). The normal absorption spectrum (a) of AX spin system, (b) simple DR spectrum with $\nu_2 = X_1$, and (c) the corresponding INDOR spectrum: B_0 and $\nu_2 = X_1$ are constant. ν_1 is varied in the CW experiment; (d) SPI carried out by FT technique by inverting the X_1 line.

to the nucleus of frequency ν_2 increase or decrease, respectively. Therefore, when the normal spectrum is subtracted, from it a PFT result identical with the CW-INDOR spectrum is obtained. The ratio ζ_{PFT}/ζ_{CW} is, of course, less by a factor of $1/\sqrt{2}$, since two spectra yield one INDOR spectrum.

SPI is another PFT analogue of INDOR.[262,1351] If, by means of a 180° pulse, a line is selectively inverted during normal PFT measurement, an INDOR spectrum is obtained in which the individual transitions have reversed roles. In the simplest AX case (see Figure 100a), upon excitation of line X_1 (see Figure 100b) ($\nu X_1 > \nu X_2$), the progressive transition A_2 ($\nu A_2 < \nu A_1$) gives a positive INDOR signal, and its regressive pair A_1 gives a negative one (see Figure 100c). Line X_1 becomes, of course, saturated and $\mathbf{X}_1$ remains unperturbed. In the SPI experiment, the line inversion of line $\mathbf{X}_1$ produces a negative signal, whereas its unperturbed pair X_2 produces a positive one (i.e., they assume the roles of lines A_1 and A_2 in the CW-INDOR experiment). The A_1 line is missing from the spectrum, since, owing to the inversion of the population of levels **1** and **2**, level **1** is overpopulated and the A_1 transition (i.e., **3** → **1**) does not occur. Thus, line A_1 in the SPI spectrum corresponds to line X_1 in CW-INDOR. On the other hand, the **4** → **2** transition, A_2, has now double intensity (see Figure 100d), but this shows no analogy with the INDOR experiment. Successful SPI may be accomplished if the conditions $\gamma B_2 \ll J_{AX}$ and $t_p \ll T_1$ are fulfilled.

2.2.5.3. *The Overhauser Effect*[1045]

The term Overhauser effect[1] was first applied on the nuclear polarization phenomenon observed when the electron spin levels of metals were saturated by radiation.[1076] In wider sense, however, all phenomena connected with changes in the population of magnetic levels induced by electromagnetic radiation are called Overhauser effect. In this sense INDOR spectra may also be regarded as a general representation of Overhauser effect.

The essence of nuclear Overhauser effect (NOE) is that upon saturation of a nucleus, the relaxation processes of another nucleus being in some connection with the former are accelerated, and thus the resonance line of the latter becomes stronger: it is able to absorb more energy from field B_1 without becoming saturated. The term "connection" is applied here in a very broad sense: it may be most frequently a spin-lattice interaction. In NOE the relaxation process according to the dipole-dipole (DD) mechanism* is speeded up; in the first NOE experiment by Kaiser[742] this was shown on the chloroform signal of a chloroform-cyclohexane mixture. Here, an increase in intensity was observed in the ^1H NMR spectrum upon saturating the hydrogens of cyclohexane. This intensity enhancement for protons may be 50% theoretically, but in practice it is less; in the experiment mentioned it was 34%. Concurrent processes (e.g., the paramagnetic molecular oxygen content of the solution) decrease the effect.**

The NOE is a consequence of a nonequilibrium nuclear polarization. The changes in signal intensities due to NOE may be determined by comparing the normal intensities of an A_nX_m spin system to those observed with NOE. This requires the determination of allowed transitions and their relative probabilities, which is a function of the occupancies and energies of levels.

In the simplest *AX* case, the saturation of nuclei *X* causes a pairwise equilibration between the occupancies of the four original levels (1, αα; 2, αβ; 3, βα; and 4, ββ). Therefore, the originally forbidden **1** → **4** and **2** → **3** transitions have now finite probabilities, W_o and W_2. If the identical probability of the further transitions is denoted by W_1, it can be shown[242,1346] that

$$\frac{I_{NOE}}{I} = 1 + \left(\frac{W_2 - W_0}{W_0 + 2W_1 + W_2}\right)\frac{\gamma_X}{\gamma_A} = 1 + \eta \qquad (242)$$

where η is the "NOE factor" characteristic of the extent of intensity enhancement. The value of η is increased by DD relaxation; all other relaxation processes have opposite effect. If, like in ^{13}C NMR, the DD mechanism is the dominant, $\eta = \gamma_H/2\gamma_C = 1.988$, and thus NOE causes a maximum threefold intensity enhancement. By a comparison of the intensities in the coupled and proton decoupled ^{13}C NMR spectra, therefore, the contribution of DD mechanism may be determined. It is, however, important to take into account that the NOE factor $\eta = \gamma_H/2\gamma_C$ pertains to the limiting case of fast molecular motions (at the extreme narrowing condition), and if the condition $\nu_c\tau_c \ll 1$ is not met, η is frequency dependent.***[372,790]

Since η contains the gyromagnetic factors of the nuclei involved, NOE may also be negative, e.g., for ^{15}N and ^{29}Si. If $\eta \approx -1$, the signal disappears, e.g., from the proton decoupled ^{15}N NMR spectrum.† In such cases NOE must be eliminated by chemical or physical methods.‡

The NOE-affected signal intensity develops at a finite rate; the signal reaches its maximum intensity after a certain time. The time constant of the process is the spin-lattice relaxation time T_1. On the other hand, decoupling immediately develops or ceases upon switching the decoupling field on or off.[470] This allows the two phenomena to be separated in time, which enables one

1. To utilize the ζ-enhancing effect of NOE without removing the multiplicities of the signal in the meantime
2. To measure absolute line intensities by eliminating NOE in ^{13}C NMR spectra
3. To determine the contribution of DD mechanism to the relaxations

These aims may be reached by means of gated decoupling (time sharing) experiments discussed in the next section.

* See Section 2.2.6.2.
** Compare p. 204, 205 and Volume II, p. 227.
***Compare p. 201.
† Compare Volume II, p. 259.
‡ Compare p. 205 and Volume II, p. 227.

NOE may be utilized not only in ^{13}C NMR PFT technique; it has significant role in the practice of structure elucidation, particularly in applications for the solution of stereochemical problems.[1492b] NOE is inversely proportional to the sixth power of the internuclear distance; therefore it is very sensitive to differences in stereochemistry.[112] The intensity changes of the formyl signal upon irradiating the methyl signals of the DMFA were examined.[44] Upon saturating the signal of the more shielded methyl protons, an intensity increase 17% was found for the signal of the formyl proton, while the saturation of the other methyl signal has not resulted in intensity changes. Thus, it was concluded[44] that the former is in *cis* relation with the formyl proton, while the latter is in *trans* relation.*

Similarly, upon saturating the *cis* methyl signal of $Me_2C; = CHCOOH$, the olefinic septet is simplified into a quartet, and its intensity is increased by 17%.[44] For $CH_2 = CMeCOOMe$, the analogous change is only 9%[533] because the relaxation of the olefinic hydrogen *cis* to the *C*-methyl group is also promoted by the *geminal* hydrogen.

In the ^1H NMR spectrum of the mixture of isomers **55a** and **b**,[1048] the signals of the bridgehead and the methyl protons, respectively, are separated, but the small shift difference does not permit unequivocal assignment. On irradiating the more intense methyl signal, an intensity increase of the stronger bridgehead proton signal was observed (31%), while the weaker bridgehead proton signal did not show NOE. Upon saturating the other methyl signal, NOE was not observed either. Thus, the Structure **55a** could be assigned to the main product, in which the methyl group is located closer to the bridgehead proton. In order to improve sensitivity in ^{13}C FT NMR spectroscopy, the NOE is utilized routinely.

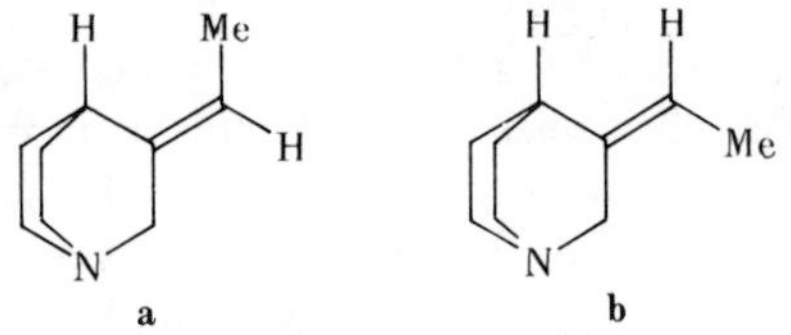

55

In the ^{13}C NMR spectrum of DMFA (see Figure 101a), the NOE can be observed in much the same way as in the ^1H DR experiments. This explains the higher intensity of the methyl signal at 36.2 ppm *cis* to the formyl hydrogen, as compared to its *trans* pair at 31.1 ppm.

In the ^{13}C NMR spectrum of $Me_2C^3(OH)–C^2 \equiv C^1H$ molecule (see Figure 101b), the intensity of the methyl signal increases more than the expected factor of 2. The intensities of the other signals decrease in the order 1-2-3 as one hydrogen is bonded to C-1, and none are bonded to C-2 and C-3. Relaxation of C-2 is, however, accelerated by DD interaction with H-1, which results in a small but observable intensity increase due to NOE.

2.2.5.4. Gated Decoupling. Determination of NOE and the Absolute Intensities of ^{13}C NMR Signals[470,521,523]

By inserting a $t_d > 3T_1$ delay time after the acquisition period, the complete pulse sequence is $(t_{pi} = t_p + t_{aq} + t_d)_n$. If the decoupling field B_2 is applied only over the t_d period, decoupling may not be observed in the spectrum, i.e., in the ^{13}C NMR spectrum, the signals of primary, secondary, and tertiary carbon atoms have quartet, triplet, and doublet structures, respectively. However, NOE is intact, since it develops during t_d, and at the first moment of t_p, when B_2 is switched off, it is maximum (see Figure 102b). From this moment it decreases exponentially, at a rate determined by T_1, but this has no effect on signal intensities, since the integral is determined by the FID value corresponding to $t_{aq} = 0$.** Thus, with the retention of multiplicities, maximum NOE, and thus much better ζ, may be attained.[521]

* Concerning the *cis-trans* isomery compare Structures *41a-c*, p. 71.
** Compare p. 170.

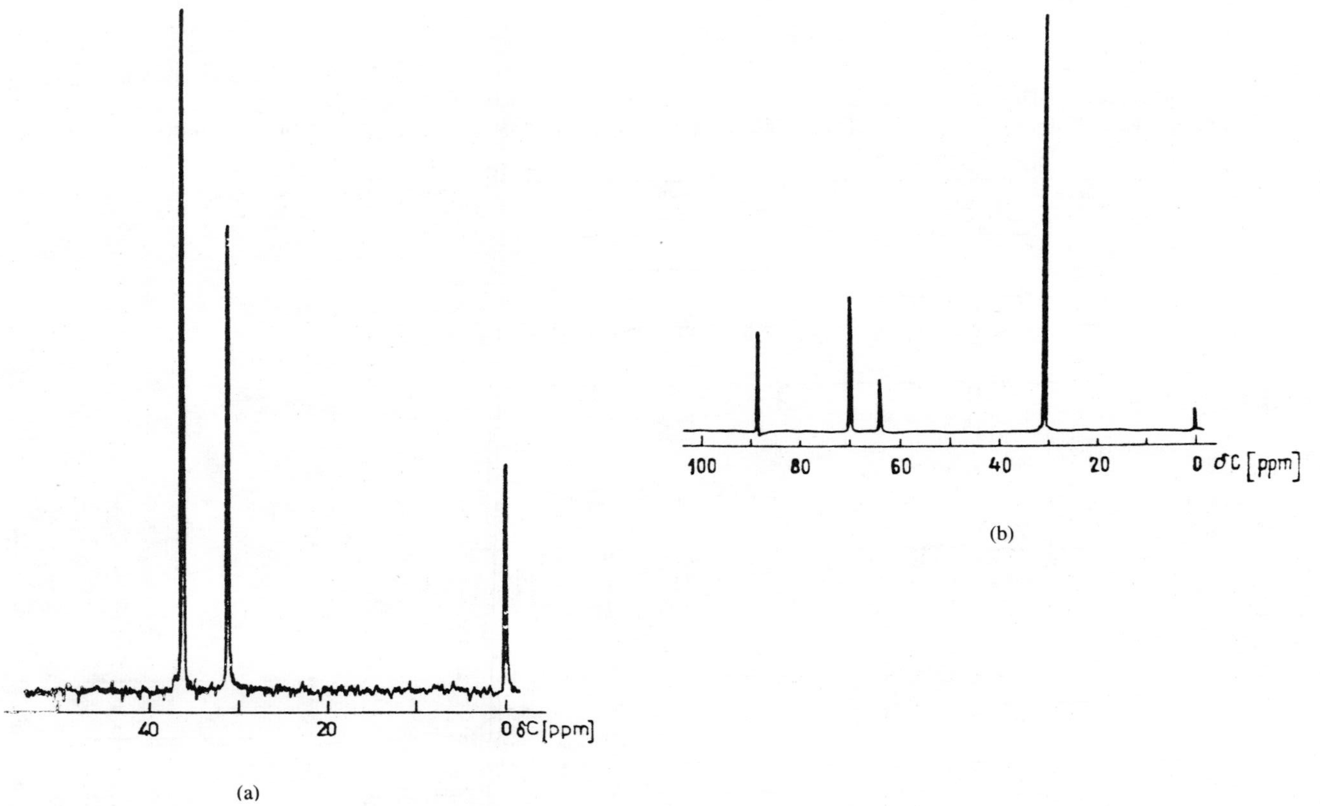

FIGURE 101. (a) Part of the ^{13}C NMR spectrum of dimethylformamide with the methyl signals [729] and (b) the ^{13}C NMR spectrum of 3-hydroxy-3-methyl-butine-1 (HC≡C–Me_2OH). (From Levy, G. C. and Nelson, G. L., *Carbon-13 Nuclear Magnetic Resonance for Organic Chemists*, John Wiley & Sons, New York, 1972. With permission.)

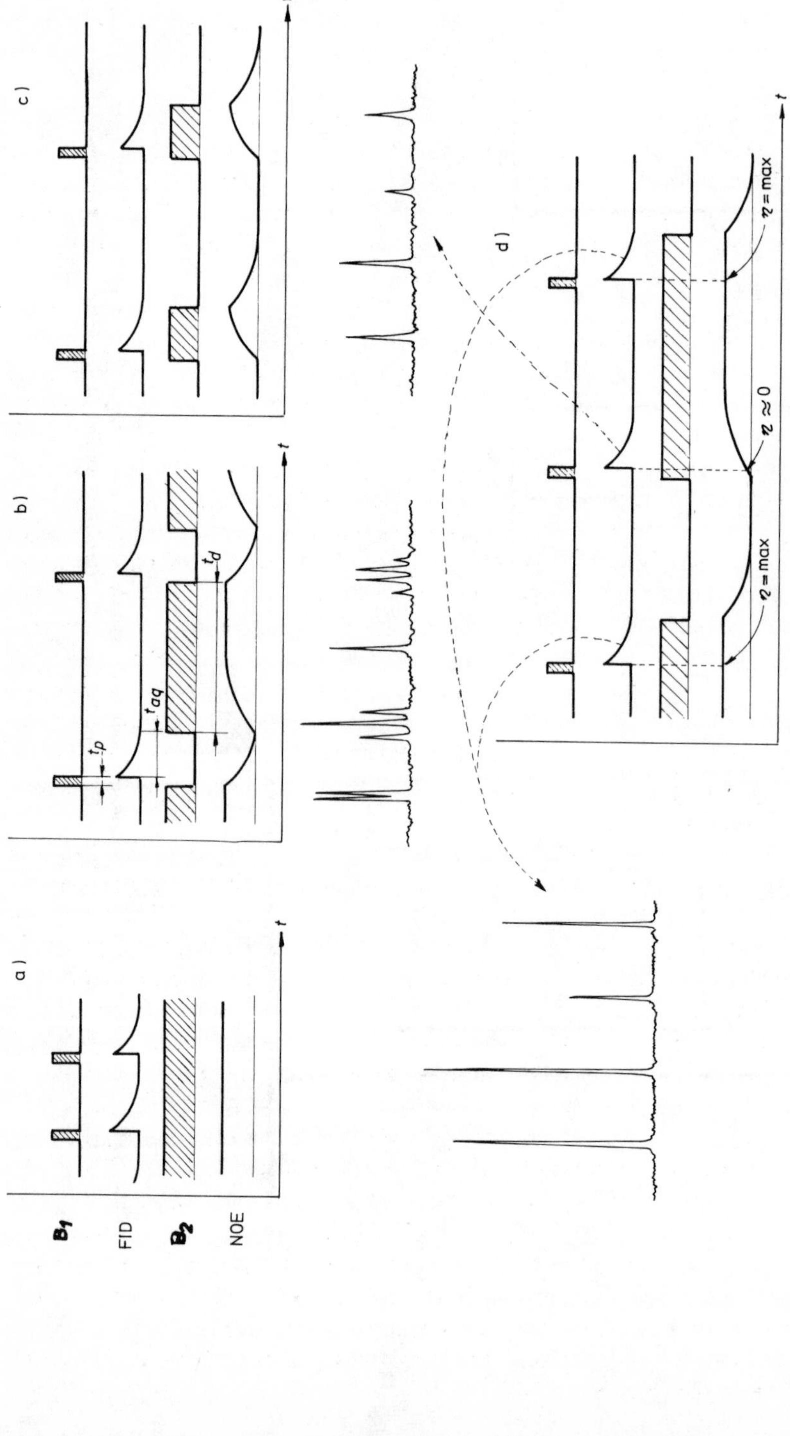

FIGURE 102. Gated decoupling. Maximum use of NOE (b) for the measurement of multiplet (coupled), (a) decoupled spectrum: the measurement of NOE, (c) proton-decoupled and NOE-free spectrum: measurement of absolute intensities, and (d) determination of absolute intensities and NOE in one experiment.

If the decoupling field B_2 is on for the $t_p + t_{aq}$ period and off during t_d (see Figure 102c), a decoupled spectrum is obtained in which the intensities of singlet peaks are not influenced by NOE, and thus they are proportional to the number of the corresponding nuclei (e.g., carbon atoms in ^{13}C NMR). During t_d NOE disappears in the absence of B_2, dropping to minimum in the moment $t_{aq} = 0$, determining signal intensities. Thus, NOE may be eliminated despite the presence of decoupling,[523] and the absolute intensities of signals may be determined. By comparing these intensities to the ones measured in the normal spectrum (see Figure 102a), one may determine the NOE factor, η, and this parameter when compared to the maximum NOE gives the contribution of DD interactions to the relaxation of the atoms studied.

Experiments represented by Figure 102a and c may be combined by switching off the B_2 field for only every second t_d period (see Figure 102d). By collecting and transforming even and odd FIDs separately, the spectra containing correct absolute intensities and those containing NOE-enhanced intensities may be obtained in a single experiment,[1492c] thereby eliminating the errors arising from the different conditions of the two measurements (e.g., different sample temperature).

2.2.5.5. Noise Modulation Proton Decoupling. Stochastic Excitation[452,743]

Broad-band double resonance (BBDR) of protons (the simultaneous heteronuclear decoupling of all hydrogens of a molecule with different chemical shifts) is one of the standard techniques of ^{13}C FT-NMR spectroscopy, which owing to the coalescence of multiplets corresponding to ^1J(C,H) couplings and to the intensity enhancement caused by NOE improves the sensitivity of measurement (increases ζ by one to two orders).

In PFT experiments the spin system is excited by consecutive, identical pulses (or pulse sequences). Excitation may, however, also be performed by pulses of varying phase, amplitude, and duration; this is stochastic or random) excitation.[452,743] The random excitation functions may be taken from electronic noise or from a synthetic random noise. The spectrum of the spin system excited by means of these functions may be calculated by the correlation of the input and output (i.e., the response of the spin system) noise functions and a subsequent FT. Excitation by the noise function, similar to pulsed excitation, gives, for statistical reasons, an even energy distribution function, and if its mathematical form is known, continuous correlation is unnecessary; it is sufficient to correlate the final time domain function obtained as a time average of excitation cycles.

Random excitation function of known mathematical form (i.e., pseudorandom) may be obtained, e.g., by an appropriately coded binary sequence. An n-bit shift register may produce 2^n one-bit pseudorandom pulse sequences. If this takes t_{si} time and the "1" states of the binary sequence represent a pulse of duration t_p, the mean (average) energy function of the sequence is identical to that of a pulse sequence characterized by parameters t_p and $t_{pi} \equiv t_{si}$; there is a difference in the phase only, with stochastic excitation the phase of side bands is statistically distributed. Therefore, the efficiency of pulsed excitation is much higher, but random excitation raises lower demands against the phase sensitivity of the instrument. In order to have a really random excitation, the condition $t_{si} = t_{aq} > T_1$ must be met.

The theory of stochastic excitation[452] is rather complicated; it may not be discussed here. It is noted that the maximum sensitivity attainable by random excitation is the same as with pulsed excitation, and as there is no favored moment of time within the sequence (like the point $t = 0$ in pulsed excitation), sensitivity and resolution may be optimized independently. The latter is determined by t_{si}, whereas the former is determined by the magnitude of mean energy, on which, however, the half-bandwidth does not depend, unlike CW and like PFT.

The bandwidth of noise modulation (frequency range of excitation) is determined by the audio frequency fed into the shift register. The output phase of this AF is turned by 180° if there is a "1" bit in the register.

At 100-MHz measuring frequency, the decoupling power is 2 to 5 kHz for protons,[1286f] which corresponds to 10 to 20 W or 50 to 100 μT. Without noise modulation, decoupling is incomplete already in a narrow interval around ν_2; gradually increasing splitting (off-resonance) may be observed.

If the phase of the carrier $\nu_1 = \nu_c$ frequency is modulated by a regular series of rectangular functions instead of pseudorandom noise, this also causes decoupling in a broad frequency range. The method is called broad-band (BB) decoupling or broad-band double resonance.[599] In the case of the BB decoupling of protons, the ^{13}C signals are sharper and the decoupling power is also higher.

2.2.6. Relaxation Mechanisms[1,100,464,563,1286g,1309]

The PFT technique has enabled relaxation times to be determined routinely and selectively (i.e., separate T_1 and T_2 values for the individual lines or nuclei). Thus, relaxation times may be added to the other, fundamental kinds of information obtained from the NMR spectra and utilized in chemical structure elucidation (chemical shifts, coupling constants, and signal intensities). A further reason for the renaissance of relaxation measurements is that due to the PFT method, the ^{13}C NMR spectra are now easier to obtain and for carbon nuclei there is a simpler and more easily interpreted relationship between relaxation and chemical structure than for protons. In order to utilize these relationships for structure elucidation, one should have at least a general view of the mechanism of relaxation processes.

2.2.6.1. The Properties of Molecular Motions Causing Relaxation

Relaxations are caused by local magnetic fields $\mathbf{B}_l$ induced by random molecular motions — tumbling* — of widely varying frequency. From the aspects of relaxation the frequency components in the megahertz range are important. The faster molecular vibrations and electronic motions are negligible. The motions may be characterized by correlation time τ_c, which is the mean lifetime of the given type of motion (i.e., frequency). For translation it is the time between two collisions; for rotation it is the time of one full revolution. For small molecules τ_c is in the order of 10^{-12} to 10^{-13} s, for molecules of medium size (ca. 100 to 300 M wt), it is around 10^{-10} s. Symmetric molecules move faster. It follows from the FT theory that motions with mean life time τ_c may have frequency components in the $\nu_c \pm 1/\tau_c$ range, i.e., between 0 and approximately 10^{12} Hz in our case.** The components with Larmor frequency ν_0 are able to induce spin-lattice relaxation, whereas the spin-spin relaxation may have contribution from any local field (including the zero frequency one). With respect to spins precessing with frequency ν_0, only the local fields of the same frequency may have static components in the x,y directions, and the M_z component of magnetization is affected only by these, $B_{x,y}$ components (see Equation 52). The $\mathbf{B}_l$ local fields of zero frequency have a static component in direction z, and thus they may affect only the $M_{x,y}$ components of magnetization, i.e., the spin-spin relaxations. Of course, the latter are also affected by the fields of frequency ν_0, which thereby decrease both T_1 and T_2. Consequently, $T_1 \geqslant T_2$ always hold.

The local fields $\mathbf{B}_l$, being a result of random molecular tumbling, have zero time average: $\overline{B_l(t)} = 0$, and thus they may be characterized only by properties with nonzero time average. Such a property is the time average of the square of the local field B_l:

$$\overline{B_l^2} = \lim_{T\to\infty} \frac{1}{T} \int_{-T/2}^{T/2} B_l^*(t) B_l(t)\,dt \tag{243}$$

* Compare p. 15.

** See Figure 88d and p. 177.

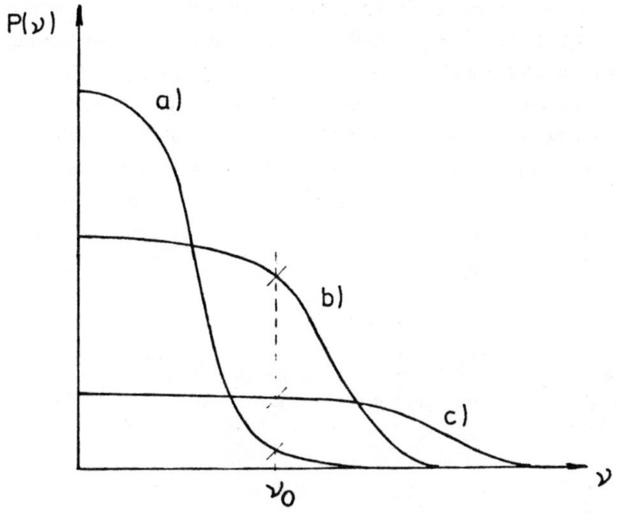

FIGURE 103. The energy distribution of molecular motions as a function of frequency, for conditions $\nu_0 > 1/\tau_c$ (a), $\nu_0 \approx 1/\tau_c$, (b),and (c) $\nu_0 < 1/\tau_c$ (slow, intermediate, and fast motions).

Equation 243 is uninformative in regards to the frequency distribution of the local field B_l. Autocorrelation function* of B_l, if known, is enlightening in this respect,[1471] being the measure of the repetition of some event or phenomenon (here the molecular motions). Following from the definition of τ_c, the probability of a change in a form of motion decreases exponentially with τ_c. Thus, the autocorrelation function may be given as:

$$R_{ff}(\tau) = \overline{B_l^*(t)B_l(t+\tau)} = \lim_{T\to\infty} \frac{1}{T}\int_{-T/2}^{T/2} B_l^*(t)B_l(t+\tau)e^{-t/\tau_c}dt =$$

$$B_l^*(t) \circledast B_l(t)e^{-t/\tau_c} \qquad (244)$$

The meaning of Equation 244 is that that the effect of a given state of molecular motion vanishes exponentially in time. After a time $\tau \gg t$ the correlation of values, $B_l(t)$ and $B_l(t + \tau)$ reduces to minimum. The Fourier transform of the autocorrelation function is the mean energy or the spectral density function:

$$P(\nu) = \mathcal{F}[R_{ff}(\tau)] = \frac{2\tau_c}{1+\omega^2\tau_c^2}\overline{B_l^2} \qquad (245)$$

This mean energy function shows how the $\nu = \omega/2\pi$ frequencies are distributed in the region of 0 to $1/\tau_c$ Hz as a function of τ_c, while the total energy of the system is constant [independent of frequency, i.e., $\int_{-\infty}^{\infty} P(\nu)d\nu$ = constant]. Since T_1 is exclusively determined and T_2 is primarily determined by the relative magnitude of the ν_0 Fourier component of molecular motions, it is worth regarding the τ_c dependence of $P(\nu)$ under the condition of $\omega = \omega_0$. Namely, it follows from the condition $\int_{-\infty}^{\infty} P(\nu)d\nu$ = constant that the frequency has an optimum as a function of τ_c. If the molecular motions are fast (τ_c is short, i.e., $\omega_0\tau_c \ll 1$), the frequency components are distributed evenly in most part of the 0 to $1/\tau_c$ region (see Figure 103), and if τ_c is long (slow motions), the ratio of low-frequency components increases, and the high-frequency ones may be neglected. As the $\omega = \omega_0$ components have

* See Reference 265, p. 175 and Equation 219.

a low probability in both cases, T_1 is long. The occurrence of an B_1 component fluctuating with the Larmor frequency is most probable[154] when $\omega_0\tau_c \approx 1$ (medium correlation times). In this case, both T_1 and T_2 are short. Since T_2 is also decreased by the $\omega < \omega_0$ components, it is short for the slow motions as well, and owing to the $\omega \approx 0$ components, the lowest T_2 values may be expected to occur in this case. This agrees with the relationship found in the qualitative interpretation of relaxation processes between T_1, T_2, and viscosity* and points out that τ_D used in the Debye theory of liquids[343] and τ_c are analogous parameters:

$$\tau_c = \frac{1}{3}\tau_D = \frac{4\pi r^3}{3} \cdot \frac{\eta}{kT} \tag{246}$$

where r is the "radius" of molecule.

Local $\mathbf{B}_l$ fields may occur for several reasons in solutions, each creating a relaxation mechanism, an energy exchange possibility, for the spin system. These interactions obey the relationship

$$T_{1,2}^{-1} \equiv R_{1,2} = \overline{B_l^2} f(\tau_c) \tag{247}$$

where R is the rate of relaxation. Since the components of R of different origin are additive, they are easier to handle than relaxation times. However, spectroscopists still think in terms of relaxation times which have historical and practical reasons: it is T that one masures.

Relaxation times determined by CW and PFT methods may be different (the latter are higher). This is due in part to differences in signal saturation and in part to the simultaneous excitation applied in the PFT measurement. Coupled nuclei may also relax transmitted by their spin-spin interaction which is eliminated by simultaneous excitation or, for ^{13}C relaxation, by the BBDR of protons.

2.2.6.2. Dipole-Dipole Relaxation Mechanism

The most important source of relaxation is the local field induced in the neighborhood of nucleus I of a given molecule by other magnetic nuclei (e.g., S) of the molecule. This dipole-dipole (DD) interaction depends on the mutual position of nuclei (distance r and the angle between r and the direction of $\mathbf{B}_0$ and on the magnetic moment of nucleus S inducing the local field:

$$B_l^{DD} = \frac{\gamma_S I_S h}{r_{IS}^3}(3\cos^2\theta_{IS} - 1) \tag{248}$$

Depending on the direction of spin I_S (α or β), B_1 is added to or subtracted from B_0. The corresponding splitting is very high for protons, approximately 2 mT, which is equivalent to approximately 80 kHz.[464c] However, solutions do not show this splitting, since B_1^{DD} is averaged by tumbling ($\int_0^{2\pi} \cos\theta d\theta = 1/3$). In solid phase or in liquid crystals, this averaging is incomplete, causing observable DD splitting. Relaxation is ensured by the fluctuation of B_1^{DD}, owing to the tumbling of molecules (variation of θ). Since according to Equation 247 and 248, B_1^{DD} and $R_1^{DD} \equiv (T_1^{DD})^-$ is proportional to the magnetic moment $\gamma_S I_S h$ of nucleus S, and to the square of it, respectively, hydrogen is the nuclear species that gives rise to the most efficient relaxation. (In radicals, of course, the dipolar relaxations are even more efficient, owing to the moment of the free electron, approximately 860 μ_H in magnitude.)

Using Equations 245, 247, and 248, R_1 and R_2 may be expressed as a function of the

* Compare p. 15 and Figure 6.

correlation time τ_c belonging to the rotation of the molecule, and for nuclei with spin 1/2, the following simplified formulas may be derived:

$$R_1 = \frac{3}{40}\frac{\gamma_I^2\gamma_S^2\hbar^2}{r^6}\left[\frac{\tau_c}{1+\Delta\omega_{IS}^2\tau_c^2} + \frac{3\tau_c}{1+\omega_I^2\tau_c^2} + \frac{6\tau_c}{1+\omega_I+\omega_S)^2\tau_c^2}\right] \tag{249}$$

and

$$R_2 = \frac{3}{80}\frac{\gamma_I^2\gamma_S^2\hbar^2}{r^6}\left[4\tau_c + \frac{3\tau_c}{1+\omega_I^2\tau_c^2} + \frac{\tau_c}{1+\Delta\omega_{IS}^2\tau_c^2} + \frac{6\tau_c}{1+\omega_S^2\tau_c^2} + \frac{6\tau_c}{1+(\omega_I+\omega_S)^2\tau_c^2}\right] \tag{250}$$

where R_2 also contains a frequency-independent term, in accordance with the earlier discussion. It can be seen from the equations that in the case of fast tumbling (in dilute, nonviscous liquids and in gases), when $\omega_c\tau_c \ll 1$, $R_1 = R_2$ (and thus $T_1 = T_2$), and they are independent of frequency. This is the "extreme narrowing condition".[248b] With large molecules, τ_c is high enough for the frequency dependence of T_1^{DD} to become measurable.[372] Thus, the situation of the minimum of T_1 depends on the size of the molecule and the measuring frequency of the instrument. At higher frequencies, the condition of "fast" motion is not met already for smaller molecules, causing line broadening. In addition, e.g., in the case of ^{13}C NMR measurement, ζ also decreases, for two reasons. First, since T_1^{DD} increases (the efficiency of DD mechanism decreases), η_{NOE} decreases,*[372] and second, since ζ increases with the ratio T_2/T_1, the increase in T_1 and a simultaneous decrease in T_2 cause automatically lower sensitivity.

Besides the intramolecular DD relaxations arising from the rotation of molecules, an analogous intermolecular interaction also exists. Since B_I^{DD} decreases with the sixth power of distance (extremely short-range effect), measurable interaction may be expected only in the case of very high B_I^{DD} fields. This is probable in the presence of paramagnetic impurities or, less frequently, through the effect of solvent molecules. The relationship between the corresponding correlation time τ_c of the translational motion and the relaxation times is similar to that of molecular rotations.[154]

2.2.6.3. *Spin Rotational Relaxation Mechanism*

Arising from the electron cloud, the entire molecule also has a magnetic moment, which produces a fluctuating magnetic field, owing to the fast rotation of molecules. This field provides the second most important relaxation mechanism, particularly for small and symmetrical molecules. On the ^{13}C and ^{15}N nuclei of them, this effect is competitive with DD relaxation. If a "molecule" consisting of a nucleus of mass m_N and an electron at a distance r_N only rotates with frequency

$$\nu_{rot} = \hbar J/2\pi m_N r_N^2 \tag{251}$$

where J is the rotational quantum number, the magnetic moment belonging to the motion of the electron is

$$\mu = \pi r_N^2 TI = e\hbar J/2m_N c = \beta_N J \sim \mu_N J \tag{252}$$

* NOE arises primarily from DD relaxations.

since current intensity is $I = e\,\nu_{rot}/c$ and the nuclear magneton, β_N, is proportional with the nuclear moment. In the vicinity of the relaxing nucleus, the latter induces a local field inversely proportional to the third power of distance r_N:

$$B_l^{SR} \sim \mu_N J r_N^{-3} \tag{253}$$

For the hydrogen molecule, for example, B_l^{SR} is approximately 3.5 mT.[464d]

It is easy to see that small molecules (small m_N; large J) induce higher B_l^{SR} fields, further enhanced by symmetry (the intermolecular interactions, e.g., hydrogen bond, van der Waals' interactions, etc., are smaller, increasing the statistical weight of rotational states corresponding to higher J quantum numbers), and thus B_l^{SR} has a larger contribution to the relaxation.

The correlation time τ_J assigned to molecular rotations may be interpreted as the time elapsed between two collisions during which the angular momentum remains unchanged. The fluctuation of field B_l^{SR} is, namely, due to the collisions in which the direction or the direction and magnitude (i.e., J) of angular moment changes. The contribution of B_l^{SR} to the relaxation rate is

$$R_1^{SR} = r_N^2 m_N kT(6\pi)^{-1}\hbar^{-2}(2C_\parallel^2 + C_\perp^2)\tau_J \tag{254}$$

where the C coefficients are the constants of spin rotation (SR) coupling.

It was shown[214,689] that for liquids below their boiling points, there is a relationship between the correlation times τ_J and τ_c:

$$\tau_c = m_N r_N^2 (6kT\tau_J)^{-1} \tag{255}$$

As can be seen, τ_J and τ_c have opposite temperature and viscosity dependences; while τ_c decreases with increasing temperature (molecular motions become faster) and increases with viscosity, τ_J has an opposite behavior (the chance of collisions decreases in gas state, at higher temperatures). Consequently, in the region of fast motions, the temperature dependence of T_1 is nonlinear, permitting one to prove the contribution of SR mechanism to relaxations (with increasing temperature τ_c decreases, whereby T_1 increases; meanwhile τ_J and consequently, R_1^{SR} increases, which has a decreasing effect on T_1). For nuclei with a wide range of chemical shifts (^{19}F, ^{13}C, ^{15}N, ^{31}P, etc.), R_1^{SR} is high, too (both depend on the electron distribution),[355,941] and it has a particularly important role in ^{19}F resonance.

2.2.6.4. *Relaxation by Chemical Shift Anisotropy*

Molecules in a field $\mathbf{B}_0$ are affected by a local field $B_l = B_0 - B_i$, where field $\mathbf{B}_i$ is induced by electronic motions. Chemical shift is a consequence of this effect. Field $\mathbf{B}_l$, and thus the chemical shift, is anisotropic property (the shielding constant σ is a tensor), with a magnitude depending on the position of the molecule in the magnetic field. In solution, the fast molecular motions produce an average of the tensor components, and thus a single line appears at a chemical shift corresponding to the local field $B_l = \overline{\sigma}\,B_0$, where

$$\overline{\sigma} = (\sigma_{xx} + \sigma_{yy} + \sigma_{zz})/3 \tag{256}$$

Consequently, the anisotropy of shielding causes no line splitting. However, on the fast time scale of relaxations, the fluctuation of the local field B_l is already observable, ensuring a further relaxation mechanism for the nuclear spins. The rate of the relaxation produced by the chemical shift anisotropy (SA, CSA) is a function of $\mathbf{B}_0$:

$$R_1^{SA} = \frac{2}{15}\,\omega_0^2(\sigma_\parallel - \sigma_\perp)^2\,\frac{\tau_c}{1 + \omega^2\tau_c^2} \tag{257}$$

and

$$R_2^{SA} = \frac{1}{45}\,\omega_0^2(\sigma_\parallel - \sigma_\perp)^2\left[4\tau_c + \frac{3\tau_c}{1+\omega^2\tau_c^2}\right] \tag{258}$$

where, assuming axial symmetry, $\sigma_\parallel$ and $\sigma_\perp$ are the shielding constants parallel with and perpendicular to the symmetry axis. For the general case, more complicated equations can be derived.[1] It is interesting to note that R_1 and R_2 do not become equal in the region of fast motions either.

$R_{1,2}^{SA}$ contributions could be detected so far in only a few cases (e.g., Reference 857), but with the spread of high field spectrometers, the number of experimental data may be expected to increase. Theoretically, in fields higher than 2.5 T, the contribution of SA mechanism becomes significant, and at approximately 6 T, it is already commensurable with the DD component.

2.2.6.5. The Scalar Relaxation Mechanism[1]

Scalar spin-spin coupling transmitted by chemical bonds may give rise to relaxation when the coupling constant J(I,S) of the interacting nuclei I and S, or the spin of one of them, changes fast with a frequency of the order of Larmor frequencies. Coupling J(I,S) may vary, e.g., in exchange processes. When nucleus S is involved in an exchange process, $\mathbf{B}_l$ induced by it fluctuates in the neighborhood of nucleus I, causing therefore the relaxation of I. The spins of quadrupole nuclei change, on the other hand, owing to their fast relaxation; if, therefore, S has a quadrupole moment, this also represents a relaxation possibility for nucleus I coupled with it.

In the first case, during the exchange processes, the field induced by nucleus S around I is $B_l = \pm JI_S/2\gamma_I$ when there is a covalent bond between the atoms, and $B_l = 0$ for the other species of the exchange, where no chemical bond exists between I and S. In the case of fast quadrupole relaxation, the spin of nucleus S is time dependent, owing to a fast exchange between the α and β spin states, which again amounts to a fluctuation of $\mathbf{B}_l$. If this process is fast and $1/\tau_{SC} \gg J$, the mean lifetime of the interconverting states (species or spin states), $T_{1,2}$(S), is shorter than $1/2\pi J$ or the relaxation times, the lines of the multiplet corresponding to coupling $J_{I,S}$ coalesce, and the relaxation contribution arising from the fluctuation of $\mathbf{B}_l$ depends on τ_{SC}:

$$R_1^{SC} = \frac{8}{3}\,\pi^2 J(\mathrm{I,S})^2 I_S\,(I_S+1)\left[\frac{\tau_{SC}}{1+\Delta\omega^2\tau_{SC}^2}\right] \tag{259}$$

and

$$R_2^{SC} = \frac{4}{3}\,\pi^2 J(\mathrm{I,S})^2 I_S(I_S+1)\left[\tau_{SC} + \frac{\tau_{SC}}{1+\Delta\omega^2\tau_{SC}^2}\right] \tag{260}$$

where $\Delta\omega = \omega_I - \omega_S$.

It follows from the above that in the case of heteronuclear coupling ($\Delta\omega$ is high), R_1^{SC} is small, and the value of T_1 is not affected significantly by SC relaxation. An exception is when T_2 of nucleus S is so short that it is commensurable with $\Delta\omega$, which is low in itself. Such a case occurs, e.g., with ^{13}C, ^{81}Br coupling,[519] where T_2(Br) is extremely short since Br has quadrupole moment, and $\Delta\omega$ is not too high (approximately 2 MHz at 2.4 T). The R_2^{SC} contribution, however, is always large, corresponding to the term not containing the resonance frequency. This is the reason for the line broadening of the signals of protons coupled with nitrogen (the ^{14}N nucleus has quadrupole moment). When $T_1(S) = \tau_{SC}$ is not very short, approximately 10^1 to 10^2 ms, the R_2^{SC} contribution is relatively large, and line broadening occurs (e.g., in the case of S = ^{14}N, ^{11}B, etc.). Then, the first term of R_2^{SC} may be large enough for the SC mechanism to make a significant contribution to the spin-spin relaxations. When, however, τ_{SC} is even smaller, as in the case of halogens where T_1(Br, Cl) $\leqslant 10^{-5}$ s, the $R_{1,2}^{SC}$ contributions are so small that no line broadening occurs.

The T_1 and T_2 relaxation times of the unsubstituted carbon atoms of *ortho*-dichlorobenzene are practically the same, in agreement with the dominant nature of the DD contribution; for the C-3,6 atoms, they are 7.8 (T_1) and 7.7 s ($T_2 = T_{1\rho}$), while for C-4,5, they are 6.3 (T_1) and 6.4 s (T_2), respectively. In contrast, for the chlorine-substituted carbons, T_1 is much longer (66 s), since there is no hydrogen immediately attached, and thus the DD relaxation is inefficient. T_2 is, however, much smaller (4.2 s) due to the vicinity of chlorine with quadrupole moment, giving rise to large R_2^{SC} contribution.[518]

Owing to the R_2^{SC} contribution, generally $T_1(^{13}C) > T_2(^{13}C)$ holds. The relation $T_2(H) > T_2(^{13}C)$ always holds, and therefore the lines in ^{13}C NMR spectrum are broader than the corresponding signals in the ^{1}H NMR spectrum. The reason is that $R_2^{SC}(^{13}C) = \pi^2 J^2(C,H)T_2(H)$, and since $J^2(C,H) \gg 1/[T_2(H)]$, $1/T_2^{SC} \gg \pi^2$.[1297] When the amplitude of ν_2 for BBDR of protons is insufficient, $1/[T_2(H)] > J(C,H)$ and the ^{13}C line broadens. To avoid this situation, the decoupling field must be increased to the level where the $1/[T_2(H)] \ll J(C,H)$ condition is met.[1297] The significant contribution of the SC mechanism to relaxations may be proved experimentally by the frequency dependence of $T_{1,2}$.

2.2.6.6. *Quadrupole Relaxation*

The nuclei with $I > 1/2$ have quadrupole moments, i.e., their charge distribution has no spherical symmetry. By changing the orientation of the electric field gradient around the nucleus, tumbling produces a fluctuating electric field in the vicinity of the nucleus, owing to the high frequency of which (τ_c is in the millisecond to microsecond range) the relaxation of quadrupole nuclei is determined primarily by the mechanism made possible by this phenomenon. According to the very short correlation times, the extreme narrowing condition is generally met ($\tau_c = T_1 = T_2 \ll 1/\omega_0$) and

$$R_1^Q = R_2^Q = \frac{3}{40}\left[\frac{2I+3}{I(2I-1)}\right]\left(1+\frac{\eta^2}{3}\right)\left(\frac{eq}{\hbar}\,eQ\right)^2\tau_c \tag{261}$$

where η is the asymmetry parameter and q is the electric field gradient. The $e^2qQ/\hbar$ factor is the quadrupole coupling constant,[1] which may vary over wide limits depending on q.

Thus, for acetonitrile the coupling constant is 4 MHz, whereas it is 0 for the ammonium cation of tetrahedral symmetry. Accordingly, the corresponding values of T_1 are 22 ns and approximately 50 s, respectively.[989] Even faster is the relaxation in the case of chlorine and bromine (order of μs) where quadrupole coupling may be very high, exceeding approximately 100 MHz. The intramolecular and dominant character of quadrupole interaction makes possible the very accurate determination of τ_c by the measurement of T_1, when quadrupole coupling is known (e.g., from ESR or NQR data).

2.2.6.7. *Nuclear-Electron Relaxations*

Unpaired electrons, as magnetic dipoles, induce anisotropic local magnetic fields which fluctuate, owing to the molecular tumbling. This provides a very efficient relaxation mechanism for the neighboring nuclei, since B_l induced by electrons is approximately 10^6 times higher than the field induced by protons, although the latter have the highest nuclear magnetic moment and induce, therefore, the highest field. (B_l is proportional to the square of the magnetic moment, and $\mu_e \approx 10^3\ \mu_H$.) Therefore, paramagnetic substances, e.g., dissolved molecular oxygen, often cause, even in very low concentrations, substantial line broadening in the signals of protons and quaternary carbons despite the larger average distance from the nuclei. If the extreme narrowing condition is met, i.e., τ_c, $\tau_e << 1/\omega_0$, then[689]

$$R_1^e = (4/3)\gamma_e^2\gamma_N^2\hbar^{-2}r^{-6}J(J+1)\tau_e \tag{262}$$

and

$$R_2^e = R_1^e + A^2\hbar^{-2} \quad J(J+1)\tau_e \tag{263}$$

where A is the electron spin-nuclear spin hyperfine coupling constant, J is the quantum number representing the total angular moment of the electron, and τ_e is the correlation time related to the electron relaxation time and also to the mean lifetime τ_{ex} of the complex between the paramagnetic substance and the molecule containing the relaxing N nucleus ($1/\tau_e = 1/I_{1e} + 1/\tau_{ex}$).[248]

Rates $R_{1,2}$ are directly proportional to the concentration of the paramagnetic substance, enabling one, for example, to determine the metal ion binding ability of biological materials by measuring the T_1 relaxation time of water molecules used as solvent. The tris-acetonylacetonate complexes of Cr^{3+} and Fe^{3+} ions, being paramagnetic substances, may be used as "relaxation reagent", or "T_2 reagent",* to accelerate the PFT measurement, enhance the intensity of ^{13}C signals insignificant owing to saturation, or eliminate NOE. The latter is necessary for the measurement of the absolute intensities of ^{13}C lines (e.g., in quantitative analysis) or in ^{15}N resonance for "developing" the signals eliminated by negative NOE.**

2.2.7. Measurement of Relaxation Times

T_1 relaxation times may also be measured by CW technique, using adiabatic fast sweep to excite a nucleus with sufficient energy to invert the population of levels and then applying normal, slow sweep in regular intervals to record the gradually decreasing negative signal and then the absorption signal of an intensity reaching its maximum exponentially. The accurate determination of spin-lattice relaxation times, although simple by using PFT instrument, requires careful work. Changes in concentration and temperature may influence the accuracy of measurement through the viscosity dependence of relaxations; therefore the application of dilute solutions and nonviscous solvents is important. Dissolved molecular oxygen and eventual paramagnetic impurities may cause completely false results, primarily when T_1 is long ($\gtrsim$ 60 s). Molecular diffusion may also lead to significant errors if $\mathbf{B}_l$ is inhomogeneous and the measuring time is long or when, owing to large sample volume, previously unexcited molecules pass into the measured volume. Even more serious discrepancies may arise from the exchange between vapor and liquid phase molecules, since the former have much shorter T_1 relaxation times owing to the SR mechanism. Therefore, the measurement is carried out preferably at temperatures where the vapor pressure of the substance is low.

2.2.7.1. Measurement of T_1 by Inversion Method[515,1460]

The most frequent, practical method of T_1 measurements is the application of a 180° - τ - 90° pulse sequence. The 180° pulse inverts the population of the levels of excited nuclei and thereby the resulting macroscopic $M_e = M_z$ magnetization along the z axis ($-M_z$, see Figure 104a). Then, through the relaxation of nuclei, $-M_z$ returns gradually into the original M_e value, passing the zero point. This process is monitored after time τ by a 90° pulse, which takes a "snapshot" of the actual magnitude of M_z (see Figure 104b). The $M_z(t)$ component remaining after time τ is turned into direction y by the 90° pulse of direction x and the $M_y(t)$ component, which is proportional to the actual $M_z(t)$, given by the initial value of FID, which may be recorded as an absorption signal after FT.[1460] If τ is small in comparison with T_1, a negative signal can be observed, the signal disappears for $\tau = T_1 \ln 2 = 0.69\ T_1$, and with longer τ intervals, already positive signals emerge (see Figure 104c). The intensities remain practically constant after τ has reached four to five times T_1 (for $\tau = 5T_1$ time $M_z(t) = 0.993\ M_e$).

* Compare Volume II, p. 227.
** Compare Volume II, p. 259.

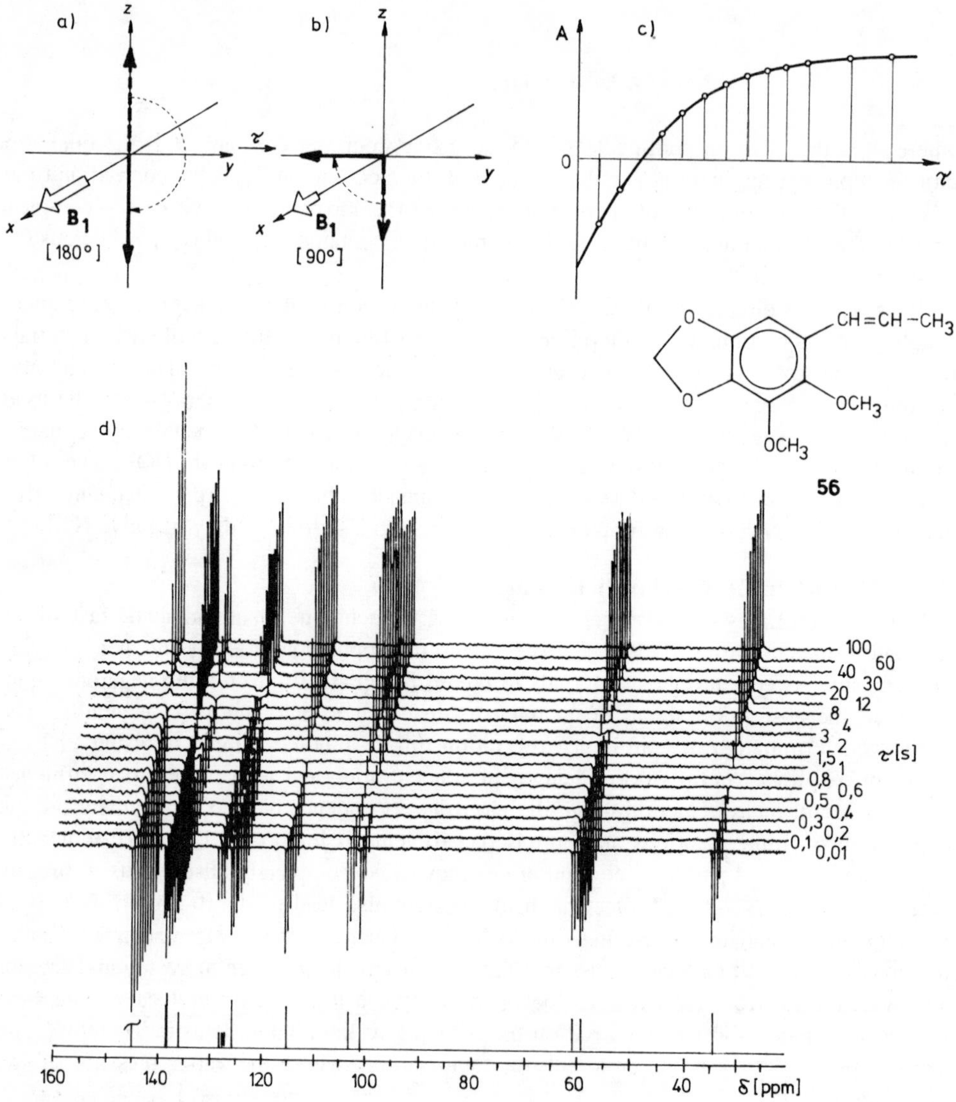

FIGURE 104. Measurement of T_1 with the inversion method, applying pulse sequence $[180° - \tau - 90° - t_{aq} - t_d]_n$. The effect of (a) 180° and (b) 90° pulses on the magnetization vector, (c) the intensities measured as a function of time τ, and (d) the proton-decoupled ^{13}C NMR spectra of Compound **56** obtained by averaging 32 transients in benzene-d_6 solution on a BRUKER WM-250 FT-instrument for the determination of T_1 by the method of inversion pulse sequence and varying the value of τ between 0.01 and 100 s. (Reproduced by courtesy of BRUKER AG, Karlsruhe, Rheinstetten, West Germany.)

The measurement is, therefore, a series of experiments with the application of pulse sequences of successively longer τ. The pulse trains must follow one another at a rate allowing a full equilibration of the spin system in the meantime. Thus, following the 90° pulse and acquisition time t_{aq} delay time t_d allowing for complete relaxation should be inserted so that $t_{aq} + t_d \geqslant 4T_1$. The complete pulse sequence is therefore $(180° - \tau - 90° - t_{aq} - t_d)_n$. Signal intensity (see Figure 104c) varies according to the relationship $M_\tau = M_e(1 - 2e^{-\tau/T_1})$. Figure 104d shows a such experiment, using the inversion method to determine T_1 for the carbon nuclei of Compound **56**.

In a modified version of the method, the difference of a FID following the above pulse sequence and a FID obtained after a normal 90° pulse is formed, whereupon the signal

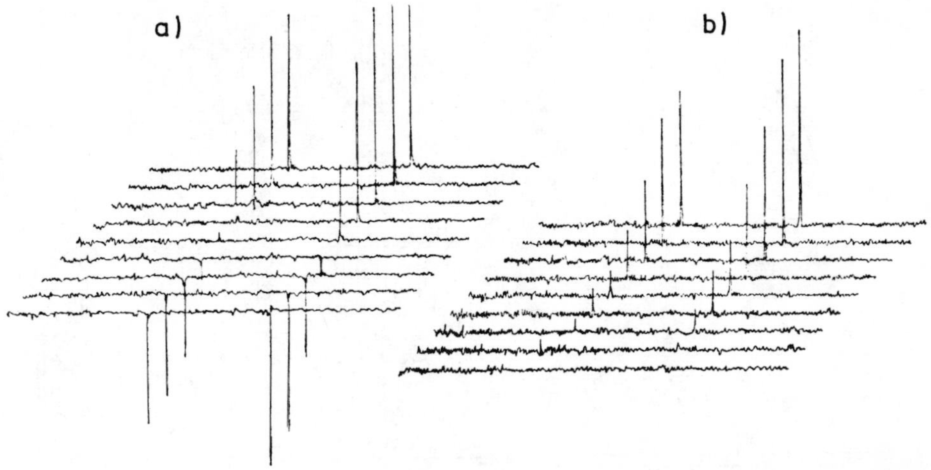

FIGURE 105. The ^{13}C NMR spectra of 1:1 mixture of benzene and dioxane for various values of τ obtained from one pulse sequence for the determination of T_1 applying (a) inversion and (b) progressive saturation technique. (From Shaw, D., *Fourier Transform NMR Spectroscopy,* Elsevier, Amsterdam, 1976, 141. With permission.)

decreases from $2M_e$ to zero. With this technique, by measuring $\Delta M = M_e - M_\tau$ differences, the errors arising from the variation of instrumental parameters during measurement are greatly eliminated.[517] The $\ln\Delta M$ vs. τ plot gives a straight line with slope T_1. When the relaxation of ^{13}C or other insensitive nuclei is measured, the method is very time consuming, since on the one hand for a sufficient ζ the FID belonging to each τ must be measured several times and on the other hand t_d is long.

2.2.7.2. *Measurement of T_1 by Progressive Saturation*[517]

If the spin system is subjected to a sequence of 90° pulses with intervals $\tau \leqslant 3T_1$, partial saturation occurs. Thus, the $[90° - t_{aq} - \tau]_n$ sequences produce increasingly stronger signals with increasing τ, provided that at the beginning of the experiment τ is short enough for the approximate or complete saturation of the signal. In this case $M_\tau = M_0[1 - e^{-(\tau + t_{aq})/T_1}]$, and the value of T_1 can be determined again from the slope of the straight line obtained by an $\ln M_\tau$ vs. τ plot. The first couple of FIDs yield false results, since the partial saturation of the signal reaches a constant, steady value only after some sequence repetitions. Since the period t_p of a 90° pulse is shorter than that of a 180° one, the measured FID is closer to the ideal $t = 0$ time, permitting more accurate measurements to be performed. On the other hand, τ may not be decreased more than t_{aq}, and the method gives accurate results only under the condition of $T_1 \gg T_2$, otherwise the second 90° pulse turns the remaining M_y component into direction z, thereby tampering the measured value of M_z. Consequently, the inversion method is more accurate, but progressive saturation is much faster, since the lengthy t_d period is omitted. This method is particularly advantageous for long T_1 and unfavorable ζ. Figure 105 shows, for a comparison of the saturation and inversion methods, the ^{13}C NMR spectra of a 1:1 mixture of benzene and dioxane after a single sequence for each τ value. It is easy to see that ζ is better in the inversion method, corresponding to the two-fold signal intensity.[1286g]

Figure 106 shows the ^{13}C NMR spectra of 3,5-dimethylcyclohexen-2-one-1 (**57**); for the determination of T_1 by the progressive saturation method.[517] Increasing signal intensities correspond to increasing τ values. The increase in intensity is slower for the signals of the more slowly relaxing quaternary C-1 and C-3 atoms.

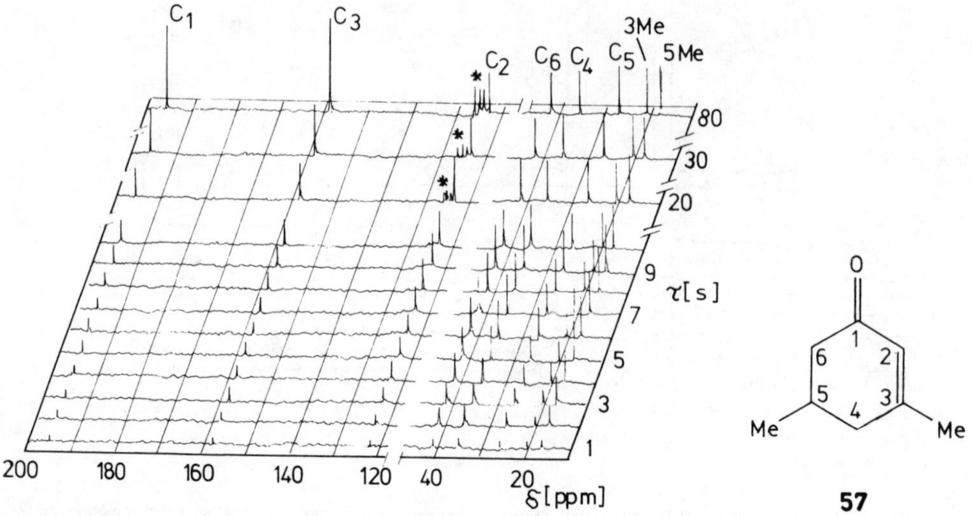

57

FIGURE 106. The determination of T_1 for the ^{13}C NMR signals of 3,5-dimethylcyclohexen-2-one-1 (**57**) in the BBDR spectrum by applying the method of progressive saturation, i.e., the pulse sequence $[90° - \tau - t_{aq} - 90°]_n$. The spectra corresponding to the various values of τ were obtained from the average of 64 FIDs at 25.1 MHz measuring frequency in benzene-d_6 solution.[517]

2.2.7.3. Measurement of T_1 by Saturation Method

Spin-lattice relaxation time may also be determined by saturating the system and following then the exponential process of the recovery of equilibrium, by gradually varying time τ elapsed between saturation and the collection of FID. In this case the signal intensity obtained after FT from the FID is determined by the following equation $M_\tau = M_e(1 - e^{-\tau/T_1})$. Saturation may be carried out by means of some (5 to 10) 90° pulses rapidly following one another[926] or by a single 90° pulse. In the latter case, field homogeneity must be spoiled for some μs, ensuring thereby the complete elimination of $M_{x,y}$ components via the very fast spin-spin relaxation. Inhomogeneity may be produced by the shim-coils.* The name HSP of the technique comes from this homogeneity spoiling pulse.[962] The pulse train is therefore $[90°, \text{HSP} - \tau - 90° - t_{aq}, \text{HSP}]_n$. T_1 can be determined again from the slope of the straight line obtained in the $\ln M_\tau$ vs. τ plot.

2.2.7.4. Measurement of NOE[523]

T_1 may also be determined by the measurement of NOE, for nuclei directly attached to hydrogen atoms (primarily for ^{13}C). This effect reaches its maximum exponentially, and the time constant of the process is T_1:

$$\eta_{NOE} = (\eta_{NOE})_{max}(1 - e^{-\tau/T_1}) \tag{264}$$

If the signal intensity is measured τ time after the application of field ν_2 used for the BB decoupling of protons (where the signal is recorded as a FID produced by a 90° pulse), and then after a time $t_d \geq 4T_1$ the sequence is repeated with another τ, the signal intensity, starting from the normal (without NOE) value ($\tau = 0$), will approach exponentially the attainable maximum intensity gradually increased by NOE. Thus, by a single although lengthy experiment (because of the long t_d delay time), both $(\eta_{NOE})_{max}$ and T_1^{DD} may be determined.

2.2.7.5. Measurement of T_2 by the Spin-Echo Method

Spin-spin relaxation time T_2 yields information on the low and zero frequency fields, too,

* Compare p. 136.

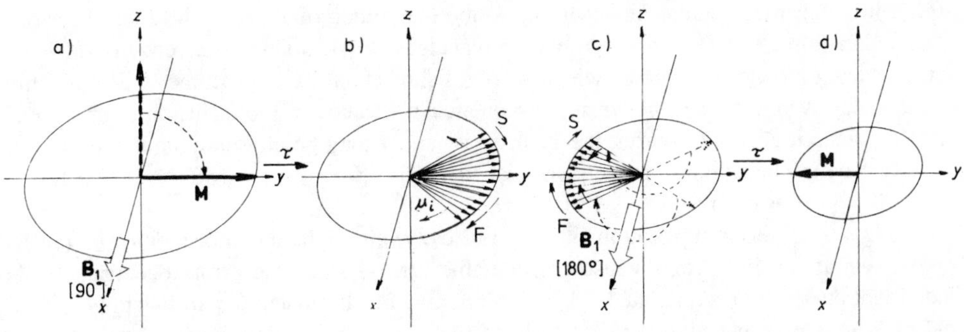

FIGURE 107. Measurement of spin-spin relaxation time, T_2 by means by the spin-echo method, using the pulse sequence $[90° - \tau - 180° - \tau - t_d]_n$. The effect of the 90° pulse (a) on the $\mathbf{M} = \Sigma\, \mu_i$ resultant magnetization vector, (b) the individual μ_i moments spread owing to field inhomogeneities, i.e., precessing faster (F) and slower (S), (c) the effect of 180° pulse inverting the moments around the x axis, and (d) the resulting magnetic moment refocused along the y axis, the spin echo. The decreasing size of the figures, i.e., the shortening of vectors in sequence (a) to (d), illustrates the exponential decrease of the individual μ_i and resulting **M** magnetization moments during measurement (under time 2τ) due to the spin-spin relaxations.[517]

but due to this fact, its measurement is a harder task than that of the T_1 spin-lattice relaxation times. Owing to the inevitable field inhomogeneities, ΔB_0, the experimentally determined T_2^* value is always lower than T_2, except when $T_2 \ll 1/\gamma\Delta B_0$), since $1/T_2^* = 1/T_2 + \gamma\Delta B_0$. ($T_2^*$ may be obtained from the spectrum without special measurement, as the $\Delta\nu_{1/2}$ half-width of the lines, since $\Delta\nu_{1/2} = 1/\pi T_2^*$).*

An elegant solution of this problem by Hahn in 1950[617] is the method known as spin-echo which applies a $90° - \tau - 180°$ pulse sequence for determination of T_2, and the $M_{x,y}$ components of magnetization are measured after 2τ time (see Figure 107). The 90° pulse of direction x turns the resultant magnetization vector into direction y. The components of this vector, arising from the various spins, spread in the $[x,y]$ plane after time τ owing to the inhomogeneities (the spin vectors for which $\nu_i > \nu_0$, will be "early" whereas others for which $\nu_i < \nu_0$ will be "late"). When these spin vectors are inverted by the 180° pulse about the x axis, the "faster" vectors will assume the positions of "slower" ones and vice versa. It is easy to see that after time 2τ from the start of experiment the moment vectors of all nuclei will be refocused in direction $-y$, but by now with a decreased intensity determined by T_2 (reduced by factor e after T_2) and with inverted sign. In such a way, the T_2 relaxation process (i.e., dephasing) may be reverted by the 180° pulse. This is the "echo" of magnetization. Before the pulse sequence may be repeated with a different τ value, a delay of $t_d \geq 5T_1$ is necessary for restoring the equilibrium state. The complete sequence is, therefore, $[90° - \tau - 180° - \tau\,(-y\text{-echo}) - t_d]_n$.

The applicability of the method is restricted by molecular diffusion; during time 2τ the molecules move to other locations in the inhomogeneous field, and owing to a change in their ν_i frequencies, the refocusing of individual spin vectors upon the 180° pulse is not perfect. A field gradient of 2.8 mT/m causes a decrease of 0.2 to 2 s in the measured $T_2^* \neq T_2$ value.[247] For the elimination of this restriction, a method was suggested by Carr and Purcell.

2.2.7.6. *Measurement of T_2 by the Carr-Purcell Method*[247]

In the Carr-Purcell sequence, the first, negative ($-y$) echo observed after time 2τ is followed by a second 180° pulse after time τ, which produces a new, positive echo 4τ time after the start of the experiment (along the $+y$ axis). Similarly, by 180° pulses applied at times 5τ, 7τ, ..., etc., further echoes of alternating sign are produced at times 6τ, 8τ, ...,

* See Equation 231.

etc., with intensities decreasing according to the magnitude of T_2, until the echo fades out.

The Carr-Purcell method has two main advantages. First, a single sequence is sufficient for producing a series of echoes, which involves substantial saving of time (the application of a t_d delay is unnecessary between the sequences). Moreover, the diffusion error is much smaller, since it acts only during the 2τ periods, and τ may be chosen quite short (approximately 1 ms). The single pulse sequence is, therefore, [$90° - \tau - 180° - \tau(-$ echo) $- \tau - 180° - \tau(+$ echo) $- \tau - 180° - \tau(-$ echo) $- \ldots$].

Owing to the serial application of 180° pulses, even slight inhomogeneities of the RF pulses (variations in t_p) may cause considerable errors, since the errors occurring in the individual pulses are accumulated. This problem may be eliminated in part by applying 180° pulses with alternating phases (alternately in the $+x$ and $-x$ directions). Thus, the even pulses are exactly in-phase, and the odd ones may cause a constant, but small, noncumulative phase error.

2.2.7.7. Measurement of T_2 by the Meiboom-Gill Method[972]

The error arising from the inhomogeneity of the RF field is easy to correct experimentally by the Meiboom-Gill method, in which the Carr-Purcell sequence is modified so that the 180° pulses are applied along the y axis (see Figure 108). If the 180° pulse is accurate, i.e., turns the magnetization vectors M exactly by 180°, the "fast" components, owing to the incoherence (see Figure 108b), will be set back, and the "slow" ones will be set forward in the [x,y] plane (see Figure 108c), and the echo produced after time 2τ (see Figure 108d) will occur exactly along the $+y$ axis. If the RF field is inhomogeneous and the magnetisation vector, M is turned, e.g., instead of 180°, by only $180° - \beta$, the defocused (see Figure 108e) and, consequently, the refocused vectors as well (see Figure 108f) point above the [x,y] plane, The vectors defocused above the [x,y] plane (see Figure 108g) are, however, turned by the next $180° - \beta$ pulse exactly into the [x,y] plane (see Figure 108h), and thus the $+y$ echo built up τ time later (see Figure 108d) will be exactly in the [x,y] plane. Thus, the second and every even echo will occur with accurate amplitude in the $+y$ direction, whereas the $+y$ amplitude of the odd ones will be slightly smaller, but its effect is not cumulative and remains constant during the measurement. The sequence is, therefore, [$90^\circ_x - \tau - 180^\circ_y - \tau(+y$-echo) $- \tau - 180^\circ_y - \tau(+y$-echo) $- \tau - 180^\circ_y - \ldots]_n$. Figure 109 shows two classical results[464e] obtained with the Carr-Purcell and the Meiboom-Gill methods.

2.2.7.8. Measurement of $T_2\rho$ (Rotating-Frame) Relaxation Time

The nutational resonance (transient nutation) method introduced by Torrey[1422] is a combination of CW and pulsed techniques. It is based on monitoring the time dependence of the magnetization induced by pulse $\mathbf{B}_1$ while the field $\mathbf{B}_0$ acts continuously on the spin system.

Upon the effect of field $\mathbf{B}_1$ in direction x, $\mathbf{M}$ rotates in the [z,y] plane at an angular velocity ω_1, and simultaneously processes around $\mathbf{B}_0$ at angular velocity ω_0. The corresponding absorption spectrum is a decaying signal at frequency ν_0, modulated by ν_1. The nutational time constant of the decay is, according to the movement of vector $\mathbf{M}$ in the [z,y] plane, $1/T_n = (1/T_1 + 1/T_2)/2$.

If ΔB_0 and ΔB_1 (inhomogeneities) are negligible, the determination of T_1 by other methods makes it possible to calculate T_2 from T_n measured here. For this measurement, however, the inhomogeneities should be eliminated.

If $B_1 \gg \Delta B_0$, the latter inhomogeneity may be neglected, since in the case of resonance $B_{eff} \approx B_1$.* However, in this case ΔB_1 is large, because it increases with B_1. This difficulty

* See p. 177.

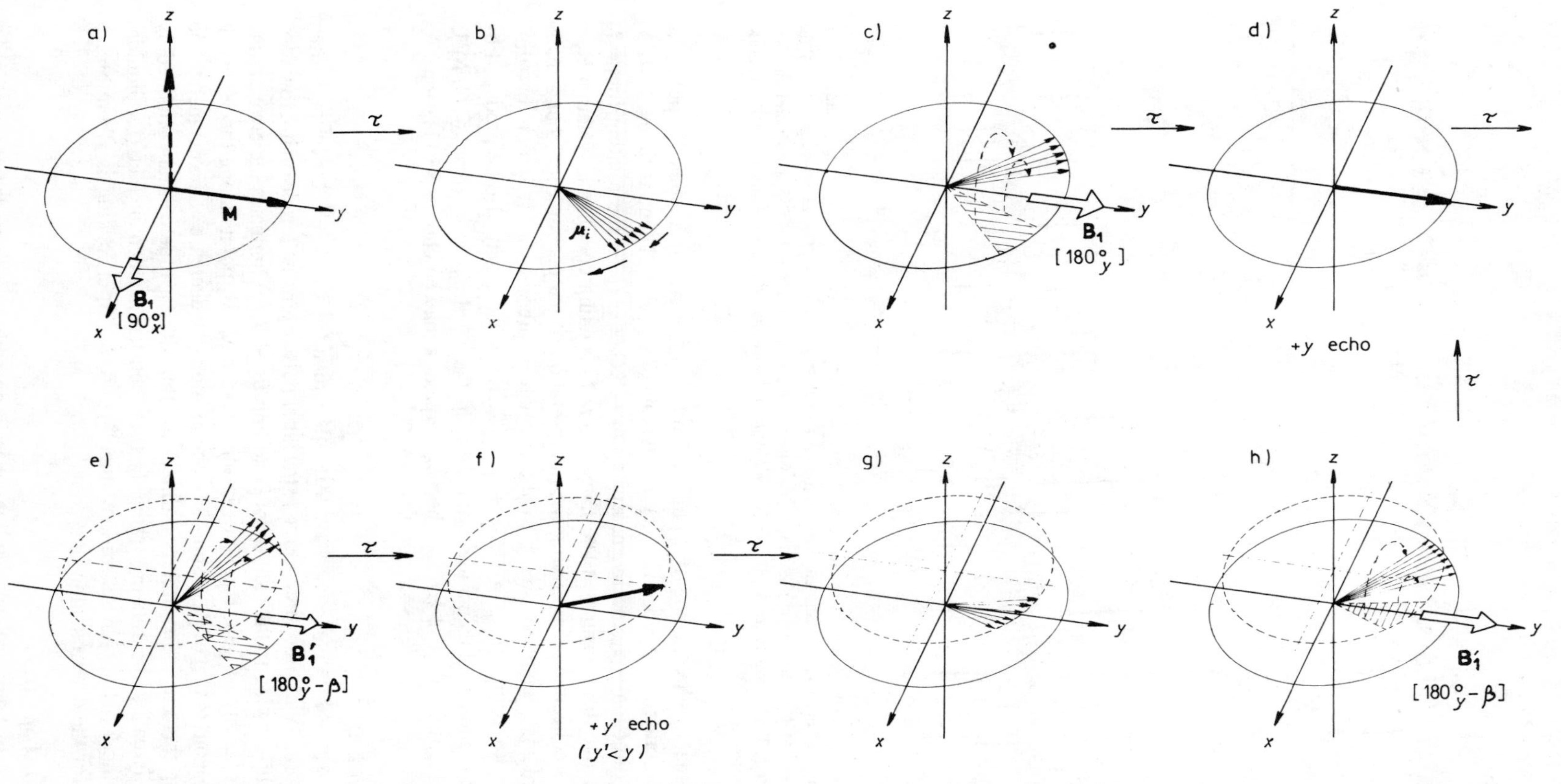

FIGURE 108. Measurement of T_2 by means of the Meiboom-Gill pulse sequence: [90°_x − τ − 180°_y − τ(echo) − τ − ...]. (a) The effect of the 90°_v pulse on the resultant magnetization vector **M** (b) the nuclear moments defocused owing to inhomogeneities, (c) the effect of a perfect 180°_v pulse on the spread individual moments μ_i (d) the resultant moment refocused in the [*xy*] plane, along the axis, (e) the effect of an imperfect 180° − β pulse ($\mathbf{B}_1$) on the spread nuclear moments, (f) the resultant vector **M** refocused after time above the [*xy*] plane: "*y*-echo", (g) the defocusing of the latter above the [*xy*] plane, and (h) the flip (imperfect inversion) of the nuclear moments spread above the [*xy*] plane into the [*xy*] plane upon the effect of the next 180° − β pulse ($\mathbf{B}'_1$) which after a further τ period become refocused in the *y* direction.

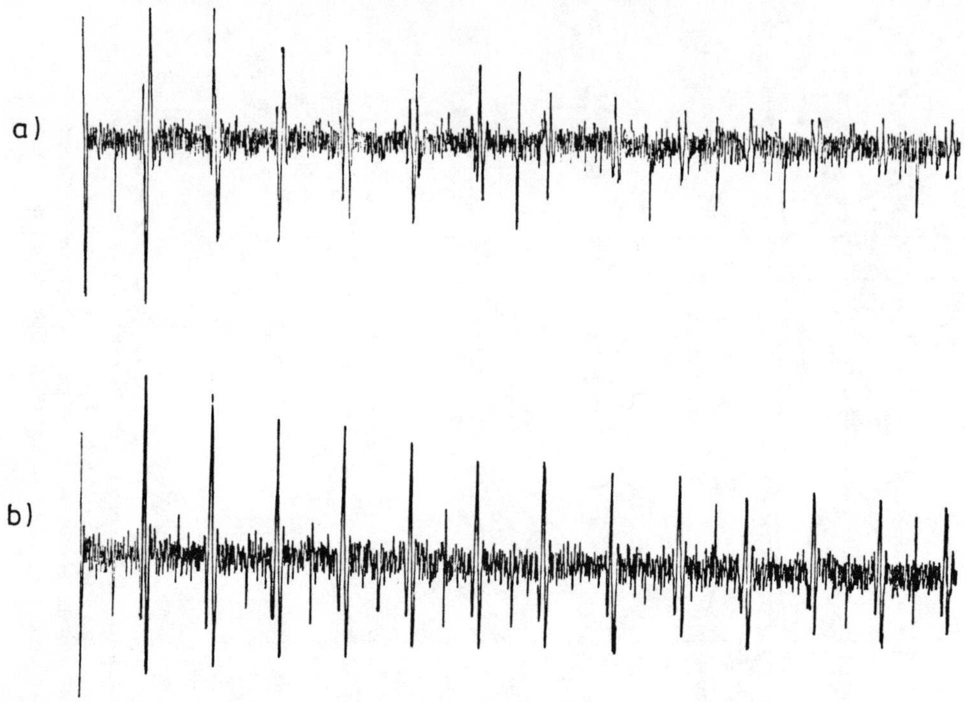

FIGURE 109. The ^{13}C NMR spectrum of acetic acid labeled on the carboxy carbon atom (60% enrichment) obtained by applying (a) the Carr-Purcell and (b) Meiboom-Gill pulse sequences. The alternating signs of the subsequent echoes are well observable in the first case, whereas the signs of the echoes in the second case are identical. (From Farrar, T. C. and Becker, E. D., *Pulse and Fourier Transform NMR*, Academic Press, New York, 1971, 26. With permission.)

may be overcome by the method of spin-echo rotation,[1348] which applies $\mathbf{B}_1$ pulses of alternating phase (alternately $+x$ and $-x$) at intervals $(2n - 1)\tau$. Then, at times $2n\tau$, $\mathbf{M}$ is refocused in direction $+z$, and by measuring these echoes, $T_{2\rho}$, "the rotating-frame T_2 time constant", may be measured. (The name refers to rotating coordinate system.) Here this is not only a convention simplifying the interpretation of the phenomenon, but a really existing experimental condition owing to the permanent presence of $\mathbf{B}_1$.) For this time constant it really holds that $1/T_{2\rho} = (1/T_1 + 1/T_2)/2$. Refocusing may also be produced by an $\mathbf{b}_0$ pulse of direction z, thus parallel with the field $\mathbf{B}_0$,[1506] if $b_0 \gg B_1$ (if $\nu_0 = 60$ MHz and $\Delta\nu_0 \approx 0.3$ Hz, then $\nu_1 \approx 2$ Hz and the γb_0 frequency may be chosen 35 kHz for $t_p \approx$ 15 μs).

2.2.7.9. *Spin Locking*[1347]

This method is based on two successive 90° pulses along $+x$ and $+y$. Since the former turns magnetization $\mathbf{M}$ just into direction $+y$, the latter pulse exerts no torque on it. However, if $B_1 \gg \Delta B_0$, this, owing to the validity of condition $B_1 \approx B_{eff}$, eliminates the error arising from the inhomogeneity of field $\mathbf{B}_0$. Since in this experiment $\mathbf{B}_1$ points along the y axis, it plays a role analogous to that of $\mathbf{B}_0$ in spin-lattice relaxation (the decrease of the magnetization component along this axis is monitored in time), and thus the $T_{1\rho}$ time constant obtained by the experiment is called the rotating-frame T_1 time constant, which in fact (for liquids) corresponds to the T_2 spin-spin relaxation time: $T_{1\rho} = T_2$. Spin locking is, therefore, a further method for the determination of T_2.

2.2.7.10. *J-Echo, J-Spectrum*

In the presence of homonuclear coupling, it is impossible to refocus the magnetization

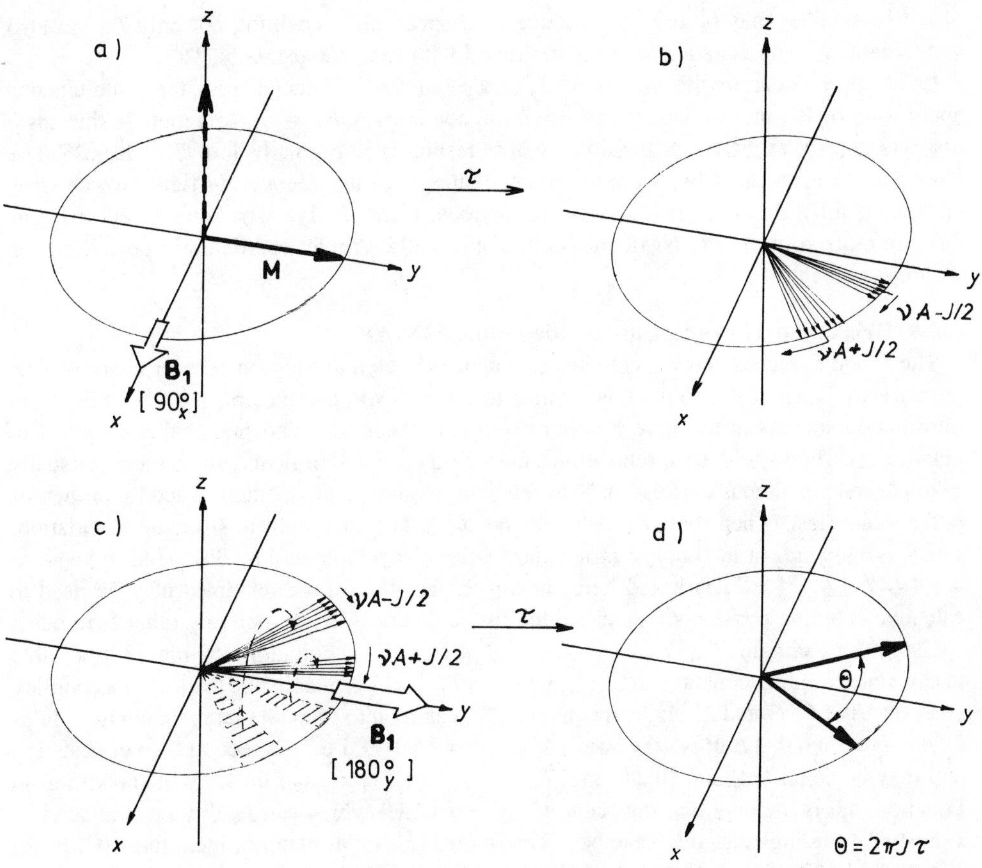

FIGURE 110. Spin echo in the case of homonuclear coupling. The flip of magnetization **M** (a) into direction y by pulse 90°_y, (b) the spread of the individual nuclear moments around frequencies $\nu A + J/2$ and $\nu A - J/2$, owing to inhomogeneities, (c) the inversion of the nuclear moments, divided into two groups due to coupling, about the y axis by a 180°_y pulse, and (d) the split resultant magnetization vectors refocused with a phase difference of $2\pi J$.

vector. This is easy to see on the example of the simplest case, an AX system. Owing to the coupling J_{AX}, the A signal is a doublet, and the frequencies of the two lines are $\nu A + J/2$ and $\nu A - J/2$. If nuclei A are excited, e.g., according to the Meiboom-Gill method, by a 90°_x pulse (see Figure 110a), the spin vectors are spread by the ΔB_0 inhomogeneities around the given two line frequencies as centers (see Figure 110b). If now, after time τ the spin vectors are inverted by a 180°_y pulse, the "slower", low-frequency components (spread around $\nu A - J/2$) remain closer to the Y axis, i.e., get ahead from the aspects of precession. Therefore, if they moved further with unchanged frequency, they would be exactly refocused after time 2τ, and by that time the "faster" components (spread around $\nu A + J/2$) would also reach them, to build up a spin echo. However, the 180°_y pulse, together with nuclei A, also inverts the spins of nuclei X, and thus the components with frequency $\nu A + J/2$ before the pulse continue the precession with frequency $\nu A - J/2$, and their pairs with frequency $\nu A - J/2$ carry on with frequency $\nu A + J/2$ (see Figure 110c). Therefore, the precession of the spins set back by the 180°_y pulse becomes slower and the precession of those turned ahead accelerates, and thus the phase difference after time 2τ is not eliminated, but increased to $J\tau$ (see Figure 110d). When there is only one coupling, the problem may be overcome,[970] e.g., by the appropriate choice of τ, since for $1/\tau = nJ$ the phase shift is $n{\cdot}360^\circ$. The difficulty may also be eliminated by means of spin locking, applying 180°_y pulses with very short intervals, or by recording the absolute-value spectrum. The J-modulated echo series may be used to produce a "*J*-spectrum", enabling not only T_2, but also very small coupling constants to be determined with extreme accuracy.[516]

Errors may occur in the values of T_2 measured for ^{13}C nuclei with the simultaneous application of BB proton decoupling when the condition $\gamma_C^2 B_1 = \gamma_H^2 B_2$ is met. In this case, a strong energy exchange is possible, which results in abnormally low T_2 values.[633] The error may be eliminated by the appropriate "gating" of the decoupling field (switching it on and off for the locking and acquisition periods, respectively). The same effect may, in turn, be utilized in the ^{13}C NMR measurement of solid samples (particularly polymers) for the improvement of ζ.[1247]

2.2.8. Relaxation Times in Cancer Diagnosis; FONAR

The T_1 and T_2 relaxation times of various magnetic nuclei, in addition to being characteristic properties of chemical compounds similar to other NMR parameters, may be utilized as information sources in the investigation of various theoretical and practical problems. No doubt, the carbon spin-lattice relaxation times are the most significant, since there is usually a simple and unanimous correlation between their magnitude and certain structural properties of the molecules.* Therefore, T_{1C} data may be utilized equally well for structure elucidation, signal assignment, or in study of molecular motion. This topic will be discussed in Sections 4.1.4.6 and 4.1.4.7. It is noted here, however, that the relaxation times may be used to calculate coupling constants which can hardly be determined, if at all, by other methods.

Using, for example, T_{1H}, T_{2H}, and $T_{1Cl} = T_{2Cl}$ of $CHCl_3$, Equation 260 makes it possible to calculate coupling constant $^2J(Cl,H)$, whereas T_{1C} and T_{2C} enable one to calculate coupling constant $^1J(Cl,C)$ (7 and 23 Hz, respectively).[464f] Taking into account that for small molecules $T_1^{DD} = T_2^{DD}$, and that $\Delta\omega^2$ is very large ($\Delta\omega_{Cl,H} \approx 40$ MHz), i.e., the second term of Equation 260 may be neglected, and finally that $T_{1Cl} = T_{2Cl} = T_{Cl}^{Q} = \tau_{SC}$, the only unknown left in Equation 260 is the coupling constant $J(I,S) = J(Cl,H)$ which can readily be evaluated.

Several coupling constants have been determined in similar manner, including $J(Cl,F)$ for ^{35}Cl and ^{37}Cl (27.5 and 23 Hz) in $CHFCl_2$ molecule.[1268] From the value of T_{1C} measured for the methyl carbon of toluene, the SR interaction constants were determined by using its relation to the paramagnetic contribution of shielding anisotropy,[489,490,1163] and from T_{2N} data, quadrupole coupling constants were evaluated.[367]

Due to the enormous importance of the subject, the rest of this chapter will devoted to a brief review of the application of relaxation times in the diagnosis of cancer.

It was observed first by Damadian in 1971 that cancerous tissues may be distinguished from healthy ones by means of NMR spectroscopy.[329] The ^1H spin-lattice relaxation times measured in malignant tissues proved to be significantly higher than in normal tissues.[304,329,527,641,669,694,1501] Increase of T_1 was also detected in noncancerous tissues of cancerous animals,[527,668,701] as well as in human tissues.[304,335,701,1167]

As an illustration, the data of Weisman[1501] may be mentioned. In NMR spectra taken from the tails of living mice, T_1 was found to increase to 0.7 s in cancerous tissue from the value of 0.3 s obtained normally (see Figure 111). The value of T_1 measured by the inversion method was determined in the usual manner, from the slope of the straight line obtained in the semilogarithmic plot of the exponential change in intensity as a function of τ. Tissue samples existing in the initial stage of the disease gave a nonexponential curve (nonlinear in the semilog plot), for lower values of τ the data corresponding to T_1 of healthy tissues, for higher values, those characteristic of cancerous tissues could be measured.

The idea of Damadian concerning the distinction of cancerous tissues by NMR methods is based on an interpretation of the physiological role of water,[694] which, amounting to 70 to 90% of living tissues, is in the living organism not simply a biological solvent, but is in

* See Volume II, p. 221-222 and Volume I, p. 198.

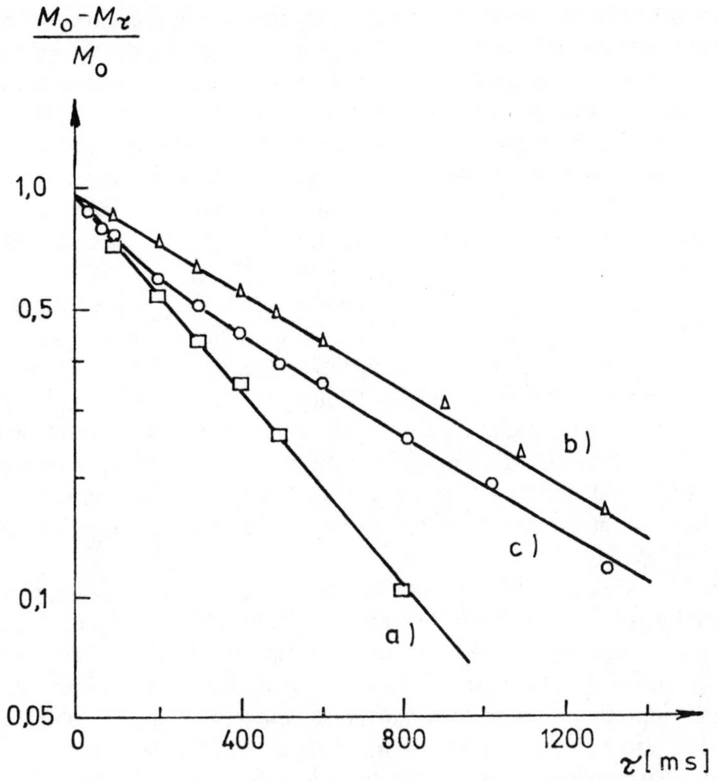

FIGURE 111. Determination of relaxation time. For normal (a) and for cancerous tissues (b) and for tissues in the initial stage of cancer (c).[1501] Lines (a) and (b) are the semilogarithmic plots of the equation $(M_e - M_\tau)/M_e = 2\exp[-\tau/T_1]$, from which T_1 (a) = 0.3 s and T_1(b) = 0.7 s may be obtained. Curve c may be approached by the following equation $M_e - M_\tau = M_a\exp[-\tau/T_{1a}] + M_b\exp[-\tau/T_{1b}]$.

a strong interaction with the macromolecules of the cells. Consequently, the physical state of the water content of cells differs from that of common water.[439-442,643,869,1430] If the theory of ion-exchange resins is applied on biological systems,[330,403] the fact that relaxation times T_1 of normal and cancerous tissues are different may be attributed to differences in chemical structure. Of course, different chemical environments assume different T_1 values. It was shown, for example, that malignant neoplasm has higher potassium content[403] and higher water content,[1167] and the ratio of hydrated and nonhydrated water molecules is also higher.[329,527,534,641]

Times T_1 measured for the tissue samples are the weighted averages of all, chemically different protons of the tissue studied. Since the proton concentration of tissues is determined primarily by the hydrogens of water molecules, the experimentally determined values T_1 are characteristic of the water molecules of healthy and cancerous tissues. This is the reason why the data of chemically widely different types of tissues are comparable at all and specific enough.

In order to assess the diagnostic values of longer T_1 times, how specific the increase in relaxation times is for cancerous diseases had to be answered, and the reason for these changes had to be found. It has been long known that water in tissues has characteristically shorter T_1 relaxation time than pure water, which is attributed to a hindered motion of cellular water molecules. Assuming that the different T_1 times measured in various tissues correspond to the mean of at least two different states, their decrease may be interpreted as an equilibrium of intra- and extracellular (less or more accessible to paramagnetic ions),[531] hydrated and nonhydrated,[479] ordered and nonordered,[297] hindered and less hindered in diffusion,[627] adsorbed and free,[1580] etc., water molecules. The change in these equilibria in cancerous tissues may account for the longer relaxation times.

The increase in T_1 was interpreted by the higher potassium ion concentration of malignant tissues.[329,335] Other authors attribute this simply to the increase in total water content,[153,668,700,701] but opposite experiences were also published.[642,701] Still others considered protein concentration as determining the magnitude of T_1,[290,1167] since cellular water, in several polarized layers, adsorbs onto cellular proteins.[870]

Longer relaxation times are also shown by some species (e.g. mice varieties),[642] certain kinds of tissues (e.g., References 700 and 1086), some young animals in rapid development,[643] and by tissues edematous for reasons other than cancer,[566,642,1158] and the value of T_1 varies within the cell.[710] It may cause difficulties in that the threshold of the method is occasionally high, and thus a tumor may be indicated only in a very developed state.[290,668] Nevertheless, the efforts directed towards the application of NMR spectroscopy in cancer diagnostics are of great importance and worth continuing. The method is fast and simple, requires no surgical intervention, and causes no change in either the tissue sample or the living organism investigated.

The increase in relaxation time T_1 was, in some cases, found to be proportional to the size or growth rate of a tumor,[290,527,668,669] and thus the results of NMR investigation permit one to draw conclusions on the stage of cancer, its rate of development, or any changes in the latter, e.g., upon treatment. It has also been published that the precancerous tissues (preneoplastic nodules), which are not yet malignant but have a high probability of becoming cancerous, may also be distinguished on the basis of NMR data, which are intermediate between those characteristic of healthy tissues and malignant tumors.[642]

Since the NMR parameters are determined directly by the chemical structure, a more accurate distinction among various cancerous diseases may be expected from the NMR method. So far, classification has been based on cell morphologic investigations, and morphologic differences are obviously less specific than chemical structure.[335]

In addition to spin-lattice relaxation time T_1, spin-spin relaxation time T_2 and diffusion constant D may also be determined by NMR measurements,[641,642,1466] as well as further parameters determined by molecular motions,* and thus they may also be utilized for diagnostic purposes. Already the first investigations of Damadian[329] were extended to the determination of T_2, and an increase analogous to that of T_1 was observed with cancerous rat tissues T_1 increased from 0.3 to 0.8 whereas T_2 increased from 0.05 to 0.1 s with respect to healthy tissue samples). In studying the application of NMR methods in cancer research, it was plausible to extend the investigations to nuclei other than protons, primarily to nuclei for which the physiological function may be expected to change upon illness and which are also accessible from the aspects of NMR measurement. This has led to the NMR investigation of ^{39}K ions[299,332] and then ^{23}Na ions[564] in cancerous tissues. (In cancerous tissues the concentrations of Na^+ and K^+ ions change.)

According to the results for these nuclei, too, the spin-lattice relaxation slows down in malignant tissues with respect to healthy ones, but the change in T_1 is not too large (approximately 25% increase), giving no firm basis for diagnosis. The first experiments with ^{29}K nuclei led to the interesting result that the intensities determined for the measurement of T_1 do not change exponentially; the method of progressive saturation produced a characteristic oscillation curve for cancerous tissues.[332] With healthy tissues, the usual exponential curve was obtained. This oscillation phenomenon was observed later for some healthy but

* From the measured values of T_1, T_2, and D, the τ_c correlation time of molecular motions may also be determined.[290,1466]

rapidly growing tissues (e.g., for newborn living mice) as well.[299] The theoretical interpretation of oscillation was given by Cope.[298]

Thus, although the small increase in the value of T_1 of ^{39}K has no diagnostic value because of the oscillation specific for cancerous cells, ^{39}K NMR measurements have proved to be a valuable source of information in cancer research. It has already emerged earlier that the diagnostic value of various NMR measurements (connected with different nuclei, or with the different NMR parameters, as T_1, T_2, D, etc., of the same nucleus) greatly increases if their results are collected into a joint ''malignance index''.[332]

Isotope ^{17}O is insensitive of NMR detection, owing to its quadrupole moment. However, the same quadrupole moment causes short relaxation times, and thus it becomes possible to distinguish[1295] or determine the relative population[1296] of water molecules in different chemical environments (e.g., intra- and extracellular water). This is impossible by means of ^1H NMR measurement, owing to the slower relaxation of protons (since the exchange of water molecules is faster than the relaxation of protons, measured values of T_1 correspond only to an average). By means of ^{17}O resonance, however, the more slowly relaxing water molecules of the nucleus (cytoblast) and those of the cytoplasm may be distinguished, and their ratio may also be determined (approximately 2:1). The nonexponential character of the T_1 curve obtained by the inversion method is due to the different relaxation rates of the two types of water molecule. Since the relaxation of oxygen is determined by the interaction of water molecules and cell proteins,[776] it is easy to understand that in dead tissues the difference disappears: the T_1 and T_2 curves become equally exponential. In cancerous tissues the nonexponential run of the T_1 and T_2 curves remains unchanged, and the magnitudes of both components of both relaxations (T_1 and T_2) increase.[1296]

The key physiological role of phosphorus derivatives, the sharp ^{31}P signals and the wide range of ^{31}P chemical shifts, explain that after the proton, ^{31}P is the most frequently studied nucleus. The ^{31}P NMR investigation of cancerous tissues[1578] indicated that similar to other nuclei T_1 is higher than in normal tissues, but the difference is greater, and thus the diagnostic value is higher.[780,1579]

The longer T_1 times may be interpreted by assuming that in the aqueous medium, less regular in cancerous tissues, phosphor-containing molecules may move more rapidly.[955,1579] Measurements have also shown that in malignant tissues phosphorus concentration is lower.[1579] The high sensitivity of FT spectrometers made it possible to determine not only relaxation time T_1, but also ^{31}P chemical shifts with a sufficient resolution. The varying chemical environment of the phosphorus atom gives rise to line-rich spectra, which may thus be regarded as an ''NMR fingerprint'' of the cell and which have high cancer diagnostic value. It was shown[780] that in cancerous tissues some of the lines of the ^{31}P NMR spectrum of normal tissues are missing, and the remaining lines are shifted and their relative intensities change.

The availability of high-resolution spectra taken from living samples suggested the idea of the application of NMR in the therapy of cancer.[1577] The characteristic absorption frequencies may become a tool in the selective destruction of malignant cells without the need of localizing these cells in the organism.

It has been known for more than a century that higher temperature cause a regression of cancer. In cell cultures, a 2-hr heat treatment at 42.5°C was shown to cause 95% destruction of cancerous cells, whereas only 43% of normal tissues are destroyed under the same conditions.[558] On this basis, an attempt was made to destroy cancerous tissues by means of RF radiation. By placing electrodes into the tumor of mice, the cell temperature was increased by means of 27 MHz RF at a power of 1 W. By maintaining temperature at 43.5°C for 1 hr, 25%, at 41.5°C for 4 hr, 40% healing could be attained. If the absorption of energy could be made selective, even 100% recovery would be possible. The selectivity of energy absorption may be ensured just by NMR resonance by placing the cancerous tissue into a magnetic field and irradiating it with the resonance frequencies of the magnetic nuclei of various chemical shifts. Unfortunately, the transfer of the required amount of energy into the cell by means of NMR appears so far to be impossible. According to rough calculations, 4.2 kJ are necessary for raising the temperature of 1 ℓ of water by 1°C, whereas the RF power absorbed by the protons of the same amount of water in the case of ^1H resonance is 0.04 J.[1577] At any rate, the idea of cancer therapy based on the principle of selective NMR is worthy of attention, and the utilization of this idea deserves further efforts.

The NMR investigation, as a nondestructive cancer diagnostic method, raised the demand towards the "NMR imaging" of voluminous "samples" (even humans) of strong magnetic inhomogeneity, i.e., toward the determination of NMR spectrum parameters (spin density, i.e., the occurrence of nuclei with given chemical environment, relaxation times T_1, etc.) in macro samples for a given limited volume or cross section. Such a "spin map" or "NMR phantom image" of macro samples (living organism, humans) is determined by the s.c. tomography or zeugmatography.[333,656,685,795,826,827,923]

The essence of the method is that a field gradient is superimposed on the homogeneous, static magnetic field, $\mathbf{B}_0$, and thus the condition of magnetic resonance given by the Larmor equation is satisfied only in a definite part of the field which changes from point to point. The required field gradient may be created by "shaping" the magnetic field using field-correction (shim) coils.* By the simultaneous control of $\mathbf{B}_0$ and $\mathbf{B}_1$, the resonating part of the sample may be located in three dimensions and restricted to arbitrary parts of space. This is called FONAR, field focusing nuclear magnetic resonance.[331,334,565]

The resonance aperture may be localized by changing the current in the shim coils. By defining the magnitude of the z component of field $\mathbf{B}_0$ in polar coordinates, the required field gradient may be given theoretically. The size of resonating volume may be determined by calibration measurements. For a solution giving an arbitrary resonance signal, line intensities are measured in two sample tubes different in size, while varying the current led into the shim coils. By plotting the intensity ratios for the two tubes as a function of current, one can determine the "normalized" current, for which this ratio is just 1:1. This current produces a magnetic field which restricts the resonating volume in the larger tube just to the size of the smaller one. By determining the normalized current for several pairs of sample tubes, the current corresponding to any resonance aperture (e.g., 1 mm^3 in volume) may be determined via the extrapolation of tube diameters vs. normalized current plot. When the aperture or the sample is moved, an NMR image or series of section recordings may be prepared from the surface or any layer of the body.[334]

Arising from the nature of the measurement, the experimental difficulties are enormous[685] and the elimination of artifact or misleading results requires great care. Both theoretical considerations and experience have shown that for cancer diagnostic purposes, lower (1 to 10 MHz) frequencies are more favorable, and the optimum must be determined separately in each case.[290,685]

2.2.9. Further Use of Pulse Sequences. Solvent Peak Elimination; Magic Angle Spinning; Two-dimensional FT-NMR Spectroscopy

In Sections 2.2.5 and 2.2.7, the application of pulse sequences in DR experiments and for measurement of relaxation times was discussed. Pulse sequences, however, can be utilized for sensitivity enhancement, suppression of strong solvent signals, and measurement of high-resolution spectra of solid samples. Two-dimensional technique, an excellent tool in the interpretation of complex spectra, may also be based on pulse sequences.

2.2.9.1. Improvement of Sensitivity: DEFT, SEFT

The advantages of PFT spectroscopy may be utilized fully for the improvement of ζ when

* Compare p. 136.

the next pulse may be applied immediately after the acquisition of FID induced by the previous pulse, i.e., the time of the measuring sequence is $t_p + t_{aq}$. *Since* t_p is very short (in the order of μs), in this case $t_{aq} \approx \pi T_2^{**}$ is the measurement time. This is possible only if $T_1 \approx T_2^*$. In practice T_2 is usually about 1 s (due to field inhomogeneities), whereas T_1 is 10 to 100 s for ^{13}C nuclei. Under such conditions, during $t_p + t_{aq}$, the magnetization component M_z is recovered to only a small extent, and if the next pulse is applied without delay, the signal disappears (saturation). The decrease in intensity due to saturation is proportional to $t/2T_1$, where $t = t_p + t_{aq} \approx \pi T_2^*$, i.e., the extent of saturation is a function of the T_2^*/T_1 ratio.[1482] The decrease in intensity may be eliminated by inserting a delay time (t_d), but this would substantially increase the duration of measurements.

Measurement time may be decreased by means of "T_1 reagents"**, or by decreasing pulse width, t_p.*** In the latter case, M_z decreases so little [$\Delta M_z = M_0(1 - \cos\theta)$] that the equilibrium is reset very quickly, and the time gain arising from this — assuming the same total measurement time — involves greater improvement in ζ (enabling one to accumulate more FIDs) than the adverse effect of the smaller $M_{(x,y)}$ components arising from the shorter pulses (namely, $M_{x,y} = M_0 \sin\theta$; and for small θ angles, $\sin\theta \gg 1 - \cos\theta$).[343] The optimum of pulse width, i.e., of θ, is a function of the T_2^*/T_1 ratio, and it must be found empirically.

Pulse sequences may also be used to improve ζ. By adding the echoes of the Carr-Purcell sequence [90° − τ − 180° − τ − 180° − τ...] to the FID recorded after the first pulse, the fraction of signal intensity lost owing to the inhomogeneities may be recovered (SEFT, spin echo FT).[25]

A more successful version of this method is DEFT (driven equilibrium FT),[105] in which the magnetization induced by a 90° pulse is refocused τ time later by a 180° pulse, and the echo appearing at time 2τ is turned back by means of a new 90° pulse into the original +z direction, more quickly than it may occur through T_1 relaxation. The sequence is thus [90° − τ − 180° − τ − 90° − t_d]$_n$, where $t_d = 4\tau$ is the optimum delay time between the sequences. FID may be acquired not only after the 90° pulse, but also after the 180° one, and the two FIDs may be coadded with reversed sign (super-DEFT). None of these methods has practical importance, since the attainable increase in sensitivity is not proportional to the complexity of experiments and the demands raised against the spectrometers.[1298]

2.2.9.2. Recording the High-Resolution Spectra of Solid Samples by Means of Pulse Sequences; The "Magic Angle"

The dipole-dipole interactions, which are averaged in gas and liquid phase owing to the fast molecular motions, cause very serious line broadening in the spectra of solid samples which had prevented the recording of high-resolution NMR spectra of solid samples and is still the main obstacle of development in this area.

The Hamilton operator describing the dipolar interaction is[434a]

$$\mathcal{H}^{DD} = \sum_{i<j}\sum \frac{1 - 3\cos^2\theta_{ij}}{2r_{ij}^3}(\mu_i\mu_j - 3\mu_{zi}\mu_{zj}) \qquad (265)$$

In liquid and gas phase $\mathcal{H}_{DD} = 0$, since[464c]

$$\frac{\int_0^{2\pi}\int_0^{\pi}(3\cos^2\Theta - 1)\sin\theta \, d\theta \, d\varphi}{\int_0^{2\pi}\int_0^{\pi}\sin\theta \, d\theta \, d\varphi} = 0 \qquad (266)$$

* Compare p. 185.
** See p. 205 and Volume II, p. 227.
***See p. 188.

i.e., the average of $1 - 3\cos^2\theta$ over all possible values of θ is zero. Operator $\mathcal{H}_{DD}$ also vanishes if $\theta = 54°44'$, i.e., when this is the angle between the individual magnetic moments and the polarizing field.[38] The angle 54°44′ is, therefore, often mentioned as "magic angle" in the literature. This condition may also be reproduced experimentally by spinning the sample at a very high rate (approximately 10 kHz) about an axis forming the magic angle with the static field. The experimental difficulties are, however, very serious.

It can be seen from Equation 265 that the condition $\mathcal{H}_{DD} = 0$ is also fulfilled if the magnetic moments themselves are precessing at the magic angle, and this may be accomplished by means of an appropriate pulse sequence.[615,1484] The simplest sequence for this purpose consists of four successive 90° pulses. The first pulse of direction x turns magnetization **M** into direction y where it spends time 2τ. Then **M** is turned back into the original (z) direction by means of a $-x$ pulse. After time τ it is turned into direction x by a $+y$ pulse where it is left for time 2τ, and finally by means of a $-y$ pulse **M** is returned into the z direction. The next sequence starts after time τ. The complete pulse train is therefore $[90^\circ_x - 2\tau - 90^\circ_{-x} - \tau - 90^\circ_y - 2\tau - 90^\circ_{-y} - \tau -]_n$. Upon the effect of this sequence, the macroscopic magnetization vector spends identical time in the three coordinate directions. Since the expression $\boldsymbol{\mu}_i \cdot \boldsymbol{\mu}_j$ depends on the relative position of the two moments only, it is not affected by the pulses, but the $\mu_{zi} \cdot \mu_{zj}$ components must be averaged for the three positions:

$$\overline{\mu_{zi}\,\mu_{zj}} = (\mu_{x'i}\,\mu_{x'j} + \mu_{y'i}\,\mu_{y'j} + \mu_{zi}\,\mu_{zj})/3 = \mu_i\mu_j/3 \tag{267}$$

(In this expression, x' and y' refer to rotating frame of reference, since the static $\mathcal{H}_{DD}$ operator acts in the fixed frame. Of course, $z' = z$, since the z axes of the fixed and rotating frames of reference coincide.) It can be seen that in this case $\mathcal{H}_{DD}$ is, indeed, zero, provided that $t_p \ll \tau$ and 6τ (the duration of one sequence) $\ll T_2$. Since in solids the typical value of T_2 is about 10 to 100 μs, the condition of the success of experiment is that $t_p \leq 1$ μs.

FID may be recorded equally for the x or y position of moment **M**. The transformation yields the usual high-resolution spectrum, in which the splittings caused by scalar couplings also appear, since the pulse sequence does not affect them (compare Figure 183c).

The method does not eliminate DD interactions completely, only decreases them by a factor of $1/\sqrt{3}$ (approximately 0.6). Line broadening is reduced by the same factor. Simultaneously, all chemical shifts decrease by $I/\sqrt{3}$, too, since the component of **M** along $\mathbf{B}_0$ is $M\cos 54°\,44/\text{ft} = M/\sqrt{3}$.

2.2.9.3. Elimination of Strong Solvent Signals: WEFT

When the NMR spectrum contains an extremely strong peak, this may suppress coincident and neighboring signals and also hinders the identification of other signals, since they hardly, if at all, emerge from the noise. (In order to prevent the ADC from overflowing, which would distort the spectrum,* the gain of the receiver must be adjusted so that only the strongest signal should fill the register of ADC completely.)

This problem is the most frequent with solvent peaks, primarily the signal of water, used almost exclusively in the ^{1}H NMR investigation of biological problems, since with respect to the 110 *M* concentration of protons in water, the proton concentration of solutes, usually dissolved in a m*M* to 10 μ*M* concentration, is 5 to 7 orders lower.

The simplest elimination of the difficulties caused is the use of isotope-substituted solvents, in this case D_2O. This solves, however, only a part of the problem (the proton concentration decreases only by approximately two orders, since heavy water is hygroscopic and thus always contains light isotopes). We have mentioned another method for the elimination of solvent signals[1404] according to which the solvent protons are selectively excluded from

* See p. 183.

excitation by means of an appropriate window function (see Section 2.2.4.3). The intensity suppression of solvent peaks is also possible by means of various pulse sequences.

The solvent signal may be saturated selectively by the application of the inversion method, i.e., the sequence consisting of two pulses, utilizing that T_1 values are different for the solvent protons and the protons of the solute.[114,1090] The moments of all protons are inverted by a 180° pulse, and after time τ they are rotated by a 90° pulse into plane $[x,y]$, where detection takes place. If the 90° pulse is applied exactly when the magnetization vector of the solvent passes zero (when $\tau = T_{1S}\ln 2$), the solvent peak will be suppressed in the spectrum (T_{1S} is the spin-lattice relaxation time of the solvent). With appropriate t_d pauses, the pulse sequence may be repeated. It is important to meet the condition $t_d \geqslant 5T_{1S}$, and thus for the success of the experiment T_{1S} must be known precisely. The method is not very efficient, since $T_1(\mathrm{HOD}) \approx 20$ s, and therefore much time is required for a good τ. Efficiency may be improved by building up the steady state, i.e., satisfying the equation[114]

$$e^{-(T_{1S}\ln 2 + t_d)/T_{1S}} + e^{-(T_{1S}\ln 2 + \tau)/T_{1S}} = 1 \qquad (268)$$

The pulse sequence is $[180° - \tau - 90° - t_d]_n$, and the name of the method is WEFT (Water Elimination FT) spectroscopy. The accuracy of the experiment may be greatly increased by removing the $M_{x,y}$ component due to the imperfect 180° pulse immediately after this pulse by means of a HSP.*[114,992]

In an analogous manner, the DEFT pulse sequence (see Section 2.2.9.1) is also applicable for the elimination of solvent signals by the appropriate timing of the 90° pulse preceding the detection after the 180° pulse.[660,661] It is noted that the solvent signal may also be eliminated by selective saturation, but in this case the neighboring signals are also suppressed (i.e., saturated), and the NOE may lead to a distortion in signal intensities. However, the result of the experiment does not depend on the relative magnitudes of the T_1 times of the solvent and the solute.[241,721]

The method of progressive saturation (series of 90° pulses) may also be used for the elimination of solvent signals.[114] As the protons of HOD molecule have long T_{1S} relaxation time by applying the pulses frequently enough, it may be attained that the intensities of the signals to be measured reach a maximum (complete relaxation for these protons during the time between the pulses), while the solvent signals become partly saturated. This method only reduces the intensity of solvent signals and does not eliminate them, but less time is required by the experiment and better ζ-values may be attained.

It is noted here that the selective exclusion of solvent protons from excitation may be performed not only by Fourier-synthesized window functions, but also by long, weak (approximately 45°) pulses,[1179,1180] exciting only the spectrum region on one or the other side of the water signal (a long, low-amplitude pulse excites only the frequency region $\Delta\nu = 2/\tau$). The problems of solvent signal elimination have been dealt with in several reviews.[1178,1284,1488]

2.2.9.4. Two-Dimensional NMR Spectroscopy (2DFTS)

Classical NMR spectroscopy investigates the spin systems in linear approximation, assuming that in a static magnetic field $\mathbf{B}_0$ the spectrum of a given spin system obtained upon the effect of the same $\mathbf{B}_1$ excitation (RF) field is always the same, i.e., the spectrum of a spin system depends on the excitation frequency alone. This assumed linearity between the excitation frequency and the response of the spin system (i.e., the spectrum) is valid under certain conditions: in CW spectroscopy, e.g., slow sweep and the application of appropriate RF field (and, of course, identical experimental conditions such as solvent, concentration, and temperature). If the sweep rate is too high, distortions occur, and if the amplitude B_1

* See p. 208.

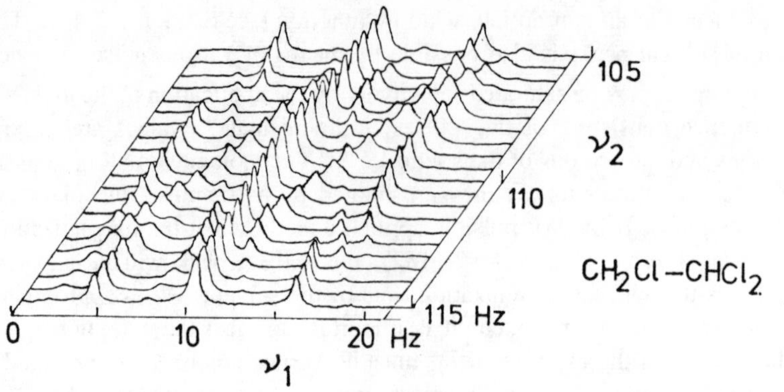

FIGURE 112. The methine triplet of 1,1,2-trichloroethane ($CHCl_2$–CH_2Cl) in the partially decoupled (tickling) spectrum, obtained by irradiating the methylene doublet with varying ν_2 frequencies ($\Delta\nu_2$ = 10 Hz).[61]

is too large, partial or complete saturation takes place; the spectrum is, therefore, a function of several variables.

The availability of computers and the introduction of FT technique allows large data sets to be collected and processed, and thus the NMR spectra may be investigated as a function of more variables. A means of the easier interpretation of the resulting lot of information (e.g., series of spectra) is two- or multidimensional NMR spectroscopy, in which the spectra of the same spin system obtained as a function of two different variables are separated into a "two-dimensional" picture, where the two axes are the variables. In this sense, the spectrum series obtained in the measurement of various relaxation times by pulse sequences plotted against the spacing τ of the individual pulses (Compare Figure 104) may also be regarded as a two-dimensional spectrum, where the two independent variables are τ and the usual frequency ν_1. A series of spectra recorded at different temperatures can be regarded also as a two-dimensional measurement, where frequency ν_1 and temperature are the changing parameters (the two axes). Zeugmatography or tomography,[826] the two- or three-dimensional NMR imaging of living organisms, the body's spin-density map* may also be classified into this group. In this case, one of the independent variable of the plot is an NMR parameter (relaxation time T_1 or T_2 pro volume unit, or the total signal intensity of a nucleus, e.g., proton), and the other (two) is (are) the spatial coordinate(s). In the above cases, the spectra are always measured as a function of only one frequency, and the second and further variables are other parameters (time, space, temperature, etc.).

The possibility of recording an NMR spectrum as *a function of two frequencies* was first suggested by Jeener.[715] It is evident that by "spreading" the NMR signals along two frequency axes, i.e., in a two-dimensional frequency plane, particularly complex spectra, like those of biopolymers or other macromolecules, may be simplified or transformed into much more easily interpretable form. The term[61] 2DFTS hence refers to a special, measurement technique.[453]

A series of DR experiments in which the second excitation frequency, ν_2, is varied is the simplest case. Figure 112 shows the methine triplet of 1,1,2-trichloroethane ($CHCl_2$–CH_2Cl) in a tickling experiment (Section 2.1.6.3) as a function of frequency ν_2 used for irradiating the methylene doublet.[61]

The series of off-resonance ^{13}C NMR spectra recorded as a function of offset $\Delta\nu$ (compare Figure 98) may also be included into the group of 2DFTS in the strict sense. This method allows one to select the corresponding ^{13}C and ^1H signals on the basis of the sequence of

* Compare p. 218.

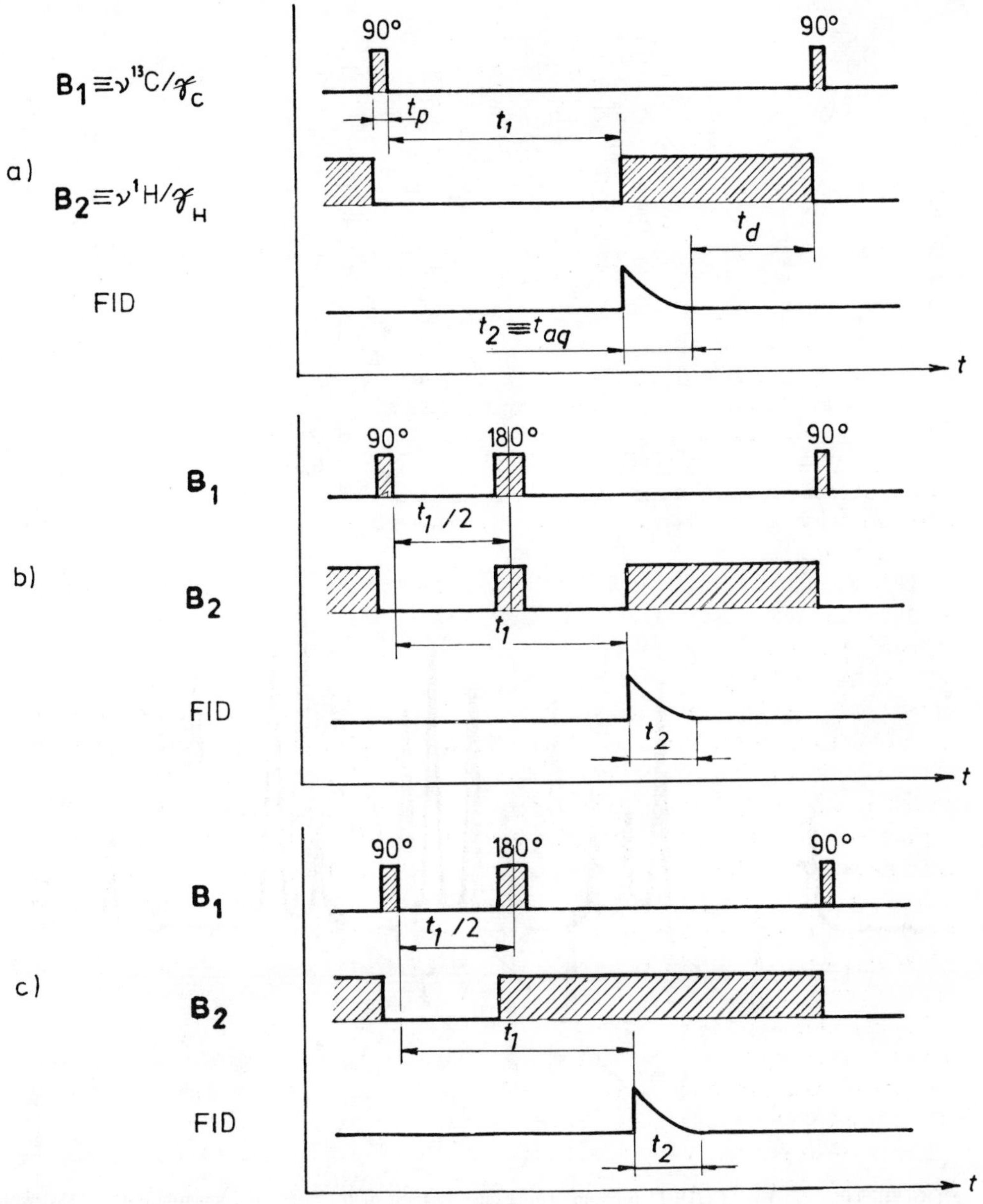

FIGURE 113. Pulse sequences for the production of ^{13}C 2DFT spectra. Decomposition of proton-coupled and decoupled ^{13}C NMR spectra into two dimensions (a), and (b) and (c) the complete separation of chemical shifts (the individual multiplets) and coupling constants (multiplicities of signals) into two dimensions: the separation of decoupled and J spectrum.[796,1016]

$\Delta\nu_i$ values belonging to the coalescence ($J_i^* = 0$) of the various multiplets (compare Figure 99).

The 2DFTS method may also be utilized in the presentation of undecoupled ^{13}C NMR spectra.[164,1016,1017] The overlapping multiplets due to closely spaced signals may often prove inseparable by off-resonance, too, and thus the order of carbon atoms cannot be determined, not to mention the loss of information represented by the true values of direct ^{13}C, ^{1}H coupling constants which become inaccessible upon the application of BBDR or off-resonance.

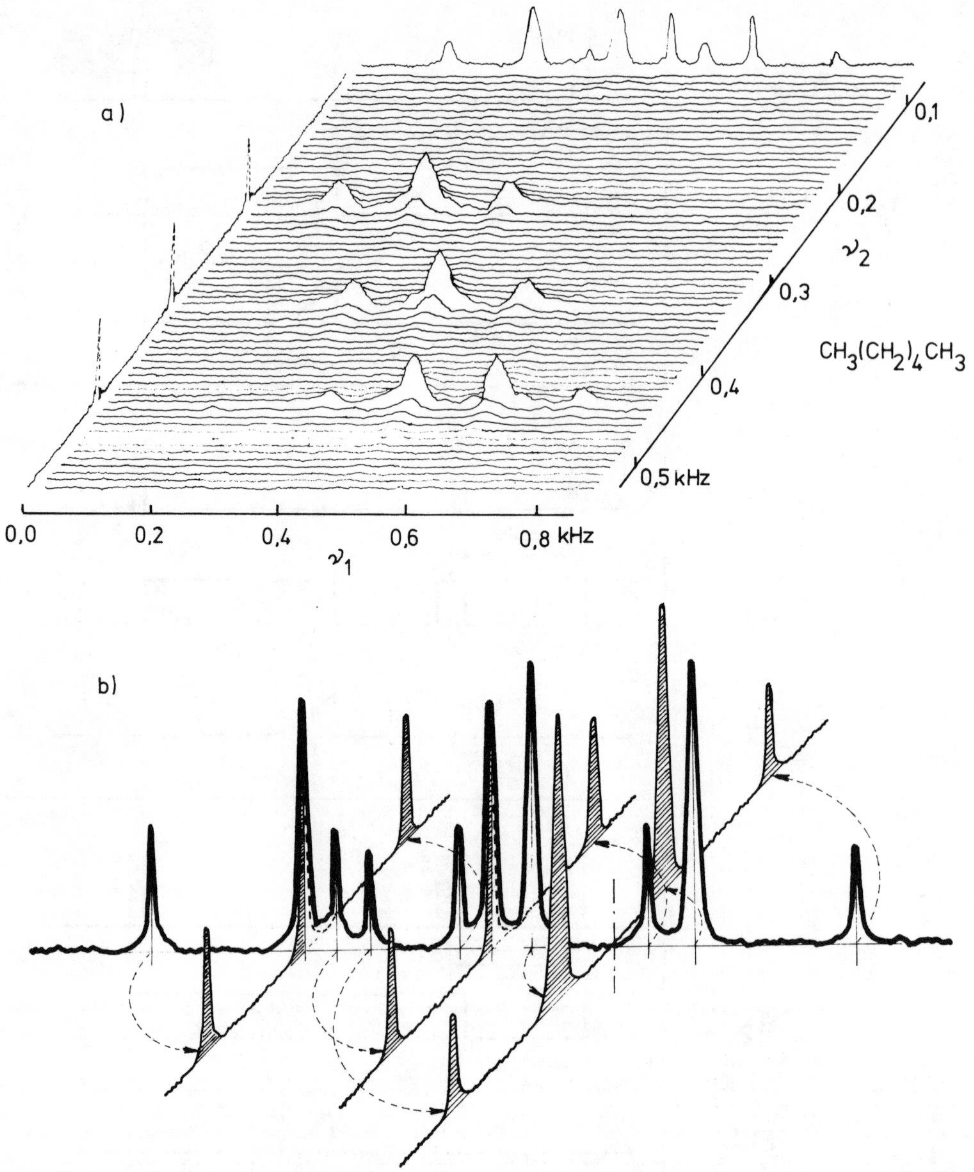

FIGURE 114. (a) The [13]C 2DFT spectrum of *n*-hexane obtained by the technique shown in Figure 113a, with the multiplets spread along the ν_2 axis. The 2DFT diagram consists of 64 spectra measured with different t_1 (0 to 35 ms), each obtained from the average of 22 FIDs[1016] Schematic representation of the two-dimensional spectrum of *n*-hexane (b) obtained by the 90° rotation of the multiplets appearing in the one-dimensional spectrum (dispersed in the frequency plane according to *J*).

The essence of a 2D experiment is that following the ν_1 excitation RF pulse of 90°, the excitation RF field ν_2 of protons is switched off for period t_1, and the $t_2 = t_{aq}$ acquisition period follows only after this time, during which the field ν_2 causing the BB decoupling of protons is present (see Figure 113a). The series of FIDs measured at changing t_1 is a function of t_1 and t_2, and the double FT of this $f(t_1,t_2)$ function yields a 2D spectrum in the frequency domain[1016]

$$\mathcal{F}_1 \mathcal{F}_2 [f(t_1,t_2)] = \int_0^\infty \int_0^\infty \cos \omega_1 t_1 \cos \omega_2 t_2 f(t_1,t_2) dt_1 dt_2 = f(\nu_1,\nu_2) \tag{269}$$

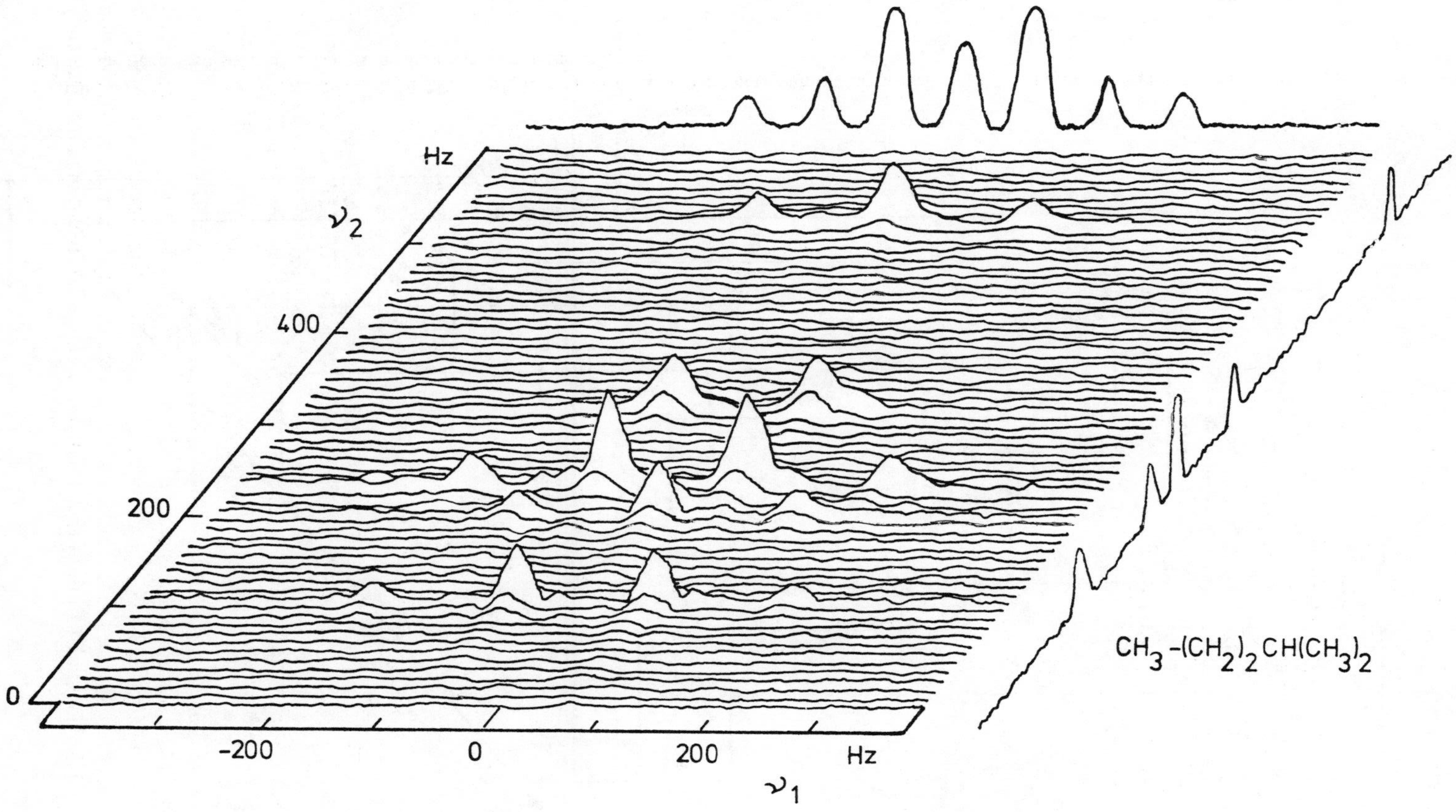

FIGURE 115. The [13]C 2DFT spectrum of 2-methylpentane obtained by the technique shown in Figure 113c,[1017] with the *J* and decoupled spectra obtained as the ν_1 and ν_2 projections.

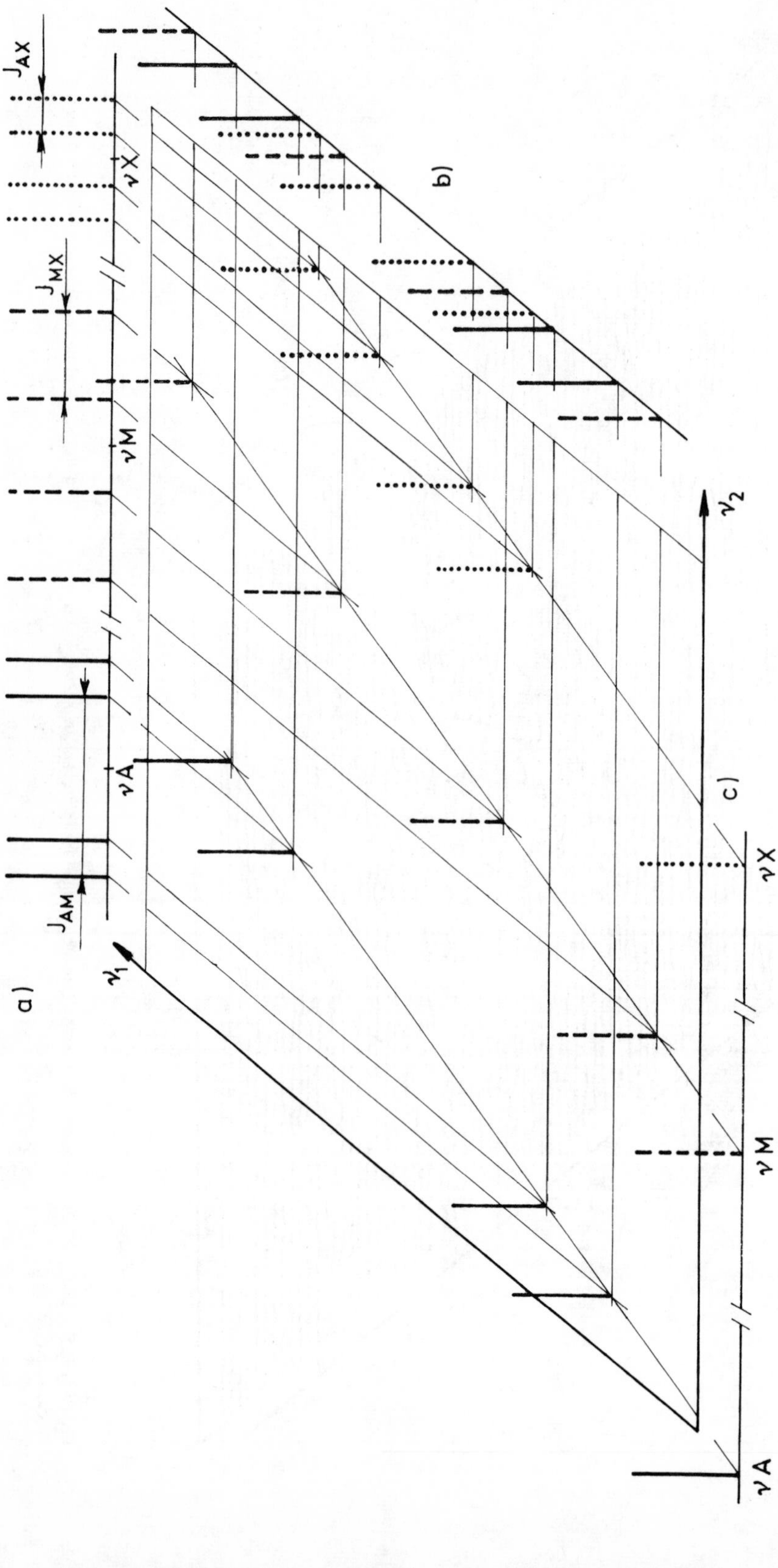

FIGURE 116. The *J*-resolved, schematic ¹H 2DFT spectrum of an *AMX* spin system, with (a) the normal and (b) *J*-spectrum obtained as the ν_1 and ν_2 projections and (c) the decoupled spectrum obtained as the projection in the diagonal.

The ^{13}C NMR spectrum of *n*-hexane is given as illustration. The projections on axis ν_1 give the coupled spectra; at $\nu_2 = 0$ the complete proton coupled spectrum may be observed. The projection along ν_2 is the decoupled spectrum (see Figure 114a). The effect of this technique is to turn the multiplets of the one-dimensional (coupled) spectrum by 90° around their centers in the 2D frequency plane, thus making it possible to separate the originally overlapping multiplets (see Figure 114b). Since in this case spin-spin coupling "allows" the spectrum to be observed in two dimensions, these spectra are called *J*-resolved 2DFT spectra.[62,1020,1021] The reason is that in period t_1 the spin system is described by the complete $\mathcal{H}^\circ + \mathcal{H}'$ Hamiltonian, whereas in period t_2, owing to BB proton decoupling, term $\mathcal{H}'$ disappears.

If the spins are inverted by a 180° pulse at the half of period t_1, the chemical shifts and spin-spin couplings may be split completely into two dimensions.[162,164,792,1017] This may be accomplished either by a 180° pulse applied on both the ^1H and the ^{13}C nuclei at point $t_1/2$ and a continuous BB decoupling over period t_2 only or by inverting ^{13}C nuclei only and applying simultaneous BB decoupling until the end of period t_2 (see Figure 113b,c). In this case, the ν_1 projection is the *J*-spectrum,* and axis ν_2 contains the decoupled normal spectrum, consisting of singlets. Figure 115 shows such a 2DFT spectrum of 2-methylpentane.[1017]

By analogous methods, *J*-resolved ^1H 2DFT spectra of first-order spin systems may also be obtained. If the spin system excited by a 90° pulse is inverted at the half of period t_1, the ν_1 projection is the *J*-spectrum, the ν_2 axis contains the normal spectrum, and the projection along the 45° diagonal axis is the spectrum consisting of singlets.[792] The technique is very suitable for the simplification and interpretation of complex, line-rich ^1H NMR spectra.

Figure 116 shows the schematic *J*-resolved 2DFT spectrum of an *AMX* spin system, containing the *J*-normal (split), and decoupled (singlet) spectra along the axes ν_1, ν_2 and the diagonal, respectively. Further discussion of 2D and MDFT (multidimensional FT) is not possible here, however, the practical value of the method may be assessed from the above. It is also clear that there is ample room for further variants of the method (e.g., see References 71,161,163,793,794,1019) The method is also applicable for the investigation of theoretical problems, e.g., for the determination of spin-level connectivities[61] or forbidden quantum transitions.[1553]

* This is such a form of the completely coupled spectrum in which the centers of all multiplets coincide: all signals appear at the same chemical shift (cf., e.g., Figure 116b).

ABBREVIATIONS IN NMR THEORY AND METHODOLOGY

ADC	analog to digital converter
AF	audio frequency
ASIS	aromatic solvent induced shift
BBDR	broad band double resonance
CAT	computer of average transients
CIDEP	chemically induced dynamic electron polarization
CIDNP	chemically induced dynamic nuclear polarization
CW	continuous wave
DAC	digital to analog converter
DD	dipole-dipole, dipolar, relaxation mechanism or interaction
DEFT	driven-equilibrium FT
2DFTS	two-dimensional FT spectroscopy
DNMR	dynamic NMR
DNP	dynamic nuclear polarization
DRDS	double resonance difference spectroscopy
ESR	electron spin resonance
e.s.u.	electrostatic unit
FFT	fast Fourier transform
FID	free induction decay
FONAR	field focusing nuclear magnetic resonance
FT	Fourier transformation
HSP	homogeneity-spoiling pulse
INDOR	internuclear double resonance
IR	infrared
LCAO	linear combination of atomic orbitals (in quantum theory)
MO	— molecular orbital (in quantum theory) — master oscillator
NMR	nuclear magnetic resonance
NOE	nuclear Overhauser effect
NQR	nuclear quadrupole resonance
PFT	pulse Fourier transformation
ppm	part per million
RF	radio frequency
SA	chemical shift anisotropy (CSA), relaxation mechanism
SC	scalar, relaxation mechanism or interaction
SEFT	spin-echo FT
SPI	selective population inversion
SR	— spin rotation, relaxation mechanism or interaction — shift reagent
VCO	voltage controlled oscillator
WEFT	water-eliminated FT

CHEMICAL ABBREVIATIONS — COMPOUNDS AND FUNCTIONAL GROUPS

Ac	acetyl, CH_3CO–
ACAC	acetylacetone, $CH_3COCH_2COCH_3$
Ar	aryl (aromatic ring containing groups)
BHC	3-*t*-butylhydroxymethylene-*d*-camphor ≡ TBC

*n*Bu	*normal* butyl, $CH_3CH_2CH_2CH_2$–
*i*Bu	*iso*-butyl, $(CH_3)_2CHCH_2$–
*s*Bu	*sec.*-butyl, $CH_3CH_2CH(CH_3)$–
*t*Bu	*tert.*-butyl, $(CH_3)_3C$–
Bz	benzoyl, C_6H_5–CO–
Bzl	benzyl, C_6H_5–CH_2–
DMF, DMF-d_7	dimethyl-formamide, heptadeuterio-DMF
DMSO, DMSO-d_6	dimethyl-sulfoxide, hexadeuterio-DMSO
DPM	dipivaloyl methane, $(CH_3)_3C$–CO–CH_2–CO–$C(CH_3)_3$
DSS	2,2-dimethyl-2-silapentane-5-sulfonic acid sodium, $(CH_3)_3Si(CH_2)_3SO_3Na$
Et	ethyl, CH_3CH_2–
FACAM	3-trifluoroacetyl-*d*-camphor
FOD	1,1,1,2,2,3,3-heptafluoro-7,7-dimethyl-octane-4,6-dione
HFBC	3-heptafluorobutyryl-*d*-camphor ≡ HFC
HFC	see HFBC
HMPA (HMPT)	hexamethylphosphoramide
LIS	lanthanide (shift reagent) induced shift
LS	shift reagent — substrate complex
LSR	lanthanide shift reagent
Me	methyl, CH_3–
Ms	mesyl, CH_3SO_2–
Ph	phenyl, C_6H_5–
*n*Pr	*normal* propyl, $CH_3CH_2CH_2$–
*i*Pr	*iso*-propyl, $(CH_3)_2CH$–
Py, αPy, βPy, γPy	pyridyl, α-, β- and γ-pyridyl, C_5H_5N–
PVC	poly(vinyl chloride), $[CH_2Cl–CHCl_2]_n$
TBC	see BHC
TFA, TFA-d	trifluoro-acetic acid (CF_3COOH), deuterio-TFA (CF_3COOD)
TFC	see FACAM
TFM	trichloro fluoromethane, $CFCl_3$
THF	tetrahydrofurane, C_4H_8O
TMHD	2,2,6,6-tetramethylheptane-3,5-dione ≡ DPM
TMS	tetramethylsilane, $(CH_3)_4Si$
Ts	tosyl, *p*CH_3–C_6H_4–SO_2–

LIST OF NOTATIONS

Basic Physical Quantities

c	$(2.997925 \pm 0.000003) \cdot 10^{8}$ [m s^{-1}], velocity of light
e	$4.8024 \cdot 10^{-10}$ [e.s.u., CGS] = $1.60210 \cdot 10^{-19}$ [A s], [C], charge of electron
h	$6.6256 \cdot 10^{-34}$ [J s], Planck's constant
$\hbar$	$h/2\pi = 1.0545 \cdot 10^{-34}$ [J s]
k	$1.38054 \cdot 10^{-23}$ [J K^{-1}], Boltzmann's constant
m_e	$(9.1091 \pm 0.0004) \cdot 10^{-31}$ [kg], static electron mass
m_{H}	$(1.67252 \pm 0.00008) \cdot 10^{-27}$ [kg], static mass of proton
N	$6.02252 \cdot 10^{23}$ [M^{-1}], Avogadro's number
R	8.3143 [J K^{-1} M^{-1}], Rydberg's constant (universal molar gas constant)
β	$9.2732 \cdot 10^{-24}$ [J T^{-1}], Bohr magneton
β_{N}	$5.0505 \cdot 10^{-27}$ [J T^{-1}], nuclear magneton
μ_{H}	$2.67519 \cdot 10^{8}$ [s^{-1} T^{-1}], magnetogyric (gyromagnetic) ratio of proton
$_{\mathrm{H}}$	$1.41049 \cdot 10^{-26}$ [J T^{-1}], magnetic moment of proton

Symbols and Their Meanings

a	antisymmetric spin state, basic function transition, etc.
a, a'	*axial, quasi axial*
A	— signal intensity — hyperfine electron-nucleus interaction constant
$\mathbf{B}$	magnetic induction field (magnetic flux density), T [m kg s^{-2} A^{-2}], [m^{-2} s V]
$\mathbf{B}_0$	static external (polarizing) field of the spectrometer
$\mathbf{B}_l$	local magnetic field
$\mathbf{B}_\ell$	induced magnetic field
$\mathbf{B}_{1,2\ldots}$	first, second, etc. exciting (RF) magnetic fields associated with ν_1, ν_2, etc.
C	constant, linear combination constant
$C_{\parallel}$, $C_{\perp}$	spin rotational coupling constants
D	diffusion coefficient
e, e'	*equatorial, quasi equatorial*
E	— energy, N [m^2 kg s^{-2}] — electronegativity
$\mathbf{E}$	electric field
E_z	z component of the electric field
$\mathcal{F}$	Fourier transformation operator
f	mole fraction
g	nuclear or electronic Lande- (g-) factor
$\mathcal{H}$	Hamilton operator (in energy or frequency unit)
$\mathcal{H}^0$, $\mathcal{H}'$	the spin-external field and spin-spin interaction terms in the Hamiltonian
$\mathbf{H}_{ij}$	element of matrix representation of $\mathcal{H}$
H	Hamilton function
$\mathcal{I}$	spin vector operator
$\mathcal{I}_w$	components of the spin vector operator $\mathcal{I}$

$\mathcal{I}^+$	raising ("absorption") spin operator
$\mathcal{I}^-$	lowering ("emission") spin operator
$\mathcal{I}^2$	square of the spin vector operator
I_w	eigenvalue of $\mathcal{I}_w$
I^2	eigenvalue of $\mathcal{I}^2$
$\mathbf{I}_{ij}$	elements of matrix representation of $\mathcal{I}^2$
i	$\sqrt{-1}$
I	— current intensity [A]
	— moment of inertia
	— nuclear spin quantum number
$\pm I$	inductive effect of atoms and functional groups
J	nuclear spin-spin coupling constant [Hz]
^nJ(X,Y) or $^nJ_{XY}$	coupling constant between X and Y nuclei through n bonds [Hz]. Subscripts are given only as algebraic symbols for any interacting pairs of nuclei. Brackets are used for indicating the coupled nuclei, e.g., $J(^{19}F, {}^{1}H)$ or J(F, H) or the coupling path, e.g., J(NCCH) ≡ ^3J(N, H).
J^*	reduced coupling constant (in off-resonance)
J	— atomic quantum number representing the total angular moment
	— rotational quantum number
^nK(X,Y)	reduced nuclear spin-spin coupling constant [$m^{-3}\ A^{-2}\ N$]
K	— combination transition
	— constant, equilibrium constant of conformational motions and other exchange processes
$\mathcal{L}$	angular moment vector operator of electron orbitals
ℓ	— distance
ℓ	— line of force
$\mathbf{L}$	torque, N [m^2 kg s^{-2}]
m	— mass of particles (electrons, protons, etc.) [kg]
	— eigenvalue of I_z of a nucleus
m_T	total magnetic quantum number for a spin system: eigenvalues of I_{zT}
M	mole
$\pm M$	mesomeric effect of atoms and functional groups
$\mathbf{M}$	macroscopic magnetization of a spin system, T
$\mathbf{M}_0$	macroscopic magnetization in field $\mathbf{B}_0$
$\boldsymbol{M}_e$	equilibrium macroscopic magnetization
M_w	components of $\boldsymbol{M}$
n	population difference of magnetic quantum states
n_e	equilibrium population difference
N	— Newton
	— nucleus
N	number of magnetic nuclei
$\mathcal{O}$	operator, representing any physical parameters
$\mathcal{P}$	permutation operator
p	— p-orbital
	— mole fraction fractional population of rotamers, etc.
$\mathbf{P}$	angular impulse moment [kg m^2 s^{-1}]
P	nuclear polarization
q	electric field gradient

Q	— nuclear quadrupole moment — various quantities and line distances in spectra of different spin systems, consisting of chemical shifts and coupling constants of the system — contributions of σ- and π-electrons to shielding component σ^P in quantum theoretical calculations of the shielding
R_w	element of matrix representation of transition moment
r	— polar coordinate — distance of atoms and groups, atomic radius, bond length, radius of ring currents, etc. [m], [Å]
R	rate of relaxations
R_1, R_2, R^{DD}, R^{SC}, R^{SA}, R^{SR}, R^Q, see at T_1, etc.	
s	s-orbital
s	symmetric spin state, basic function, transition, etc.
S	electron density around a nucleus
S	singlet electron state
t	time [s]
t_p	pulse time
t_{aq}	acquisition time
t_d	delay time
t_{pi}	period of repetitive pulses
t_{si}	period of repetitive pulse sequences
T	— forbidden transition — triplet electron state — temperature [°C, °K]
T_c	coalescence temperature
T_s	spin temperature
T_1, T_{1N}	spin-lattice relaxation times (for nucleus N), [s]
T_2	spin-spin relaxation times [s]
$T_{1\rho}$, $T_{2\rho}$	spin-lattice and spin-spin relaxation times in the rotating frame
T_2*	measurable spin-spin dephasing time shortened by field inhomogeneities
V	volume [m^3]
w	the common notation of Cartesian coordinates
W	— transition probability — signal width [Hz] for unresolved multiplet
Z	saturation factor
Z_0	saturation factor corresponding to maximal value of signal shape function
α, β	basic eigenfunction of spin corresponding $m = +1/2$ and $m = -1/2$
γ, γ_N	magnetogyric ratio of N nucleus [T^{-1} s^{-1}]
ɣ	$\gamma/2\pi$
δN	chemical shift of nucleus N ppm
δ_{ij}	Krönecker-delta
Δ	difference
ΔΔ	difference of differences
ΔB_0, ΔB_1, etc.	inhomogeneity in fields B_0, B_1, etc.

Δ_c, Δ_d, Δ_p	Fermi-contact, diamagnetic and pseudocontact shifts in SR-technique
ΔE	— differences in electronegativity — average electronic excitation energy
$\Delta G‡$	free enthalpy of activation [J M^{-1}]
Δ_i, Δ_{Ti}	SR's induced shift in the complex LS
$\Delta\nu$, $\Delta\nu XY$, $\Delta\delta$, $\Delta\delta XY$	chemical shift difference for X and Y nuclei, $\Delta\nu XY = \nu X - \nu Y$ [Hz], $\Delta\delta XY = \delta X - \delta Y$ [ppm]
$\Delta\nu_{1/2}$	half band width (band width at the half height of a signal)
$\Delta\nu_i$	spectral width
$\Delta\sigma$	anisotropy in σ, $\Delta\sigma = \sigma_{\parallel} - \sigma_{\perp}$
$\Delta\chi$	anisotropy in χ, $\Delta\chi = \chi_{\parallel} - \chi_{\perp}$
ϵ	sign-factor in CIDNP referring to reaction path
ζ	signal-to-noise ratio (''NMR sensitivity'', receptivity)
η	— viscosity, P [m^{-1} kg s^{-1}] — NOE-factor — assymmetry factor in $e^2qQ/\hbar$
θ, Θ	— dihedral or valence bond angle — polar coordinate — pulse angle
κ	rate constant of exchange processes
λ	— wavelength [m] — eigenvalues of various operators
$\boldsymbol{\mu}$	magnetic dipole moment, T
μ	— magnetic dipole moment in a given direction (e.g., μ_{B0} or μ_w, etc.) — sign-factor in CIDNP referring to electronic state of radical pair precursor — rationalized permeability
μ_0	permeability of vacuum
μ'	nonrationalized permeability
ν	frequency, Hz [s^{-1}]
ν_0	measuring frequency (operating frequency of an NMR spectrometer)
ν_1, ν_2,	first, second, etc. excitation frequencies
ν_N	Larmor (precession) frequency of a nucleus N
ν_c	carrier frequency
π	π-electrons
ξ	steric correction substituent factors [ppm]
ρ	— electron density [C], [A s] — substituent constant [ppm] — L/S ratio in SR measurements
σ	— shielding constant — LS/L ratio in SR measurements — σ-electrons — sign-factor in CIDNP, referring to interacting nuclei in a common radical, or in different radicals
σ_N	shielding constant of a nucleus N
$\boldsymbol{\sigma}$	shielding constant tensor
σ_{vw}	shielding constant tensor components
$\sigma_{\parallel}$, $\sigma_{\perp}$	shielding constant components parallel or perpendicular to

	the molecular axis
σ^d, σ^p	diamagnetic and paramagnetic contributions to the shielding constant
τ	— time [s] — lifetime of individual species in exchange processes, and generally of a spin state — time between pulses in pulse sequences — chemical shift (thau scale) [ppm]
τ_c	correlation time of molecular tumblings
τ_J	correlation time of molecular rotation at a given ν_{rot} frequency
τ_{SC}	correlation time of a spin state in scalar relaxation
τ_e	correlation time in nuclear-electron relaxation
φ	— product eigen function — polar coordinate — dihedral or valence bond angle
χ	magnetic susceptibility
ψ	basic function of spin systems, eigenfunction of electronic quantum states
ω	angular velocity [s^{-1}]
	ω_N, ω_0, ω_1, ω_2, etc. see at ν_N,..., etc.

REFERENCES

1. **Abragam, A.,** *The Principles of Nuclear Magnetism,* Oxford University Press, Oxford, 1961; (a) chap. 2, 8; (b) chap. 9.
2. **Abraham, R. J.,** *J. Chem. Soc. B,* 1022, 1969.
3. **Abraham, R. J. and Bernstein, H. J.,** *Can. J. Chem.,* 39, 905, 1961.
4. **Abraham, R. J., Bullock, E., and Mitra, S. S.,** *Can. J. Chem.,* 37, 1859, 1959.
5. **Abraham, R. J., Jackson, A. J., and Kenner, G. W.,** *J. Chem. Soc.,* 3468, 1961.
6. **Abraham, R. J., Lapper, R. D., Smith, K. M., and Unsworth, J. F.,** *J. Chem. Soc. Perkin Trans. 2,* 1004, 1974.
7. **Abraham, R. J., MacDonald, D. B., and Pepper, E. S.,** *J. Am. Chem. Soc.,* 90, 147, 1968.
8. **Abraham, R. J. and Pachler, K. G. R.,** *Mol. Phys.,* 7, 164, 1964.
9. **Abraham, R. J. and Siverns, T. M.,** *Org. Magn. Reson.,* 5, 253, 1973.
10. **Abraham, R. J. and Thomas, W. A.,** *J. Chem. Soc.,* 3739, 1964.
11. **Abraham, R. J. and Thomas, W. A.,** *J. Chem. Soc.,* 335, 1965.
12. **Abriktsen, P.,** *Acta Chem. Scand.,* 27, 3889, 1973.
13. **Abruscato, G. J., Ellis, P. D., and Tidwell, T. T.,** *Chem. Commun.,* 988, 1972.
14. **Adler, G. and Lichter, R. L.,** *J. Org. Chem.,* 39, 3547, 1974.
15. **Ahmad, N., Bhacca, N. S., Selbin, J., and Wander, J. D.,** *J. Am. Chem. Soc.,* 93, 2564, 1971.
16. **Aime, S. and Milone, L.,** Dynamic ^{13}C NMR spectroscopy of metal carbonyls, in *Progress in Nuclear Magnetic Resonance Spectroscopy,* Vol. 11, Emsley, J. W., Feeney, J., and Sutcliffe, L. H., Eds., Pergamon Press, Oxford, 1977, 183.
17. **Aito, Y., Matsuo, T., and Aso, C.,** *Bull. Chem. Soc. Jpn.,* 40, 130, 1967 .
18. **Albert, A. and Sergeant, E. P.,** *Ionization Constants of Acids and Bases,* Methuen, London, 1962.
19. **Alder, F. and Yu, F. C.,** *Phys. Rev.,* 81, 1967, 1951.
20. **Alei, M., Florin, A. E., Litchman, W. M., and O'Brien, J. F.,** *J. Phys. Chem.,* 75, 932, 1971.
21. **Alger, T. D., Grant, D. M., and Harris, R. K.,** *J. Phys. Chem.,* 76, 281, 1972.
22. **Alger, T. D., Grant, D. M., and Lyerla, J. R., Jr.,** *J. Phys. Chem.,* 75, 2539, 1971.
23. **Alger, T. D., Grant, D. M., and Paul, E. G.,** *J. Am. Chem. Soc.,* 88, 5397, 1966.
24. **Al-Iraqi, M. A., Al-Rawi, J. M. A., and Khuthier, A. H.,** *Org. Magn. Reson.,* 14, 161, 1980.
25. **Allerhand, A. and Cochran, D. W.,** *J. Am. Chem. Soc.,* 92, 4482, 1970.
26. **Allerhand, A. and Doddrell, D.,** *J. Am. Chem. Soc.,* 93, 2777, 1971.
27. **Allerhand, A., Doddrell, D., and Komoroski, R.,** *J. Chem. Phys.,* 55, 189, 1971.
28. **Allerhand, A. and Komoroski, R. A.,** *J. Am. Chem. Soc.,* 95, 8228, 1973.
29. **Allerhand, A. and Oldfield, E.,** *Biochemistry,* 12, 3428, 1973.
30. **Allingham, Y., Cookson, R. C., and Crabb, T. A.,** *Tetrahedron,* 24, 1989, 1968.
31. **Allred, A. L. and Rochow, E. G.,** *J. Am. Chem. Soc.,* 79, 5361, 1957.
32. **Aminova, R. M. and Samitov, Yu. Yu.,** *Zh. Strukt. Khim.,* 15, 607, 1974.
33. **Ammon, R. V. and Fischer, R. D.,** *Angew. Chem.,* 84, 737, 1972.
34. **Ammon, R. V. and Fischer, R. D.,** *Angew. Chem. Int., Ed. Engl.,* 11, 675, 1972.
35. **Anderson, C. B. and Sepp, D. T.,** *Chem. Ind. (London),* 2054, 1964.
36. **Anderson, W. A., Freeman, R., and Hill, H. D. W.,** *Pure Appl. Chem.,* 32, 27, 1972.
37. **Andersson, L. O., Mason, J., and van Bronswijk, W.,** *J. Chem. Soc. A,* 296, 1970.
38. **Andrew, E. R., Bradbury, A., and Eades, R. G.,** *Arch. Sci.,* 11, 223, 1958; *Nature,* 182, 1659, 1958.
39. **Anet, F. A. L.,** *J. Am. Chem. Soc.,* 84, 747, 1962.
40. **Anet, F. A. L.,** *J. Am. Chem. Soc.,* 84, 1053, 1962.
41. **Anet, F. A. L.,** *J. Am. Chem. Soc.,* 86, 458, 1964.
42. **Anet, F. A. L., Ahmad, M., and Hall, L. D.,** *Proc. Chem. Soc.,* 145, 1964.
43. **Anet, F. A. L. and Bock, L. A.,** *J. Am. Chem. Soc.,* 90, 7130, 1968.
44. **Anet, F. A. L. and Bourn, A. J. R.,** *J. Am. Chem. Soc.,* 87, 5250, 1965.
45. **Anet, F. A. L., Bradley, C. H., and Buchanan, G. W.,** *J. Am. Chem. Soc.,* 93, 258, 1971.
46. **Anet, F. A. L., Trepka, R. D., and Cram, D. J.,** *J. Am. Chem. Soc.,* 89, 357, 1967.
47. **Anet, F. A. L. and Yavari, I.,** *J. Org. Chem.,* 41, 3589, 1976.
48. **Anet, R.,** *Can. J. Chem.,* 40, 1249, 1962.
49. **Angyal, S. J.,** *Angew. Chem.,* 81, 172, 1969.
50. **Angyal, S. J. and Pickles, V. A.,** *Aust. J. Chem.,* 25, 1695, 1972.
51. *Annual Review of NMR Spectroscopy,* Mooney, E. F., Ed., Academic Press, London, Volumes 1-7, 1968.
52. *Applied Spectroscopy Reviews,* Brame, E. G., Jr., Ed., Marcel Dekker, New York, Volumes 1-7, 1968.
53. **Apsimon, J. W., Beierbeck, H., and Saunders, J. K.,** *Can. J. Chem.,* 53, 338, 1975.
54. **Apsimon, J. W., Craig, W. G., Demarco, P. V., Mathieson, D. W., Saunders, L., and Whalley, W. B.,** *Tetrahedron,* 23, 2339, 1967.

55. **Armarego, W. L. F.,** *J. Chem. Soc. B,* 191, 1966.
56. **Armitage, I. M., Hall, L. D., Marshall, A. G., and Werbelow, L. G.,** Determination of molecular configuration from lanthanide-induced proton NMR chemical shifts, in *Nuclear Magnetic Resonance Shift Reagents,* Sievers, R. F., Ed., Academic Press, New York, 1973, 313.
57. **Arnold J. T.,** *Phys. Rev.,* 102, 136, 1956.
58. **Arnold, J. T., Dharmatti, S. S., and Packard, M. E.,** *J. Chem. Phys.,* 19, 507, 1951.
59. **Arnold, J. T. and Packard, M. E.,** *J. Chem. Phys.,* 19, 1608, 1951.
60. **Asakawa, J., Kasai, R., Yamasaki, K., and Tanaka, O.,** *Tetrahedron,* 33, 1935, 1977.
61. **Aue, W. P., Bartholdi, E., and Ernst, R. R.,** *J. Chem. Phys.,* 64, 2229, 1976.
62. **Aue, W. P., Karhan, J., and Ernst, R. R.,** *J. Chem. Phys.,* 64, 4226, 1976.
63. **Axenrod, T., Mangiaracina, P., and Pregosin, P. S.,** *Helv. Chim. Acta,* 59, 1655, 1976.
64. **Axenrod, T., Pregosin, P. S., and Milne, G. W. A.,** *Chem. Commun.,* 702, 1968.
65. **Axenrod, T., Pregosin, P. S., and Milne, G. W. A.,** *Tetrahedron Lett.,* 5293, 1968.
66. **Axenrod, T., Pregosin, P. S., Wieder, M. J., Becker, E. D., Bradley, R. B., and Milne, G. W. A.,** *J. Am. Chem. Soc.,* 93, 6536, 1971.
67. **Axenrod, T., Pregosin, P. S., Wieder, M. J., and Milne, G. W. A.,** *J. Am. Chem. Soc.,* 91, 3681, 1969.
68. **Axenrod, T. and Wieder, M. J.,** *J. Am. Chem. Soc.,* 93, 3541, 1971.
69. **Axenrod, T. and Wieder, M. J.,** *Org. Magn. Reson.,* 8, 350, 1976.
70. **Axenrod, T., Wieder, M. J., and Milne, G. W. A.,** *Tetrahedron Lett.,* 401, 1969.
71. **Bachmann, P., Aue, W. P., Müller, L., and Ernst, R. R.,** *J. Magn. Reson.,* 28, 29, 1977.
72. **Bacon, M. R. and Maciel, G. E.,** *J. Am. Chem. Soc.,* 95, 2413, 1973.
73. **Baldeschwieler, J. D. and Randall, E. W.,** *Chem. Rev.,* 63, 81, 1963.
74. **Bak, B., Dambmann, C., Nicolaisen, F., Pedersen, E. J., and Bhacca, N. S.,** *J. Mol. Spectrosc.,* 26, 78, 1968.
75. **Baker, E. B.,** *J. Chem. Phys.,* 37, 911, 1962.
76. **Baker, E. B. and Popov, A.I.,** *J. Phys. Chem.,* 76, 2403, 1972.
77. **Banks, R. E., Barlow, M. G., Davies, W. D., Haszeldine, R. N., Mullen, K., and Taylor, D. R.,** *Tetrahedron Lett.,* 3909, 1968.
78. **Banks, R. E., Barlow, M. G., and Mullen, K.,** *J. Chem. Soc. C,* 1331, 1969.
79. **Banks, R. E., Braithwaite, A., Haszeldine, R. N., and Taylor, D. R.,** *J. Chem. Soc. C,* 2593, 1968.
80. **Banwell, C. N. and Sheppard, N.,** *Mol. Phys.,* 3, 351, 1960.
81. **Barbarella, G. and Dembech, P.,** *Org. Magn. Reson.,* 13, 282, 1980.
82. **Barbarella, G., Dembech, P., Garbesi, A., and Fava, A.,** *Org. Magn. Reson.,* 8, 108, 469, 1976.
83. **Barfield, M.,** *J. Chem. Phys.,* 44, 1836, 1966.
84. **Barfield, M. and Gearhart, H. L.,** *Mol. Phys.,* 27, 899, 1974.
85. **Barfield, M. and Grant, D. M.,** *J. Am. Chem. Soc.,* 85, 1899, 1963.
86. **Bargon, J. and Fischer, H.,** *Z. Naturforsch.,* 22a, 1556, 1967.
87. **Barlow, M. G.,** *Chem. Commun.,* 703, 1966.
88. **Bartholdi, E. and Ernst, R. R.,** *J. Magn. Reson.,* 11, 9, 1973.
89. **Bartuska, V. J. and Maciel, G. E.,** *J. Magn. Reson.,* 7, 36, 1972.
90. **Batchelor, J. G.,** *J. Am. Chem. Soc.,* 97, 3410, 1975.
91. **Batchelor, J. G., Prestegard, J. H., Cushley, R. J., and Lipsky, S. R.,** *J. Am. Chem. Soc.,* 95, 6358, 1973.
92. **Battiste, D. R. and Traynham, J. G.,** *J. Org. Chem.,* 40, 1239, 1975.
93. **Bauld, N. L. and Rim, Y. S.,** *J. Org. Chem.,* 33, 1303, 1968.
94. **Beauté, C., Wolkowski, Z. W., and Thoai, N.,** *Chem. Commun.,* 700, 1971.
95. **Becconsall, J. K. and Hampson, P.,** *J. Mol. Phys.,* 10, 21, 1965.
96. **Becconsall, J. K. and Jones, R. A. Y.,** *Tetrahedron Lett.,* 1103, 1962.
97. **Becconsall, J. K., Jones, R. A. Y., and McKenna, J.,** *J. Chem. Soc.,* 1726, 1965.
98. **Beck, W., Becker, W., Nöth, H., and Wrackmeyer, B.,** *Chem. Ber.,* 105, 2883, 1972.
99. **Becker, E. D.,** *High Resolution NMR Theory and Chemical Applications,* Academic Press, New York, London, 1969.

99a. **Becker, E. D.,** *High Resolution NMR,* Appendix A, Academic Press, New York, 1969.

100. **Becker, E. D.,** Nuclei other than hydrogen, in *A Review of Some Nuclear Properties and a Discussion of Their Relaxation Mechanisms,* Axenrod, T. and Webb, G. A., Eds., Wiley-Interscience, New York, 1974, 1.
101. **Becker, E. D. and Bradley, R. B.,** *J. Chem. Phys.,* 31, 1413, 1959.
102. **Becker, E. D., Bradley, R. B., and Axenrod, T.,** *J. Magn. Reson.,* 4, 136, 1971.
103. **Becker, E. D., Bradley, R. B., and Watson, C. J.,** *J. Am. Chem. Soc.,* 83, 3743, 1961.
104. **Becker, E. D. and Farrar, T. C.,** *Science,* 178, 316, 1972.
105. **Becker, E. D., Ferretti, J. A., and Farrar, T. C.,** *J. Am. Chem. Soc.,* 91, 7784, 1969.

106. **Becker, E. D., Liddel, U., and Shoolery, J. N.,** *J. Mol. Spectrosc.*, 2, 1, 1958.
107. **Becker, G., Lutther, W., and Schrumpf, G.,** *Angew. Chem. Int. Ed. Engl.*, 12, 339, 1973.
108. **Beckett, A. H., Taylor, J. F., Casy, A. F., and Hassan, M. M. A.,** *J. Pharm. Pharmacol.*, 20, 754, 1968.
109. **Beierbeck, H., Saunders, J. K., and ApSimon, J. W.,** *Can. J. Chem.*, 55, 2813, 1977.
110. **Bell, C. L. and Danyluk, S. S.,** *J. Am. Chem. Soc.*, 88, 2344, 1966.
111. **Bell, R. A., Chan, C. L., and Sayer, B. G.,** *Chem. Commun.*, 67, 1972.
112. **Bell, R. A. and Saunders, J. K.,** *Can. J. Chem.*, 48, 1114, 1970.
113. **Bentrude. W. G. and Tan, H. W.,** *J. Am. Chem. Soc.*. 98, 1850. 1976.
114. **Benz, F. W., Feeney, J., and Roberts, G. C. K.,** *J. Magn. Reson.*, 8, 114, 1972.
115. **Berger, S. and Rieker, A.,** *Tetrahedron*, 28, 3123, 1972.
116. **Berkeley, P. J., Jr. and Hanna, M. W.,** *J. Phys. Chem.*, 67, 846, 1963.
117. **Berkeley, P. J. and Hanna, M. W.,** *J. Am. Chem. Soc.*, 86, 2990, 1964.
118. **Berlin, K. D. and Rengaraju, S.,** *J. Org. Chem.*, 36, 2912, 1971.
119. **Bernáth, G., Göndös, Gy., Kovács, K., and Sohár, P.,** *Tetrahedron*, 29, 981, 1973.
120. **Bernáth, G., Sohár, P., Láng, K. L., Tornyai, I., and Kovács, Ö. K. J.,** *Acta Chim. Acad. Sci. Hung.*, 64, 81, 1970.
121. **Bernet, W. A.,** *J. Chem. Educ.*, 44, 17, 1967.
122. **Bernheim, R. A. and Lavery, B. J.,** *J. Am. Chem. Soc.*, 89, 1279, 1967.
123. **Bernstein, H. J., Pople, J. A., and Schneider, W. G.,** *Can. J. Chem.*, 35, 65, 1957
124. **Bernstein, H. J., Schneider, W. G., and Pople, J. A.,** *Proc. R. Soc. London Ser. A*, 236, 515, 1956.
125. **Bernstein, H. J. and Sheppard, N.,** *J. Chem. Phys.*, 37, 3012, 1962.
126. **Bertrand, R. D., Grant, D. M., Allred, E. L., Hinshaw, J. C., and Strong, A. B.,** *J. Am. Chem. Soc.*, 94, 997, 1972.
127. **Besserre, D. and Coffi-Nketsia, S.,** *Org. Magn. Reson.*, 13, 235, 1980.
128. **Beugelmans, R., Shapiro, R. H., Durham, L. J., Williams, D. H., Budzikiewicz, H., and Djerassi, C.,** *J. Am. Chem. Soc.*, 86, 2832, 1964.
129. **Bhacca, N. S., Giannini, D. D., Jankowski, W. S., and Wolff, M. E.,** *J. Am. Chem. Soc.*, 95, 8421, 1973.
130. **Bhacca, N. S., Horton, D., and Paulsen, H.,** *J. Org. Chem.*, 33, 2484, 1968.
131. **Bhacca, N. S. and Williams, D. H.,** *Applications of NMR Spectroscopy in Organic Chemistry. Illustrations from the Steroid Field*, Holden-Day, San Francisco, 1964.
132. **Bidló-Iglóy, M., Méhesfalvi-Vajna, Zs., and Nagy, J.,** *Acta Chim. Acad. Sci. Hung.*, 79, 1, 1973.
133. **Biellmann, J. F. and Callot, H.,** *Bull. Soc. Chim. (France)*, 397, 1967.
134. **Binsch, G.,** The study of intramolecular rate processes by dynamic nuclear magnetic resonance, in *Topics in Stereochemistry*, Vol. 3, Eliel, E. L. and Allinger, N. L., Eds., Wiley-Interscience, New York, 1968, 97.
135. **Binsch, G., Lambert, J. B., Roberts, B. W., and Roberts, J. D.,** *J. Am. Chem. Soc.*, 86, 5564, 1964.
136. **Binsch, G. and Roberts, J. D.,** *J. Am. Chem. Soc.*, 87, 5157, 1965.
137. **Binst, G. and Tourwe, D.,** *Heterocycles*, 1, 257, 1973.
138. **Birchall, T. and Jolly, W. L.,** *J. Am. Chem. Soc.*, 87, 3007, 1965.
139. **Bjorgo, J., Boyd, D. R., Watson, C. G., and Jennings, W. B.,** *J. Chem. Soc. Perkin Trans. 2*, 757, 1974.
140. **Black, P. J. and Heffernan, M. L.,** *Aust. J. Chem.*, 16, 1051, 1963.
141. **Black, P. J. and Heffernan, M. L.,** *Aust. J. Chem.*, 18, 353, 1965.
142. **Blackman, R. B. and Tukey, J. W.,** *The Measurement of Power Spectra*, Dover, New York, 1958.
143. **Bleaney, B.,** *J. Magn. Reson.*, 8, 91, 1972.
144. **Bleaney, B., Dobson, C. M., Levine, B. A., Martin, R. B., Williams, R. J. P., and Xavier, A. V.,** *Chem. Commun.*, 791, 1972.
145. **Blizzard, A. C. and Santry, D. P.,** *Chem. Commun.*, 1085, 1970.
146. **Blizzard, A. C. and Santry, D. P.,** *J. Chem. Phys.*, 55, 950, 1971.
147. **Blizzard, A. C. and Santry, D. P.,** *J. Chem. Phys.*, 58, 4714, 1973.
148. **Bloch, F.,** *Phys. Rev.*, 70, 460, 1946.
149. **Bloch, F.,** *Phys. Rev.*, 94, 496, 1954.
150. **Bloch, F., Hansen, W. W., and Packard, M.,** *Phys. Rev.*, 69, 127, 1946.
151. **Bloch, F., Hansen, W. W., and Packard, M.,** *Phys. Rev.*, 70, 474, 1946.
152. **Bloch, F. and Siegert, A.,** *Phys. Rev.*, 57, 522, 1940.
153. **Block, R. E. and Maxwell, G. P.,** *J. Magn. Reson.*, 14, 329, 1974.
154. **Bloembergen, N., Purcell, E. M., and Pound, R. V.,** *Phys. Rev.*, 73, 679, 1948.
155. **Bloom, A. L. and Shoolery, J. N.,** *Phys. Rev.*, 151, 102, 1965.
156. **Blunt, J. W. and Munro, M. H. G.,** *Org. Magn. Reson.*, 13, 26, 1980.
157. **Blunt, J. W. and Stothers, J. B.,** *Org. Magn. Reson.*, 9, 439, 1977.

158. **Boča, R., Pelikan, P., Valko, L., and Miertus, S.,** *J. Chem. Phys.*, 11, 229, 1975.
159. **Bock, K. and Pedersen, C.,** *J. Chem. Soc. Perkin Trans. 2*, 293, 1974.
160. **Bock, K. and Wiebe, L.,** *Acta Chem. Scand.*, 27, 2676, 1973.
161. **Bodenhausen, G., Freeman, R., Morris, G., and Turner, D. L.,** *J. Magn. Reson.*, 31, 75, 1978.
162. **Bodenhausen, G., Freeman, R., Niedermeyer, R., and Turner, D. L.,** *J. Magn. Reson.*, 24, 291, 1976.
163. **Bodenhausen, G., Freeman, R., Niedermeyer, R., and Turner, D. L.,** *J. Magn. Reson.*, 26, 133, 1977.
164. **Bodenhausen, G., Freeman, R., and Turner, D. L.,** *J. Chem. Phys.*, 65, 839, 1976.
165. **Boekelheide, V. and Phillips, J. B.,** *J. Am. Chem. Soc.*, 89, 1695, 1967.
166. **Bohlmann, F. and Zeisberg, R.,** *Chem. Ber.*, 108, 1043, 1975.
167. **Bohman, O. and Allenmark, S.,** *Acta Chem. Scand.*, 22, 2716, 1968.
168. **Booth, H.,** Applications of ^{1}H NMR spectroscopy to the conformational analysis of cyclic compounds, in *Progress in Nuclear Magnetic Resonance Spectroscopy*, Vol. 5, Emsley, J. W., Feeney, J., and Sutcliffe, L. H., Eds., Pergamon Press, London, 1969, 149.
169. **Booth, H. and Little, J. H.,** *Tetrahedron*, 23, 291, 1967.
170. **Booth, G. E. and Ouelette, R. J.,** *J. Org. Chem.*, 31, 544, 1966.
171. **Bordás, B., Sohár, P., Matolcsy, Gy., and Berencsi, P.,** *J. Org. Chem.*, 37, 1727, 1972.
172. **Borgen, G.,** *Acta Chem. Scand.*, 26, 1740, 1972.
173. **Bothner-By, A. A. and Moser, E.,** *J. Am. Chem. Soc.*, 90, 2347, 1968.
174. **Bothner-By, A. A. and Naar-Colin, C.,** *J. Am. Chem. Soc.*, 84, 743, 1962.
175. **Bottini, A. T. and O'Rell, M. K.,** *Tetrahedron Lett.*, 423, 429, 1967.
176. **Bottini, A. T. and Roberts, J. D.,** *J. Am. Chem. Soc.*, 80, 5203, 1958.
177. **Bourn, A. J . R. and Randall, E. W.,** *Mol. Phys.*, 8, 567, 1964.
178. **Bovey, F. A.,** *NMR Data Tables for Organic Compounds*, Interscience, New York, 1957.
179. **Bovey, F. A.,** *Nuclear Magnetic Resonance Spectroscopy*, Academic Press, New York, 1969.
180. **Bovey, F. A.,** *High Resolution Nucler Magnetic Resonance of Macromolecules*, Academic Press, New York, 1972.
181. **Bovey, F. A., Hood, F. P., III, Anderson, E. W., and Kornegay, R. L.,** *J. Chem. Phys.*, 41, 2041, 1964.
182. **Bovey, F. A. and Tiers, G. V. D.,** *J. Am. Chem. Soc.*, 81, 2870, 1959.
183. **Bowie, J. H., Cameron, D. W., Schütz, P. E., Williams, D. H., and Bhacca, N. S.,** *Tetrahedron*, 22, 1771, 1966.
184. **Bracewell, R.,** *The Fourier Transform and Its Physical Applications*, McGraw-Hill, New York, 1955.
185. **Bradbury, E. M. and Crane-Robinson, C.,** *Nature (London)*, 220, 1079, 1968.
186. **Bradley, C. H., Hawkes, G. E., Randall, E. W., and Roberts, J. D.,** *J. Am. Chem. Soc.*, 97, 1958, 1975.
187. **Bramwell, M. R. and Randall, E. W.,** *Chem. Commun.*, 250, 1969.
188. **Braun, S. and Frey, G.,** *Org. Magn. Reson.*, 7, 194, 1975.
189. **Brederode, H. and Huysmans, G. B.,** *Tetrahedron Lett.*, 1695, 1971.
190. **Breitmaier, E., Jung, G., and Voelter, W.,** *Angew. Chem.*, 83, 659, 1971.
191. **Breitmaier, E., Jung, G., and Voelter, W.,** *Angew. Chem. Int. Ed. Engl.*, 10, 73, 1971.
192. **Breitmaier, E., Jung, G., and Voelter, W.,** *Chimia*, 26, 136, 1972.
193. **Breitmaier, E., Jung. G., Voelter, W., and Pohl, L.,** *Tetrahedron*, 29, 2485, 1973.
194. **Breitmaier, E. and Spohn, K. H.,** *Tetrahedron*, 29, 1145, 1973.
195. **Breitmaier, E. and Voelter, W.,** *Tetrahedron*, 29, 227, 1973.
196. **Breitmaier, E. and Voelter, W.,** *^{13}C NMR Spectroscopy*, Verlag Chemie, Weinheim, 1974, (a) pp. 195-202, (b) p. 98, (c) pp. 268-274, (d) p. 83.
197. **Breitmaier, E., Voelter, W., Jung, G., and Tänzer, C.,** *Chem. Ber.*, 104, 1147, 1971.
198. **Breslow, R. and Ryan, G.,** *J. Am. Chem. Soc.*, 89, 3073, 1967.
199. **Brewer, J. P. N., Heaney, H., and Marples, B. A.,** *Chem. Commun.*, 27, 1967.
200. **Brey, W. S. and Ramey, K. C.,** *J. Chem. Phys.*, 39, 844, 1963.
201. **Briggs, J. M., Farnell, L. F., and Randall, E. W.,** *Chem. Commun.*, 680, 1971.
202. **Briggs, J., Frost, G. H., Hart, F. A., Moss, G. P., and Staniforth, M. L.,** *Chem. Commun.*, 749, 1970.
203. **Briggs, J., Hart, F. A., and Moss, G. P.,** *Chem. Commun.*, 1506, 1970.
204. **Briggs, J., Hart, F. A., Moss, G. P., and Randall, E. W.,** *Chem. Commun.*, 364, 1971.
205. **Briggs, J. M., Moss, G. P., Randall, E. W., and Sales, K. D.,** *Chem. Commun.*, 1180, 1972.
206. **Briggs, J. M., Rahkamaa, E., and Randall, E. W.,** *J. Magn. Reson.*, 12, 40, 1973.
207. **Briguet, A., Duplan, J. C., and Delmau, J.,** *Mol. Phys.*, 29, 837, 1975.
208. **Brimacombe, J. S. and Tucker, L. C. N.,** *Carbohydrate Res.*, 5, 36, 1967.
209. **Brois, S. J.,** *J. Org. Chem.*, 27, 3532, 1962.
210. **Brois, S. J.,** *J. Am. Chem. Soc.*, 90, 506, 508, 1968.
211. **Brois, S. J. and Beardsley, G. P.,** *Tetrahedron Lett.*, 5113, 1966.

212. **Brookhart, M., Levy, G. C., and Winstein, S.,** *J. Am. Chem. Soc.*, 89, 1735, 1967.
213. **Brouant, P., Limouzin, Y., and Maire, J. C.,** *Helv. Chim. Acta,* 56, 2057, 1973.
214. **Brown, R. J. C., Gutowsky, H. S., and Shimomura, K.,** *J. Chem. Phys.*, 38, 76, 1963.
215. **Browne, D. T., Kenyon, G. L., Packer, E. L., Sternlicht, H., and Wilson, D. M.,** *J. Am. Chem. Soc.*, 95, 1316, 1973; *Biochem. Biophys. Res. Commun.*, 50, 42, 1973.
216. **Brownstein, S., Dunogues, J., Lindsay, D., and Ingold, K. U.,** *J. Am. Chem. Soc.*, 99, 2073, 1977.
217. **Bruce, J. M. and Knowles, P.,** *Proc. Chem. Soc.*, 294, 1964.
218. (a) Bruker: *High Power Pulsed NMR Applications,* Bruker Analytic GMBH, Karlsruhe, Rheinstettin, West-Germany; (b) Bruker WM Series: *High Field Multinuclear NMR,* Bruker Analytic GMBH, Karlsruhe, Rheinstettin, West-Germany.
219. Bruker: *High Resolution NMR in Solids by Magic Angle Spinning,* Bruker Analytic GMBH, Karlsruhe, Rheinstettin, West Germany.
220. **Bucci, P.,** *J. Am. Chem. Soc.*, 90, 252, 1968.
221. **Buchanan, G. W. and Dawson, B. A.,** *Can. J. Chem.*, 54, 790, 1976.
222. **Buchanan, G. W. and Dawson, B. A.,** *Org. Magn. Reson.*, 13, 293, 1980.
223. **Buchanan, G. W. and Stothers, J. B.,** *Can. J. Chem.*, 47, 3605, 1969.
224. **Buckingham, A. D.,** *Can. J. Chem.*, 38, 300, 1960.
225. **Buckingham, A. D. and McLauchan, K. A.,** *Proc. Chem. Soc.*, 144, 1963.
226. **Buckingham, A. D., Schaefer, T., and Schneider, W. G.,** *J. Chem. Phys.*, 32, 1227, 1960.
227. **Budhram, R. S. and Uff, B. C.,** *Org. Magn. Reson.*, 13, 89, 1980.
228. **Bulman, M. J.,** *Tetrahedron,* 25, 1433, 1969.
229. **Bulusu, S., Autera, J. R., and Axenrod, T.,** *Chem. Commun.*, 602, 1973.
230. **Burke, J. J. and Lauterbur, P. C.,** *J. Am. Chem. Soc.*, 86, 1870, 1964.
231. **Burton, R., Hall, L. D., and Steiner, P. R.,** *Can. J. Chem.*, 49, 588, 1971.
232. **Bystrov, V. F.,** Spin-spin coupling and the conformational states of peptide systems, in *Progress in Nuclear Magnetic Resonance Spectroscopy,* Vol. 10, Emsley, J. W., Feeney, J., and Sutcliffe, L. H., Eds., Pergamon Press, Oxford, 1976, 41.
233. **Bystrov, V. F., Gavrilov, Y. D., and Solkan, V. N.,** *J. Magn. Reson.*, 19, 123, 1975.
234. **Bystrov, V. F., Ivanov, V. T., Portnova, S. L., Balashova, T. A., and Ovchinnikov, Yu. A.,** *Tetrahedron,* 29, 873, 1973.
235. **Bystrov, V. F., Portnova, S. L., Csetlin, V. I., Ivanov, V. T., and Ovcsinnikov, V. A.,** *Tetrahedron,* 25, 493, 1969.
236. **Bystrov, V. F. and Stepanyants, A. U.,** *J. Mol. Spectrosc.*, 21, 241, 1966.
237. **Caddy, B., Martin-Smith, M., Norris, R. K., Reid, S. T., and Sternhell, S.,** *Aust. J. Chem.*, 21, 1853, 1968.
238. **Calder, I. C. and Sondheimer, F.,** *Chem. Commun.*, 904, 1966.
239. **Cameron, D. W., Kingston, D. G. J., Sheppard, N., and Todd, J.,** *J. Chem. Soc.*, 98, 1964.
240. **Campaigne, E., Chamberlain, N. F., and Edwards, B. E.,** *J. Org. Chem.*, 27, 135, 1962.
241. **Campbell, I. D., Dobson, C. M., and Williams, R. J. P.,** *Proc. R. Soc. London Ser. B,* 189, 485, 1975.
242. **Campbell, I. D. and Freeman, R.,** *J. Magn. Reson.*, 11, 143, 1973.
243. **Canet, C., Goulon-Ginet, C., and Marchal, J. P.,** *J. Magn. Reson.*, 22, 539, 1976 (with an erratum in 25, 397, 1977).
244. **Cantacuzène, J., Jantzen, R., Tordeux, M., and Chachaty, C.,** *Org. Magn. Reson.*, 7, 407, 1975.
245. **Caputo, J. F. and Martin, A. R.,** *Tetrahedron Lett.*, 4547, 1971.
246. **Carman, C. J., Tarpley, A. R., and Goldstein, J. H.,** *J. Am. Chem. Soc.*, 93, 2864, 1971.
247. **Carr, H. Y. and Purcell, F. M.,** *Phys. Rev.*, 94, 630, 1954.
248. **Carrington, A. and McLachlan, A. D.,** *Introduction to Magnetic Resonance,* Harper & Row, New York, 1967; (a) p. 187; (b) chap. 11.
249. **Cartledge, F. K. and Riedel, K. H.,** *J. Organomet. Chem.*, 34, 11, 1972.
250. **Caspi, E., Wittstruck, T. A., and Piatak, D. M.,** *J. Org. Chem.*, 27, 3183, 1962.
251. **Castellano, S., Sun, C., and Kostelnik, R. J.,** *J. Chem. Phys.*, 46, 327, 1967.
252. **Casu, B., Reggiani, M., Gallo, G. G., and Vigevani, A.,** *Tetrahedron,* 22, 3061, 1966.
253. **Casu, B., Reggiani, M., Gallo, G. G., and Vigevani, A.,** *Tetrahedron,* 24, 803, 1968.
254. **Casy, A. F.,** *PMR Spectroscopy in Medicinal and Biological Chemistry,* Academic Press, New York, 1971.
255. **Carey, P. R., Kroto, H. W., and Turpin, M. A.,** *Chem. Commun.*, 188, 1969.
256. **Caughey, W. S. and Iber, P. K.,** *J. Org. Chem.*, 28, 269, 1963.
257. **Cavalli, L.,** Fluorine-19 NMR spectroscopy, in *Annual Reports on NMR Spectroscopy,* Vol. 6B, Mooney, E. F., Ed., Academic Press, New York, 1976, 43.
258. **Cavanaugh, J. R.,** *J. Chem. Phys.*, 39, 2378, 1963.
259. **Cavanaugh, J. R. and Dailey, B. P.,** *J. Chem. Phys.*, 34, 1099, 1961.
260. **Celotti, J. C., Reisse, J., and Chiurdoglu, G.,** *Tetrahedron,* 22, 2249, 1966.
261. **Chadwick, D. J. and Williams, D. H.,** *J. Chem. Soc. Perkin Trans. 2,* 1202, 1974.

262. **Chalmers, A. A., Pachler, K. G. R., and Wessels, P. L.,** *J. Magn. Reson.*, 15, 419, 1974.
263. **Chambers, R. D., Hole, M., Musgrave, W. K. R., Storey, R. A., and Iddon, B.,** *J. Chem. Soc. C*, 2331, 1966.
264. **Chambers, R. D. and Palmer, A. J.,** *Tetrahedron Lett.*, 2799, 1968.
265. **Champeney, D. C.,** *Fourier Transforms and Their Physical Applications*, Academic Press, New York, 1973.
266. **Chapman, A. C., Homer, J., Mowthorpe, D. J., and Jones, R. T.,** *Chem. Commun.*, 121, 1965.
267. **Chapman, D. and Magnus, P. D.,** *Introduction to Practical High Resolution Nuclear Magnetic Resonance Spectroscopy*, Academic Press, New York, 1966.
268. **Chapman, O. L. and King, R. W.,** *J. Am. Chem. Soc.*, 86, 1256, 1964.
269. **Cheney, B. V. and Grant, D. M.,** *J. Am. Chem. Soc.*, 89, 5319, 1967.
270. **Cherry, P. C., Cottrell, W. R. T., Meakins, G. D., and Richards, E. E.,** *J. Chem. Soc. C*, 459, 1968.
271. **Chew, K. F., Derbyshire, W., Logan, N., Norbury, A. H., and Sinha, A. I. P.,** *Chem. Commun.*, 1708, 1970.
272. **Chew, K. F., Healy, M. A., Khalil, M. J., Logan, N., and Derbyshire, W.,** *J. Chem. Soc. (Dalton)*, 1315, 1975.
273. **Chiang, Y. and Whipple, E. B.,** *J. Am. Chem. Soc.*, 85, 2763, 1963.
274. **Chow, Y. L.,** *Angew. Chem. Int. Ed. Educ.*, 6, 75, 1968.
275. **Christ, H. A.,** *Helv. Phys. Acta*, 33, 572, 1960.
276. **Christ, H. A. and Diehl, P.,** *Helv. Phys. Acta*, 36, 170, 1963.
277. **Christ, H. A., Diehl, P., Schneider, H., and Dahn, H.,** *Helv. Chim. Acta*, 44, 865, 1961.
278. **Christl, M., Reich, H. J., and Roberts, J. D.,** *J. Am. Chem. Soc.*, 93, 3463, 1971.
279. **Christl, M., Warren, J. P., Hawkins, B. L., and Roberts, J. D.,** *J. Am. Chem. Soc.*, 95, 4392, 1973.
280. **Claret, P. A. and Osborne, A. G.,** *Org. Magn. Reson.*, 8, 147, 1976.
281. **Clerc, T. and Pretsch, E.,** *Kernresonanzspektroskopie*, Akademische Verlagsgesellschaft, Frankfurt am Main, 1970.
282. **Closs, G. L.,** *J. Am. Chem. Soc.*, 91, 4552, 1969.
283. **Closs, G. L.,** *23rd International Congress of Pure and Applied Chemistry*, Vol. 4, Boston, Butterworths, London, 1971, 19.
284. **Closs, G. L.,** Chemically induced dynamic nuclear polarization, in *Advances in Magnetic Resonance*, Vol. 7, Waugh, J. S., Ed., Academic Press, New York, 1974, 157.
285. **Closs, G. L. and Closs, L. E.,** *J. Am. Chem. Soc.*, 85, 2022, 1963.
286. **Closs, G. L. and Closs, L. E.,** *J. Am. Chem. Soc.*, 91, 4549, 1969.
287. **Closs, G. L. and Moss, R. A.,** *J. Am. Chem. Soc.*, 86, 4042, 1964.
288. **Closs, G. L. and Trifunac, A. D.,** *J. Am. Chem. Soc.*, 91, 4454, 1969.
289. **Cohen, A. D., Sheppard, N., and Turner, J. J.,** *Proc. Chem. Soc.*, 118, 1958.
290. **Coles, B. A.,** *J. Natl. Cancer Inst.*, 57, 389, 1976.
291. **Colli, H. N., Gold, V., and Pearson, J. E.,** *Chem. Commun.*, 408, 1973.
292. **Connes, J.,** *Rev. Opt.*, 40, 45, 116, 171, 231, 1961.
293. **Connor, T. M. and Reid, C.,** *J. Mol. Spectrosc.*, 7, 32, 1961.
294. **Conti, F., Segre, A., Pini, P., and Porri, L.,** *Polymer*, 15, 5, 1974.
295. **Cooley, J. W. and Tukey, J. W.,** *Math. Comput.*, 19, 297, 1965.
296. **Cooper, R. A., Lichter, R. L., and Roberts, J. D.,** *J. Am. Chem. Soc.*, 95, 3724, 1973.
297. **Cope, F. W.,** *Biophys. J.*, 9, 303, 1969.
298. **Cope, F. W.,** *Physiol. Chem. Phys.*, 10, 535, 541, 547, 1978.
299. **Cope, F. W. and Damadian, R.,** *Physiol. Chem. Phys.*, 11, 143, 1979.
300. **Corio, P. L.,** *Structure of High Resolution NMR Spectra*, Academic Press, New York, 1966.
301. **Corio, P. L.,** *Chem. Rev.*, 60, 363, 1960.
302. **Corio, P. L. and Dailey, B. P.,** *J. Am. Chem. Soc.*, 78, 3043, 1956.
303. **Cornwell, C. D.,** *J. Chem. Phys.*, 44, 874, 1966.
304. **Cottam, G. L., Vasek, A., and Lusted, D.,** *Res. Commun. Chem. Pathol. Pharmacol.*, 4, 495, 1972.
305. **Cowley, A. H. and Schweiger, J. R.,** *J. Am. Chem. Soc.*, 95, 4179, 1973.
306. **Cox, P. F.,** *J. Am. Chem. Soc.*, 85, 380, 1963.
307. **Cox, R. H. and Bothner-By, A. A.,** *J. Phys. Chem.*, 72, 1642, 1968.
308. **Coxon, B.,** *Tetrahedron*, 22, 2281, 1966.
309. **Coxon, B.,** *Ann. N.Y. Acad. Sci.*, 222, 953, 1973.
310. **Crabb, T. A.,** Nuclear magnetic resonance of alkaloids, in *Annual Reports on NMR Spectroscopy*, Vol. 6A, Mooney, E. F., Ed., Academic Press, New York, 1975, 250.
311. **Crabb, T. A.,** Nuclear magnetic resonance of alkaloids, in *Annual Reports on NMR Spectroscopy*, Vol. 8, Webb, G. A., Academic Press, New York, 1978, 2.
312. **Cramer, R. E. and Dubois, R.,** *Chem. Commun.*, 936, 1973.
313. **Cramer, R. E., Dubois, R., and Seff, K.,** *J. Am. Chem. Soc.*, 96, 4125, 1974.

314. **Cramer, R. E. and Seff, K.,** *Chem. Commun.*, 400, 1972.
315. **Creagh, L. T. and Truitt, P.,** *J. Org. Chem.*, 33, 2956, 1968.
316. **Crecely, K. M., Crecely, R. W., and Goldstein, J. H.,** *J. Phys. Chem.*, 74, 2680, 1970.
317. **Crecely, K. M., Crecely, R. W., and Goldstein, J. H.,** *J. Mol. Spectrosc.*, 37, 252, 1971.
318. **Crecely, K. M., Watts, V. S., and Goldstein, J. H.,** *J. Mol. Spectrosc.*, 30, 184, 1969.
319. **Cremer, S. and Srinivasan, R.,** *Tetrahedron Lett.*, 24, 1960.
320. **Crump, D. R., Sanders, J. K. M., and Williams, D. H.,** *Tetrahedron Lett.*, 4949, 1970.
321. **Crutchfield, M. M., Dungan, C. H., Letcher, J. H., Mark, V., and Van Wazer, J. R.,** *Topics in Phosphorus Chemistry,* Vol. 5, Grayson, M. and Griffith, E. J., Eds., Interscience, New York, 1967.
322. **Cudby, M. E. A. and Willis, H. A.,** The nuclear magnetic resonance spectra of polymers, in *Annual Reports on NMR Spectroscopy,* Vol. 4, Mooney, E. F., Ed., Academic Press, New York, 1971, 363.
323. **Cushley, R. J., Codington, J. F., and Fox, J. J.,** *Carbohydr. Res.*, 5, 31, 1967.
324. **Cushley, R. J., Naugler, D., and Ortig, C.,** *Can. J. Chem.*, 53, 3419, 1975.
325. **Cushley, R. J., Watanabe, K. A., and Fox, J. J.,** *J. Am. Chem. Soc.*, 89, 394, 1967.
326. **Dahlquist, K. I. and Forsén, S.,** *J. Phys. Chem.*, 69, 4062, 1965.
327. **Dalling, D. K. and Grant, D. M.,** *J. Am. Chem. Soc.*, 89, 6612, 1967.
328. **Dalling, D. K. and Grant, D. M.,** *J. Am. Chem. Soc.*, 94, 5318, 1972.
329. **Damadian, R.,** *Science,* 171, 1151, 1971.
330. **Damadian, R.,***Biophys. J.*, 11, 739, 773, 1971.
331. **Damadian, R.,** U.S. Patent, 3,789,832, Field 17.03, 1972.
332. **Damadian, R. and Cope, F. W.,** *Physiol. Chem. Phys.*, 6, 309, 1974.
333. **Damadian, R., Goldsmith, M., and Minkoff, L.,** *Physiol. Chem. Phys.*, 9, 97, 1977.
334. **Damadian, R., Minkoff, L., Goldsmith, M., Stanford, M., and Koutcher, J. A.,** *Science,* 194, 1430, 1976.
335. **Damadian, R., Zaner, K. S., Hor, D., Dimaio, T., Minkoff, L., and Goldsmith, M.,** *Ann. N.Y. Acad. Sci.*, 222, 1048, 1973.
336. **Daniel, A. and Pavia, A. A.,** *Tetrahedron Lett.*, 1145, 1967.
337. **Daunis, J., Follet, M., and Marzin, C.,** *Org. Magn. Reson.*, 13, 330, 1980.
338. **Davis, J. B., Jackman, L. M., Siddons, P. T., and Weedon, B. C. L.,** *J. Chem. Soc. C,* 2154, 1966.
339. **Davis, M. and Hassel, O.,** *Acta Chem. Scand.*, 17, 1181, 1963.
340. **Davison, A. and Rakita, P. E.,** *J. Am. Chem. Soc.*, 90, 4479, 1968.
341. **Davoust, D., Massias, M., and Molho, D.,** *Org. Magn. Reson.*, 13, 218, 1980.
342. **Dear, R. E. A. and Gilbert, E. E.,** *J. Org. Chem.*, 33, 819, 1968.
343. **Debye, D.,** *Polar Molecules,* Dover, New York, 1948.
344. **Dehmlow, E. V., Zeisberg, R., and Dehmlow, S. S.,** *Org. Magn. Reson.*, 7, 418, 1975.
345. **De'Kowalewski, D. G. and Kowalewski, V. J.,** *J. Chem. Phys.*, 37, 1009, 1962.
346. **Delbaere, L. T. J., James, M. N. G., and Lemieux, R. U.,** *J. Am. Chem. Soc.*, 95, 7866, 1973.
347. **Della, E. W.,** *Chem. Commun.*, 1558, 1968.
348. **Demarco, P. V., Elzey, T. K., Lewis, R. B., and Wenkert, E.,** *J. Am. Chem. Soc.*, 92, 5734, 1970.
349. **Demarco, P. V., Farkas, E., Doddrell, D., Mylari, B. L., and Wenkert, E.,** *J. Am. Chem. Soc.*, 90, 5480, 1968.
350. **De Mare, G. R. and Martin, S. J.,** *J. Am. Chem. Soc.*, 88, 5033, 1966.
351. **Dennison, D. M.,** *Proc. R. Soc London Ser. A,* 155, 483, 1927.
352. **Deslauriers, R. and Smith, I. C. P.,** Conformation and structure of peptides, in *Topics in Carbon-13 NMR Spectroscopy,*Vol. 2, Levy, G. C., Ed., Wiley-Interscience, New York, 1976.
353. **Deslauriers, R., Walter, R., and Smith, I. C. P.,** *FEBS Lett.*, 37, 27, 1973.
354. **Desreux, J. F., Fox, L. E., and Reilley, C. N.,** *Anal. Chem.*, 44, 2217, 1972.
355. **Deverell, C.,** *Mol. Phys.*, 18, 319, 1970.
356. **Dewey, R. S., Schönewaldt, E. F., Joshua, H., Paleveda, W. J., Schwam, H., Barkemeyer, H., Arison, B. H., Veber, D. F., Denkenwalter, R. G., and Hirschmann, R.,** *J. Am. Chem. Soc.*, 90, 3254, 1968.
357. **Deyrup, J. A. and Greenwald, R. B.,** *J. Am. Chem. Soc.*, 87, 4538, 1965.
358. **Dhami, K. S. and Stothers, J. B.,** *Can. J. Chem.*, 43, 479, 1965.
359. **Dhami, K. S. and Stothers, J. B.,** *Can. J. Chem.*, 43, 498, 1965.
360. **Dhami, K. S. and Stothers, J. B.,** *Can. J. Chem.*, 43, 510, 1965.
361. **Dhami, K. S. and Stothers, J. B.,** *Can. J. Chem.*, 44, 2855, 1966.
362. **Dhami, K. S. and Stothers, J. B.,** *Can. J. Chem.*, 45, 233, 1967.
363. **Dickerman, S. C. and Haase, R. J.,** *J. Am. Chem. Soc.*, 89, 5458, 1967.
364. **Dickinson, W. C.,** *Phys. Rev.*, 77, 736, 1950.
365. **Dickinson, W. C.,** *Phys. Rev.*, 80, 563, 1951.
366. **Diehl, P.,** *Helv. Chim. Acta,* 44, 829, 1961.
367. **Dinesh, Rogers, M. T.,** *J. Magn. Reson.*, 7, 30, 1972.

368. **Ditchfield, R. and Ellis, P. D.,** *Chem. Phys. Lett.*, 17, 342, 1972.
369. **Ditchfield, R. and Ellis, P. D.,** Theory of ^{13}C chemical shifts, in *Topics in Carbon-13 NMR Spectroscopy,* Vol. 1, Levy, G. C., Ed., Wiley-Interscience, New York, 1974, 1.
370. **Doddrell, D. M. and Allerhand, A.,** *J. Am. Chem. Soc.*, 93, 1558, 1971.
371. **Doddrell, D. M., Burfitt, J., Grutzner, J. B., and Barfield, M.,** *J. Am. Chem. Soc.*, 96, 1241, 1974.
372. **Doddrell, D. M., Glushke, V., and Allerhand, A.,** *J. Chem. Phys.*, 56, 3683, 1972.
373. **Doddrell, D. M., Khong, P. W., and Lewis, K. G.,** *Tetrahedron Lett.*, 2381, 1974.
374. **Doddrell, D. M. and Wells, P. R.,** *J. Chem. Soc. Perkin Trans. 2,* 1333, 1973.
375. **Doleschall, G. and Lempert, K.,** *Tetrahedron,* 29, 639, 1973.
376. **Doomes, E. and Cromwell, N. H.,** *J. Org. Chem.*, 34, 310, 1969.
377. **Dorman, D. E., Angyal, S. J., and Roberts, J. D.,** *J. Am. Chem. Soc.*, 92, 1351, 1970.
378. **Dorman, D. E., Bauer, D., and Roberts, J. D.,** *J. Org. Chem.*, 40, 3729, 1975.
379. **Dorman, D. E., Jautelat, M., and Roberts, J. D.,** *J. Org. Chem.*, 36, 2757, 1971.
380. **Dorman, D. E., Jautelat, M., and Roberts, J. D.,** *J. Org. Chem.*, 38, 1026, 1973.
381. **Dorman, D. E. and Roberts, J. D.,** *Proc. Natl. Acad. Sci.*, 65, 19, 1970.
382. **Dorman, D. E. and Roberts, J. D.,** *J. Am. Chem. Soc.*, 92, 1355, 1970.
383. **Dorman, D. E. and Roberts, J. D.,** *J. Am. Chem. Soc.*, 93, 4463, 1971.
384. **Dorman, D. E. and Bovey, F. A.,** *J. Org. Chem.*, 38, 2379, 1973.
385. **Douglas, A. W.,** *J. Chem. Phys.*, 40, 2413, 1964.
386. **Douglas, A. W.,** *J. Chem. Phys.*, 45, 3465, 1966.
387. **Douglas, A. W. and Dietz, D.,** *J. Chem. Phys.*, 46, 1214, 1967.
388. **Downing, A. P., Ollis, W. D., and Sutherland, I. O.,** *Chem. Commun.*, 171, 1967.
389. **Downing, A. P., Ollis, W. D., Sutherland, I. O., and Mason, J.,** *Chem. Commun.*, 329, 1968.
390. **Dörnyei, G., Bárczai-Beke, M., Majoros, B., Sohár, P., and Szántai, Cs.,** *Acta Chim. Acad. Sci. Hung.*, 90, 275, 1976.
391. **Drakenberg, T. and Carter, R. E.,** *Org. Magn. Reson.*, 7, 307, 1975.
392. **Drakenberg, T., Jost, R., and Sommer, J.,** *Chem. Commun.*, 1011, 1974.
393. **Dreeskamp, H.,** *Z. Phys. Chem.*, 59, 321, 1968.
394. **Dreeskamp, H., Hilderbrand, K., and Pfisterer, G.,** *Mol. Phys.*, 17, 429, 1969.
395. **Dreeskamp, H. and Sackman, E.,** *Z. Phys. Chem.*, 34, 273, 1962.
396. **Dreeskamp, H. and Stegmeier, G.,** *Z. Naturforsch.*, 22a, 1458, 1967.
397. **Duch, M. W. and Grant, D. M.,** *Macromolecules,* 3, 165, 1970.
398. **Dudek, E. P. and Dudek, G. O.,** *J. Org. Chem.*, 32, 823, 1967.
399. **Dudek, G. O. and Dudek, E. P.,** *J. Am. Chem. Soc.*, 86, 4283, 1964; 88, 2407, 1966.
400. **Dudek, G. O. and Dudek, E. P.,** *Tetrahedron,* 23, 3245, 1967.
401. **Dudek, G. O. and Holm, R. H.,** *J. Am. Chem. Soc.*, 83, 2099, 1961.
402. **Dungan, C. H. and Van Wazer, J. R.,** *Compilation of Reported F^{19} NMR Chemical Shifts 1951-1967,* Wiley-Interscience, New York, 1970.
403. **Dunham, L., Nichols, S., and Brunschwig, A.,** *Cancer Res.*, 6, 230, 1946.
404. **Durette, P. L. and Horton, D.,** *J. Org. Chem.*, 36, 2658, 1971.
405. **Duus, F., Jakobsen, P., and Lawesson, S. O.,** *Tetrahedron,* 24, 5323, 1968.
406. **Dvornik, D. and Schilling, G.,** *J. Med. Chem.*, 8, 466, 1965.
407. **Dyer, D. S., Cunningham, J. A., Brooks, J. J., Sievers, R. E., and Rondeau, R. F.,** Interactions of nucleophiles with lanthanide shift reagents, in *Nuclear Magnetic Resonance Shift Reagents,* Sievers, R. F., Ed., Academic Press, New York, 1973, 21.
408. **Dyer, J. and Lee, J.,** *Trans. Faraday Soc.*, 62, 257, 1966.
409. *Dynamic Nuclear Magnetic Resonance Spectroscopy,* **Jackman, L. M. and Cotton, F. A., Eds.,** Academic Press, New York, 1975.
410. **Eaton, D. R., Josey, A. D., Phillips, W. D., and Benson, R. E.,** *J. Chem. Phys.*, 39, 3513, 1963.
411. **Ebsworth, E. A. V. and Frankiss, S. G.,** *Trans. Faraday Soc.*, 59, 1518, 1963.
412. **Echols, R. E. and Levy, G. C.,** *J. Org. Chem.*, 39, 1321, 1974.
413. **Edward, J. T.,** *Chem. Ind. (London),* 1102, 1955.
414. **Eggert, H. and Djerassi, C.,** *J. Am. Chem. Soc.*, 95, 3710, 1973.
415. **Ejchart, A.,** *Org. Magn. Reson.*, 9, 351, 1977.
416. **Ejchart, A.,** *Org. Magn. Reson.*, 10, 263, 1977.
417. **Ejchart, A.,** *Org. Magn. Reson.*, 13, 368, 1980.
418. **Elguero, J., Fruchier, A., and Pardo, M. C.,** *Can. J. Chem.*, 54, 1329, 1976.
419. **Elguero, J., Jacquier, R., and Tarrago, G.,** *Tetrahedron Lett.*, 4719, 1965.
420. **Elguero, J., Jacquier, R., and Tarrago, G.,** *Bull. Soc. Chim.*, 2981, 1966.
421. **Elguero, J., Johnson, B. L., Pereillo, J. M., Pouzard, G., Rajzman, M., and Randall, E. W.,** *Org. Magn. Reson.*, 9, 145, 1977.

422. **Eliel, E. L., Allinger, N. L., Angyal, S. J., and Morrison, G. A.,** *Conformational Analysis,* Interscience, New York, 1965.
423. **Eliel, E. L., Bailey, W. F., Kopp, L. D., Willer, R. L., Grant, D. M., Bertrand, R., Christensen, K. A., Dalling, D. K., Duch, M. W., Wenkert, E., Schell, F. M., and Cochran, D. W.,** *J. Am. Chem. Soc.,* 97, 322, 1975.
424. **Eliel, E. L. and Pietrusiewicz, K. M.,** ^{13}C NMR of nonaromatic heterocyclic compounds, in *Topics in Carbon-13 NMR Spectroscopy,* Vol. 3, Levy, G. C., Ed., Wiley-Interscience, New York, 1979, chap. 3.
425. **Eliel, E. L. and Pietrusiewicz, K. M.,** *Org. Magn. Reson.,* 13, 193, 1980.
426. **Eliel, E. L., Rao, V. S., and Pietrusiewicz, K. M.,** *Org. Magn. Reson.,* 12, 461, 1979.
427. **Eliel, E. L., Rao, V. S., and Riddel, F. G.,** *J. Am. Chem. Soc.,* 98, 3583, 1976.
428. **Elleman, D. D. and Manatt, S. L.,** *J. Mol. Spectrosc.,* 9, 477, 1962.
429. **Elleman, D. D., Manatt, S. L., and Pearce, C. D.,** *J. Chem. Phys.,* 42, 650, 1965.
430. **Ellis, G. E. and Jones, R. G.,** *J. Chem. Soc. Perkin Trans. 2,* 437, 1972.
431. **Ellis, G. E., Jones, R. G., and Papadopoulos, M. G.,** *J. Chem. Soc. Perkin Trans. 2,* 1381, 1974.
432. **Ellis, J., Jackson, A. H., Kenner, G. W., and Lee, J.,** *Tetrahedron Lett.,* 23, 1960.
433. **Elvidge, J. A. and Ralph, P. D.,** *J. Chem. Soc. B,* 249, 1966.
434. **Emsley, J. W., Feeney, J., and Sutcliffe, L. H.,** *High Resolution Nuclear Magnetic Resonance Spectroscopy,* Pergamon Press, London, 1965; (a) Vol. 1, p. 31; (b) Vol. 2, pp. 1011-1031.
435. **Emsley, J. W. and Phillips, L.,** Fluorine chemical shifts, in *Progress in Nuclear Magnetic Resonance Spectroscopy,* Vol. 7, Emsley, J. W., Feeney, J., and Sutcliffe, L. H., Eds., Pergamon Press, New York, 1971, 1.
436. **Emsley, J. W., Phillips, L., and Wray, V.,** Fluorine coupling constants, in *Progress in Nuclear Magnetic Resonance Spectroscopy,* Vol. 10, Emsley, J. W., Feeney, J., and Sutcliffe, L. H., Eds., Pergamon Press, New York, 1976, 83.
437. **Engelhardt, G., Radeglia, R., Jancke, H., Lippmaa, E., and Mägi, M.,** *Org. Magn. Reson.,* 5, 561, 1973.
438. **Eremenko, L. T. and Borisenko, A. A.,** *Izv. Akad. Nauk. (SSSR) Ser. Khim.,* 675, 1968.
439. **Ernst, E.,** *Biophysics of Straited Muscle,* Akadémiai Kiadó, Budapest, 1963.
440. **Ernst, E.,** *Acta Biochem. Biophys. Acad. Sci. Hung.,* 10, 95, 1975.
441. **Ernst, E. and Hazlewood, C. F.,** *Inorg. Perspectives Biol. Med.,* 2, 27, 1978.
442. **Ernst, E., Tigyi, J., and Zahorcsek, A.,** *Acta Physiol. Acad. Sci. Hung.,* 1, 5, 1950.
443. **Ernst, L.,** *Chem. Ber.,* 108, 2030, 1975.
444. **Ernst, L.,** *Z. Naturforsch.,* 30b, 788, 1975.
445. **Ernst, L.,** *Z. Naturforsch.,* 30b, 794, 1975.
446. **Ernst, L.,** *J. Magn. Reson.,* 20, 544, 1975.
447. **Ernst, L.,** *J. Magn. Reson.,* 22, 279, 1976.
448. **Ernst, L.,** *Org. Magn. Reson.,* 8, 161, 1976.
449. **Ernst, L. and Mannschreck, A.,** *Tetrahedron Lett.,* 3023, 1971.
450. **Ernst, R. R.,** *J. Chem. Phys.,* 45, 3845, 1966.
451. **Ernst, R. R.,** *Adv. Magn. Reson.,* 2, 1, 1967.
452. **Ernst, R. R.,** *J. Magn. Reson.,* 3, 10, 1970.
453. **Ernst, R. R.,** *Chimia,* 29, 179, 1975.
454. **Ernst, R. R. and Anderson, W. A.,** *Rev. Sci. Instrum.,* 37, 93, 1966.
455. **Espersen, W. G. and Martin, R. B.,** *J. Phys. Chem.,* 80, 741, 1976.
456. **Ettinger, R., Blume, P., Patterson, A., Jr., and Lauterbur, P. C.,** *J. Chem. Phys.,* 33, 1597, 1960.
457. **Evans, D. F.,** *Proc. Chem. Soc.,* 115, 1958.
458. **Evans, D. F., Tucker, J. N., and de Villardi, G. C.,** *Chem. Commun.,* 205, 1975.
459. **Evans, D. F. and Wyatt, M.,** *Chem. Commun.,* 312, 1972.
460. **Evans, H. B., Jr., Tarpley, A. R., and Goldstein, J. H.,** *J. Phys. Chem.,* 72, 2552, 1968.
461. **Ewing, D. F.,** Two-bond coupling between protons and carbon-13, in *Annual Reports on NMR Spectroscopy,* Vol. 6A, Mooney, E. F., Ed., Academic Press, New York, 1975, 389.
462. **Faller, J. W., Adams, M. A., and La Mar, G. N.,** *Tetrahedron Lett.,* 699, 1974.
463. **Farminer, A. R. and Webb, G. A.,** *Org. Magn. Reson.,* 8, 102, 1976.
464. **Farrar, T. C. and Becker, E. D.,** *Pulse and Fourier Transform NMR,* Academic Press, New York, 1971; (a) p. 41; (b) p. 34; (c) p. 54, (d) p. 63; (e) p. 26; (f) p. 61.
465. **Farrar, T. C., Druck, S. J., Shoup, R. R., and Becker, E. D.,** *J. Am. Chem. Soc.,* 94, 699, 1972.
466. **Faure, R., Clinas, J.-R., Vincent, E.-J., and Larice, J.-L.,** *C. R. Acad. Sci. Paris, Ser. C,* 279, 717, 1974.
467. **Faure, R., Galy, J.-P., Vincent, E.-J., and Elguero, J.,** *Can. J. Chem.,* 56, 46, 1978.
468. **Feeney, J. and Partington, P.,** *Chem. Commun.,* 611, 1973.
469. **Feeney, J., Partington, P., and Roberts, G. C. K.,** *J. Magn. Reson.,* 13, 268, 1974.

470. **Feeney, J., Pauwells, P., and Shaw, D.,** *Chem. Commun.*, 554, 1970.
471. **Feeney, J. and Sutcliffe, L. H.,** *J. Chem. Soc.*, 1123, 1962.
472. **Feeney, J. and Sutcliffe, L. H.,** *Spectrochim. Acta*, 24a, 1135, 1968.
473. **Fellgett, P.,** *J. Phys. Radium*, 19, 187, 1958.
474. **Fessenden, R. W. and Waugh, J. S.,** *J. Chem. Phys.*, 31, 996, 1959.
475. **Fields, R.,** Fluorine-19 nuclear magnetic resonance spectroscopy, in *Annual Reports on NMR Spectroscopy*, Vol. 5A, Mooney, E. F., Ed., Academic Press, New York, 1972, 99.
476. **Fields, R.,** Fluorine-19 NMR spectroscopy of fluoroalklyl and fluoroaryl derivatives of transition metals, in *Annual Reports on NMR Spectroscopy*, Vol. 7, Webb, G. A., Ed., Academic Press, New York, 1977, 1.
477. **Figeys, H. P., Geerlings, P., Raeymaekers, P., van Lommen, G., and Defay, N.,** *Tetrahedron*, 31, 1731, 1975.
478. **Filleux-Blanchard, M. L., Durand, H., Bergeon, M. T., Clesse, F., Quiniou, H., and Martin, G. J.,** *J. Mol. Structure*, 3, 351, 1969.
479. **Finch, E. D., Harmon, J. F., and Muller, B. H.,** *Arch. Biochem. Biophys.*, 147, 299, 1971.
480. **Finer, E. G. and Harris, R. K.,** Spin-spin couplings between phosphorus nuclei, in *Progress in Nuclear Magnetic Resonance Spectroscopy*, Vol. 6, Emsley, J. W., Feeney, J., and Sutcliffe, L. H., Pergamon Press, New York, 1971, 61.
481. **Firl, J. R. and Runge, W.,** *Z. Naturforsch.*, 29b, 393, 1974.
482. **Firl, J. R., Runge, W., Hartmann, W., and Utikal, H. P.,** *Chem. Lett.*, 51, 1975.
483. **Fischer, H.,** New Series, Landolt-Börnstein, Group II, in *Magnetic Properties of Free Radicals*, Vol. 1, Hellwege, K. H., Ed., Springer-Verlag, Berlin, 1965.
484. **Fischer, H. and Bargon, J.,** *Acc. Chem. Res.*, 2, 110, 1969.
485. **Fleming, I., Hanson, S. W., and Sanders, J. K. M.,** *Tetrahedron Lett.*, 3733, 1971.
486. **Fleming, I. and Williams, D. H.,** *Tetrahedron*, 23, 2747, 1967.
487. **Fletcher, J. R. and Sutherland, I. O.,** *Chem. Commun.*, 1504, 1969.
488. **Flohé, L., Breitmaier, E., Günzler, W. A., Voelter, W., and Jung, G.,** *Hoppe-Seyler's Z. Physiol. Chem.*, 353, 1159, 1972.
489. **Flygare, W. H.,** *J. Chem. Phys.*, 41, 793, 1964.
490. **Flygare, W. H. and Goodisman, J.,** *J. Chem. Phys.*, 49, 3122, 1968.
491. **Fodor, L., Szabó, J., and Sohár, P.,** *Tetrahedron*, 37, 963, 1981.
492. **Foote, C. S.,** *Tetrahedron Lett.*, 579, 1963.
493. **Foreman, M. I.,** Medium effects. NMR shift reagents, in *Specialist Periodical Reports, Nuclear Magnetic Resonance*, Vol. 1, Harris, R. K., Ed., The Chemical Society, London, 1972, 310.
494. **Foreman, M. I.,** Medium effects. NMR shift reagents, in *Specialist Periodical Reports, Nuclear Magnetic Resonance*, Vol. 2, Harris, R. K., Ed., The Chemical Society, London, 1972, 355.
495. **Foreman, M. I.,** Medium effects. NMR shift reagents, in *Specialist Periodical Reports, Nuclear Magnetic Resonance*, Vol.. 3, Harris, R. K., Ed., The Chemical Society, London 1972, 342.
496. **Foreman, M. I.,** Medium effects. NMR shift reagents, in *Specialist Periodical Reports, Nuclear Magnetic Resonance*, Vol. 4, Harris, R. K., Ed., The Chemical Society, London, 1972, 294.
497. **Foreman, M. I.,** Medium effects on chemical shifts and coupling constants, in *Specialist Periodical Reports, Nuclear Magnetic Resonance*, Vol. 5, Harris, R. K., Ed., The Chemical Society, London, 1973, 292.
498. **Foreman, M. I.,** Solvent effects. NMR shift reagents, in *Specialist Periodical Reports, Nuclear Magnetic Resonance*, Vol. 6, Abraham, R. J., Ed., The Chemical Society, London, 1975, 244.
499. **Formacek, V., Desnayer, L., Kellerhals, H. P., Keller, T., and Clerk, J. T.,** *Bruker Data Bank* ^{13}C, Vol. 1, Bruker Physic Morisch OHG, Karlsruhe, 1976.
500. **Forsén, S.,** *Acta Chem. Scand.*, 13, 1472, 1959.
501. **Forsén, S. and Norin, T.,** *Tetrahedron Lett.*, 2845, 1964.
502. **Foster, A. B., Inch, T. D., Quadir, M. H., and Webber, J. M.,** *Chem. Commun.*, 1086, 1968.
503. **Fourier, J. B. J.,** *Theorie Analytique de la Chaleur*, Paris, 1822.
504. **Fraenkel, G. and Franconi, C.,** *J. Am. Chem. Soc.*, 82, 4478, 1960.
505. **Frankiss, S. G.,** *J. Phys. Chem.*, 67, 752, 1963.
506. **Franklin, N. C. and Feltkamp, H.,** *Tetrahedron*, 22, 2801, 1966.
507. **Frasca, A. R. and Dennler, E. B.,** *Chem. Ind. (London)*, 509, 1967.
508. **Fraser, M., McKenzie, S., and Reid, D. H.,** *J. Chem. Soc. B*, 44, 1966.
509. **Fraser, R. R.,** *Can. J. Chem.*, 38, 2226, 1960.
510. **Fraser, R. R., Petit, M. A., and Miskow, M.,** *J. Am. Chem. Soc.*, 94, 3253, 1972.
511. **Freeburger, M. E. and Spialter, L.,** *J. Am. Chem. Soc.*, 93, 1894, 1971.
512. **Freed, I. H. and Pedersen, J. B.,** The theory of chemically induced dynamic spin polarization in *Advances in Magnetic Resonance*, Vol. 8, Waugh, J. S., Ed., Academic Press, New York, 1976, 2.
513. **Freeman, J. P.,** *J. Org. Chem.*, 28, 2508, 1963.

514. **Freeman, R. and Anderson, W. A.,** *J. Chem. Phys.*, 39, 1518, 1963.
515. **Freeman, R. and Hill, H. D. W.,** *J. Chem. Phys.*, 51, 3140, 1969.
516. **Freeman, R. and Hill, H. D. W.,** *J. Chem. Phys.*, 54, 301, 1971.
517. **Freeman, R. and Hill, H. D. W.,** *J. Chem. Phys.*, 54, 3367, 1971.
518. **Freeman, R. and Hill, H. D. W.,** *J. Chem. Phys.*, 55, 1985, 1971.
519. **Freeman, R. and Hill, H. D. W.,** *Molecular Spectroscopy 1971,* Institute of Petroleum, London, 1971.
520. **Freeman, R. and Hill, H. D. W.,** *J. Magn. Reson.*, 4, 366, 1971.
521. **Freeman, R. and Hill, H. D. W.,** *J. Magn.Reson.*, 5, 278, 1971.
522. **Freeman, R., Hill, H. D. W., and Kaptein, R.,** *J. Magn. Reson.*, 7, 82, 1972.
523. **Freeman, R., Hill, H. D. W., and Kaptein, R.,** *J. Magn. Reson.*, 7, 327, 1972.
524. **Freeman, R., Pachler, K. G. R., and La Mar, G. N.,** *J. Chem. Phys.*, 55, 4586, 1971.
525. **Freeman, R. and Whiffen, D. H.,** *Mol. Phys.*, 4, 321, 1961.
526. **Frei, K. and Bernstein, H. J.,** *J. Chem. Phys.*, 38, 1216, 1963.
527. **Frey, H. E., Knispel, R. R., Kruuv, J., Sharp, A. R., Thompson, R. T., and Pintar, M. M.,** *J. Natl. Cancer Inst.*, 49, 903, 1972.
528. **Friedel, R. A. and Retcofsky, H. L.,** *J. Am. Chem. Soc.*, 85, 1300, 1963.
529. **Friedel, R. A. and Retcofsky, H. L.,** *Chem. Ind. (London),* 455, 1966.
530. **Frith, P. G. and McLauchlan, K. A.,** Chemically induced dynamic nuclear polarization, in *NMR Specialist Periodical Reports,* Vol. 3, Harris, R. K., Ed., The Chemical Society, London, 1974, 378.
531. **Fritz, O. G., Jr. and Swift, T. J.,** *Biophys. J.*, 7, 675, 1967.
532. **Fronza, G., Gamba, A., Mondelli, R., and Pagani, G.,** *J. Magn. Reson.*, 12, 231, 1973.
533. **Fukumi, T., Arata, Y., and Fujiwara, S.,** *J. Mol. Spectrosc.*, 27, 443, 1968.
534. **Fung, B. M.,** *Biochim. Biophys. Acta,* 362, 209, 1974.
535. **Gagnaire, D. and Vincendon, M.,** *Chem. Commun.*, 509, 1977.
536. **Gagnaire, D. and Vottero, P.,** *Bull. Soc. Chim. France,* 2779, 1963.
537. **Gainer, J., Howarth, G. A., and Hoyle, W.,** *Org. Magn. Reson.*, 8, 226, 1976.
538. **Gansow, O. A., Burke, A. R., and La Mar, G. N.,** *Chem. Commun.*, 456, 1972.
539. **Gansow, O. A., Loeffler, P. A., Willcott, M. R., and Lenkinski, E.,** *J. Am. Chem. Soc.*, 95, 3389, 1973.
540. **Gansow, O. A., Willcott, M. R., and Lenkinski, R. E.,** *J. Am. Chem. Soc.*, 93, 4297, 1971.
541. **Ganter, G., Pokras, S. M., and Roberts, J. D.,** *J. Am. Chem. Soc.*, 88, 4235, 1966.
542. **Garbisch, E. W., Jr. and Griffith, M. G.,** *J. Am. Chem. Soc.*, 90, 3590, 1968.
543. **Garbisch, E. W., Jr. and Griffith, M. G.,** *J. Am. Chem. Soc.*, 90, 6543, 1968.
544. **Garnier, R., Faure, R., Babadjamian, A., and Vincent, E.-J.,** *Bull. Soc. Chim. France,* 3, 1040, 1972.
545. **Garrat, P. J.,** *Aromaticity,* McGraw-Hill, London, 1971, 46.
546. **Gash, V. W. and Bauer, D. J.,** *J. Org. Chem.*, 31, 3602, 1966.
547. **Gassmann, P. G. and Heckert, D. C.,** *J. Org. Chem.*, 30, 2859, 1965.
548. **Gault, I., Price, B. J., and Sutherland, I. O.,** *Chem. Commun.*, 540, 1967.
549. **Georgian, V., Kerwin, J. F., Wolff, M. E., and Owings, F. F.,** *J. Am. Chem. Soc.*, 84, 3594, 1962.
550. **Gerlach, H. and Zagalak, B.,** *Chem. Commun.*, 274, 1973.
551. **Gerlach, W. and Stern, O.,** *Ann. Phys. Leipzig,* 74, 673, 1924.
552. **Ghersetti, S., Lunazzi, L., Maccagnani, G., and Mangini, A.,** *Chem. Commun.*, 834, 1969.
553. **Gibbons, W. A., Sogn, J. A., Stern, A., Craig, L. C., and Johnson, L. F.,** *Nature (London),* 227, 840, 1970.
554. **Gierki, T. D. and Flygare, W. H.,** *J. Am. Chem. Soc.*, 94, 7277, 1972.
555. **Gil, V. M. S. and Geraldes, C. F. G. C.,** *J. Magn. Reson.*, 11, 268, 1973.
556. **Gilles, D. G. and Shaw, D.,** The application of Fourier transformation to high resolution nuclear magnetic resonance spectroscopy, in *Annual Reports on NMR Spectroscopy,* Vol. 5A, Mooney, E. F., Ed., Academic Press, New York, 1972, 557.
557. **Gillespie, R. P. and Birchall, T.,** *Can. J. Chem.*, 41, 148, 1963.
558. **Giovanella, B. C., Morgan, A. C., Stehlin, J. S., and Williams, L. J.,** *Cancer Res.*, 33, 2568, 1973.
559. **Girard, P., Kagan, H., and David, S.,** *Tetrahedron,* 27, 5911, 1971.
560. **Goe, G. L.,** *J. Org. Chem.*, 38, 4285, 1974.
561. **Goering, H. L., Eikenberry, J. N., and Koermer, G. S.,** *J. Am. Chem. Soc.*, 93, 5913, 1971.
562. **Goering, H. L., Eikenberry, J. N., Koermer, G. S., and Lattimer, C. J.,** *J. Am. Chem. Soc.*, 96, 1493, 1974.
563. **Goldman, M.,** *Spin Temperature and Nuclear Magnetic Resonance in Solids,* Oxford University Press, Oxford, 1970.
564. **Goldsmith, M. and Damadian, R.,** *Physiol. Chem. Phys.*, 7, 263, 1975.
565. **Goldsmith, M., Damadian, R., Stanford, M., and Lipkowitz, M.,** *Physiol. Chem. Phys.*, 9, 105, 1977.
566. **Goldsmith, M., Koutcher, J. A., and Damadian, R.,** *Br. J. Cancer,* 36, 235, 1977.

567. **Goldstein, J. H., Watts, V. S., Rattet, L. S.,** ^{13}CH satellite NMR spectra, in *Progress in Nuclear Magnetic Resonance Spectroscopy,* Vol. 8, Emsley, J. W., Feeney, J., and Sutcliffe, L. H., Eds., Pergamon Press, Oxford, 1971, 104.
568. **Goodman, B. A. and Raynor, J. B.,** *Adv. Inorg. Chem. Radiochem.,* 13, 135, 1970.
569. **Gorin, P. A. J.,** *Can. J. Chem.,* 52, 458, 1974.
570. **Gorkom, M. and Hall, G. E.,** *Quart. Rev. Chem. Soc.,* 22, 14, 1968.
571. **Gorodetsky, M., Luz, Z., and Mazur, Y.,** *J. Am. Chem. Soc.,* 89, 1183, 1967.
572. **Gorter, C. J.,** *Physica,* 3, 995, 1936.
573. **Gorter, C. J. and Broer, L. F. J.,** *Physica,* 9, 591, 1942.
574. **Gough, J. L., Guthrie, J. P., and Stothers, J. B.,** *Chem. Commun.,* 979, 1972.
575. **Govill, G.,** *Mol. Phys.,* 21, 953, 1971.
576. **Graham, D. M. and Holloway, C. E.,** *Can. J. Chem.,* 41, 2114, 1963.
577. **Granacher, I.,** *Helv. Phys. Acta,* 34, 272, 1961.
578. **Granger, P. and Maugras, M.,** *Org. Magn. Reson.,* 7, 598, 1975.
579. **Grant, D. M. and Cheney, B. V.,** *J. Am. Chem. Soc.,* 89, 5315, 1967.
580. **Grant, D. M., Hirst, R. C., and Gutowsky, H. S.,** *J. Chem. Phys.,* 38, 470, 1963.
581. **Grant, D. M. and Paul, E. G.,** *J. Am. Chem. Soc.,* 86, 2984, 1964.
582. **Gray, G. A., Cremer, S. E., and Marsi, K. L.,** *J. Am. Chem. Soc.,* 98, 2109, 1976.
583. **Gray, G. A., Ellis, P. D., Traficante, D. D., and Maciel, G. E.,** *J. Magn. Reson.,* 1, 41, 1969.
584. **Gray, G. A., Maciel, G. E., and Ellis, P. D.,** *J. Magn. Reson.,* 1, 407, 1969.
585. **Greene, J. L., Jr. and Shevlin, P. B.,** *Chem. Commun.,* 1092, 1971.
586. **Griffith, D. L. and Roberts, J. D.,** *J. Am. Chem. Soc.,* 87, 4089, 1965.
587. **Grim, S. O. and McFarlane, W.,** *Nature (London),* 208, 995, 1965.
588. **Grinter, R.,** Nuclear spin-spin coupling, in *Specialist Periodical Reports,* Vol. 3, Harris, R. K., Ed., The Chemical Society, London, 1974, 50.
589. **Grinter, R. and Mason, J.,** *J. Chem. Soc. A.,* 2196, 1970.
590. **Grishin, Y. K., Sergeyev, N. M., Subbotin, O. A., and Ustynyuk, Y. A.,** *Mol. Phys.,* 25, 297, 1973.
591. **Grishin, Y. K., Sergeyev, N. M., and Ustynyuk, Y. A.,** *J. Organomet. Chem.,* 22, 361, 1970.
592. **Gronowitz, S., Hörnfeldt, A., Gestblom, B., and Hoffman, R. A.,** *Arkiv. Kemi,* 18, 133, 1962.
593. **Gronowitz, S., Sörlin, G., Gestblom, B., and Hoffman, R. A.,** *Arkiv. Kemi,* 19, 483, 1963.
594. **Grover, S. H., Guthrie, J. P., Stothers, J. B., and Tan, C. T.,** *J. Magn. Reson.,* 10, 227, 1973.
595. **Grover, S. H. and Stothers, J. B.,** *Can. J. Chem.,* 52, 870, 1974.
596. **Grunwald, E., Loewenstein, A., and Meiboom, S.,** *J. Chem. Phys.,* 27, 630, 1957.
597. **Grunwald, E., Loewenstein, A., and Meiboom, S.,** *J. Chem. Phys.,* 27, 641, 1957.
598. **Grutzner, J. B., Jautelat, M., Dence, J. B., Smith, R. A., and Roberts, J. D.,** *J. Am. Chem. Soc.,* 92, 7107, 1970.
599. **Grutzner, J. B. and Santini, R. E.,** *J. Magn. Reson.,* 19, 173, 1975.
600. **Gust, D., Moon, R. B., and Roberts, J. D.,** *Proc. Natl. Acad. Sci. U.S.A.,* 72, 4696, 1975.
601. **Gutowsky, H. S.,** *J. Chem. Phys.,* 37, 2196, 1962.
602. **Gutowsky, H. S., Belford, G. G., and McMahon, P. E.,** *J. Chem. Phys.,* 36, 3353, 1962.
603. **Gutowsky, H. S. and Holm, C. H.,** *J. Chem. Phys.,* 25, 1228, 1956.
604. **Gutowsky, H. S., Holm, C. H., Saika, A., and Williams, G. A.,** *J. Am. Chem. Soc.,* 79, 4596, 1957.
605. **Gutowsky, H. S., Karplus, M., and Grant, D. M.,** *J. Chem. Phys.,* 31, 1278, 1959.
606. **Gutowsky, H. S. and McCall, D. W.,** *J. Chem. Phys.,* 22, 162, 1954.
607. **Gutowsky, H. S., McCall, D. W., and Slichter, C. P.,** *J. Chem. Phys.,* 21, 279, 1953.
608. **Gutowsky, H. S. and Saika, A.,** *J. Chem. Phys.,* 21, 1688, 1953.
609. **Günther, H. and Herrig, W.,** *Chem. Ber.,* 106, 3938, 1973.
610. **Günther, H. and Keller, T.,** *Chem. Ber.,* 103, 3231, 1970.
611. **Günther, H. and Schmickler, H.,** *Angew. Chem. Int. Ed. Engl.,* 12, 243, 1973.
612. **Günther, H. and Ulmen, J.,** *Tetrahedron,* 30, 3781, 1974.
613. **Haake, P. and Miller, W. B.,** *J. Am. Chem. Soc.,* 85, 4044, 1963.
614. **Haake, P., Miller, W. B., and Tyssee, D. A.,** *J. Am. Chem. Soc.,* 86, 3577, 1964.
615. **Haeberlen, U. and Waugh, J. S.,** *Phys. Rev.,* 175, 453, 1960.
616. **Hagen, R. and Roberts, J. D.,** *J. Am. Chem. Soc.,* 91, 4504, 1969.
617. **Hahn, E. L.,** *Phys. Rev.,* 80, 580, 1950.
618. **Hahn, E. L. and Maxwell, D. E.,** *Phys. Rev.,* 88, 1070, 1952.
619. **Hall, L. D.,** *Adv. Carbohydr. Chem.,* 19, 51, 1964.
620. **Hall, L. D. and Manville, J. F.,** *Chem. Ind. (London),* 991, 1965.
621. **Hall, L. D. and Manville, J. F.,** *Chem. Commun.,* 37, 1968.
622. **Halpern, B., Nitecki, D. E., and Weinstein, B.,** *Tetrahedron Lett.,* 3075, 1967.
623. **Hamlow, H. P., Okuda, S., and Nakagawa, N.,** *Tetrahedron Lett.,* 2553, 1964.

624. **Hammel, J. C. and Smith, J. A. S.,** *J. Chem. Soc. A,* 2883, 1969.
625. **Hampson, P. and Mathias, A.,** *Mol. Phys.,* 11, 541, 1966.
626. **Hampson, P. and Mathias, A.,** *Chem. Commun.,* 371, 1967.
627. **Hansen, J. R.,** *Biochim. Biophys. Acta,* 230, 482, 1971.
628. **Hansen, P. E., Poulsen, O. K., and Berg, A.,** *Org. Magn. Reson.,* 7, 23, 1975.
629. **Hansen, P. E., Poulsen, O. K., and Berg, A.,** *Org. Magn. Reson.,* 7, 405, 1975.
630. **Hansen, P. E., Poulsen, O. K., and Berg, A.,** *Org. Magn. Reson.,* 7, 475, 1975.
631. **Hansen, P. E., Poulsen, O. K., and Berg, A.,** *Org. Magn. Reson.,* 9, 649, 1977.
632. **Harris, R. K.,** *Chem. Soc. Rev.,* 5, 1, 1976.
633. **Hartmann, S. R. and Hahn, E. L.,** *Phys. Rev.,* 128, 2042, 1962.
634. **Hatada, K., Nagata, K., and Yuki, H.,** *Bull. Chem. Soc. Jpn.,* 43, 3195, 3267, 1970.
635. **Hatada, K., Takeshita, M., and Yuki, H.,** *Tetrahedron Lett.,* 4621, 1968.
636. **Hausser, K. H. and Stehlik, D.,** Dynamic nuclear polarization in liquids, in *Advances in Magnetic Resonance,* Vol. 3, Waugh, J. S., Ed., Academic Press, New York, 1968, 79.
637. **Hawkes, G. E., Herwig, K., and Roberts, J. D.,** *J. Org. Chem.,* 39, 1021, 1974.
638. **Hawkes, G. E., Randall, E. W., Elguero, J., and Marzin, C. J.,** *J. Chem. Soc. Perkin Trans. 2,* 1024, 1977.
639. **Hawkes, G. E., Randall, E. W., Hull, W. E., Gattegno, D., and Conti, F.,** *Biochemistry,* 17, 3986, 1978.
640. **Hayamizu, K. and Yamamoto, O.,** *Org. Magn. Reson.,* 13, 460, 1980.
641. **Hazlewood, C. F., Chang, D. C., Medina, D., Cleveland, G., and Nichols, B. L.,** *Proc. Natl. Acad. Sci. U.S.A.,* 69, 1478, 1972.
642. **Hazlewood, C. F., Cleveland, G., and Medina, D.,** *J. Natl. Cancer Inst.,* 52, 1849, 1974.
643. **Hazlewood, C. F., Nichols, B. L., Chang, D. C., and Brown, B.,** *Johns Hopkins Med. J.,* 128, 117, 1971.
644. **Heap, N. and Whitham, G. H.,** *J. Chem. Soc. B,* 164, 1966.
645. **Henneike, H. F., Jr. and Drago, R. S.,** *Inorg. Chem.,* 7, 1908, 1968.
646. **Herbison-Evans, D. and Richards, R. E.,** *Mol. Phys.,* 8, 19, 1964.
647. **Hess, R. E., Schaeffer, C. D., and Yoder, C. H.,** *J. Org. Chem.,* 36, 2201, 1971.
648. **Heyd, W. E. and Cupas, C. A.,** *J. Am. Chem. Soc.,* 91, 1559, 1969.
649. *High Resolution NMR Spectra,* SADTLER Research Laboratories, Philadelphia; Heyden and Son, London, 1967-1975.
650. **Hill, E. A. and Roberts, J. D.,** *J. Am. Chem. Soc.,* 89, 2047, 1967.
651. **Hill, R. K. and Chan, T. H.,** *Tetrahedron,* 21, 2015, 1965.
652. **Hinckley, C. C.,** *J. Am. Chem. Soc.,* 91, 5160, 1969.
653. **Hinckley, C. C.,** *J. Org. Chem.,* 35, 2834, 1970.
654. **Hinckley, C. C., Boyd, W. A., and Smith, G. V.,** *Tetrahedron Lett.,* 879, 1972.
655. **Hinckley, C. C., Boyd, W. A., and Smith, G. V.,** Chemistry of lanthanide shift reagents secondary deuterium isotope effects, in *Nuclear Magnetic Resonance Shift Reagents,* Sievers, R. F., Ed., Academic Press, New York, 1973, 1.
656. **Hinshaw, W. S., Bottomley, P. A., and Holland, G. N.,** *Nature (London),* 270, 722, 1977.
657. **Hirayama, M. and Hanyu, Y.,** *Bull. Chem. Soc. Jpn.,* 46, 2687, 1973.
658. **Hirsch, J. A. and Havinga, E.,** *J. Org. Chem.,* 41, 455, 1976.
659. **Hirsike, E.,** *J. Phys. Soc. Jpn.,* 15, 270, 1960.
660. **Hochmann, J. and Kellerhals, H. P.,** *J. Magn. Reson.,* 38, 23, 1980.
661. **Hochmann, J., Rosanske, R. C., and Levy, G. C.,** *J. Magn. Reson.,* 33, 275, 1979.
662. **Hoffman, R. A. and Forsen, S.,** High resolution nuclear magnetic double and multiple resonance, in *Progress in Nuclear Magnetic Resonance Spectroscopy,* Vol. 1, Emsley, J. W., Feeney, J., and Sutcliffe, L. H., Eds., Pergamon Press, Oxford, 1966, 15.
663. **Hoffman, R. A. and Gronowitz, S.,** *Arkiv. Kemi,* 16, 501, 515, 563, 1960.
664. **Hoffmann, T. A. and Ladik, J.,** *Adv. Chem. Phys.,* 7, 94, 1969.
665. **Hofman, W., Stefaniak, L., Urbanski, T., and Witanowski, M.,** *J. Am. Chem. Soc.,* 86, 554, 1964.
666. **Hogeveen, H.,** *Rec. Trav. Chim.,* 85, 1072, 1966.
667. **Holland, C. V., Horton, D., and Jewell, J. S.,** *J. Org. Chem.,* 32, 1818, 1967.
668. **Hollis, D. P., Saryan, L. A., Economou, J. S., Eggleston, J. C., Czeisler, J. L., and Morris, H. P.,** *J. Natl. Cancer Inst.,* 53, 807, 1974.
669. **Hollis, D. P., Saryan, L. A., and Morris, H. P.,** *Johns Hopkins Med. J.,* 131, 441, 1972.
670. **Hollósi, M., Radics, L., and Wieland, T.,** *J. Peptide Protein Res.,* 10, 1286, 1977.
671. **Holly, S. and Sohár, P.,** Theoretical and technical introduction, in *Absorption Spectra in the Infrared Region,* Láng, L. and Prichard, W. H., Eds., Akadémiai Kiadó, Budapest, 1975.
672. **Homer, J.,** *Appl. Spectrosc. Rev.,* 9, 1, 1975.

673. **Homer, J.**, Intramolecular effects. NMR shift reagents, in *Specialist Periodical Reports, Nuclear Magnetic Resonance,* Vol. 7, Abraham, R. J., Ed., The Chemical Society, London, 1976, 330.
674. **Homer, J. and Thomas, L. F.,** *Trans. Faraday Soc.,* 59, 2431, 1963.
675. **Horn, R. R. and Everett, G. W., Jr.,** *J. Am. Chem. Soc.,* 93, 7173, 1971.
676. **Horrocks, W. DeW., Jr.,** *J. Am. Chem. Soc.,* 96, 3022, 1974.
677. **Horrocks, W. DeW., Jr. and Hall, D. DeW.,** *Coord. Chem. Rev.,* 6, 147, 1971.
678. **Horrocks, W. DeW., Jr., Sipe, J. P., III, and Luber, J. R.,** *J. Am Chem. Soc.,* 93, 5258, 1971.
679. **Horsley, W. J. and Sternlicht, H.,** *J. Am. Chem. Soc.,* 90, 3738, 1968.
680. **Horsley, W. J., Sternlicht, H., and Cohen, J. S.,** *Biochem. Biophys. Res. Commun.,* 37, 47, 1969.
681. **Horsley, W. J., Sternlicht, H., and Cohen, J. S.,** *J. Am. Chem. Soc.,* 92, 680, 1970.
682. **Horton, D. and Jewell, J. S.,** *Carbohydr. Res.,* 2, 251, 1966; 3, 255, 1966.
683. **Horton, D. and Turner, W. N.,** *J. Org. Chem.,* 30, 3387, 1965.
684. **Horváth, T., Sohár, P., and Ábrahám, G.,** *Carbohydr. Res.,* 73, 277, 1979.
685. **Hoult, D. I. and Lauterbur, P. C.,** *J. Magn. Reson.,* 34, 425, 1979.
686. **House, H. O., Latham, R. A., and Whitesides, G. M.,** *J. Org. Chem.,* 32, 2481, 1967.
687. **Howard, B. B., Linder, B., and Emerson, M. T.,** *J. Chem. Phys.,* 36, 485, 1962.
688. **Howarth, O. W., and Lynch, R. J.,** *Mol. Phys.,* 15, 431, 1968.
689. **Hubbard, P. S.,** *Phys. Rev.,* 131, 1155, 1963.
690. **Huggins, C. M., Pimentel, G. C., and Shoolery, J. N.,** *J. Phys. Chem.,* 60, 1311, 1956.
691. **Huggins, M. L.,** *J. Am. Chem. Soc.,* 75, 4123, 1953.
692. **Huitric, A. C., Roll, D. B., and Deboer, J. R.,** *J. Org. Chem.,* 32, 1661, 1967.
693. **Ihrig, A. M. and Marshall, J. L.,** *J. Am. Chem. Soc.,* 94, 1756, 1972.
694. **Iijima, N. and Fujii, N.,** *JEOL (Jpn. Electron Opt. Lab.) News,* 9a, 5, 1972.
695. **Imanari, M., Kohno, M., Ohuchi, M., and Ishizu, K.,** *Bull. Chem. Soc. Jpn.,* 47, 708, 1974.
696. **Inamoto, N., Kushida, K., Masuda, S., Ohta, H., Satoh, S., Tamura, Y., Tokumaru, K., Tori, K., and Yoshida, M.,** *Tetrahedron Lett.,* 3617, 1974.
697. **Inch, T. D.,** Nuclear magnetic resonance spectroscopy in the study of carbohydrates and related compounds, in *Annual Reports on NMR Spectroscopy,* Vol. 2, Mooney, E. F., Ed., Academic Press, New York, 1969, 35.
698. **Inch, T. D.,** Nuclear magnetic resonance spectroscopy in the study of carbohydrates and related compounds, in *Annual Reports on NMR Spectroscopy,* Vol. 5A, Mooney, E. F., Ed., Academic Press, New York, 1972, 305.
699. **Inch, T. D., Plimmer, J. R., and Fletcher, H. G., Jr.,** *J. Org. Chem.,* 31, 1825, 1966.
700. **Inch, W. R., McCredie, J. A., Geiger, C., and Boctor, Y.,** *J. Natl. Cancer Inst.,* 53, 689, 1974.
701. **Inch, W. R., McCredie, J. A., Knispel, R. R., Thompson, R. T., and Pintar, M. M.,** *J. Natl. Cancer Inst.,* 52, 353, 1974.
702. **Ingold, C. K.,** *Structure and Mechanism in Organic Chemistry,* G. Bell and Sons, London, 1953.
703. **Inoue, Y., Chûjô, R., and Nishioka, A.,** *Polym. J.,* 4, 244, 1973.
704. **Irving, C. S. and Lapidot, A.,** *Chem. Commun.,* 43, 1976.
705. **Jackman, L. M. and Kelley, D. P.,** *J. Chem. Soc. B,* 102, 1970.
706. **Jackman, L. M., Sondheimer, F., Amiel, Y., Ben-Efraim, D. A., Gaoni, Y., Wolovsky, R., and Bothner-By, A. A.,** *J. Am. Chem. Soc.,* 84, 4307, 1962.
707. **Jackson, J. A., Lemons, J. F., and Taube, H.,** *J. Chem. Phys.,* 32, 553, 1960.
708. **Jaeckle, H., Haeberlen, U., and Schweitzer, D.,** *J. Magn. Reson.,* 4, 198, 1971.
709. **Jakobsen, P. and Treppendahl, S.,** *Org. Magn. Reson.,* 14, 133, 1980.
710. **James, T. L. and Gillen, K. T.,** *Biochim. Biophys. Acta,* 286, 10, 1972.
711. **Jameson, C. J. and Damasco, M. C.,** *Mol. Phys.,* 18, 491, 1970.
712. **Jameson, C. J. and Gutowsky, H. S.,** *J. Chem. Phys.,* 51, 2790, 1969.
713. **Jardine, R. V. and Brown, R. K.,** *Can. J. Chem.,* 41, 2067, 1963.
714. **Jautelat, M., Grutzner, J. B., and Roberts, J. D.,** *Proc. Natl. Acad. Sci.,* 65, 288, 1970.
715. **Jeener, J.,** Ampere Int. Summer School II, *Basko Polje, Yugoslavia, 1971.*
716. **Jennings, W. B., Boyd, D. R., Watson, C. G., Becker, E. D., Bradley, R. B., and Jerina, D. M.,** *J. Am. Chem. Soc.,* 94, 8501, 1972.
717. **Jensen, F. R., Noyce, D. S., Sederholm, C. H., and Berlin, A. J.,** *J. Am. Chem. Soc.,* 84, 386, 1962.
718. **Jensen, F. R. and Smith, L. A.,** *J. Am. Chem. Soc.,* 86, 956, 1964.
719. *JEOL (Jpn. Electron Opt. Lab.) News,* 7C, 16, 1970.
720. *JEOL NMR Applications,* JEOL Co., Ltd., Tokyo, Japan.
721. **Jesson, J. P., Meakin, P., and Kneissel, G.,** *J. Am. Chem. Soc.,* 95, 618, 1973.
722. **Johns, S. R. and Willing, R. J.,** *Aust. J. Chem.,* 29, 1617, 1976.
723. **Johnson, B. F. G., Lewis, J., McArdle, P., and Norton, J. R.,** *Chem. Commun.,* 535, 1972.
724. **Johnson, C. E. and Bovey, F. A.,** *J. Chem. Phys.,* 29, 1012, 1958.

725. **Johnson, C. S., Jr., Weiner, M. A., Waugh, J. S., and Seyferth, D.,** *J. Am. Chem. Soc.*, 83, 1306, 1961.
726. **Johnson, F. P., Melera, A., and Sternhell, S.,** *Aust. J. Chem.*, 19, 1523, 1966.
727. **Johnson, L. F.,** *VARIAN Assoc. Techn. Inform. Bull.*, 6, 1965.
728. **Johnson, L. F., Heatley, F., and Bovey, F. A.,** *Macromolecules,* 3, 175, 1970.
729. **Johnson, L. F. and Jankowski, W. C.,** *C-13 NMR Spectra,* Wiley-Interscience, New York, 1972.
730. **Johnston, M. D., Jr., Shapiro, B. L., Shapiro, M. J., Proulx, T. W., Godwin, A. D., and Pearce, H. L.,** *J. Am. Chem. Soc.*, 97, 542, 1975.
731. **Jolley, K. W., Sutcliffe, L. H., and Walker, S. M.,** *Trans. Faraday Soc.*, 64, 269, 1968.
732. **Jonathan, N., Gordon, S., and Dailey, B. P.,** *J. Chem. Phys.*, 36, 2443, 1962.
733. **Jones, A. J., Alger, T. D., Grant, D. M., and Litchman, W. M.,** *J. Am. Chem. Soc.*, 92, 2386, 1970.
734. **Jones, A. J., Beeman, C. P., Hasan, M. U., Casy, A. F., and Hassan, M. M. A.,** *Can. J. Chem.*, 54, 126, 1976.
735. **Jones, A. J., Gardner, P. D., Grant, D. M., Litchman, W. M., and Boekelheide, V.,** *J. Am. Chem. Soc.*, 92, 2395, 1970.
736. **Jones, A. J. and Grant, D. M.,** *Chem. Commun.*, 1670, 1968.
737. **Jones, A. J., Grant, D. M., and Kuhlmann, K.,** *J. Am. Chem. Soc.*, 91, 5013, 1969.
738. **Jones, A. J., Grant, D. M., Winkley, M. W., and Robins, R. K.,** *J. Am. Chem. Soc.*, 92, 4079, 1970.
739. **Jones, K. and Mooney, E. F.,** Fluorine-19 nuclear magnetic resonance spectroscopy, in *Annual Reports on NMR Spectroscopy,* Vol. 3, Mooney, E. F., Ed., Academic Press, New York, 1970, 261.
740. **Jones, K. and Mooney, E. F.,** Fluorine-19 nuclear magnetic resonance spectroscopy, in *Annual Reports on NMR Spectroscopy,* Vol. 4, Mooney, E. F., Ed., Academic Press, New York, 1971, 391.
741. **Joseph-Nathan, P., Herz, J. E., and Rodriguez, V. M.,** *Can. J. Chem.*, 50, 2788, 1972.
742. **Kaiser, R.,** *J. Chem. Phys.*, 42, 1838, 1965.
743. **Kaiser, R.,** *J. Magn. Reson.*, 3, 28, 1970.
744. **Kalinowski, H. and Kessler, H.,** *Angew. Chem. Int. Ed. Engl.*, 13, 90, 1974.
745. **Kametani, T., Fukumoto, K., Ihara, M., Ujiie, A., and Koizumi, H.,** *J. Org. Chem.*, 40, 3280, 1975.
746. **Kametani, T., Ujiie, A., Ihara, M., Fukumoto, K., and Koizumi, H.,** *Heterocycles,* 3, 371, 1975.
747. **Kaptein, R.,** *Chem. Commun.*, 732, 1971.
748. **Kaptein, R. and Oosterhoff, J. L.,** *Chem. Phys. Lett.*, 4, 195, 214, 1969.
749. **Karabatsos, G. J., Graham, J. D., and Vane, F. M.,** *J. Am. Chem. Soc.*, 84, 37, 1962.
750. **Karabatsos, G. J. and Vane, F. M.,** *J. Am. Chem. Soc.*, 85, 3886, 1963.
751. **Karplus, M.,** *J. Chem. Phys.*, 30, 11, 1959.
752. **Karplus, M.,** *J. Chem. Phys.*, 33, 1842, 1960.
753. **Karplus, M.,** *J. Am. Chem. Soc.*, 84, 2458, 1962.
754. **Karplus, M. and Anderson, D. H.,** *J. Chem. Phys.*, 30, 6, 1959.
755. **Karplus, M. and Grant, D. M.,** *Proc. Natl. Acad. Sci. U.S.A.*, 45, 1269, 1959.
756. **Karplus, M. and Pople, J. A.,** *J. Chem. Phys.*, 38, 2803, 1963.
757. **Kato, H. and Yonezawa, T.,** *Bull. Chem. Soc. Jpn.*, 43, 1921, 1970.
758. **Kato, Y. and Saika, A.,** *J. Chem. Phys.*, 46, 1975, 1967.
759. **Katritzky, A. R. and Maine, F. W.,** *Tetrahedron,* 20, 299, 1964.
760. **Kawazoe, Y. and Tsuda, M.,** *Chem. Pharm. Bull.*, 15, 1405, 1967.
761. **Kawazoe, Y., Tsuda, M., and Ohniski, M.,** *Chem. Pharm. Bull.*, 15, 51, 1967.
762. **Keller, C. E. and Petitt, R.,** *J. Am. Chem. Soc.*, 88, 604, 606, 1966.
763. **Kellie, G. M. and Riddell, F. G.,** *J. Chem. Soc. B,* 1030, 1971.
764. **Kessler, H.,** *Angew. Chem.*, 82, 237, 1970.
765. **Khuong-Huu, F., Sangare, M., Chari, V. M., Bekaert, A., Devys, M., Barbier, M., and Lukacs, G.,** *Tetrahedron Lett.*, 1787, 1975.
766. **Kiefer, E. F., Gericke, W., and Amimoto, S. T.,** *J. Am. Chem. Soc.*, 90, 6246, 1968.
767. **King, M. M., Yeh, H. J. C., and Dudek, G. O.,** *Org. Magn. Reson.*, 8, 208, 1976.
768. **Kintzinger, J. P. and Lehn, J. M.,** *Helv. Chim. Acta,* 58, 905, 1975.
769. **Kitching, W., Bullpitt, M., Garsthore, D., Adcock, W., Khor, T. C., Doddrell, D., and Rae, I. D.,** *J. Org. Chem.*, 42, 2411, 1977.
770. **Kleinfelter, D. C.,** *J. Am. Chem. Soc.*, 89, 1734, 1967.
771. **Klesper, E., Johnson, A., Gronski, W., and Wehrli, F. W.,** *Macromol. Chem.*, 176, 1071, 1975.
772. **Klinck, R. E. and Stothers, J. B.,** *Can. J. Chem.*, 40, 2329, 1962.
773. **Knight, S. A.,** *Org. Magn. Reson.*, 6, 603, 1973.
774. **Knight, W. D.,** *Phys. Res.*, 76, 1259, 1949.
775. **Knox, L. H., Velarde, E., and Cross, A. D.,** *J. Am. Chem. Soc.*, 85, 2533, 1963.
776. **Koenig, S. H., Hallenga, K., and Shporer, M.,** *Proc. Natl. Acad. Sci. U.S.A.*, 72, 2667, 1975.
777. **Koer, F. J., de Hoog, A. J., and Altona, C.,** *Recl. Trav. Chim. Pays-Bas,* 94, 75, 1975.

778. **Komoroski, R. A., Peat, I. R., and Levy, G. C.,** ^{13}C NMR studies of biopolymers, in *Topics in Carbon-13 NMR Spectroscopy,* Vol. 2, Levy, G. C., Ed., Wiley-Interscience, New York, 1976, 180.
779. **Korver, P. K., Haak, P. J., Steinberg, H., and Deboer, T. J.,** *Recl. Trav. Chim.*, 84, 129, 1965.
780. **Koutcher, J. A. and Damadian, R.,** *Physiol. Chem. Phys.*, 9, 181, 1977.
781. **Kowalsky, A. and Cohn, M.,** *Annu. Rev. Biochem.*, 33, 481, 1964.
782. **Kraus, W. and Suhr, H.,** *Annalen,* 695, 27, 1966.
783. **Kreishman, G. P., Witkowski, J. T., Robins, R. K., and Schweizer, M. P.,** *J. Am. Chem. Soc.*, 94, 5894, 1972.
784. **Kristiansen, P. and Ledaal, T.,** *Tetrahedron Lett.*, 4457, 1971.
785. **Kroschwitz, J. I., Winokur, M., Reich, H. J., and Roberts, J. D.,** *J. Am. Chem. Soc.*, 91, 5927, 1969.
786. **Krow, G. R. and Ramey, K. C.,** *Tetrahedron Lett.*, 3143, 1971.
787. **Kruczynski, L., Lishingman, L. K. K., and Takats, J.,** *J. Am. Chem. Soc.*, 96, 4006, 1974.
788. **Krueger, G. L., Kaplan, F., Orchin, M., and Faul, W. H.,** *Tetrahedron Lett.*, 3979, 1965.
789. **Kuhlmann, K. F. and Grant, D. M.,** *J. Chem. Phys.*, 55, 2998, 1971.
790. **Kuhlmann, K. F., Grant, D. M., and Harris, R. K.,** *J. Chem. Phys.*, 52, 3439, 1970.
791. **Kulka, M.,** *Can. J. Chem.*, 42, 2791, 1964.
792. **Kumar, A., Aue, W. P., Bachmann, P., Karhan, J., Müller, L., and Ernst, R. R.,** Two-dimensional spin-echo spectroscopy. A means to resolve proton and carbon NMR spectra, in *Magnetic Resonance and Related Phenomena. Proc. XIXth Congr. Ampère, Heidelberg, 1976,* Brunner, H., Hausser, K. H., and Schweitzer, D., Eds., Groupement Ampère, Heidelberg, 1976, 473.
793. **Kumar, A. and Ernst, R. R.,** *Chem. Phys. Lett.*, 37, 162, 1976.
794. **Kumar, A. and Ernst, R. R.,** *J. Magn. Reson.*, 24, 425, 1976.
795. **Kumar, A., Welti, D., and Ernst, R. R.,** *J. Magn. Reson.*, 18, 69, 1975.
796. **Kurland, R. J., Rubin, M. B., and Wyse, W. B.,** *J. Chem. Phys.*, 40, 2426, 1964.
797. **Kuszmann, J. and Sohár, P.,** *Carbohydr. Res.*, 14, 415, 1970.
798. **Kuszmann, J. and Sohár, P.,** *Carbohydr. Res.*, 21, 19, 1972.
799. **Kuszmann, J. and Sohár, P.,** *Carbohydr. Res.*, 27, 157, 1973.
800. **Kuszmann, J. and Sohár, P.,** *Carbohydr. Res.*, 35, 97, 1974.
801. **Kuszmann, J. and Sohár, P.,** *Acta Chim. Acad. Sci. Hung.*, 83, 373, 1974.
802. **Kuszmann, J., Sohár, P., and Horváth, Gy.,** *Tetrahedron,* 27, 5055, 1971.
803. **Kuszmann, J., Sohár, P., Horváth, Gy., and Méhesfalvi-Vajna, Zs.,** *Tetrahedron,* 30, 3905, 1974.
804. **Kuszmann, J. and Vargha, L.,** *Carbohydr. Res.*, 17, 309, 1971.
805. **La Mar, G. N. and Faller, J. W.,** *J. Am. Chem. Soc.*, 95, 3818, 1973.
806. **La Mar, G. N., Horrocks, W. DeW., Jr., and Allen, L. C.,** *J. Chem. Phys.*, 41, 2126, 1964.
807. **Lamb, W. E.,** *Phys. Rev.*, 60, 817, 1941.
808. **Lambert, J. B., Binsch, G., and Roberts, J. D.,** *Proc. Natl. Acad. Sci. U.S.A.*, 51, 735, 1964.
809. **Lambert, J. B. and Keske, R. G.,** *J. Am. Chem. Soc.*, 88, 620, 1966.
810. **Lambert, J. B., Netzel, D. A., Sun, H.-N., and Lilianstrom, K. K.,** *J. Am. Chem. Soc.*, 98, 3778, 1976.
811. **Lambert, J. B., Roberts, B. W., Binsch, G., and Roberts, J. D.,** ^{15}N Magnetic resonance spectroscopy, in *Nuclear Magnetic Resonance in Chemistry,* Pesce, B., Ed., Academic Press, New York, 1965, 269.
812. **Lambert, J. B. and Roberts, J. D.,** *J. Am. Chem. Soc.*, 85, 3710, 1963.
813. **Lambert, J. B. and Roberts, J. D.,** *J. Am. Chem. Soc.*, 87, 3884, 3891, 1965.
814. **Lambert, J. B. and Roberts, J. D.,** *J. Am. Chem. Soc.*, 87, 4087, 1965.
815. **László, P.,** Solvent effects and NMR, in *Progress in Nuclear Magnetic Resonance Spectroscopy,* Vol. 3, Emsley, J. W., Feeney, J., and Sutcliffe, L. H., Eds., Pergamon Press, New York, 1967, 231.
816. **Lauterbur, P. C.,** *J. Chem. Phys.*, 26, 217, 1957.
817. **Lauterbur, P. C.,** *Ann. N. Y. Acad. Sci.*, 70, 841, 1958.
818. **Lauterbur, P. C.,** *Tetrahedron Lett.*, 274, 1961.
819. **Lauterbur, P. C.,** *J. Am. Chem. Soc.*, 83, 1838, 1961.
820. **Lauterbur, P. C.,** *J. Am. Chem. Soc.*, 83, 1846, 1961.
821. **Lauterbur, P. C.,** Nuclear magnetic resonance spectra of elements other than hydrogen and fluorine, in *Determination of Organic Structures by Physical Methods,* Vol. 2, Nachod, F. C. and Phillips, W. D., Eds., Academic Press, New York, 1962, 465.
822. **Lauterbur, P. C.,** *J. Chem. Phys.*, 38, 1406, 1963.
823. **Lauterbur, P. C.,** *J. Chem. Phys.*, 38, 1415, 1963.
824. **Lauterbur, P. C.,** *J. Chem. Phys.*, 38, 1432, 1963.
825. **Lauterbur, P. C.,** *J. Chem. Phys.*, 43, 360, 1965.
826. **Lauterbur, P. C.,** *Nature (London),* 242, 190, 1973.
827. **Lauterbur, P. C.,** *NMR in Biology,* Dwek, R. A., Campbell, I. D., Richards, R. E., and Williams, R. J. P., Academic Press, London, 1977, 323.
828. **Lauterbur, P. C. and King, R. B.,** *J. Am. Chem. Soc.*, 87, 3266, 1965.

829. **Lawler, R. G.,** *J. Am. Chem. Soc.*, 89, 5519, 1967.
830. **Lawler, R. G.,** *Acc. Chem. Res.*, 5, 25, 1972.
831. **Lawler, R. G.,** Chemically induced dynamic nuclear polarization, in *Progress in NMR Spectroscopy,* Vol. 9, Emsley, J. W., Feeney, J., and Sutcliffe, L. H., Eds., Pergamon Press, Oxford, 1973, 145.
832. **Lee, K. and Anderson, W. A.,** *Handbook of Chemistry and Physics,* 55th ed., Weast, R. C., Ed., Chemical Rubber Co., Cleveland, 1974, E69.
833. **Le Févre, C. G. and Le Févre, R. J. W.,** *J. Chem. Soc.*, 3549, 1956.
834. **Lehn, J. M. and Riehl, J. J.,** *J. Mol. Phys.*, 8, 33, 1964.
835. **Lehn, J. M. and Wagner, J.,** *Chem. Commun.*, 148, 1968.
836. **Lemieux, R. U.,** in *Molecular Rearrangements,* De Mayo, A., Ed., Interscience, New York, 1963.
837. **Lemieux, R. U.,** *Ann. N. Y. Acad. Sci.*, 222, 915, 1973.
838. **Lemieux, R. U., Kullnig, R. K., Bernstein, H. J., and Schneider, W. G.,** *J. Am. Chem. Soc.*, 79, 1005, 1957.
839. **Lemieux, R. U. and Stevens, J. D.,** *Can. J. Chem.*, 44, 249, 1966.
840. **Lempert-Sréter, M. and Sohár, P.,** *Acta Chim. Acad. Sci. Hung.*, 54, 203, 1967.
841. **Lepley, A. R. and Closs, G. L.,** *Chemically Induced Magnetic Polarization,* Wiley-Interscience, New York, 1973.
842. **Levin, R. H. and Roberts, J. D.,** *Tetrahedron Lett.*, 135, 1973.
843. **Levy, G. C.,** *Chem. Commun.*, 352, 1972.
844. **Levy, G. C.,** *J. Magn. Reson.*, 8, 122, 1972.
845. **Levy, G. C.,** *Tetrahedron Lett.*, 3709, 1972.
846. **Levy, G. C.,** *Acc. Chem. Res.*, 6, 161, 1973.
847. **Levy, G. C. and Cargioli, J. D.,** *J. Magn. Reson.*, 6, 143, 1972.
848. **Levy, G. C. and Cargioli, J. D.,** *J. Magn. Reson.*, 10, 231, 1973.
849. **Levy, G. C., Cargioli, J. D., and Anet, F. A. L.,** *J. Am. Chem. Soc.*, 95, 1527, 1973.
850. **Levy, G. C. and Dittmer, D. C.,** *Org. Magn. Reson.*, 4, 107, 1972.
851. **Levy, G. C. and Edlund, U.,** *J. Am. Chem. Soc.*, 97, 5031, 1975.
852. **Levy, G. C. and Komoroski, R. A.,** *J. Am. Chem. Soc.*, 96, 678, 1974.
853. **Levy, G. C., Komoroski, R. A., and Echols, R. E.,** *Org. Magn. Reson.*, 7, 172, 1975.
854. **Levy, G. C., Komoroski, R. A., and Halstead, J. A.,** *J. Am. Chem. Soc.*, 96, 5456, 1974.
855. **Levy, G. C. and Nelson, G. L.,** *Carbon-13 Nuclear Magnetic Resonance for Organic Chemists,* Wiley-Interscience, New York, 1972, (a) p. 110, 121, 123; (b) p. 125; (c) p. 81; (d) pp. 176-198.
856. **Levy, G. C. and Nelson, G. L.,** *J. Am. Chem. Soc.*, 94, 4897, 1972.
857. **Levy, G. C., White, D. M., and Anet, F. A. L.,** *J. Magn. Reson.*, 6, 453, 1972.
858. **Lewis, W. B., Jackson, J. A., Lemons, J. F., and Taube, H.,** *J. Chem. Phys.*, 36, 694, 1962.
859. **Lichtenthaler, F. W.,** *Chem. Ber.*, 96, 845, 2047, 1963.
860. **Lichter, R. L.,** *Determination of Organic Structures by Physical Methods,* Vol. 4, Nachod, F. C. and Zuckerman, J. J., Eds., Academic Press, New York, 1971, 195.
861. **Lichter, R. L., Dorman, D. E., and Wasylishen, R.,** *J. Am. Chem. Soc.*, 96, 930, 1974.
862. **Lichter, R. L. and Roberts, J. D.,** *J. Chem. Phys.*, 74, 912, 1970.
863. **Lichter, R. L. and Roberts, J. D.,** *J. Am. Chem. Soc.*, 93, 5218, 1971.
864. **Lichter, R. L. and Roberts, J. D.,** *J. Am. Chem. Soc.*, 94, 2495, 1972.
865. **Lichter, R. L. and Roberts, J. D.,** *J. Am. Chem. Soc.*, 94, 4904, 1972.
866. **Lillien, I. and Doughty, R. A.,** *J. Am. Chem. Soc.*, 89, 155, 1967.
867. **Lindeman, L. P. and Adams, J. Q.,** *Anal. Chem.*, 43, 1245, 1971.
868. **Lindström, G.,** *Phys. Res.*, 78, 1817, 1950.
869. **Ling, G. N.,** *A Physical Theory of the Living State,* Blaisdel, New York, 1962.
870. **Ling, G. N.,** *Ann. N. Y. Acad. Sci.*, 125, 401, 1965.
871. **Lippert, E. and Prigge, H.,** *Ber. Bunsenges Phys. Chem.*, 67, 415, 1963.
872. **Lippmaa, E., Magi, M., Novikov, S. S., Khmelnitski, L. I., Prihodko, A. S., Lebedev, O. V., and Epishina, L. V.,** *Org. Magn. Reson.*, 4, 153, 1972.
873. **Lippmaa, E., Magi, M., Novikov, S. S., Khmelnitski, L. I., Prihodko, A. S., Lebedev, O. V., and Epishina, L. V.,** *Org. Magn. Reson.*, 4, 197, 1972.
874. **Lippmaa, E. and Pehk, T.,** *Kem. Teollisuus,* 24, 1001, 1967.
875. **Lippmaa, E. and Pehk, T.,** *Eesti NSV Tead. Akad. Toim. Keem. Geol.*, 17, 210, 1968.
876. **Lippmaa, E. and Pehk, T.,** *Eesti NSV Tead. Akad. Toim. Keem. Geol.*, 17, 287, 1968.
877. **Lippmaa, E., Pehk, T., Andersson, K., and Rappe, C.,** *Org. Magn. Reson.*, 2, 109, 1970.
878. **Lippmaa, E., Pehk, T., and Past, J.,** *Eesti NSV Tead. Akad. Toim. Fuus. Mat.*, 16, 345, 1967.
879. **Lippmaa, E., Saluvere, T., and Laisaar, S.,** *Chem. Phys. Lett.*, 11, 120, 1971.
880. **Liska, K. J., Fentiman, A. F., Jr., and Foltz, R. L.,** *Tetrahedron Lett.*, 4657, 1970.
881. **Litchman, W. M. and Grant, D. M.,** *J. Am. Chem. Soc.*, 89, 6775, 1967.

882. **Litchman, W. M. and Grant, D. M.,** *J. Am. Chem. Soc.*, 90, 1400, 1968.
883. **Llinas, J.-R., Vincent, E.-J., and Peiffer, G.,** *Bull. Soc. Chim. France,* 3209, 1973.
884. **Looney, C. E., Phillips, W. D., and Reilly, E. L.,** *J. Am. Chem. Soc.*, 79, 6136, 1957.
884a. **Löw, M., Kisfaludy, L., and Sohár, P.,** *Z. Physiol. Chem.*, 359, 1643, 1978.
885. **Lukacs, G., Khuong-Huu, F., Bennett, C. R., Buckwalter, B. L., and Wenkert, E.,** *Tetrahedron Lett.*, 3515, 1972.
886. **Lunazzi, L., Macciantelli, D., and Boicelli, C. A.,** *Tetrahedron Lett.*, 1205, 1975.
887. **Lunazzi, L., Macciantelli, D., and Taddei, F.,** *Mol. Phys.*, 19, 137, 1970.
888. **Lustig, E., Benson, W. R., and Duy, N.,** *J. Org. Chem.*, 32, 851, 1967.
889. **Lustig, E., Ragelis, E. P., and Duy, N.,** *Spectrochim. Acta,* 23a, 133, 1967.
890. **Lyerla, J. R., Jr., Grant, D. M., and Bertrand, R. D.,** *J. Phys. Chem.*, 75, 3967, 1971.
891. **Lyerla, J. R., Jr., Grant, D. M., and Harris, R. K.,** *J. Phys. Chem.*, 75, 585, 1971.
892. **Lyerla, J. R., Jr. and Levy, G. C.,** Carbon-13 nuclear spin relaxation, in *Topics in Carbon-13 NMR Spectroscopy,* Vol. 1, Levy, G. C., Ed., Wiley-Interscience, New York, 1974, 79; (a) p. 88; (b) p. 96.
893. **Lynch, D. M. and Cole, W.,** *J. Org. Chem.*, 31, 3337, 1966.
894. **Lynden-Bell, R. M. and Sheppard, N.,** *Proc. R. Soc. London Ser. A,* 269, 385, 1962.
895. **Maciel, G. E.,** *J. Phys. Chem.*, 69, 1974, 1965.
896. **Maciel, G. E.,** ^{13}C-^{13}C coupling constants, in *Nuclear Magnetic Resonance Spectroscopy of Nuclei Other Than Protons,* Axenrod, T. and Webb, G. A., Eds., Wiley-Interscience, New York, 1974, 187.
897. **Maciel, G. E. and Beatty, D. A.,** *J. Phys. Chem.*, 69, 3920, 1965.
898. **Maciel, G. E., Dallas, J. L., and Miller, D. P.,** *J. Am. Chem. Soc.*, 98, 5074, 1976.
899. **Maciel, G. E., Ellis, P. D., and Hofer, D. C.,** *J. Phys. Chem.*, 71, 2160, 1967.
900. **Maciel, G. E., Ellis, P. D., Natterstad, J. J., and Savitsky, G. B.,** *J. Magn. Reson.*, 1, 589, 1969.
901. **Maciel, G. E. and James, R. V.,** *J. Am. Chem. Soc.*, 86, 3893, 1964.
902. **Maciel, G. E., McIver, J. W., Jr., Ostlund, N. S., and Pople, J. A.,** *J. Am. Chem. Soc.*, 92, 1, 11, 1970.
903. **Maciel, G. E. and Natterstad, J. J.,** *J. Chem. Phys.*, 42, 2752, 1965.
904. **Maciel, G. E. and Savitsky, G. B.,** *J. Phys. Chem.*, 69, 3925, 1965.
905. **Maciel, G. E. and Traficante, D. D.,** *J. Am. Chem. Soc.*, 88, 220, 1966.
906. **MacNicol, D. D.,** *Tetrahedron Lett.*, 3325, 1975.
907. **Magi, M., Erashko, V. I., Shevelev, S. A., and Fainzil'berg, A. A.,** *Eesti Nsv. Tead. Toim. Keem. Geol.*, 20, 297, 1971.
908. **Maia, H. L., Orrell, K. G., and Rydon, H. N.,** *Chem. Commun.*, 1209, 1971.
909. **Maksić, Z. B., Eckert-Maksić, M., and Randic, M.,** *Theor. Chim. Acta,* 22, 70, 1971.
910. **Malinowski, E. R.,** *J. Am. Chem. Soc.*, 83, 4479, 1961.
911. **Malinowski, E. R., Pollara, L. Z., and Larmann, J. P.,** *J. Am. Chem. Soc.*, 84, 2649, 1962.
912. **Malinowski, E. R., Vladimirov, T., and Tavares, R. F.,** *J. Phys. Chem.*, 70, 2046, 1966.
913. **Manatt, S. L.,** *J. Am. Chem. Soc.*, 88, 1323, 1966.
914. **Manatt, S. L. and Bowers, J.,** *J. Am. Chem. Soc.*, 91, 4381, 1969.
915. **Manatt, S. L., Elleman, D. D., and Brois, S. J.,** *J. Am. Chem. Soc.*, 87, 2220, 1965.
916. **Mandel, F. S., Cox, R. H., and Taylor, R. C.,** *J. Magn. Reson.*, 14, 235, 1974.
917. **Mandel, M.,** *J. Biol. Chem.*, 240, 1586, 1965.
918. **Mann, B. E.,** Dynamic ^{13}C NMR spectroscopy, in *Progress in Nuclear Magnetic Resonance Spectroscopy,* Vol. 11, Emsley, J. W., Feeney, J., and Sutcliffe, L. H., Eds., Pergamon Press, Oxford, 1977, 95.
919. **Mann, B. E.,** The common nuclei. Fluorine ^{19}F, in *NMR and the Periodic Table,* Harris, R. K. and Mann, B. E., Eds., Academic Press, New York, 1978, 98.
920. **Mann, B. E.,** The common nuclei. Phosphorus ^{31}P, in *NMR and the Periodic Table,* Harris, R. K. and Mann, B. E., Eds., Academic Press, New York, 1978, 100.
921. **Mannschreck, A.,** *Tetrahedron Lett.*, 1341, 1965.
922. **Mannschreck, A., Seitz, W., and Staab, H. A.,** *Ber. Bunsenges Phys. Chem.*, 67, 471, 1963.
923. **Mansfield, P. and Maudsley, A. A.,** *Br. J. Radiol.*, 50, 188, 1977.
924. **Marchal, J. P. and Canet, D.,** *J. Am. Chem. Soc.*, 97, 6581, 1975.
925. **Marcus, S. H. and Miller, S. I.,** *J. Am. Chem. Soc.*, 88, 3719, 1966.
926. **Markley, J. L., Horsley, W. J., and Klein, M. P.,** *J. Chem. Phys.*, 55, 3604, 1971.
927. **Markowski, V., Sullivan, G. R., and Roberts, J. D.,** *J. Am. Chem. Soc.*, 99, 714, 1977.
928. **Marr, D. H. and Stothers, J. B.,** *Can. J. Chem.*, 43, 596, 1965.
929. **Marr, D. H. and Stothers, J. B.,** *Can. J. Chem.*, 45, 225, 1967.
930. **Marshall, J. L., Faehl, L. G., Ihrig, A. M., and Barfield, M.,** *J. Am. Chem. Soc.*, 98, 3406, 1976.
931. **Marshall, J. L. and Ihrig, A. M.,** *Org. Magn. Reson.*, 5, 235, 1971.
932. **Marshall, J. L., Ihrig, A. M., and Miiller, D. E.,** *J. Magn. Reson.*, 16, 439, 1974.
933. **Marshall, J. L. and Miiller, D. E.,** *J. Am. Chem. Soc.*, 95, 8305, 1973.

934. **Marshall, J. L., Miiller, D. E., Conn, S. A., Seitwell, R., and Ihrig, A. M.,** *Acc. Chem. Res.*, 7, 333, 1974.
935. **Marshall, T. W. and Pople, J. A.,** *Mol. Phys.*, 1, 199, 1958.
936. **Marshall, T. W. and Pople, J. A.,** *Mol. Phys.*, 3, 339, 1960.
937. **Martin, G. J., Gouesnard, J. P., Dorie, J., Rabilles, C., and Martin, M. L.,** *J. Am. Chem. Soc.*, 99, 1381, 1977.
938. **Martin, J. S. and Dailey, B. P.,** *J. Chem. Phys.*, 39, 1722, 1963.
939. **Martin, L. L., Chang, C. J., Floss, H. G., Mabe, J. A., Hagaman, E. W., and Wenkert, E.,** *J. Am. Chem. Soc.*, 94, 8942, 1972.
940. **Martin, R. H., Defay, N., Figeys, H. P., F.-Barbieux, M., Cosyn, J. P., Gelbcke, M., and Schurter, J. J.,** *Tetrahedron,* 25, 4985, 1969.
941. **Maryott, A. A., Farrar, T. C., and Malmberg, M. S.,** *J. Chem. Phys.*, 54, 64, 1971.
942. **Masamune, S., Kemp-Jones, A. V., Green, J., Rabenstein, D. L., Yasunami, M., Takase, K., and Nozoe, T.,** *Chem. Commun.*, 283, 1973.
943. **Mason, J.,** *J. Chem. Soc. A*, 1038, 1971.
944. **Mason, J. and van Bronswijk, W.,** *J. Chem. Soc. A,* 1763, 1970.
945. **Mason, J. and van Bronswijk, W.,** *J. Chem. Soc. A,* 791, 1971.
946. **Mason, J. and Vinter, J. G.,** *J. Chem. Soc. (Dalton),* 2522, 1975.
947. **Mathias, A.,** *Tetrahedron,* 22, 217, 1966.
948. **Matsuo, T. and Shosenji, H.,** *Chem. Commun.*, 501, 1969.
949. **Matsuura, S. and Goto, T.,** *Tetrahedron Lett.*, 1499, 1963.
950. **Matsuzaki, K., Ito, H., Kawamura, T., and Uryu, T.,** *J. Polym. Sci.*, 11, 971, 1973.
951. **Matsuzaki, K., Kanai, T., Kawamura, T., Matsumoto, S., and Uryu, T.,** *J. Polym. Sci.*, 11, 961, 1973.
952. **Matter, U. E., Pascual, C., Pretsch, E., Pross, A., Simon, W., and Sternhell, S.,** *Tetrahedron,* 25, 691, 1969.
953. **Matusch, R.,** *Angew. Chem.*, 87, 283, 1975.
954. **Mavel, G.,** NMR studies of phosphorus compounds, in *Annual Reports on NMR Spectroscopy,* Vol. 5B, Mooney, E. F., Ed., Academic Press, New York, 1973, 1.
955. **Maxwell, L. R. and Bennett, L. H.,** *Physiol. Chem. Phys.*, 10, 59, 1978.
956. **Mayo, R. E. and Goldstein, J. H.,** *J. Mol. Spectrosc.*, 14, 173, 1964.
957. **McConnel, H. M.,** *J. Chem. Phys.*, 24, 460, 1956.
958. **McConnel, H. M.,** *J. Chem. Phys.*, 27, 226, 1957.
959. **McConnel, H. M.,** *J. Chem. Phys.*, 28, 430, 1958.
960. **McConnel, H. M. and Robertson, R. E.,** *J. Chem. Phys.*, 29, 1361, 1958.
961. **McCreary, M. D., Lewis, D. W., Wernick, D. L., and Whitesides, G. M.,** *J. Am. Chem. Soc.*, 96, 1038, 1974.
962. **McDonald, G. G. and Leigh, J. S., Jr.,** *J. Magn. Reson.*, 9, 358, 1973.
963. **McFarlane, H. C. E. and McFarlane, W.,** *Org. Magn. Reson.*, 4, 161, 1972.
964. **McFarlane, H. C. E., McFarlane, W., and Rycroft, D. S.,** *J. Chem. Soc. (Faraday II),* 68, 1300, 1972.
965. **McFarlane, W.,** *Quart. Rev. Chem. Soc.*, 23, 187, 1969.
966. **McInnes, A. G., Walter, J. A., Wright, J. L. C., and Vining, L. C.,** ^{13}C NMR biosynthetic studies, in *Topics in Carbon-13 NMR Spectroscopy,* Vol. 2, Levy, G. C., Ed., Wiley-Interscience, New York, 1976, 123.
967. **McKeever, L. D., Waack, R., Doran, M. A., and Baker, E. B.,** *J. Am. Chem. Soc.*, 91, 1057, 1969.
968. **McKenna, J., McKenna, J. M., Tulley, A., and White, J.,** *J. Chem. Soc.*, 1726, 1965.
969. **McLachlan, A. D.,** *J. Chem. Phys.*, 32, 1263, 1960.
970. **McLaughlin, A. C., McDonald, G. G., and Leigh, J. S.,** *J. Magn. Reson.*, 11, 107, 1973.
971. **Mechin, B., Richer, J. C., and Odiot, S.,** *Org. Magn. Reson.*, 14, 79, 1980.
972. **Meiboom, S. and Gill, D.,** *Rev. Sci. Instrum.*, 29, 688, 1958.
973. **Meiboom, S. and Snyder, L. C.,** *J. Am. Chem. Soc.*, 89, 1038, 1967.
974. **Meinwald, J. and Meinwald, Y. C.,** *J. Am. Chem. Soc.*, 85, 2541, 1963.
975. **Merrill, J. R.,** *J. Phys. Chem.*, 65, 2023, 1961.
976. **Merrit, R. F.,** *J. Am. Chem. Soc.*, 89, 609, 1967.
977. **Méhesfalvi-Vajna, Zs.,** *Ph.D. thesis,* Budapest, 1971.
978. **Méhesfalvi-Vajna, Zs., Neszmélyi, A., Baicz, E., and Sohár, P.,** *Acta Chim. Acad. Sci. Hung.*, 86, 159, 1975.
979. **Mislow, K. and Raban, M.,** Stereoisomeric relationships of groups in molecules, in *Topics in Stereochemistry,* Vol. 1, Allinger, N. L. and Eliel, E. L., Eds., Interscience, New York, 1967, 1.
980. **Mison, P., Chaabouni, R., Diab, Y., Martino, R., Lopez, A., Lattes, A., Wehrli, F. W., and Wirthlin, T.,** *Org. Magn. Reson.*, 8, 79, 1976.
981. **Mitchell, R. H., Klopfenstein, C. E., and Boekelheide, V.,** *J. Am. Chem. Soc.*, 91, 4931, 1969.

982. **Miyajima, G. and Takahashi, K.,** *J. Phys. Chem.*, 75, 331, 1971.
983. **Miyajima, G. and Takahashi, K.,** *J. Phys. Chem.*, 75, 3766, 1971.
984. **Miyajima, G., Takahashi, K., and Nishimoto, K.,** *Org. Magn. Reson.*, 6, 413, 1974.
985. **Miyajima, G., Utsumi, Y., and Takahashi, K.,** *J. Phys. Chem.*, 73, 1370, 1969.
986. **Mochel, V. D.,** *Rev. Macromol. Chem.*, C8, 289, 1972.
987. **Mollere, P. D., Houk, K. N., Bomse, D. S., and Morton, T. H.,** *J. Am. Chem. Soc.*, 98, 4732, 1976.
988. **Mondelli, R. and Merlini, L.,** *Tetrahedron,* 22, 3253, 1966.
989. **Moniz, W. B. and Gutowsky, H. S.,** *J. Chem. Phys.*, 38, 1155, 1963.
990. **Moniz, W. B., Poranski, C. F., Jr., and Sojka, S. A.,** ^{13}C CIDNP as a mechanistic and kinetic probe, in *Topics in Carbon-13 NMR Spectroscopy,* Vol. 3, Levy, G. C., Ed., Wiley-Interscience, New York, 1979, 362.
991. **Montaudo, G., Librando, V., Caccamese, S., and Maravigna, P.,** *J. Am. Chem. Soc.*, 95, 6365, 1973.
992. **Mooberry, E. S. and Krugh, T. R.,** *J. Magn. Reson.*, 17, 128, 1975.
993. **Mooney, E. F.,** *An Introduction to* ^{19}*F NMR Spectroscopy,* Heyden, London, 1970.
994. **Mooney, E. F. and Winson, P. H.,** Fluorine-19 nuclear magnetic resonance spectroscopy, in *Annual Reports on NMR Spectroscopy,* Vol. 1, Mooney, E. F., Ed., Academic Press, London, 1968, 243.
995. **Mooney, E. F. and Winson, P. H.,** Nitrogen magnetic resonance spectroscopy, in *Annual Reports on NMR Spectroscopy,* Vol. 2, Mooney, E. F., Ed., Academic Press, London, 1969, 125.
996. **Mooney, E. F. and Winson, P. H.,** Carbon-13 nuclear magnetic resonance spectroscopy, carbon-13 chemical shifts and coupling constants, in *Annual Reports on NMR Spectroscopy,* Vol. 2, Mooney, E. F., Ed., Academic Press, New York, 1969, 153; (a) p. 176.
997. **Moore, G. G. I., Kirk, A. R., and Newmark, R. A.,** *J. Heterocyclic Chem.*, 16, 789, 1979.
998. **Moreland, C. G. and Carroll, F. I.,** *J. Magn. Reson.*, 15, 596, 1974.
999. **Mori, N., Omura, S., Yamamoto, O., Suzuki, T., and Tsuzuki, Y.,** *Bull. Chem. Soc. Jpn.*, 36, 1401, 1963.
1000. **Moriarty, R. M. and Kliegman, J. M.,** *J. Org. Chem.*, 31, 3007, 1966.
1001. **Morishima, I., Mizuno, A., Yonezawa, T., and Goto, K.,** *Chem. Commun.*, 1321, 1970.
1002. **Morishima, I., Yoshikawa, K., Okada, K., Yonezawa, T., and Goto, K.,** *J. Am. Chem. Soc.*, 95, 165, 1973.
1003. **Morrill, T. C., Opitz, R. J., and Mozzer, R.,** *Tetrahedron Lett.*, 3715, 1973.
1004. **Morris, D. G. and Murray, A. M.,** *J. Chem. Soc. Perkin Trans. 2,* 1579, 1976.
1005. **Mortimer, F. S.,** *J. Mol. Spectrosc.*, 5, 199, 1960.
1006. **Moy, D. and Young, A. R.,** *J. Am. Chem. Soc.*, 87, 1889, 1965.
1007. **Neszmélyi, A.,** ^{13}C Data Bank of Central Research Institute of Chemistry of the Hungarian Academy of Sciences, Budapest, Hungary.
1008. **Muller, N.,** *J. Chem. Phys.*, 36, 359, 1962.
1009. **Muller, N. and Pritchard, D. E.,** *J. Chem. Phys.*, 31, 768, 1959.
1010. **Muller, N. and Pritchard, D. E.,** *J. Chem. Phys.*, 31, 1471, 1959.
1011. **Muller, N. and Rose, P. I.,** *J. Am. Chem. Soc.*, 85, 2173, 1963.
1012. **Murrell, J. N.,** *Progr. Nucl. Magn. Reson. Spectrosc.*, 6, 1, 1970.
1013. **Murthy, A. S. N. and Rao, C. N. R.,** Spectroscopic studies of the hydrogen bond, in *Applied Spectroscopy Reviews,* Vol. 2, Brame, E. G., Jr., Ed., Marcel Dekker, New York, 1969, 69.
1014. **Musher, J. I.,** *J. Chem. Phys.*, 34, 594, 1961.
1015. **Musher, J. I. and Corey, E. J.,** *Tetrahedron,* 18, 791, 1962.
1016. **Müller, L., Kumar, A., and Ernst, R. R.,** *J. Chem. Phys.*, 63, 5490, 1975.
1017. **Müller, L., Kumar, A., and Ernst, R. R.,** *J. Magn. Reson.*, 25, 383, 1977.
1018. **Nagal, Y., Ohtsuki, M. A., Nakano, T., and Watanabe, H.,** *J. Organometal. Chem.*, 35, 81, 1972.
1019. **Nagayama, K., Bachmann, P., Wüthrich, K., and Ernst, R. R.,** *J. Magn. Reson.*, 31, 133, 1978.
1020. **Nagayama, K., Wüthrich, K., Bachmann, P., and Ernst, R. R.,** *Biochem. Biophys. Res. Commun.*, 78, 99, 1977.
1021. **Nagayama, K., Wüthrich, K., Bachmann, P., and Ernst, R. R.,** *Naturwissenschaften,* 64, 581, 1977.
1022. **Nakagawa, N. and Saito, S.,** *Tetrahedron Lett.*, 1003, 1967.
1023. **Nakanishi, H. and Yamamoto, O.,** *Tetrahedron Lett.*, 1803, 1974.
1024. **Narasimhan, P. T. and Rogers, M. T.,** *J. Am. Chem. Soc.*, 82, 5983, 1960.
1025. **Narasimhan, P. T. and Rogers, M. T.,** *J. Chem. Phys.*, 34, 1049, 1961.
1026. **Nasfay Scott, K.,** *J. Am. Chem. Soc.*, 94, 8564, 1972.
1027. **Naulet, N., Beljean, M., and Martin, G. J.,** *Tetrahedron Lett.*, 3597, 1976.
1028. **Negrebetskii, V. V., Bogdanov, V. S., and Kessenikh, A. V.,** *Z. Strukt. Khim.*, 12, 716, 1971.
1029. **Neiman, Z. and Bergmann, F.,** *Chem. Commun.*, 1002, 1968.
1030. **Nelson, G. L., Levy, G. C., and Cargioli, J. D.,** *J. Am. Chem. Soc.*, 94, 3089, 1972.
1031. **Neszmélyi, A., Lipták, A., Nánási, P., and Szejtli, J.,** *J. Am. Chem. Soc.*, in press.

1032. **Neudert, W. and Röpke, H.,** *Atlas of Steroid Spectra,* Springer-Verlag, Berlin, 1965.
1033. **Neuss, N., Nash, C. N., Lemke, P. A., and Grutzner, J. B.,** *J. Am. Chem. Soc.*, 93, 2337, 1971.
1034. **Newmark, R. A. and Hill, J. R.,** *Org. Magn. Reson.*, 13, 40, 1980.
1035. **Newmark, R. A. and Sederholm, C. H.,** *J. Chem. Phys.*, 39, 3131, 1963.
1036. **Newmark, R. A. and Sederholm, C. H.,** *J. Chem. Phys.*, 43, 602, 1965.
1037. **Newton, M. D., Schulman, J. M., and Manns, M. M.,** *J. Am. Chem. Soc.*, 96, 17, 1974.
1038. **Ng, S. and Sederholm, C. H.,** *J. Chem. Phys.*, 40, 2090, 1964.
1039. **Nickon, A., Castle, M. A., Harada, R., Berkoff, C. E., and Williams, R. O.,** *J. Am. Chem. Soc.*, 85, 2185, 1963.
1040. **Witanowski, M. and Webb, G. A., Eds.,** *Nitrogen NMR,* Plenum Press, London, 1973.
1041. **Nixon, J. F. and Pidcock, A.,** Phosphorus-31 nuclear magnetic resonance spectra of co-ordination compounds, in *Annual Reports on NMR Spectroscopy,* Vol. 2, Mooney, E. F., Ed., Academic Press, New York, 1969, 346.
1042. *NMR Basic Principles and Progress, Grundlagen und Fortschritte 1-8,* Diehl, P., Fluck, E., and Kosfeld, R., Eds., Springer-Verlag, Berlin, 1969.
1043. *NMR Specialist Periodical Reports,* Harris, R. K. and Abraham, R. J., Eds., Pergamon Press, Oxford; (a) Vol. 3, p. 79; (b) Vol. 1, p. 281.
1044. *NMR Spectra Catalog,* Compiled by Bhacco, N. S., Hollis, D. P., Johnson, L. F., Pier, G. A., and Shoolery, J. N., VARIAN Associates, Palo Alto, Calif., 1962-1963.
1045. **Noggle, J. H. and Schirmer, R. E.,** *The Nuclear Overhauser Effect; Chemical Applications,* Academic Press, New York, 1971.
1046. **Nógrádi, M., Ollis, W. D., and Sutherland, I. O.,** *Chem. Commun.*, 158, 1970.
1047. **Nonhebel, D. C.,** *Tetrahedron,* 24, 1869, 1968.
1048. **Nouls, J. C., Binst, G., and van Martin, R. H.,** *Tetrahedron Lett.*, 4065, 1967.
1049. **Nöth, H. and Wrackmeyer, B.,** *Chem. Ber.*, 107, 3070, 1974.
1050. **Nöth, H. and Wrackmeyer, B.,** *Chem. Ber.*, 107, 3089, 1974.
1051. *Nuclear Magnetic Resonance for Organic Chemists,* Mathieson, D. W., Ed., Academic Press, New York, 1967.
1052. *Nuclear Magnetic Resonance Spectra and Chemical Structure,* Brügel, W., Ed., Academic Press, New York, 1967.
1053. **Axenrod, T. and Webb, G. A., Eds.,** *Nuclear Magnetic Resonance Spectroscopy of Nuclei Other than Protons,* Wiley-Interscience, New York, 1974.
1054. **Ogg, R. A., Jr. and Diehl, P.,** *Helv. Phys. Acta,* 30, 251, 1957.
1055. **Ogg, R. A. and Ray, J. D.,** *J. Chem. Phys.*, 26, 1340, 1957.
1056. **Ogg, R. A. and Ray, J. D.,** *J. Chem. Phys.*, 26, 1515, 1957.
1057. **Ohtsuru, M. and Tori, K.,** *J. Mol. Spectrosc.*, 27, 296, 1968.
1058. **Ohtsuru, M. and Tori, K.,** *Tetrahedron Lett.*, 4043, 1970.
1059. **Okamura, W. H. and Sondheimer, F.,** *J. Am. Chem. Soc.*, 89, 5991, 1967.
1060. **Oki, M., Iwamura, H., and Hayakava, N.,** *Bull. Chem. Soc. Jpn.*, 36, 1542, 1963.
1061. **Oláh, G. A., Bollinger, J. M., and Brinich, J.,** *J. Am. Chem. Soc.*, 90, 2587, 1968.
1062. **Oláh, G. A., DeMember, J. R., and Schlosberg, R. H.,** *J. Am. Chem. Soc.*, 91, 2112, 1969.
1063. **Oláh, G. A., Denis, J.-M., and Westerman, P. W.,** *J. Org. Chem.*, 39, 1206, 1974.
1064. **Oláh, G. A. and Kiovsky, T. E.,** *J. Am. Chem. Soc.*, 90, 4666, 1968.
1065. **Oláh, G. A. and Kreienbühl, P.,** *J. Am. Chem. Soc.*, 89, 4756, 1967.
1066. **Oláh, G. A. and Matescu, G. D.,** *J. Am. Chem. Soc.*, 92, 1430, 1970.
1067. **Oláh, G. A. and Pittman, C. U., Jr.,** *Advances in Physical Organic Chemistry,* Vol. 4, Gold, V., Ed., Academic Press, New York, 1966.
1068. **Oláh, G. A. and White, A. M.,** *J. Am. Chem. Soc.*, 89, 7072, 1967.
1069. **Oláh, G. A. and White, A. M.,** *J. Am. Chem. Soc.*, 91, 5801, 1969.
1070. **Ollis, W. D. and Sutherland, I. O.,** *Chem. Commun.*, 402, 1966.
1071. **O'Reilly, D. E.,** *J. Chem. Phys.*, 36, 274, 1962.
1072. **Oth, J. F. M., Merényi, R., Röttele, H., and Schröder, G.,** *Tetrahedron Lett.*, 3941, 1968.
1073. **Oth, J. F. M., Müllen, K., Gilles, J. M., and Schröder, G.,** *Helv. Chim. Acta,* 57, 1415, 1974.
1074. **Ouelette, R. J.,** *J. Am. Chem. Soc.*, 86, 4378, 1964.
1075. **Overberger, C. G., Kurtz, T., and Yaroslavsky, S.,** *J. Org. Chem.*, 30, 4363, 1965.
1076. **Overhauser, A.,** *Phys. Rev.*, 89, 689, 1953; 92, 411, 1953.
1077. **Ozubko, R. S., Buchanan, G. W., and Smith, I. C. P.,** *Can. J. Chem.*, 52, 2493, 1974.
1078. **Padwa, A., Shefter, E., and Alexander, E.,** *J. Am. Chem. Soc.*, 90, 3717, 1968.
1079. **Page, J. E.,** Nuclear magnetic resonance spectra of steroids, in *Annual Reports on NMR Spectroscopy,* Vol. 3, Mooney, E. F., Ed., Academic Press, New York, 1970, 149.
1080. **Page, T. F., Alger, T., and Grant, D. M.,** *J. Am. Chem. Soc.*, 87, 5333, 1965.

1081. **Paolillo, L. and Becker, E. D.,** *J. Magn. Reson.*, 2, 168, 1970.
1082. **Paolillo, L., Tancredi, T., Temussi, P. A., Trivellone, E., Bradbury, E. M., and Crane-Robinson, C.,** *Chem. Commun.*, 335, 1972.
1083. **Parker, R. G. and Roberts, J. D.,** *J. Am. Chem. Soc.*, 92, 743, 1970.
1084. **Parker, R. G. and Roberts, J. D.,** *J. Org. Chem.*, 35, 996, 1970.
1085. **Parmigiani, A., Perotti, A., and Riganti, V.,** *Gazz. Chim. Ital.*, 91, 1148, 1961.
1086. **Parrish, R., Kurland, D., Janese, W., and Bakay, L.,** *Science*, 183, 438, 1974.
1087. **Pascual, C., Meier, J., and Simon, W.,** *Helv. Chim. Acta*, 49, 164, 1966.
1088. **Patel, D. J., Howden, M. E. H., and Roberts, J.D.,** *J. Am. Chem. Soc.*, 85, 3218, 1963.
1089. **Patrick, T. B. and Patrick, P. H.,** *J. Am. Chem. Soc.*, 95, 6230, 1972.
1090. **Patt, S. L. and Sykes, D. B.,** *J. Chem. Phys.*, 56, 3182, 1972.
1091. **Patterson, A., Jr. and Ettinger, R.,** *Z. Elektrochem.*, 64, 98, 1960.
1092. **Paudler, W. W. and Blewitt, H. L.,** *J. Org. Chem.*, 30, 4081, 1965.
1093. **Paudler, W. W. and Dunham, D. E.,** *J. Heterocyclic Chem.*, 2, 410, 1965.
1094. **Paudler, W. W. and Kress, T. J.,** *Chem. Ind. (London)*, 1557, 1966.
1095. **Paudler, W. W. and Kuder, J. E.,** *J. Heterocyclic Chem.*, 3, 33, 1966.
1096. **Paudler, W. W. and Kuder, J. E.,** *J. Org. Chem.*, 31, 809, 1966.
1097. **Paul, E. G. and Grant, D. M.,** *J. Am. Chem. Soc.*, 85, 1701, 1963.
1098. **Pauli, W.,** *Naturwissenschaften*, 12, 741, 1924.
1099. **Peake, A. and Thomas, L. F.,** *Trans. Faraday Soc.*, 62, 2980, 1966.
1100. **Pehk, T. and Lippmaa, E.,** *Eesti NSV Tead. Akad. Toim. Keem. Geol.*, 17, 291, 1968.
1101. **Pehk, T. and Lippmaa, E.,** *Org. Magn. Reson.*, 3, 679, 1971.
1102. Perkin-Elmer Model R-10 60 MHz Spectrometer, Perkin-Elmer Corp., Norwalk, Conn.
1103. *Perkin-Elmer NMR Q.*, 2, 11, 1971.
1104. *Perkin-Elmer NMR Q.*, 5, 12, 1972.
1105. **Perlin, A. S., Casu, B., and Koch, H. J.,** *Can. J. Chem.*, 48, 2599, 1970.
1106. **Perlin, A. S., Cyr, N., Koch, H. J., and Korsch, B.,** *Ann. N. Y. Acad. Sci.*, 222, 935, 1973.
1107. **Perlin, A. S. and Koch, H. J.,** *Can. J. Chem.*, 48, 2639, 1970.
1108. **Petrakis, L. and Sederholm, C. H.,** *J. Chem. Phys.*, 35, 1174, 1961.
1109. **Pfeffer, H. U. and Klessinger, M.,** *Org. Magn. Reson.*, 9, 121, 1977.
1110. **Philipsborn, W.,** *Angew. Chem.*, 83, 470, 1971.
1111. **Piette, L. H., Ray, J. D., and Ogg, R. A., Jr.,** *J. Mol. Spectrosc.*, 2, 66, 1958.
1112. **Pihlaja, K. and Pasanen, P.,** *Suom. Kemisbil.*, 46, 273, 1973.
1113. **Pimentel, G. C. and McClennan, A. L.,** *The Hydrogen Bond*, W. H. Freeman, San Francisco, 1960.
1114. **Pinhey, J. T. and Sternhell, S.,** *Tetrahedron Lett.*, 275, 1963.
1115. **Pirkle, W. H.,** *J. Am. Chem. Soc.*, 88, 1837, 1966.
1116. **Pirkle, W. H. and Beare, S. D.,** *J. Am. Chem. Soc.*, 89, 5485, 1967.
1117. **Pirkle, W. H. and Beare, S. D.,** *J. Am. Chem. Soc.*, 91, 5150, 1969.
1118. **Pitcher, E., Buckingham, A. D., and Stone, F. G. A.,** *J. Chem. Phys.*, 36, 124, 1962.
1119. **Pitzer, K. S. and Donath, W. E.,** *J. Am. Chem. Soc.*, 81, 3213, 1959.
1120. **Pohland, A. E., Badger, R. C., and Cromwell, N. H.,** *Tetrahedron Lett.*, 4369, 1965.
1121. **Pomerantz, M. and Fink, R.,** *Chem. Commun.*, 430, 1975.
1122. **Pomerantz, M., Fink, R., and Gray, G. A.,** *J. Am. Chem. Soc.*, 98, 291, 1976.
1123. **Pomerantz, M. and Hillenbrand, D. F.,** *J. Am. Chem. Soc.*, 95, 5809, 1973.
1124. **Pomerantz, M. and Hillenbrand, D. F.,** *Tetrahedron*, 31, 217, 1975.
1125. **Pople, J. A.,** *Proc. R. Soc. London Ser. A*, 239, 541, 1957.
1126. **Pople, J. A.,** *Dis. Faraday Soc.*, 34, 7, 1962.
1127. **Pople, J. A.,** *J. Chem. Phys.*, 37, 53, 60, 1962.
1128. **Pople, J. A.,** *Proc. R. Soc. London Ser. A*, 239, 550, 1957.
1129. **Pople, J. A. and Bothner-By, A. A.,** *J. Chem. Phys.*, 42, 1339, 1965.
1130. **Pople, J. A. and Gordon, M. S.,** *J. Am. Chem. Soc.*, 89, 4253, 1967.
1131. **Pople, J. A. and Santry, D. P.,** *Mol. Phys.*, 8, 1, 1964.
1132. **Pople, J. A., Schneider, W. G., and Bernstein, H. J.,** *High-Resolution Nuclear Magnetic Resonance*, McGraw-Hill, New York, 1959.
1133. **Pople, J. A. and Untch, K. G.,** *J. Am. Chem. Soc.*, 88, 4811, 1966.
1134. **Poranski, C. F. and Moniz, W. B.,** *J. Phys. Chem.*, 71, 1142, 1967.
1135. **Porte, A. L., Gutowsky, H. S., and Hunsberger, I. M.,** *J. Am. Chem. Soc.*, 82, 5057, 1960.
1136. **Porter, R., Marks, T. J., and Shriver, D. F.,** *J. Am. Chem. Soc.*, 95, 3548, 1973.
1137. **Portoghese, P. S. and Telang, V. G.,** *Tetrahedron*, 27, 1823, 1971.
1138. **Posner, T. B., Markowski, V., Loftus, P., and Roberts, J. D.,** *Chem. Commun.*, 769, 1975.
1139. **Pouchoulin, G., Llinas, J. R., Buomo, G., and Vincent, E.-J.,** *Org. Magn. Reson.*, 8, 518, 1976.

1140. **Pregosin, P. S. and Randall, E. W.,** *Chem. Commun.,* 399, 1971.
1141. **Pregosin, P. S. and Randall, E. W.,** ^{13}C Nuclear magnetic resonance, in *Determination of Organic Structures by Physical Methods,* Vol. 4, Nachod, F. C. and Zuckerman, J. J., Eds., Academic Press, New York, 1971, chap. 6.
1142. **Pregosin, P. S., Randall, E. W., and White, A. I.,** *J. Chem. Soc. Perkin Trans. 2,* 1, 1972.
1143. **Pretsch, E., Clerc, T., Seibl, J., and Simon, W.,** *Tabellen zur Strukturaufklärung organischer Verbindungen mit spektroskopischen Methoden,* Springer-Verlag, Berlin, 1976; (a) p. C250; (b) p. C10; (c) p. C15; (d) p. C75; (e) p. C50; (f) p. C40; (g) p. C70; (h) p. C100; (i) p. C90; (j) p. C110; (k) p. C170; (l) p. C175, p. C180, p. C185; (m) p. B10; (n) p. C120, p. C125; (o) p. C150; (p) p. C140, p. C145; (r) p. C160; (s) p. C220; (t) p. C230.
1144. **Price, D., Suschitzky, H., and Hollies, J. I.,** *J. Chem. Soc. C,* 1967, 1966.
1145. **Proctor, W. G. and Yu, F. C.,** *Phys. Rev.,* 77, 717, 1950.
1146. **Proctor, W. F. and Yu, F. C.,** *Phys. Rev.,* 81, 20, 1951.
1147. *Progress in Nuclear Magnetic Resonance Spectroscopy,* Emsley, J. W., Feeney, J., and Sutcliffe, L. H., Eds., Pergamon Press, New York, 1966, (a) Vol. 11, 1977, p. 2, p. 3.
1148. **Pugmire, R. J. and Grant, D. M.,** *J. Am. Chem. Soc.,* 90, 697, 1968.
1149. **Pugmire, R. J. and Grant, D. M.,** *J. Am. Chem. Soc.,* 90, 4232, 1968.
1150. **Pugmire, R. J. and Grant, D. M.,** *J. Am. Chem. Soc.,* 93, 1880, 1971.
1151. **Pugmire, R. J., Grant, D. M., Robins, M. J., and Robins, R. K.,** *J. Am. Chem. Soc.,* 91, 6381, 1969.
1152. **Pugmire, R. J., Robins, M. J., Grant, D. M., and Robins, R. K.,** *J. Am. Chem. Soc.,* 93, 1887, 1971.
1153. **Purcell, E. M., Torrey, H. C., and Pound, R. V.,** *Phys. Rev.,* 69, 37, 1946.
1154. **Raban, M. and Mislow, K.,** *Tetrahedron Lett.,* 3961, 1966.
1155. **Rabenstein, D. L.,** *Anal. Chem.,* 43, 1599, 1971.
1156. **Rabenstein, D. L. and Sayer, T. L.,** *J. Magn. Reson.,* 24, 27, 1976.
1157. **Rabi, I. I., Millman, S., Kusch, P., and Zacharias, J. R.,** *Phys. Rev.,* 55, 526, 1939.
1158. **Rácz, P., Tompa, K., and Pocsik, I.,** *Exp. Eye Res.,* 28, 129, 1979; 29, 601, 1979.
1159. **Radeglia, R., Storek, W., Engelhardt, G., Ritschil, F., Lippmaa, E., Pehk, T., Magi, M., and Martin, D.,** *Org. Magn. Reson.,* 5, 419, 1973.
1160. **Rader, C. P.,** *J. Am. Chem. Soc.,* 88, 1713, 1966.
1161. **Ramey, K. C., Lini, D. C., and Krow, G.,** General review of nuclear magnetic resonance, in *Annual Reports on NMR Spectroscopy,* Vol. 6A, Mooney, E. F., Ed., Academic Press, New York, 1975, 147.
1162. **Ramsey, N. F.,** *Phys. Rev.,* 77, 567, 1950.
1163. **Ramsey, N. F.,** *Phys. Rev.,* 78, 699, 1950.
1164. **Ramsey, N. F.,** *Phys. Rev.,* 86, 243, 1952.
1165. **Ramsey, N. F.,** *Phys. Rev.,* 91, 303, 1953.
1166. **Ramsey, N. F. and Purcell, E. M.,** *Phys. Rev.,* 85, 143, 1952.
1167. **Ranade, S. S., Shah, S., Korgaonkar, K. S., Kasturi, S. R., Chaughule, R. S., and Vijayaraghavan, R.,** *Physiol. Chem. Phys.,* 8, 131, 1976.
1168. **Randall, E. W. and Gilles, D. G.,** Nitrogen nuclear magnetic resonance, in *Progress in Nuclear Magnetic Resonance Spectroscopy,* Vol. 6, Emsley, J. W., Feeney, J., and Sutcliffe, L. H., Eds., Pergamon Press, Oxford, 1971, 119.
1169. **Randall, E. W. and Shaw, D.,** *Spectrochim. Acta,* 23a, 1235, 1967.
1170. **Ranft, J.,** *Ann. Physik.,* 10, 399, 1963.
1171. **Rao, C. N. R.,** *Can. J. Chem.,* 40, 963, 1962.
1172. **Rattet, L. S., Williamson, A. D., and Goldstein, J. H.,** *J. Phys. Chem.,* 72, 2954, 1968.
1173. **Read, J. M., Mayo, R. E., and Goldstein, J. H.,** *J. Mol. Spectrosc.,* 22, 419, 1967.
1174. **Reddy, G. S. and Goldstein, J. H.,** *J. Am. Chem. Soc.,* 83, 2045, 1961.
1175. **Reddy, G. S. and Goldstein, J. H.,** *J. Phys. Chem.,* 65, 1539, 1961..
1176. **Reddy, G. S., Hobgood, R. T., Jr., and Goldstein, J. H.,** *J. Am. Chem. Soc.,* 84, 336, 1962.
1177. **Reddy, G. S., Mandell, L., and Goldstein, J. H.,** *J. Am. Chem. Soc.,* 83, 1300, 1961.
1178. **Redfield, A. G.,** *Methods Enzymol.,* 49, 253, 1978.
1179. **Redfield, A. G. and Gupta, R. K.,** *J. Chem. Phys.,* **54,** 1418, 1971.
1180. **Redfield, A. G., Kunz, S. D., and Ralph, E. K.,** *J. Magn. Reson.,* 19, 114, 1975.
1181. **Reeves, L. W.,** *Trans. Faraday Soc.,* 55, 1684, 1959.
1182. **Reeves, L. W.,** *Adv. Phys. Org. Chem.,* 3, 187, 1965.
1183. **Reeves, L. W., Riveros, J. M., Spragg, R. A., and Vavin, J. A.,** *Mol. Phys.,* 25, 9, 1973.
1184. **Reeves, L. W. and Schneider, W. G.,** *Can. J. Chem.,* 36, 793, 1958.
1185. **Reeves, L. W. and Schneider, W. G.,** *Trans. Faraday Soc.,* 54, 314, 1958.
1186. **Reich, H. J., Jautelat, M., Messe, M. T., Weigert, F. J., and Roberts, J. D.,** *J. Am. Chem. Soc.,* 91, 7445, 1969.
1187. **Reilly, C. A. and Swalen, J. D.,** *J. Chem. Phys.,* 32, 1378, 1960.

1188. **Reiter, J., Sohár, P., Lipták, J., and Toldy, L.,** *Tetrahedron Lett.*, 1417, 1970.
1189. **Reiter, J., Sohár, P., and Toldy, L.,** *Tetrahedron Lett.*, 1411, 1970.
1190. **Retcofsky, H. L. and Friedel R. A.,** *J. Phys. Chem.*, 71, 3592, 1967; 72, 290, 2619, 1968.
1191. **Retcofsky, H. L. and Friedel, R. A.,** *J. Magn. Reson.*, 8, 398, 1972.
1192. **Retcofsky, H. L. and McDonald, F. R.,** *Tetrahedron Lett.*, 2575, 1968.
1193. **Reuben, J.,** Effects of chemical equilibrium and adduct stoichiometry in shift reagent studies, in *Nuclear Magnetic Resonance Shift Reagents,* Sievers, R. F., Ed., Academic Press, New York, 1973, 341.
1194. **Reuben, J.,** *J. Am. Chem. Soc.*, 95, 3534, 1973.
1195. **Reuben, J.,** *J. Magn. Reson.*, 11, 103, 1973.
1196. **Reuben, J.,** Paramagnetic lanthanide shift reagents in NMR spectroscopy; principles, methodology and applications, in *Progress in Nuclear Magnetic Resonance Spectroscopy,* Vol. 9, Emsley, J. W., Feeney, J. and Sutcliffe, L. H., Eds., Pergamon Press, Oxford, 1973, 1.
1197. **Reuben, J. and Fiat, D.,** *J. Chem. Phys.*, 51, 4909, 1969.
1198. **Reuben, J. and Leight, J. S., Jr.,** *J. Am. Chem. Soc.*, 94, 2789, 1972.
1199. **Reuben, J., Tzalmona, A., and Samuel, D.,** *Proc. Chem. Soc.*, 353, 1962.
1200. **Reutov, O. A., Shatkina, T. N., Lippmaa, E. T., and Pehk, T. I.,** *Dokl. Akad. Nauk. SSSR,* 181, 1400, 1968.
1201. **Reutov, O. A., Shatkina, T. N., Lippmaa, E. T., and Pehk, T. I.,** *Tetrahedron,* 25, 5757, 1969.
1202. **Revel, M., Roussel, J., Navech, J., and Mathis, J.,** *Org. Magn. Reson.*, 8, 399, 1976.
1203. **Ricca, G. S., Danieli, B., Palmisano, G., Duddeck, H., and Elgamal, M. H. A.,** *Org. Magn. Reson.*, 11, 163, 1978.
1204. **Richard, C. and Granger, P.,** Chemically induced dynamic nuclear and electron polarizations — CIDNP and CIDEP, in *NMR Basic Principles and Progress,* Vol. 8, Diehl, P., Fluck, E., and Kosfeld, R., Eds., Springer-Verlag, Berlin, 1974.
1205. **Richards, R. E. and Schaefer, T. P.,** *Trans. Faraday Soc.*, 54, 1280, 1958.
1206. **Richards, R. E. and Thomas, N. A.,** *J. Chem. Soc. Perkin Trans. 2,* 368, 1974.
1207. **Riddell, F. G.,** *J. Chem. Soc. B,* 331, 1970.
1208. **Riddell, F. G. and Lehn, J. M.,** *J. Chem. Soc. B,* 1224, 1968.
1209. **Roberts, B. W., Lambert, J. B., and Roberts, J. D.,** *J. Am. Chem. Soc.*, 87, 5439, 1965.
1210. **Roberts, J. D.,** *J. Am. Chem. Soc.*, 78, 4495, 1956.
1211. **Roberts, J. D.,** *An Introduction to the Analysis of Spin-Spin Splitting in High-Resolution Nuclear Magnetic Resonance Spectra,* Benjamin, New York, 1962.
1212. **Roberts, J. D., Lutz, R. P., and Davis, D. R.,** *J. Am. Chem. Soc.*, 83, 246, 1961.
1213. **Roberts, J. D., Weigert, F. J., Kroschwitz, J. I., and Reich, H. J.,** *J. Am. Chem. Soc.*, 92, 1338, 1970.
1214. **Roberts, R. T. and Chachaty, C.,** *Chem. Phys. Lett.*, 22, 348, 1973.
1215. **Rodger, C., Sheppard, N., McFarlane, C., and McFarlane, W.,** Group VI — oxygen, sulphur, selenium and tellurium, in *NMR and the Periodic Table,* Harris, R. K. and Mann, B. E., Eds., Academic Press, New York, 1978, 383.
1216. **Rogers, E. H.,** *NMR-EPR Workshop Notes,* VARIAN Associates, Palo Alto, Calif.,
1217. **Rogers, M. T. and LaPlanche, L. A.,** *J. Phys. Chem.*, 69, 3648, 1965.
1218. **Ronayne, J. and Williams, D. H.,** *Chem. Commun.*, 712, 1966.
1219. **Ronayne, J. and Williams, D. H.,** *J. Chem. Soc. C,* 2642, 1967.
1220. **Ronayne, J. and Williams, D. H.,** Solvent effects in proton magnetic resonance spectroscopy, in *Annual Reports on NMR Spectroscopy,* Vol. 2, Mooney, E. F., Ed., Academic Press, New York, 1969, 83.
1221. **Rondeau, R. E., Berwick, M. A., Steppel, R. N., and Servé, M. P.,** *J. Am. Chem. Soc.*, 94, 1096, 1972.
1222. **Rondeau, R. E. and Sievers, R. E.,** *J. Am. Chem. Soc.*, 93, 1522, 1971.
1223. **Rosenberg, D., DeHaan, J. W., and Drenth, W.,** *Rec. Trav. Chim. Pays-Bas,* 87, 1387, 1968.
1224. **Rosenberg, D. and Drenth, W.,** *Tetrahedron,* 27, 3893, 1971.
1225. **Rousselot, M. M.,** *C. R. Acad. Sci. France C,* 262, 26, 1966.
1226. **Rowe, J. J. M., Hinton, J., and Rowe, K. L.,** *Chem. Rev.*, 70, 1, 1970.
1227. **Royden, V.,** *Phys. Rev.*, 96, 543, 1954.
1228. **Runsink, J. and Günther, H.,** *Org. Magn. Reson.*, 13, 249, 1980.
1229. **Sackmann, E. and Dreeskamp, H.,** *Spectrochim. Acta,* 21, 2005, 1965.
1230. *SADTLER Nuclear Magnetic Resonance Spectra,* SADTLER Research Laboratories, Philadelphia, 1965.
1231. **Saika, A. and Slichter, C. P.,** *J. Chem. Phys.*, 22, 26, 1954.
1232. **Saito, H. and Smith, I. C. P.,** *Arch. Biochem. Biophys.*, 158, 154, 1973.
1233. **Saito, H., Tanaka, Y., and Nagata, S.,** *J. Am. Chem. Soc.*, 95, 324, 1973.
1234. **Sanders, J. K. M. and Williams, D. H.,** *Chem. Commun.*, 422, 1970.
1235. **Sanders, J. K. M. and Williams, D. H.,** *J. Am. Chem. Soc.*, 93, 641, 1971.

1236. **Sanders, J. K. M. and Williams, D. H.,** *Tetrahedron Lett.*, 2813, 1971.
1237. **Sasaki, Y., Kawaki, H., and Okazaki, Y.,** *Chem. Pharm. Bull. (Jpn.),* 21, 2488, 1973.
1238. **Saunders, M. and Hyne, J. B.,** *J. Chem. Phys.*, 29, 253, 1319, 1958.
1239. **Savitsky, G. B.,** *J. Phys. Chem.*, 67, 2723, 1963.
1240. **Savitsky, G. B., Ellis, P. D., Namikawa, K., and Maciel, G. E.,** *J. Chem. Phys.*, 49, 2395, 1968.
1241. **Savitsky, G. B., Namikawa, K., and Zweifel, G.,** *J. Phys. Chem.*, 69, 3105, 1965.
1242. **Savitsky, G. B., Pearson, R. M., and Namikawa, K.,** *J. Phys. Chem.*, 60, 1425, 1965.
1243. **Schaefer, J.,** *Macromolecules,* 2, 210, 1969.
1244. **Schaefer, J.,** *Macromolecules,* 4, 105, 1971.
1245. **Schaefer, J.,** The carbon-13 NMR analysis of synthetic high polymers, in *Topics in Carbon-13 NMR Spectroscopy,* Vol. 1, Levy, G. C., Ed., Wiley-Interscience, New York, 1974, 149.
1246. **Schaefer, J. and Stejskal, E. O.,** High-resolution ^{13}C NMR of solid polymers, in *Topics in Carbon-13 NMR Spectroscopy,* Vol. 3, Levy, G. C., Ed., Wiley-Interscience, New York, 1979, 284.
1247. **Schaefer, J., Stejskal, E. O., and Buchdahl, R.,** *Macromolecules,* 8, 291, 1979.
1248. **Schaefer, T.,** *J. Chem. Phys.*, 36, 2235, 1962.
1249. **Schaefer, T., Reynolds, W. F., and Yonemoto, T.,** *Can. J. Chem.*, 41, 2969, 1963.
1250. **Schaefer, T. and Schneider, W. G.,** *Can. J. Chem.*, 38, 2066, 1960.
1251. **Schaefer, T. and Schneider, W. G.,** *J. Chem. Phys.*, 32, 1224, 1960.
1252. **Scheiner, P. and Litchman, W. M.,** *Chem. Commun.*, 781, 1972.
1253. **Schiemenz, G. P.,** *Tetrahedron,* 29, 741, 1973.
1254. **Schmidbaur, H.,** *Chem. Ber.*, 97, 1639, 1964.
1255. **Schmidt, C. F. and Chan, S. I.,** *J. Chem. Phys.*, 55, 4670, 1971.
1256. **Schmidt, C. F. and Chan, S. I.,** *J. Magn. Reson.*, 5, 151, 1971.
1257. **Schmutzler, R.,** *J. Chem. Soc.*, 4551, 1964.
1258. **Schneider, H. J., Freitag, W., and Hoppen, V.,** *Org. Magn. Reson.*, 13, 266, 1980.
1259. **Schneider, H. J., Price, R., and Keller, T.,** *Angew. Chem. Int. Ed. Engl.*, 10, 730, 1971.
1260. **Schneider, W. G., Bernstein, H. J., and Pople, J. A.,** *J. Chem. Phys.*, 28, 601, 1958.
1261. **Schraml, J., Duc-Chuy, N., Chvalovsky, V., Mägi, M., and Lippmaa, E.,** *Org. Magn. Reson.*, 7, 379, 1975.
1262. **Schröder, G.,** *Angew. Chem. Int. Ed. Engl.*, 4, 752, 1965.
1263. **Schulman, J. M. and Venanzi, T.,** *J. Am. Chem. Soc.*, 98, 4701, 1976.
1264. **Schultheiss, H. and Fluck, E.,** *Z. Naturforsch.*, 32b, 257, 1977.
1265. **Schwarcz, J. A. and Perlin, A. S.,** *Can. J. Chem.*, 50, 3667, 1972.
1266. **Schwarz, R. M. and Rabjohn, N.,** *Org. Magn. Reson.*, 13, 9, 1980.
1267. **Schwyzer, R. and Ludescher, U.,** *Helv. Chim. Acta,* 52, 2033, 1969.
1268. **Sears, R. E. J.,** *J. Chem. Phys.*, 56, 983, 1971.
1269. **Seel, F., Hartmann, V., and Gombler, W.,** *Z. Naturforsch.*, 27b, 325, 1972.
1270. **Seel, H., Aydin, R., and Günther, H.,** *Z. Naturforsch.*, 33, 353, 1978.
1271. **Segre, A. and Musher, J. I.,** *J. Am. Chem. Soc.*, 89, 706, 1967.
1272. **Sen, B. and Wu, W. C.,** *Anal. Chim. Acta,* 46, 37, 1969.
1273. **Senda, Y., Ishiyama, J.-I., and Imaizumi, S.,** *Bull. Chem. Soc. Jpn.*, 50, 2813, 1977.
1274. **Seo, S., Tomita, Y., and Tori, K.,** *Chem. Commun.*, 270, 1975.
1275. **Sepulchre, A. M., Septe, B., Lukacs, G., Gero, S. D., Voelter, W., and Breitmaier, E.,** *Tetrahedron,* 30, 905, 1974.
1276. **Sergeyev, N. M. and Solkan, V. N.,** *Chem. Commun.*, 12, 1975.
1277. **Servis, K. L. and Fang, K. N.,** *J. Am. Chem. Soc.*, 90, 6712, 1968.
1278. **Sewell, P. R.,** The nuclear magnetic resonance spectra of polymers, in *Annual Reports on NMR Spectroscopy,* Vol. 1, Mooney, E. F., Ed., Academic Press, New York, 1968, 165.
1279. **Shadowitz, A.,** *The Electromagnetic Field,* McGraw-Hill, New York, 1975.
1280. **Shafer, P. R., Davis, D. R., Vogel, M., Nagarajan, K., and Roberts, J. D.,** *Proc. Natl. Acad. Sci. U.S.A.*, 47, 49, 1961.
1281. **Shapiro, B. L., Hlubucek, J. R., Sullivan, G. R., and Johnson, L. F.,** *J. Am. Chem. Soc.*, 93, 3281, 1971.
1282. **Shapiro, B. L. and Johnston, M. D., Jr.,** *J. Am. Chem. Soc.*, 94, 8185, 1972.
1283. **Shapiro, B. L., Johnston, M. D., Jr., and Towns, R. L. R.,** *J. Am. Chem. Soc.*, 94, 4381, 1972.
1284. **Shaw, D.,** Fourier transform NMR, in *NMR Specialist Periodical Reports,* Vol. 3, Harris, R. K., Ed., The Chemical Society, London, 1974, 249.
1285. **Shaw, D.,** Fourier transform NMR, in *NMR Specialist Periodical Reports,* Vol. 5, Harris, R. K., Ed., The Chemical Society, London, 1976, 188.
1286. **Shaw, D.,** *Fourier Transform NMR Spectroscopy,* Elsevier, Amsterdam, 1976, (a) p. 191; (b) pp. 131-134; (c) p. 189; (d) p. 195; (e) p. 184; (f) p. 141; (g) p. 335.

1287. **Sheinblatt, M.,** *J. Am. Chem. Soc.*, 88, 2845, 1966.
1288. **Sheppard, N. and Turner, J. J.,** *Proc. R. Soc. London Ser. A*, 252, 506, 1959.
1289. **Sheppard, N. and Turner, J. J.,** *Mol. Phys.*, 3, 168, 1960.
1290. **Shimanouchi, T.,** *Pure Appl. Chem.*, 12, 287, 1966.
1291. **Shoolery, J. N.,** *VARIAN Tech. Inform. Bull.*, 2, 3, 1959.
1292. **Shoolery, J. N. and Alder, B.,** *J. Chem. Phys.*, 23, 805, 1955.
1293. **Shoolery, J. N. and Rogers, M. T.,** *J. Am. Chem. Soc.*, 80, 5121, 1958.
1294. **Shoppee, C. W., Johnson, F. P., Lack, R. E., Shannon, J. S., and Sternhell, S.,** *Tetrahedron Suppl.*, 8, 421, 1966.
1295. **Shporer, M. and Civan, M. M.,** *Biochim. Biophys. Acta*, 385, 81, 1975.
1296. **Shporer, M., Haas, M., and Civan, M. M.,** *Biophys. J.*, 16, 601, 1976.
1297. **Shouf, R. R. and Van der Hart, D. L.,** *J. Am. Chem. Soc.*, 93, 2053, 1971.
1298. **Shoup, R. R., Becker, E. D., and Farrar, T. C.,** *J. Magn. Reson.*, 8, 290, 1972.
1299. **Siddall, T. H.,** *J. Phys. Chem.*, 70, 2249, 1966.
1300. **Siddall, T. H., III.,** *Chem. Commun.*, 452, 1971.
1301. **Siemion, I. Z. and Sucharda-Sobczyk, A.,** *Tetrahedron*, 26, 191, 1970.
1302. **Sievers, R. E.,** *Nuclear Magnetic Resonance Shift Reagents*, Academic Press, New York, 1973.
1303. **Silver, B. L. and Luz, Z.,** *Q. Rev. Chem. Soc.*, 21, 458, 1967.
1304. **Simmons, H. E. and Park, C. H.,** *J. Am. Chem. Soc.*, 90, 2428, 2429, 2431, 1968.
1305. **Simon, W. and Clerc, T.,** *Strukturaufklärung organischer Verbindungen mit spektroskopischen Methoden*, Vol. 7, Akademische Verlagsgesellschaft, Frankfurt am Main, 1967.
1306. **Simonnin, M. P., Lecourt, M. J., Terrier, F., and Dearing, C. A.,** *Can. J. Chem.*, 50, 3558, 1972.
1307. **Singh, R. D. and Singh, S. N.,** *J. Magn. Reson.*, 16, 110, 1974.
1308. **Sinha, S. P.,** *Europium*, Springer-Verlag, Berlin, 1967.
1309. **Slichter, C. P.,** *Principles of Magnetic Resonance*, Harper, New York, 1963, (a) pp. 16-22.
1310. **Slomp, G. and McKellar, F.,** *J. Am. Chem. Soc.*, 82, 999, 1960.
1311. **Smith, G. V., Boyd, W. A., and Hinckley, C. C.,** *J. Am. Chem. Soc.*, 93, 6319, 1971.
1312. **Smith, G. V. and Kriloff, H.,** *J. Am. Chem. Soc.*, 85, 2016, 1963.
1313. **Smith, I. C. and Schneider, W. G.,** *Can. J. Chem.*, 39, 1158, 1961.
1314. **Smith, W. B.,** Carbon-13 NMR spectroscopy of steroids, in *Annual Reports on NMR Spectroscopy*, Vol. 8, Webb, G. A., Ed., Academic Press, New York, 1978, 199; (a) p. 212, 216; (b) p. 215, pp. 218-220; (c) p. 205; (d) p. 202.
1315. **Snyder, E. I., Altman, L. J., and Roberts, J. D.,** *J. Am. Chem. Soc.*, 84, 2004, 1962.
1316. **Snyder, E. I. and Roberts, J. D.,** *J. Am. Chem. Soc.*, 84, 1582, 1962.
1317. **Sogn, J. A., Gibbons, W. A., and Randall, E. W.,** *Biochemistry*, 12, 2100, 1973.
1318. **Sohár, P.,** *Magy. Kém. Foly.*, 76, 577, 1970, (in Hungarian).
1319. **Sohár, P. and Bernáth, G.,** *Acta Chim. Acad. Sci. Hung.*, 87, 285, 1975.
1320. **Sohár, P.,** *Magy. Kém. Lapja*, 30, 100, 309, 1975 (in Hungarian).
1321. **Sohár, P.,** unpublished results.
1322. **Sohár, P. and Bernáth, G.,** *Org. Magn. Reson.*, 5, 159 1973.
1323. **Sohár, P., Fehér, G., and Toldy, L.,** *Org. Magn. Reson.*, 11, 9, 1978.
1324. **Sohár, P., Fehér, Ö., and Tihanyi, E.,** *Org. Magn. Reson.*, 12, 205, 1979.
1325. **Sohár, P., Gera, L., and Bernáth, G.,** *Org. Magn. Reson.*, 14, 204, 1980.
1326. **Sohár, P. and Hajós, A.,** unpublished results.
1327. **Sohár, P., Horváth, T., and Ábrahám, G.,** *Acta Chim. (Budapest)*, 103, 95, 1980.
1328. **Sohár, P., Kosáry, J., and Kasztreiner, E.,** *Acta Chim. Acad. Sci. Hung.*, 84, 201, 1975.
1329. **Sohár, P. and Kuszmann, J.,** *Org. Magn. Reson.*, 3, 647, 1971.
1330. **Sohár, P. and Kuszmann, J.,** *Org. Magn. Reson.*, 6, 407, 1974.
1331. **Sohár, P. and Kuszmann, J.,** *Acta Chim. Acad. Sci. Hung.*, 86, 285, 1975.
1332. **Sohár, P., Kuszmann, J., Horváth, Gy., and Méhesfalvi-Vajna, Zs.,** *Kém. Közlemények*, 46, 481, 1976 (in Hungarian).
1333. **Sohár P., Kuszmann, J., Ullrich, E., and Horváth, Gy.,** *Acta Chim. Acad. Sci. Hung.*, 79, 457, 1973.
1334. **Sohár, P. and Lázár, J.,** *Acta Chim. Acad. Sci. Hung.*, 105, 105, 1980.
1335. **Sohár, P., Mányai, Gy., Hideg, K., Hankovszky, H. O., and Lex, L.,** *Org. Magn. Reson.*, 14, 125, 1980.
1336. **Sohár, P., Medgyes, G., and Kuszmann, J.,** *Org. Magn. Reson.*, 11, 357, 1978.
1337. **Sohár, P., Méhesfalvi-Vajna, Zs., and Bernáth, G.,** *Kém. Közlemények*, 46, 487, 1976 (in Hungarian).
1338. **Sohár, P., Nyitrai, J., Zauer, K., and Lempert, K.,** *Acta Chim. Acad. Sci. Hung.*, 65, 189, 1970.
1339. **Sohár, P., Ocskay, Gy., and Vargha, L.,** *Acta Chim. Acad. Sci. Hung.*, 84, 381, 1975.
1340. **Sohár, P., Reiter, J., and Toldy, L.,** *Org. Magn. Reson.*, 3, 689, 1971.
1341. **Sohár, P. and Sipos, Gy.,** *Acta Chim. Acad. Sci. Hung.*, 67, 365, 1971.

1342. **Sohár, P., Széll, T., and Dudás, T.,** *Acta Chim. Acad. Sci. Hung.*, 70, 355, 1971.
1343. **Sohár, P., Széll, T., Dudás, T., and Sohár, I.,** *Tetrahedron Lett.*, 1101, 1972.
1344. **Sohár, P. and Toldy, L.,** *Acta Chim. Acad. Sci. Hung.*, 75, 99, 1973.
1345. **Sohár, P. and Varsányi, Gy.,** *Acta Chim. Acad. Sci. Hung.*, 55, 189, 1968.
1346. **Solomon, I.,** *Phys. Rev.*, 99, 559, 1955.
1347. **Solomon, I.,** *C. R. Acad. Sci. Paris*, 248, 92, 1959.
1348. **Solomon, I.,** *Phys. Rev. Lett.*, 2, 301, 1959.
1349. **Sólyom, S., Sohár, P., Toldy, L., Kálmán, A., and Párkányi, L.,** *Tetrahedron Lett.*, 48, 4245, 1977.
1350. **Sorensen, S., Hansen, M., and Jakobsen, H. J.,** *J. Magn. Reson.*, 12, 340, 1973.
1351. **Sorensen, S., Hansen, R. S., and Jakobsen, H. J.,** *J. Magn. Reson.*, 14, 243, 1974.
1352. *Spectrometry of Fuels,* Friedel, R. A., Ed., Plenum Press, New York, 1970, 90.
1353. **Spiesecke, H.,** *Z. Naturforsch.*, 23a, 467, 1968.
1354. **Spiesecke, H. and Schneider, W. G.,** *J. Chem. Phys.*, 35, 722, 1961.
1355. **Spiesecke, H. and Schneider, W. G.,** *J. Chem. Phys.*, 35, 731, 1961.
1356. **Spiesecke, H. and Schneider, W. G.,** *Tetrahedron Lett.*, 468, 1961.
1357. **Spiess, H. W., Schweitzer, D., Haeberlen, U., and Hausser, K. H.,** *J. Magn. Reson.*, 5, 101, 1971.
1358. **Spotswood, T. M. and Tänzer, C. I.,** *Tetrahedron Lett.*, 911, 1967.
1359. **Springer, C. S., Bruder, A. H., Tanny, S. R., Pickering, M., and Rockefeller, H. A.,** $Ln(fod)_3$ complexes as NMR shift reagents, in *Nuclear Magnetic Resonance Shift Reagents,* Academic Press, New York, 1973, 283.
1360. **Srinivasan, P. R. and Lichter, R. L.,** *Org. Magn. Reson.*, 8, 198, 1976.
1361. **Staab, H. A., Brettschneider, H., and Brunner, H.,** *Chem. Ber.*, 103, 1101, 1970.
1362. **Stájer, G., Szabó, E. A., Pintye, J., Klivényi, F., and Sohár, P.,** *Chem. Ber.*, 107, 299, 1974.
1363. **Staley, S. W. and Kingsley, W. G.,** *J. Am. Chem. Soc.*, 95, 5805, 1973.
1364. **Stefaniak, L.,** *Spectrochim. Acta,* 32a, 345, 1976.
1365. **Stefaniak, L.,** *Tetrahedron,* 32a, 1065, 1976.
1366. **Stefaniak, L.,** *Tetrahedron,* 33, 2571, 1977.
1367. **Stefaniak, L. and Grabowska, A.,** *Bull. Acad. Polon. Sci. Ser. Sci. Chim.*, 22, 267, 1974.
1368. **Sternhell, S.,** *Q. Rev. Chem. Soc.*, 23, 236, 1969.
1369. **Steur, R., Van Dongen, J. P. C. M., De Bie, M. J. A., Drenth, W., De Haan, J. W., and Van de Ven, L. J. M.,** *Tetrahedron Lett.*, 3307, 1971.
1370. **Stolow, R. D. and Gallo, A. A.,** *Tetrahedron Lett.*, 3331, 1968.
1371. **Stothers, J. B.,** *Q. Rev. Chem. Soc.*, 19, 144, 1965.
1372. **Stothers, J. B.,** *Carbon-13 NMR Spectroscopy,* Academic Press, New York, 1972, (a) p. 105; (b) p. 118; (c) p. 134; (d) p. 135; (e) p. 168; (f) p. 153; (g) p. 272; (h) p. 70, p. 71; (i) p. 145, p. 151; (j) p. 151; (k) p. 291; (l) p. 442; (m) pp. 433-439; (n) p. 371; (o) p. 372; (p) pp. 362-370, pp. 375-381; (r) p. 432.
1373. **Stothers, J. B.,** ^{13}C NMR studies of reaction mechanisms and reactive intermediates, in *Topics in Carbon-13 NMR Spectroscopy,* Vol. 1, Levy, G. C., Ed., Wiley-Interscience, New York, 1974, 229; (a) pp. 238-244.
1374. **Stothers, J. B. and Lauterbur, P. C.,** *Can. J. Chem.*, 42, 1563, 1964.
1375. **Sterehlow, H.,** *Magnetische Kernresonanz und chemische Struktur,* Vol. 7, (Fortschritte der physikalischen Chemie), Steinkoff-Verlag, Darmstadt, 1968.
1376. **Su, J.-A., Siew, E., Brown, E. V., and Smith, S. L.,** *Org. Magn. Reson.*, 10, 122, 1977.
1377. **Suhr, H.,** *Chem. Ber.*, 96, 1720, 1963.
1378. **Suhr, H.,** *J. Mol. Phys.*, 6, 153, 1963.
1379. **Suhr, H.,** *Anwendungen der Kernmagnetischen Resonanz in der Organischen Chemie,* Springer-Verlag, Berlin, 1965.
1380. **Sullivan, G. R. and Roberts, J. D.,** *J. Org. Chem.*, 42, 1095, 1977.
1381. **Sunners, B., Piette, L. H., and Schneider, W. G.,** *Can. J. Chem.*, 38, 681, 1960.
1382. **Sutherland, I. O.,** The investigation of the kinetics of conformational changes by nuclear magnetic resonance spectroscopy, in *Annual Reports on NMR Spectroscopy,* Vol. 4, Mooney, E. F., Ed., Academic Press, New York, 1971, 71.
1383. **Szántay, Cs., Novák, L., and Sohár, P.,** *Acta Chim. Acad. Sci. Hung.*, 57, 335, 1968.
1384. **Szarek, W. A., Vyas, D. M., Sepulchre, A. M., Gero, S. D., and Lukacs, G.,** *Can. J. Chem.*, 52, 2041, 1974.
1385. **Sztraka, L.,** *Basic Principles of Fourier-Transform IR Spectrometry,* A kémia ujabb eredményei, Vol. 36, Csákvári, B., Ed., Akadémiai Kiadó, Budapest, 1977 (in Hungarian).
1386. **Szymanski, S., Witanowski, M., Gryff-Keller, A.,** Problems in theory and analysis of dynamic nuclear magnetic spectra, in *Annual Reports on NMR Spectroscopy,* Vol. 8, Webb, G. A., Ed., Academic Press, New York, 1978, 227.
1387. **Tadokoro, S., Fujiwara, S., and Ichihara, Y.,** *Chem. Lett.*, 849, 1973.

1388. **Taillandier, M., Liquier, J., and Taillandier, E.,** *J. Mol. Struct.,* 2, 437, 1968.
1389. **Takahashi, K.,** *Bull. Chem. Soc. Jpn.,* 39, 2782, 1966.
1390. **Takahashi, K., Sone, T., and Fujieda, K.,** *J. Phys. Chem.,* 74, 2765, 1970.
1391. **Takeuchi, Y.,** *Org. Magn. Reson.,* 7, 181, 1975.
1392. **Takeuchi, Y., Chivers, P. J., and Crabb, T. A.,** *Chem. Commun.,* 210, 1974.
1393. **Takeuchi, Y. and Dennis, N.,** *J. Am. Chem. Soc.,* 96, 3657, 1974.
1394. **Tanabe, M. and Detre, G.,** *J. Am. Chem. Soc.,* 88, 4515, 1966.
1395. **Tangerman, A. and Zwannenburg, B.,** *Tetrahedron Lett.,* 5195, 1973.
1396. **Tarpley, A. R. and Goldstein, J. H.,** *J. Mol. Spectrosc.,* 37, 432, 1971.
1397. **Tarpley, A. R. and Goldstein, J. H.,** *J. Mol. Spectrosc.,* 39, 275, 1971.
1398. **Temple, C., Thorpe, M. C., Coburn, W. C., and Montgomery, J. A.,** *J. Org. Chem.,* 31, 935, 1966.
1399. **Terui, Y., Aono, K., and Tori, K.,** *J. Am. Chem. Soc.,* 90, 1069, 1968.
1400. **Thétaz, C., Wehrli, F. W., and Wentrup, C.,** *Helv. Chim. Acta,* 59, 259, 1976.
1401. **Thomas, H. A.,** *Phys. Rev.,* 80, 901, 1950.
1402. **Thomas, W. A.,** NMR spectroscopy in conformational analysis, in *Annual Reports on NMR Spectroscopy 1-5,* Vol. 1, Mooney, E. F., Ed., Academic Press, New York, 1968.
1403. **Thomas, W. A.,** NMR and conformations of amino acids, peptides and proteins, in *Annual Reports on NMR Spectroscopy,* Vol. 6B, Mooney, E. F., Ed., Academic Press, New York, 1976, 1.
1404. **Thomlinson, B. L. and Hill, H. D. W.,** *J. Chem. Phys.,* 59, 1775, 1973.
1405. **Thorpe, M. C., Cobrun, W. C., and Montgomery, J. A.,** *J. Magn. Reson.,* 15, 98, 1974.
1406. **Tiers, G. V. D.,** *J. Phys. Chem.,* 62, 1151, 1958.
1407. **Tiers, G. V. D.,** *Proc. Chem. Soc.,* 1960, 1960.
1408. **Toldy, L., Sohár, P., Faragó, K., Tóth, I., and Bartalits, L.,** *Tetrahedron Lett.,* 2167, 1970.
1409. **Toldy, L., Sohár, P., Faragó, K., Tóth, I., and Bartalits, L.,** *Tetrahedron Lett.,* 2177, 1970.
1410. *Topics in Carbon-13 NMR Spectroscopy,* Levy, G. C., Ed., Wiley-Interscience, New York, Vol. 1 to 3, 1974 to 1979, a) Vol. 2, p. 433.
1411. **Torchia, D. A. and Vanderhart, D. L.,** High-Power double-resonance studies in fibrous proteins, proteoglycans, and model membranes, in *Topics in Carbon-13 NMR Spectroscopy,* Vol. 3, Levy, G. C., Ed., Wiley-Interscience, New York, 1979, 325.
1412. **Tori, K., Kitahonoki, K., Takano, Y., Tanida, H., and Tsuji, T.,** *Tetrahedron Lett.,* 559, 1964.
1413. **Tori, K., Kitahonoki, K., Takano, Y., Tanida, H., and Tsuji, T.,** *Tetrahedron Lett.,* 869, 1965.
1414. **Tori, K., Komeno, T., Sangare, M., Septe, B., Delpech, B., Ahand, A., and Lukacs, G.,** *Tetrahedron Lett.,* 1157, 1974.
1415. **Tori, K. and Kondo, E.,** *Tetrahedron Lett.,* 645, 1963.
1416. **Tori, K., Muneyuki, R., and Tanida, H.,** *Can. J. Chem.,* 41, 3142, 1963.
1417. **Tori, K. and Nakagawa, T.,** *J. Phys. Chem.,* 68, 3163, 1964.
1418. **Tori, K. and Ogata, M.,** *Chem. Pharm. Bull.,* 12, 272, 1964.
1419. **Tori, K. and Ohtsuru, M.,** *Chem. Commun.,* 886, 1966.
1420. **Tori, K., Ohtsuru, M., Aono, K., Kawazoe, Y., and Ohnishi, M.,** *J. Am. Chem. Soc.,* 89, 2765, 1967.
1421. **Tori, K. and Yoshimura, Y.,** *Tetrahedron Lett.,* 3127, 1973.
1422. **Torrey, H. C.,** *Phys. Rev.,* 76, 1059, 1949.
1423. **Towl, A. D. C. and Schaumburg, K.,** *Mol. Phys.,* 22, 49, 1971.
1424. **Traficante, D. D. and Maciel, G. E.,** *J. Phys. Chem.,* 69, 1348, 1965.
1425. **Trager, W. F., Nist, B. J., and Huitric, A. C.,** *Tetrahedron Lett.,* 2931, 1965.
1426. **Tran-Dihn, S., Fermandjian, S., Sala, E., Mermet-Bouvier, R., Cohen, M., and Fromageot, P.,** *J. Am. Chem. Soc.,* 96, 1484, 1974.
1427. **Tran-Dihn, S., Fermandjian, S., Sala, E., Mermet-Bouvier, R., and Fromageot, P.,** *J. Am. Chem. Soc.,* 97, 1267, 1975.
1428. **Triplett, J. W., Digenis, G. A., Layton, W. J., and Smith, S. L.,** *Spectrosc. Lett.,* 10, 141, 1977.
1429. **Tronchet, J. M. J., Barbalat-Rey, F., and Le-Hong, N.,** *Helv. Chim. Acta,* 54, 2615, 1971.
1430. **Troshin, A. S.,** *Problems of Cell Permeability,* Pergamon Press, New York, 1966.
1431. **Truce, W. E. and Brady, D. G.,** *J. Org. Chem.,* 31, 3543, 1966.
1432. **Tulloch, A. P.,** *Can. J. Chem.,* 55, 1135, 1977.
1433. **Turner, A. B., Heine, H. W., Irwing, J., and Bush, J. B., Jr.,** *J. Am. Chem. Soc.,* 87, 1050, 1965.
1434. **Turner, T. E., Fiora, V. C., and Kendrick, W. M.,** *J. Chem. Phys.,* 23, 1966, 1955.
1435. **Uhlenbeck, G. E. and Goudsmit, S.,** *Naturwissenschaften,* 13, 953, 1925; *Nature,* 117, 264, 1926.
1436. **Untch, K. G. and Wysocki, D. C.,** *J. Am. Chem. Soc.,* 88, 2608, 1966.
1437. **Vanasse, G. A. and Sakai, H.,** Fourier spectroscopy, in *Progress in Optics,* Vol. 6, Wolf, E., Ed., North-Holland, Amsterdam, 1967, 261.
1438. **Van Der Veen, J. M.,** *J. Org. Chem.,* 28, 564, 1963.
1439. **Van Vleck, J. H.,** *The Theory of Electric and Magnetic Susceptibilities,* Oxford University Press, Oxford, 1932.

1440. **Van' Wazer, J. R., Callis, C. F., Shoolery, J. N., and Jones, R. C.,** *J. Am. Chem. Soc.*, 78, 5715, 1956.
1441. **Vargha, L., Kuszmann, J., and Sohár, P.,** *Magy. Kém. Lapja,* 356, 1972 (in Hungarian).
1442. **Vargha, L., Kuszmann, J., Sohár, P., and Horváth, Gy.,** *J. Heterocyclic Chem.*, 9, 341, 1972.
1443. *VARIAN Instrum. Appl.*, 2, 4, 1968.
1444. **Vegar, M. R. and Ewlis, R. J.,** *Tetrahedron Lett.*, 2847, 1971.
1445. **Venien, F.,** *C. R. Acad. Sci. France C,* 269, 642, 1969.
1446. **Verchére, C., Rousselle, D., and Viel, C.,** *Org. Magn. Reson.*, 13, 110, 1980.
1447. **Verkade, J. G., McCarley, R. E., Hendricker, D. G., and King, R. W.,** *Inorg. Chem.*, 4, 228, 1965.
1448. **Versmold, H. and Yoon, C.,** *Ber. Bunsenges. Phys. Chem.*, 76, 1164, 1972.
1449. **Vincent, E.-J. and Metzger, J.,** *C. R. Acad. Sci. Paris,* 261, 1964, 1965.
1450. **Vinkler, E., Németh, P., Stájer, G., Sohár, P., and Jerkovich, Gy.,** *Arch. Pharm.*, 309, 265, 1976.
1451. **Voelter, W. and Breitmaier, E.,** *Org. Magn. Reson.*, 5, 311, 1973.
1452. **Voelter, W., Breitmaier, E., and Jung, G.,** *Angew. Chem.*, 83, 1011, 1961; *Angew. Chem. Int. Ed. Engl.*, 10, 935, 1961.
1453. **Voelter, W., Breitmaier, E., Jung, G., Keller, T., and Hiss, D.,** *Angew. Chem. Int. Ed. Engl.*, 9, 803, 1970.
1454. **Voelter, W., Breitmaier, E., Price, R., and Jung, G.,** *Chimia,* 25, 168, 1971.
1455. **Voelter, W., Jung, G., Breitmaier, E., and Bayer, E.,** *Z. Naturforsch.*, 26b, 213, 1971.
1456. **Voelter, W., Jung, G., Breitmaier, E., and Price, R.,** *Hoppe-Seyler's Z. Physiol. Chem.*, 352, 1034, 1971.
1457. **Voelter, W. and Oster, O.,** *Z. Naturforsch.*, 28b, 370, 1973.
1458. **Voelter, W., Zech, K., Grimminger, W., Breitmaier, E., and Jung, G.,** *Chem. Ber.*, 105, 3650, 1972.
1459. **Vo-Kim-Yen, Papoušková, Z., Schraml, J., and Chvalovský, V.,** *Collect. Czech. Chem. Commun.*, 38, 3167, 1973.
1460. **Vold, R. L., Waugh, J. S., Klein, M. P., and Phelps, D. E.,** *J. Chem. Phys.*, 48, 3831, 1968.
1461. **Vögeli, U. and von Philipsborn, W.,** *Org. Magn. Reson.*, 5, 551, 1973.
1462. **Vögeli, U. and von Philipsborn, W.,** *Org. Magn. Reson.*, 7, 617, 1975.
1463. **Vögtle, F., Mannschreck, A., and Staab, H. A.,** *Liebigs Ann. Chem.*, 708, 51, 1967.
1464. **Wallach, D.,** *J. Chem. Phys.*, 47, 5258, 1967.
1465. **Wallach, D.,** *J. Phys. Chem.*, 73, 307, 1969.
1466. **Walter, J. A. and Hope, A. B.,** Nuclear magnetic resonance and the state of water in cells, in *Progress in Biophysics and Molecular Biology,* Vol. 23, Pergamon Press, Oxford, 1971, 1.
1467. **Walter, R., Havran, R. T., Swartz, I. L., and Johnson, L. F.,** *Peptides 1969,* Scoffone, E., Ed., North-Holland, Amsterdam, 1971.
1468. **Walter, R. and Johnson, L. F.,** *Biophys. J.*, 9a, 159, 1969.
1469. **Wang, C. and Kingsbury, C. A.,** *J. Org. Chem.*, 40, 3811, 1975.
1470. **Wang, C. H., Grant, D. M., and Lyerla, J. R., Jr.,** *J. Chem. Phys.*, 55, 4674, 1971.
1471. **Wang, M. C. and Uhlenbeck, G. E.,** *Rev. Mod. Phys.*, 17, 323, 1945.
1472. **Ward, H. R.,** *Acc. Chem. Res.*, 5, 18, 1972.
1473. **Ward, H. R. and Lawler, R. G.,** *J. Am. Chem. Soc.*, 89, 5518, 1967.
1474. **Warren, J. P. and Roberts, J. D.,** *J. Phys. Chem.*, 78, 2507, 1974.
1475. **Wasserman, H. H. and Keehn, P. M.,** *J. Am. Chem. Soc.*, 91, 2374, 1969.
1476. **Wasylishen, R. E.,** Spin-spin coupling between carbon-13 and the first row nuclei, in *Annual Reports on NMR Spectroscopy,* Vol. 7, Webb, G. A., Ed., Academic Press, New York, 1977, 245; (a) pp. 287-291.
1477. **Wasylishen, R. E. and Schaefer, T.,** *Can. J. Chem.*, 50, 2710, 1972.
1478. **Wasylishen, R. E. and Schaefer, T.,** *Can. J. Chem.*, 50, 2989, 1972.
1479. **Watts, V. S. and Goldstein, J. H.,** *J. Phys. Chem.*, 70, 3887, 1966.
1480. **Watts, V. S. and Goldstein, J. H.,** *J. Chem. Phys.*, 46, 4165, 1967.
1481. **Watts, V. S., Loemker, J., and Goldstein, J. H.,** *J. Mol. Spectrosc.*, 17, 348, 1965.
1482. **Waugh, J. S.,** *J. Mol. Spectrosc.*, 35, 298, 1970.
1483. **Waugh, J. S. and Fessenden, R. W.,** *J. Am. Chem. Soc.*, 79, 846, 1957.
1484. **Waugh, J. S., Huber, L. M., and Haeberlen, U.,** *Phys. Rev. Lett.*, 20, 180, 1968.
1485. **Webb, R. G., Haskell, M. W., and Stammer, C. H.,** *J. Org. Chem.*, 34, 576, 1969.
1486. **Webster, D. E.,** *J. Chem. Soc.*, 5132, 1960.
1487. **Wehrli, F. W.,** *Chem. Commun.*, 379, 1973.
1488. **Wehrli, F. W.,** *Solvent Suppression Technique in Pulsed NMR,* VARIAN Application Note NMR, 75, 2, 1975.
1489. **Wehrli, F. W.,** Organic structure assignments using ^{13}C spin-relaxation data, in *Topics in Carbon-13 NMR Spectroscopy,* Vol. 2, Levy, G. C., Ed., Wiley-Interscience, New York, 1976, 343. (a) pp. 362-364; (b) p. 373.

1490. **Wehrli, F. W., Giger, W., and Simon, W.,** *Helv. Chim. Acta,* 54, 229, 1971.
1491. **Wehrli, F. W., Jeremic, D., and Mihailovic, M. L., and Milosavljevic, S.,** *Chem. Commun.,* 302, 1978.
1492. **Wehrli, F. W. and Wirthlin, T.,** *Interpretation of Carbon-13 NMR Spectra,* Heyden, London, 1976; (a) p. 68; (b) p. 248, p. 249; (c) p. 145; (d) p. 37; (e) pp. 129-151.
1493. **Weigert, F. J. and Roberts, J. D.,** *J. Am. Chem. Soc.,* 89, 2967, 1967.
1494. **Weigert, F. J. and Roberts, J. D.,** *J. Am. Chem. Soc.,* 90, 3543, 1968.
1495. **Weigert, F. J. and Roberts, J. D.,** *J. Am. Chem. Soc.,* 90, 3577, 1968.
1496. **Weigert, F. J. and Roberts, J. D.,** *J. Am. Chem. Soc.,* 92, 1347, 1970.
1497. **Weigert, F. J. and Roberts, J. D.,** *J. Am. Chem. Soc.,* 94, 6021, 1972.
1498. **Weigert, F. J., Winokur, M., and Roberts, J. D.,** *J. Am. Chem. Soc.,* 90, 1566, 1968.
1499. **Weiler, L.,** *Can. J. Chem.,* 50, 1975, 1972.
1500. **Weinstein, B.,** *Peptides,* Marcel Dekker, New York, 1970.
1501. **Weisman, I. D., Bennett, L. H., Maxwell, L. R., Woods, M. W., and Burke, D.,** *Science,* 178, 1288, 1972.
1502. **Weissman, S. I.,** *J. Am. Chem. Soc.,* 93, 4928, 1971.
1503. **Weitkamp, H. and Korte, F.,** *Chem. Ber.,* 95, 2896, 1962.
1504. **Weitkamp, H. and Korte, F.,** *Tetrahedron Suppl.,* 75, 1966.
1505. **Wellman, K. M. and Bordwell, F. G.,** *Tetrahedron Lett.,* 173, 1963.
1506. **Wells, E. J. and Abramson, K. H.,** *J. Magn. Reson.,* 1, 378, 1969.
1507. **Wells, P. R., Arnold, D. P., and Doddrell, D.,** *J. Chem. Soc. Perkin Trans. 2,* 1745, 1974.
1508. **Wenkert, E. and Buckwalter, B. L.,** *J. Am. Chem. Soc.,* 94, 4367, 1972.
1509. **Wenkert, E., Buckwalter, B. L., Burfitt, I. R., Gasic, M. J., Gottlieb, H. E., Hagaman, E. W., Schell, F. M., Wovkulich, P. M., and Zheleva, A.,** Carbon-13 NMR Spectroscopy, in *Topics in Carbon-13 NMR Spectroscopy,* Vol. 2, Levy, G. C., Ed., Wiley-Interscience, New York, 1976.
1510. **Wenkert, E., Clouse, A. O., Cochran, D. W., and Doddrell, D.,** *Chem. Commun.,* 1433, 1969.
1511. **Wenkert, E., Cochran, D. W., Hagaman, E. W., Lewis, R. B., and Schell, F. M.,** *J. Am. Chem. Soc.,* 93, 6271, 1971.
1512. **Whipple, E. B., Goldstein, J. H., and Mandell, L.,** *J. Chem. Phys.,* 30, 1109, 1959.
1513. **Whipple, E. B., Goldstein, J. H., and Stewart, W. E.,** *J. Am. Chem. Soc.,* 81, 4761, 1959.
1514. **White, D. M. and Levy, G. C.,** *Macromolecules,* 5, 526, 1972.
1515. **Whitesides, G. M. and Fleming, J. S.,** *J. Am. Chem. Soc.,* 89, 2855, 1967.
1516. **Whitesides, G. M. and Lewis, D. W.,** *J. Am. Chem. Soc.,* 92, 6979, 1970.
1517. **Whitlock, H. W.,** *J. Am. Chem. Soc.,* 84, 3412, 1962.
1518. **Whitman, D. R.,** *J. Mol. Spectrosc.,* 10, 250, 1963.
1519. **Wiberg, K. B. and Nist, B. J.,** *J. Am. Chem. Soc.,* 83, 1226, 1961.
1520. **Wieland, T. and Bende, H.,** *Chem. Ber.,* 98, 504, 1965.
1521. **Williams, D. H. and Bhacca, N. S.,** *Tetrahedron,* 21, 2021, 1965.
1522. **Williams, D. H. and Fleming, I.,** *Spektroskopische Methoden in der Organischen Chemie,* Thieme-Verlag, Stuttgart, 1968.
1523. **Williams, D. H., Ronayne, J., Moore, H. W., and Shelden, H. R.,** *J. Org. Chem.,* 33, 998, 1968.
1524. **Williams, F., Sears, B., Allerhand, A., and Cordes, E. H.,** *J. Am. Chem. Soc.,* 95, 4871, 1973.
1525. **Williamson, K. L., Howell, T., and Spencer, T. A.,** *J. Am. Chem. Soc.,* 88, 325, 1966.
1526. **Williamson, K. L., Lanford, C. A., and Nicholson, C. R.,** *J. Am. Chem. Soc.,* 86, 762, 1964.
1527. **Williamson, K. L., Li Hsu, Y.-F., Hall, F. H., Swager, S., and Coulter, M. S.,** *J. Am. Chem. Soc.,* 90, 6717, 1968.
1528. **Williamson, K. L. and Roberts, J. D.,** *J. Am. Chem. Soc.,* 98, 5082, 1976.
1529. **Williamson, M. P., Kostelnik, R. J., and Castellano, S. M.,** *J. Chem. Phys.,* 49, 2218, 1968.
1530. **Wilson, N. K. and Zehr, R. D.,** *J. Org. Chem.,* 43, 1768, 1978.
1531. **Wing, R. M., Uebel, J. J., and Anderson, K. K.,** *J. Am. Chem. Soc.,* 95, 6046, 1973.
1532. **Witanowski, M.,** *Tetrahedron,* 23, 4299, 1967.
1533. **Witanowski, M.,** *J. Am. Chem. Soc.,* 90, 5683, 1968.
1534. **Witanowski, M.,** *Pure Appl. Chem.,* 37, 225, 1974.
1535. **Witanowski, M. and Januszewski, H.,** *J. Chem. Soc. B,* 1063, 1967.
1536. **Witanowski, M. and Januszewski, H.,** *Can. J. Chem.,* 47, 1321, 1969.
1537. **Witanowski, M. and Stefaniak, L.,** *J. Chem. Soc. B,* 1061, 1967.
1538. **Witanowski, M., Stefaniak, L., Januszewski, H., Bahadur, K., and Webb, G. A.,** *J. Cryst. Mol. Struct.,* 5, 137, 1975.
1539. **Witanowski, M., Stefaniak, L., Januszewski, H., Grabowski, Z., and Webb, G. A.,** *Bull. Acad. Polon. Sci. Ser. Sci. Chim.,* 20, 917, 1972.
1540. **Witanowski, M., Stefaniak, L., Januszewski, H., and Piotrowska, H.,** *Bull. Acad. Polon. Sci. Ser. Sci. Chim.,* 23, 333, 1975.

1541. **Witanowski, M., Stefaniak, L., Januszewski, H., Szymansky, S., and Webb, G. A.,** *Tetrahedron,* 29, 2833, 1973.

1542. **Witanowski, M., Stefaniak, L., Januszewski, H., Voronkov, M. G., and Tandura, S. N.,** *Bull. Acad. Polon. Sci. Ser. Sci. Chim.,* 24, 281, 1976.

1543. **Witanowski, M., Stefaniak, L., Januszewski, H., and Webb, G. A.,** *Bull. Acad. Polon. Sci . Ser. Sci. Chim.,* 21, 71, 1973.

1544. **Witanowski, M., Stefaniak, L., Szymanski, S., Grabowski, Z., and Webb, G. A.,** *J. Magn. Reson.,* 21, 185, 1976.

1545. **Witanowski, M., Stefaniak, L., Szymanski, S., and Januszewski, H.,** *J. Magn. Reson.,* 28, 217, 1977.

1546. **Witanowski, M., Stefaniak, L., Szymanski, S., and Webb, G. A.,***Tetrahedron,* 32, 2127, 1976.

1547. **Witanowski, M., Stefaniak, L., and Webb, G. A.,** *J. Chem. Soc. B,* 1065, 1967.

1548. **Witanowski, M., Stefaniak, L., and Webb, G. A.,** Nitrogen NMR spectroscopy, in *Annual Reports on NMR Spectroscopy,* Vol. 7, Webb, G. A., Ed., Academic Press, London, 1977, 117.

1549. **Witanowski, M., Urbanski, T., and Stefaniak, L.,** *J. Am. Chem. Soc.,* 86, 2569, 1964.

1550. **Witanowski, M. and Webb, G. A.,** Nitrogen NMR spectroscopy, in *Annual Reports on NMR Spectroscopy,* Vol. 5A, Mooney, E. F., Ed., Academic Press, London, 1972, 395.

1551. **Wittstruck, T. A. and Cronan, J. F.,** *J. Phys. Chem.,* 72, 4243, 1968.

1552. **Wittstruck, T. A., Malhotra, S. K., and Ringold, H. J.,** *J. Am. Chem. Soc.,* 85, 1699, 1963.

1553. **Wokaun, A. and Ernst, R. R.,** *Chem. Phys. Lett.,* 52, 407, 1977.

1554. **Wolkowski, Z. W.,** *Tetrahedron Lett.,* 825, 1971.

1555. **Woods, W. G. and Strong, P. L.,** *J. Am. Chem. Soc.,* 88, 4667, 1966.

1556. **Woodward, R. B. and Skarie, V.,** *J. Am. Chem. Soc.,* 83, 4676, 1961.

1557. **Woolfenden, W. R. and Grant, D. M.,** *J. Am. Chem. Soc.,* 88, 1496, 1966.

1558. **Wright, G. E.,** *Tetrahedron Lett.,* 1097, 1973.

1559. **Wright, G. E. and Tang-Wei, T. Y.,** *J. Pharm. Sci.,* 61, 299, 1972.

1560. **Wright, G. E. and Tang-Wei, T. Y.,** *Tetrahedron,* 29, 3775, 1973.

1561. **Wüthrich, K.,** *FEBS Lett.,* 25, 104, 1972.

1562. **Wüthrich, K., Meiboom, S., and Snyder, L. C.,** *J. Chem. Phys.,* 52, 230, 1970.

1563. **Yamagishi, T., Hayashi, K., Mitsuhashi, H., Imanari, M., and Matsushita, K.,** *Tetrahedron Lett.,* 3527, 1973.

1564. **Yamagishi, T., Hayashi, K., Mitsuhashi, H., Imanari, M., and Matsushita, K.,** *Tetrahedron Lett.,* 3531, 1973.

1565. **Yamaguchi, I.,** *Bull. Chem. Soc. Jpn.,* 34, 353, 1961.

1566. **Yamamoto, O., Watabe, M., and Kikuchi, O.,** *Mol. Phys.,* 17, 249, 1969.

1567. **Yamazaki, M., Usami, T., and Takeuchi, T.,** *Nippon Kagaku Kaishi,* 11, 2135, 1973; *Chem. Abstr.,* 80, 47030b, 1974.

1568. **Yee, K. C. and Bentrude, W. G.,** *Tetrahedron Lett.,* 2775, 1971.

1569. **Yeh, H. J. C., Ziffer, H., Jerina, D. M., and Body, D. R.,** *J. Am. Chem. Soc.,* 95, 2741, 1973.

1570. **Yoder, C. H., Griffith, D. R., and Schaeffer, C. D.,** *J. Inorg. Nucl. Chem.,* 32, 3689, 1970.

1571. **Yoder, C. H., Sheffy, F. K., Howell, R., Hess, R. E., Pacala, L., Schaeffer, C. D., Jr., and Zuckerman, J. J.,** *J. Org. Chem.,* 41, 1511, 1976.

1572. **Yoder, C. H., Tuck, R. H., and Hess, R. E.,** *J. Am. Chem. Soc.,* 91, 539, 1969.

1573. **Yonezawa, T. and Morishima, I.,** *J. Mol. Spectrosc.,* 27, 210, 1968.

1574. **Yonezawa, T., Morishima, I., and Fukuta, K.,** *Bull. Chem. Soc. Jpn.,* 41, 2297, 1968.

1575. **Yonezawa, T., Morishima, I., and Kato, H.,** *Bull. Chem. Soc. Jpn.,* 39, 1398, 1966.

1576. **Young, J. A., Grasselli, J. G., and Ritchey, W. M.,** *J. Magn. Reson.,* 14, 194, 1974.

1577. **Zaner, K. S. and Damadian, R.,** *Physiol. Chem. Phys.,* 7, 437, 1975.

1578. **Zaner, K. S. and Damadian, R.,** *Science,* 189, 729, 1975.

1579. **Zaner, K. S. and Damadian, R.,** *Physiol. Chem. Phys.,* 9, 473, 1977.

1580. **Zimmerman, J. R. and Brittin, W. E.,** *J. Phys. Chem.,* 61, 1328, 1957.

1581. **Zürcher, R. F.,** *Helv. Chim. Acta,* 46, 2054, 1963.

INDEX

A

D

E

F

G

H

I

J

N

O

P

Q

R

T

U

V

W

X

Z